Orthogonal Sets and Polar Methods in Linear Algebra

PURE AND APPLIED MATHEMATICS

A Wiley-Interscience Series of Texts, Monographs, and Tracts

Founded by RICHARD COURANT
Editor Emeritus: PETER HILTON and HARRY HOCHSTADT
Editors: MYRON B. ALLEN III, DAVID A. COX, PETER LAX,
JOHN TOLAND

A complete list of the titles in this series appears at the end of this volume.

Orthogonal Sets and Polar Methods in Linear Algebra

Applications to Matrix Calculations, Systems of Equations, Inequalities, and Linear Programming

ENRIQUE CASTILLO, ANGEL COBO,
FRANCISCO JUBETE, and ROSA EVA PRUNEDA

A Wiley-Interscience Publication
JOHN WILEY & SONS, INC.
New York / Chichester / Weinheim / Brisbane / Singapore / Toronto

For ordering and customer service, call 1-800-CALL-WILEY.

Library of Congress Cataloging in Publication Data:

Orthogonal sets and polar methods in linear algebra : applications to matrix calculations, systems of equations, inequalities, and linear programming / Enrique Castillo . . . [et al.].
p. cm. — (Pure and applied mathematics)
" A Wiley-Interscience publication."
Includes index.
ISBN 0-471-32889-8 (acid-free paper)
1. Algebras, Linear. 2. Orthogonalization methods. I. Castillo, Enrique, 1946– . II. Series: Pure and applied mathematics (John Wiley & Sons : Unnumbered)
QA184.077 1999
512'.5—dc21 98-38391
CIP

10 9 8 7 6 5 4 3 2 1

Preface

Inverting matrices, calculating determinants, solving linear systems of equations and/or inequalities, and linear programming problems are mathematical problems one has to deal with very frequently. One is prepared to solve them using many available tools, procedures, or computer packages. In fact, these simple problems are studied in detail in standard courses. However, one may not be prepared to solve other frequently appearing problems such as:

1. Calculating the inverse of a matrix after modifying, removing, and/or adding rows or columns, based on the inverse of the initial matrix.

2. Calculating the inverse of a symbolic matrix.

3. Solving systems of linear equations in some selected variables.

4. Obtaining the solution of a system of linear equations when some equations and/or variables are changed, removed, or added, without starting from scratch, that is, using the solution of the initial system.

5. Analyzing the compatibility of systems of linear equations and inequalities with restricted variables.

In this book we give methods that allow solving a much wider range of problems, which, apart from the above, include systems of linear equations and inequalities with unrestricted, restricted, or strictly restricted variables. This allows us to solve very interesting practical problems.

In this book we do not deal with the standard methods of analysis, which are very well covered in many other books. Instead, we present new methods and algorithms for solving standard and new problems based on the concept of orthogonality.

When analyzing and discussing several problems in linear algebra involving linear spaces, one possibility is to analyze them using the appropriate definitions and concepts directly, and stating the conditions to be satisfied by the problem under consideration. This is referred to as the *primal approach*. Alternatively, one can deal with orthogonal sets and state the corresponding conditions using the concept of orthogonality. This, the *dual approach*, which initially seems to be much more involved but is computationally simpler and more elegant and appealing, will be used in this book.

Thus, this book does not deal with standard methods or solutions. The reader looking for standard methods or references to published works with this orientation should consult one of the very many books on the topic (see References). On the contrary, in this book the above problems are discussed from a different point of view and the corresponding methods are given.

In addition to obtaining solutions, mathematicians and engineers are interested in analyzing conditions leading to well-defined problems. In this context, the problems of compatibility and uniqueness of solutions play a central role. In fact, they lead to very interesting physical or engineering conclusions, which relate the physical conditions, obtained from reality, and the corresponding conditions behind the mathematical models.

This book is organized in four parts. In Part I we deal with linear spaces, matrices, and systems of equations (including the problems of compatibility and uniqueness) and obtaining their solutions.

Part II deals with cones and systems of linear equations and inequalities in unrestricted, restricted, and strictly restricted variables, where again the problems of compatibility, uniqueness of solution, and obtaining solutions are treated.

We also show that a linear space of finite dimension is simply a particular case of a cone. Thus, the cone concept is broader than the linear space concept, and, what is more important, it offers the possibility of replacing the usual methods of classical linear algebra with the specific methods of cones and polarity, which are more efficient.

Part III is devoted to linear programming problems. However, instead of describing well-known methods, such as the simplex, the primal–dual, and the interior point methods, we present the new exterior point method, which progresses from outside the feasible region.

In addition to this innovative method, we also discuss how the solution of a linear programming problem can be obtained when removing or adding constraints and/or variables, without starting from the very beginning and thus eliminating a huge amount of computation.

Part IV includes a wide collection of examples, which are treated by a mixture of mathematics, physics, and engineering points of view. Examples such as the matrix analysis of structures, the transportation and production–

scheduling problems, the input–output tables, the diet problem and the network flow problems are used to illustrate concepts and motivate applications.

All of the above models have arisen from real life. In other words, they are mathematical models of physical, economic, or engineering problems. Selection of the adequate model reproducing the reality is a crucial step for a satisfactory solution to a real problem. The mathematical structures are not arbitrary, but a consequence of reality itself. In this book we make a great effort to connect physical and mathematical realities. We show the reader the reasoning that leads to the analysis of the different structures and concepts. This becomes apparent in the illustrative examples, which show the connection between model and reality.

This book can be used as a reference or consulting book and as a textbook in upper-division undergraduate courses or in graduate-level courses. Included in the book are numerous illustrative examples and end-of-chapter exercises. We have also developed some computer programs in Java and Mathematica to implement the various algorithms and methodologies presented in this book. The current version of these programs, together with a brief User's Guide, can be obtained from the World Wide Web site

`http://ccaix3.unican.es/~AIGroup/Orthogonal.html.`

This book is addressed to a wide audience, including mathematicians, engineers, and applied scientists. There are few prerequisites for the reader, though a previous knowledge of linear algebra and some familiarity with matrices are essential.

Several colleagues read earlier versions of the manuscript for this book and have provided us with valuable comments and suggestions. Their contributions have given rise to the current substantially improved version. In particular, we acknowledge the help of Ali S. Hadi and Juan Angel Díaz Hernando, who made a very careful reading of the manuscript and gave many suggestions.

Special thanks are given to the University of Cantabria, Iberdrola and José Antonio Garrido for their finantial support.

ENRIQUE CASTILLO, ANGEL COBO,
FRANCISCO JUBETE, AND ROSA EVA PRUNEDA

Santander, Spain
January 1999

Contents

Part I Linear Spaces and Systems of Equations

1 *Basic Concepts* 5
- 1.1 *Introduction* 5
- 1.2 *Linear space* 6
- 1.3 *The Euclidean Space E^n* 11
- 1.4 *Orthogonal Sets and Decompositions* 14
- 1.5 *Matrices* 16
- 1.6 *Systems of Linear Equations* 20
- *Exercises* 21

2 *Orthogonal Sets* 23
- 2.1 *Introduction and Motivation* 23
- 2.2 *Orthogonal Decompositions* 24
- 2.3 *The Orthogonalization Module* 31
- 2.4 *Mathematica Program* 36
- *Exercises* 39

3 *Matrix Calculations Using Orthogonal Sets* 41

3.1 Introduction 41
3.2 Inverting a Matrix 42
3.3 The Rank of a Matrix 43
3.4 Calculating the Determinant of a Matrix 44
3.5 Algorithm for Matrix Calculations 46
3.6 Complexity 50
3.7 Inverses and Determinants of Row-Modified Matrices 51
3.8 Inverses of Symbolic Matrices 55
3.9 Extensions to Partitioned Matrices 56
3.10 Inverses of Modified Matrices 60
3.11 Mathematica Programs 67
Exercises 69

4 More Applications of Orthogonal Sets 73
4.1 Intersection of Two Linear Subspaces 73
4.2 Reciprocals Images in Linear Transformations 77
4.3 Other Applications 79
4.4 Mathematica Programs 81
Exercises 83

5 Orthogonal Sets and Systems of Linear Equations 85
5.1 Introduction 85
5.2 Compatibility of a System of Linear Equations 86
5.3 Solving a System of Linear Equations 90
5.4 Complexity 95
5.5 Checking Systems Equivalence 96
5.6 Solving a System in Some Selected Variables 99
5.7 Modifying Systems of Equations 101
5.8 Applications 124
5.9 Mathematica Programs 135
Exercises 137
Appendix: Proof of Lemma 5.2 140

Part II Cones and Systems of Inequalities

6 Polyhedral Convex Cones 147
6.1 Introduction 147

6.2 Convex Sets 147
6.3 Types of Linear Combinations 148
6.4 Polyhedral Convex Cones 151
6.5 The Γ-Process 157
6.6 The Complete Γ-Algorithm 182
6.7 Mathematica Program 184
Exercises 189

7 Polytopes and Polyhedra 191
7.1 Introduction 191
7.2 Polytopes 191
7.3 Polyhedra 207
Exercises 212

8 Cones and Systems of Inequalities 215
8.1 Introduction 215
8.2 A Discussion of Linear Systems 216
8.3 Solving Linear Systems 226
8.4 Applications to Linear Programming 247
Exercises 248

Part III Linear Programming

9 An Introduction to Linear Programming 255
9.1 Introduction 255
9.2 Problem Statement and Basic Definitions 255
9.3 Linear Programming Problem in Standard Form 261
9.4 Basic Solutions 264
9.5 Duality 264
Exercises 271

10 The Exterior Point Method 275
10.1 Introduction 275
10.2 The Exterior Point Method 275
10.3 Making the EPM More Efficient 299
10.4 Complexity 318
10.5 Recovering the Final Tableau from the Solution 320
10.6 Modifying a Linear Programming Problem 321

Exercises 339

Part IV Applications

11 Applications 345
11.1 Introduction 345
11.2 Matrix Analysis of Engineering Structures 346
11.3 The Transportation Problem 365
11.4 Production–Scheduling Problems 369
11.5 The Input–Output Tables 376
11.6 The Diet Problem 378
11.7 Network Flow Problems 380
Exercises 387

Part V Appendices

Appendix A: A Java Application 395
A.1 How to Use the Program 396

Appendix B: List of Notation 407

References 415

Index 419

Orthogonal Sets and Polar Methods in Linear Algebra

Part I

Linear Spaces and Systems of Equations

Introduction to Part I

Linear spaces, matrices, and systems of equations are topics covered by most science curricula. However, only some particularly simple problems and solutions are studied. For example, standard books, in general, deal with different problems related to matrices and systems of linear equations in unrestricted variables, but not with problems such as:

1. Calculating the inverse of a matrix after modifying, removing, and/or adding rows or columns, based on the inverse of the initial matrix.

2. Calculating the inverse of a symbolic matrix, that is, a matrix whose elements are not numbers but symbols.

3. Solving systems of linear equations in a selected subset of variables.

4. Obtaining the solution of a system of linear equations when some equations and/or variables are changed, removed, or added, without starting from scratch, that is, using the solution of the initial system to eliminate a huge amount of computation.

5. Giving practical ways of analyzing the compatibility of systems of linear equations.

6. Checking the equivalence of two systems of linear equations, without solving them.

7. Obtaining the inverse image of a linear space by a given linear transformation.

After a first introductory chapter in which we review some elementary concepts, such as linear and Euclidean spaces, orthogonal decompositions, and matrices. Chapter 2 introduces methods and algorithms to obtain the sets orthogonal to a given vector or linear space. These algorithms are used throughout the book to solve a wide collection of problems. In fact, we can almost say that one single algorithm allows all of the above stated problems to be solved.

In Chapter 3 we deal with matrices and related problems and use this algorithm to solve some standard problems, such as finding the rank or determinant of a matrix, inverting a matrix in numeric or symbolic form, and other non-standard problems, such as finding the inverse of a matrix after changing, adding, or removing rows or columns, based on the inverse of the initial matrix.

In Chapter 4 we analyze some interesting problems such as how to obtain the intersection of two linear spaces and the reciprocal image of a linear space by a linear transformation.

In Chapter 5 we deal with systems of linear equations. Linear systems of equations appear in important areas of science, such as mathematics, economics, and engineering. Historically, linear systems of equations underwent a very important changes with the appearance of problems in calculus of structures or more precisely, with the pioneering advances in this area during the nineteenth century.

The development of mechanics and its contributions to the analysis of the new structures being built in that initial period unavoidably led to systems of linear equations with many unknowns.

This problem of solving systems of equations occupied the engineers of several generations who tried to reduce the huge amount of calculations. Some illustrative examples are the methods of Cremona and Hennenberg and, later, with the appearance of reinforced concrete structures, those of Cros and Kani.

Simultaneously, mathematicians, encouraged by the same purpose, discovered many processes and algorithms of general application. As simple examples it is worthwhile mentioning the work of Chió, Choleski, and Picone and, in the beginning of the twentieth century, the iterative method of Zadunaiski.

Certainly, the appearance of the computer in the mid–1940s obscured most of these methods and only those with a simple and essential concept survived, such as the Gauss elimination method. In fact, thanks to the computer, today the problem of solving a system of linear equations can be considered a simple routine problem.

From a theoretical point of view, the largest impact on linear systems of equations is due to economics, first with Quesnay (eighteenth century) and his "Tableau Economique" and afterwards, and more clearly, with the Lausanne School (middle of the nineteenth century), when Walras and Pareto created the marginal utility theory. This theory establishes some interdependence between different economic variables, prices, demand, and supply of different items, by means of systems of equations with a very large number of unknowns.

In Chapter 5 we deal with systems of equations and discuss the problems of compatibility and uniqueness of solutions. We also give methods of solution and check the equivalence of two systems of equations, solve the systems in some selected variables, and obtain the solution of a system of equations after modifying, adding, or removing equations based on the solution of the initial system of equations.

1

Basic Concepts

1.1 INTRODUCTION

In this chapter we introduce some concepts and definitions that are needed for the following chapters, such as linear spaces, subspaces, Euclidean space, orthogonal decompositions, matrices, and systems of linear equations. The reader familiar with these concepts can skip this chapter and go directly to Chapter 2, where the basics of the algorithms to be used in this book are given. For a more detailed treatment of these topics, see Bloom (1979), Mansfield (1976), Díaz–Hernando (1985), Samelson (1974), or Strong (1980).

In Section 1.2 we review the classical axiomatic definition of linear spaces and some related concepts, such as subspaces, linear combinations, set of generators, and basis.

In Section 1.3 we introduce the concepts of dot product and Euclidean spaces.

Section 1.4 deals with orthogonal sets and decompositions, which are the most important concepts for the rest of the book, since the methods and algorithms are based on them.

Section 1.5 is devoted to some elemental concepts of matrices, including inverse, rank, and determinant.

Finally, Section 1.6 introduces systems of linear equations and the important concepts of uniqueness of solution, and general and particular solutions.

1.2 LINEAR SPACE

Definition 1.1 (Linear space). *Let $\mathcal{K}$ be a commutative field. A linear space over $\mathcal{K}$ is a triplet $(\mathcal{V}, +, *)$, where:*

- *$\mathcal{V}$ is a set,*
- *$+ : \mathcal{V} \times \mathcal{V} \to \mathcal{V}$ is a function (internal operation), and*
- *$* : \mathcal{K} \times \mathcal{V} \to \mathcal{V}$ is another function (external operation),*

satisfying the following conditions:

1. *$(\mathcal{V}, +)$ is a commutative group; that is, the internal operation + satisfies:*
   (a) *The commutative property: $\mathbf{u} + \mathbf{v} = \mathbf{v} + \mathbf{u}$; $\forall \mathbf{u}, \mathbf{v} \in \mathcal{V}$.*
   (b) *The associative property: $(\mathbf{u}+\mathbf{v})+\mathbf{w} = \mathbf{u}+(\mathbf{v}+\mathbf{w})$; $\forall \mathbf{u}, \mathbf{v}, \mathbf{w} \in \mathcal{V}$.*
   (c) *There is a vector $\mathbf{0} \in \mathcal{V}$ such that $\mathbf{0} + \mathbf{u} = \mathbf{u} + \mathbf{0} = \mathbf{u}$; $\forall \mathbf{u} \in \mathcal{V}$.*
   (d) *Each $\mathbf{u} \in \mathcal{V}$ has an inverse $\check{\mathbf{u}}$ such that $\check{\mathbf{u}} + \mathbf{u} = \mathbf{u} + \check{\mathbf{u}} = \mathbf{0}$.*
2. *Distributive property I: $\alpha * (\mathbf{u} + \mathbf{v}) = \alpha * \mathbf{u} + \alpha * \mathbf{v}$; $\forall \alpha \in \mathcal{K}; \forall \mathbf{u}, \mathbf{v} \in \mathcal{V}$.*
3. *Distributive property II: $(\alpha + \beta) * \mathbf{u} = \alpha * \mathbf{u} + \beta * \mathbf{u}$; $\forall \alpha, \beta \in \mathcal{K}; \forall \mathbf{u} \in \mathcal{V}$.*
4. *$(\alpha\beta) * \mathbf{u} = \alpha * (\beta * \mathbf{u})$; $\forall \alpha, \beta \in \mathcal{K}; \forall \mathbf{u} \in \mathcal{V}$.*
5. *$1 * \mathbf{u} = \mathbf{u}$; $\forall \mathbf{u} \in \mathcal{V}$, where $1 \in \mathcal{K}$ is the unit element.*

*The elements of $\mathcal{V}$ are called vectors; the elements of $\mathcal{K}$, scalars; the + function, sum of vectors; and the * function, multiplication by a scalar.*

Consider the set $\mathbb{R}^n$ of n-tuples of real numbers, that is,

$$\mathbb{R}^n \equiv \left\{ \mathbf{x} = (x_1, \ldots, x_n)^T | x_i \in \mathbb{R}; i = 1, \ldots, n \right\}$$

In this book, boldfaced letters will be used to represent elements of $\mathbb{R}^n$ and their components will be referred to by subscripts as in $\mathbf{x} = (x_1, \ldots, x_n)^T$, where T denotes the transpose of $(x_1, \ldots, x_n)$. The equality of sets is denoted in this book as $\equiv$.

The following example shows that $\mathbb{R}^n$ is a linear space.

Example 1.1 ($\mathbb{R}^n$ is a linear space). In $\mathbb{R}^n$ we can define two operations:

1. Sum of vectors:

$$\mathbf{x} + \mathbf{y} = \begin{pmatrix} x_1 + y_1 \\ \vdots \\ x_n + y_n \end{pmatrix}; \ \forall \mathbf{x}, \mathbf{y} \in \mathbb{R}^n.$$

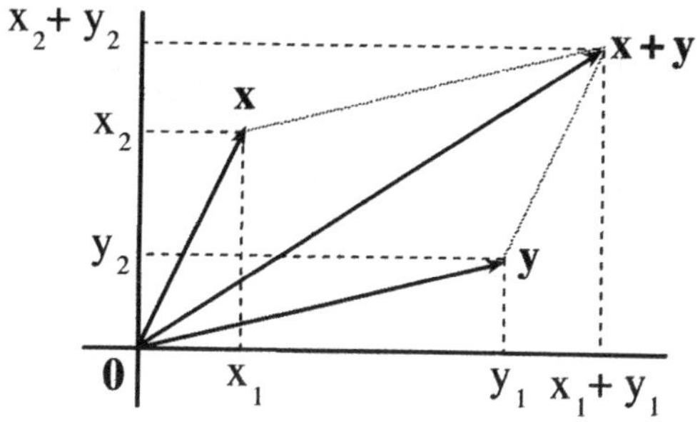

Fig. 1.1 Graphical illustration of the sum of vectors (parallelogram law).

2. Multiplication by a scalar:

$$\alpha \mathbf{x} = \begin{pmatrix} \alpha x_1 \\ \vdots \\ \alpha x_n \end{pmatrix}; \ \forall \alpha \in \mathbb{R}; \forall \mathbf{x} \in \mathbb{R}^n.$$

The notions of vector sum and multiplication by a scalar determine the linear-space structure of $\mathbb{R}^n$ over $\mathbb{R}$ with zero vector $\mathbf{0} = (0, \ldots, 0)^T$. It is easy to verify that they satisfy the conditions in Definition 1.1. ■

Example 1.2 (The set of free vectors). The set of all "free" vectors in the real plane is a linear space, where two quantities determine a vector: a length and a direction. A vector is represented by an arrow of a given length pointing to a given direction.

Choosing a given point as origin, the set of all "position" vectors is obtained. In this case, a single point determines a vector, represented by an arrow going from the fixed origin to the given point. Then, there is a one to one correspondence between vectors of $\mathbb{R}^2$ and points of the Cartesian plane $\mathbb{R} \times \mathbb{R}$. In this book, we shall refer indistinctly to both spaces.

Given two such vectors, their sum can be found by the familiar parallelogram law, as shown in Fig. 1.1.

The scalar multiple $\alpha \mathbf{x}$ of a vector $\mathbf{x}$ has the same (if $\alpha > 0$) or opposite (if $\alpha < 0$) direction as the vector $\mathbf{x}$, and its length or module, $\ell(\mathbf{x}) = \sqrt{x_1^2 + x_2^2}$, is $|\alpha|$ times the length of $\mathbf{x}$ (see Fig. 1.2).

■

Example 1.3 (The set of polynomials of degree smaller than or equal to n). Let $\mathcal{V} = \mathcal{P}_n(\mathbb{R})$ be the set of polynomials of degree smaller than or equal to n with coefficients in $\mathbb{R}$. That is,

$$\mathcal{P}_n(\mathbb{R}) \equiv \{a_0 + a_1 x + \cdots + a_n x^n \mid a_i \in \mathbb{R}, i = 0, 1, \ldots, n\}.$$

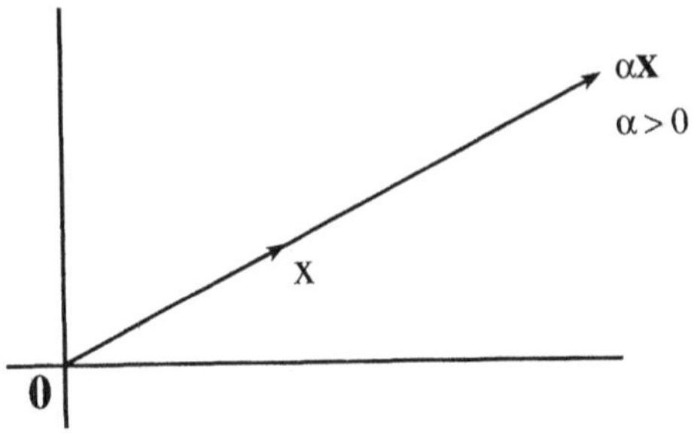

Fig. 1.2 Graphical illustration of the multiplication of a vector by a scalar.

$(\mathcal{P}_n(\mathbb{R}), +, *)$ is a linear space over $\mathbb{R}$, where $+$ and $*$ are the sum of polynomials and product of a polynomial by a real number, respectively; that is, if

$$p_a(x) = a_0 + a_1 x + \cdots + a_n x^n,$$

and

$$p_b(x) = b_0 + b_1 x + \cdots + b_n x^n,$$

we have

$$p_a(x) + p_b(x) = (a_0 + b_0) + (a_1 + b_1)x + \cdots + (a_n + b_n)x^n$$

and

$$\alpha p_a(x) = (\alpha a_0) + (\alpha a_1)x + \cdots + (\alpha a_n)x^n.$$

The set, $(\mathcal{P}(\mathbb{R}), +, *)$, of all polynomials with coefficients in $\mathbb{R}$ is also a linear space. ■

Definition 1.2 (Subspace). *A subset of $\mathcal{V}$ such that it is itself a linear space, with respect to addition and multiplication by scalars of $\mathcal{V}$, is called a subspace of $(\mathcal{V}, +, *)$.*

Remark 1.1 *A subset $\mathcal{W}$ of a linear space is a subspace iff $\forall \mathbf{u}, \mathbf{v} \in \mathcal{W}$ and $\forall \alpha \in \mathcal{K}$*

$$\mathbf{u} + \mathbf{v} \in \mathcal{W} \quad \textit{and} \quad \alpha \mathbf{u} \in \mathcal{W}. \tag{1.1}$$

Example 1.4 (Subsets defined by equations). Consider the following subsets of $\mathbb{R}^n$:

$$\begin{aligned} \mathcal{W}_1 &\equiv \{\mathbf{x} \in \mathbb{R}^n \mid x_1 + \cdots + x_n = 0\}, \\ \mathcal{W}_2 &\equiv \{\mathbf{x} \in \mathbb{R}^n \mid x_1 = \cdots = x_n\}, \\ \mathcal{W}_3 &\equiv \{\mathbf{x} \in \mathbb{R}^n \mid x_n = 1\}. \end{aligned}$$

According to Remark 1.1, $\mathcal{W}_1$ and $\mathcal{W}_2$ are subspaces. However, $\mathcal{W}_3$ is not, because vectors $\mathbf{u} = (0, \ldots, 0, 1)^T$ and $\mathbf{v} = (1, \ldots, 1, 1)^T$ belong to $\mathcal{W}_3$ but $\mathbf{u} + \mathbf{v} \notin \mathcal{W}_3$. ■

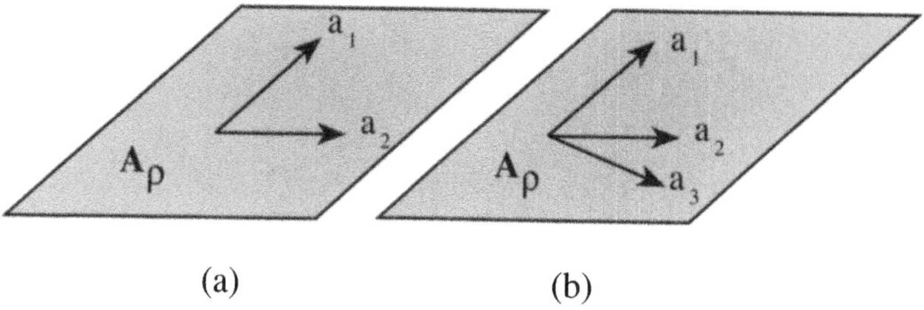

Fig. 1.3 Examples of sets $\mathbf{A}_\rho$ in $\mathbb{R}^2$.

Example 1.5 (The subset of polynomials having r as a root). Let $\mathcal{W}_r$ be the subset of polynomials that have r as a root, that is,

$$\mathcal{W}_r \equiv \{p(x) \in \mathcal{P}(\mathbb{R}) \mid p(r) = 0\}.$$

The set $\mathcal{W}_r$ is a subspace, because if $p(x), q(x) \in \mathcal{W}_r$ and $\lambda \in \mathbb{R}$, $(p+q)(r) = p(r) + q(r) = 0$ and $(\lambda p)(r) = \lambda p(r) = 0$, and then $p(x) + q(x), \lambda p(x) \in \mathcal{W}_r$. ■

Definition 1.3 (Linear combination). *Let $\mathbf{A} \equiv \{\mathbf{a}_1, \ldots, \mathbf{a}_k\}$ be an ordered set of k vectors of $\mathbb{R}^n$. A vector $\mathbf{x} \in \mathbb{R}^n$ is said to be a linear combi nation of the vectors $\mathbf{a}_1, \ldots, \mathbf{a}_k$, if it can be expressed in the following form:*

$$\mathbf{x} = \rho_1 \mathbf{a}_1 + \cdots + \rho_k \mathbf{a}_k, \tag{1.2}$$

where $\rho_i \in \mathbb{R}$, $i = 1, \ldots, k$.

In this book, the greek letter ρ is used to refer to unrestricted real numbers, and the set of all linear combinations of vectors of $\mathbf{A} \equiv \{\mathbf{a}_1, \ldots, \mathbf{a}_k\}$, which is a subspace of $\mathbb{R}^n$, will be denoted by $\mathbf{A}_\rho$, that is,

$$\mathbf{A}_\rho \equiv \{\rho_1 \mathbf{a}_1 + \cdots + \rho_k \mathbf{a}_k \mid \rho_i \in \mathbb{R},\ i = 1, \ldots, k\}. \tag{1.3}$$

The vectors $\mathbf{a}_1, \ldots, \mathbf{a}_k$ are known as the generators of $\mathbf{A}_\rho$.

Example 1.6 (The set $\mathbf{A}_\rho$). Fig. 1.3 shows the sets $\mathbf{A}_\rho$ associated with two and three vectors in $\mathbb{R}^2$. In both cases, the set coincides with $\mathbb{R}^2$, because every vector can be written as a linear combination of two of the generators. ■

Definition 1.4 (Non-null linear combination). *Let $\mathbf{A} \equiv \{\mathbf{a}_1, \ldots, \mathbf{a}_k\}$ be an ordered set of k vectors in E^n (Euclidean space of dimension n). A vector $\mathbf{x} \in E^n$ is said to be a non-null linear combination of vectors of $\mathbf{A}$, iff*

$$\mathbf{x} = \xi_1 \mathbf{a}_1 + \cdots + \xi_k \mathbf{a}_k, \tag{1.4}$$

where $\exists i \in \{1, \ldots, k\} \mid \xi_i \in \mathbb{R} \setminus \{0\}$.

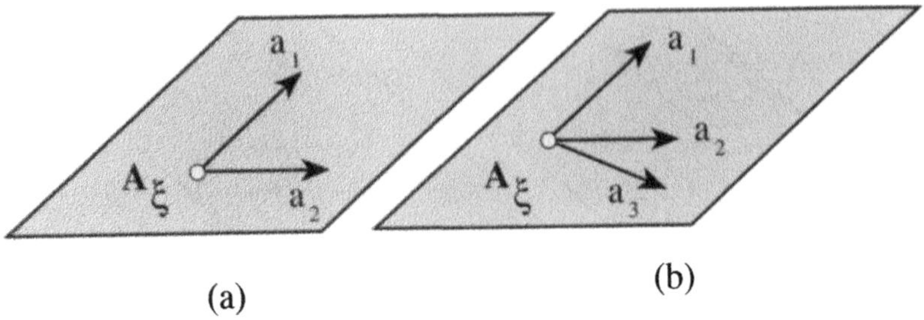

Fig. 1.4 Examples of sets $\mathbf{A}_\xi$ in E^2.

In this book, the greek letter ξ is used to refer to non-null real numbers, and $\mathbf{A}_\xi$ to denote the set of all non-null linear combinations of vectors in $\mathbf{A}$.

Example 1.7 (The set $\mathbf{A}_\xi$). Fig. 1.4 shows the sets $\mathbf{A}_\xi$ associated with two and three vectors in E^2. The white circle indicates that the origin doesn't belong to $\mathbf{A}_\xi$. Now the two sets $\mathbf{A}_\xi$ are equal again because every vector can be written as a non-null linear combination of two generators. ■

Definition 1.5 (Linear dependence). *Vectors in $\mathbf{A} \equiv \{\mathbf{a}_1, \ldots, \mathbf{a}_k\}$ are said to be linearly dependent if the zero vector, $\mathbf{0}$, can be written as a non-null linear combination of those vectors, that is, iff*

$$\mathbf{0} = \xi_1 \mathbf{a}_1 + \cdots + \xi_k \mathbf{a}_k \quad \text{with} \quad \sum_{i=1}^{k} |\xi_i| \neq 0. \tag{1.5}$$

In other words, vectors in $\mathbf{A}$ are linearly dependent iff $\mathbf{0} \in \mathbf{A}_\xi$. Otherwise, the vectors are said to be linearly independent.

Example 1.8 (Linearly dependent vectors). The vectors

$$\begin{aligned} \mathbf{a} &= (1, -2, 3, 5)^T, \\ \mathbf{b} &= (0, -1, 1, 2)^T, \\ \mathbf{c} &= (2, -1, 3, 4)^T \end{aligned}$$

in $\mathbb{R}^4$ are linearly dependent because $\mathbf{0} = (0,0,0,0)^T$ can be written as a non-null linear combination of them: $2\mathbf{a} - 3\mathbf{b} - \mathbf{c} = \mathbf{0}$. ■

Remark 1.2 *The set of vectors $\mathbf{A} \equiv \{\mathbf{a}_1, \ldots, \mathbf{a}_k\}$ is linearly dependent if it contains $\mathbf{0}$ or if one of the vectors is a linear combination of the remaining ones, and it is linearly independent iff*

$$\rho_1 \mathbf{a}_1 + \cdots + \rho_k \mathbf{a}_k = \mathbf{0} \Rightarrow \rho_1 = \cdots = \rho_k = 0. \tag{1.6}$$

Definition 1.6 (Set of generators). *A set of vectors $\{\mathbf{a}_1, \ldots, \mathbf{a}_k\}$ is said to be a set of generators of a linear space iff every vector in it can be expressed as a linear combination of them.*

Definition 1.7 (Basis). *A set of vectors $\{\mathbf{a}_1, \ldots, \mathbf{a}_k\}$ is a basis of $\mathbb{R}^n$ iff*

(i) it is a set of generators of $\mathbb{R}^n$, and

(ii) they are linearly independent.

Example 1.9 (Canonical basis). The unit coordinate vectors

$$\mathbf{e}_1 = \begin{pmatrix} 1 \\ 0 \\ \vdots \\ 0 \end{pmatrix}, \quad \mathbf{e}_2 = \begin{pmatrix} 0 \\ 1 \\ \vdots \\ 0 \end{pmatrix}, \ldots, \quad \mathbf{e}_n = \begin{pmatrix} 0 \\ 0 \\ \vdots \\ 1 \end{pmatrix}$$

form a basis of $\mathbb{R}^n$, which is known as the *canonical basis*. ■

Remark 1.3 *Every basis of $\mathbb{R}^n$ has exactly n elements, so $\mathbb{R}^n$ is said to be an n-dimensional space. Every set of generators of $\mathbb{R}^n$ has at least n vectors and n is also the maximum number of linearly independent vectors in $\mathbb{R}^n$.*

Example 1.10 (Polynomials). Any polynomial $p(x) \in \mathcal{P}_n(\mathbb{R})$ can be expressed as a linear combination of the polynomials in $\{1, x, \ldots, x^n\}$, that is, $\{1, x, \ldots, x^n\}$ is a set of generators of $\mathcal{P}_n(\mathbb{R})$. Moreover, these polynomials are linearly independent because

$$\xi_0 1 + \xi_1 x + \cdots + \xi_n x^n = 0 \Leftrightarrow \xi_i = 0, \quad i = 1, 2, \ldots, n,$$

so, they form a basis of $\mathcal{P}_n(\mathbb{R})$. ■

1.3 THE EUCLIDEAN SPACE E^N

The notions of vector sum and multiplication by scalars determine the linear space structure of $\mathbb{R}^n$, but they are not sufficient to define concepts such as distance, orthogonality, and angle. To define these concepts we introduce a dot product in $\mathbb{R}^n$.

Definition 1.8 (Inner or dot product). *An inner or dot product in $\mathbb{R}^n$ is a function*

$$\begin{array}{ccc} \mathbb{R}^n \times \mathbb{R}^n & \rightarrow & \mathbb{R} \\ (\mathbf{x}, \mathbf{y}) & \rightarrow & \mathbf{x} \bullet \mathbf{y} \end{array} \tag{1.7}$$

satisfying the following properties:

1. *Positive definiteness:* $\mathbf{x} \bullet \mathbf{x} \geq 0, \ \forall\, \mathbf{0} \neq \mathbf{x} \in \mathbb{R}^n$ *and* $\mathbf{x} \bullet \mathbf{x} = 0$ *iff* $\mathbf{x} = \mathbf{0}$.

2. *Symmetry:* $\mathbf{x} \bullet \mathbf{y} = \mathbf{y} \bullet \mathbf{x},\ \forall\ \mathbf{x}, \mathbf{y} \in \mathbb{R}^n$.
3. *Linearity:* $\mathbf{x} \bullet (\alpha\mathbf{y} + \beta\mathbf{z}) = \alpha(\mathbf{x} \bullet \mathbf{y}) + \beta(\mathbf{x} \bullet \mathbf{z}),\ \forall\ \mathbf{x}, \mathbf{y}, \mathbf{z} \in \mathbb{R}^n$ *and* $\alpha, \beta \in \mathbb{R}$.

An example of dot product in $\mathbb{R}^n$ is

$$\mathbf{x} \bullet \mathbf{y} = \sum_{i=1}^{n} x_i y_i, \tag{1.8}$$

which is known as the *Euclidean dot product.*

Remark 1.4 *Obviously, a dot product can be defined in any linear space. For example, the following expression defines a dot product in* $\mathcal{P}(\mathbb{R})$:

$$p(x) \bullet q(x) = \int_0^1 p(x)q(x)dx; \quad \forall p(x), q(x) \in \mathcal{P}(\mathbb{R}).$$

It is easy to check that this definition satisfies the positive definiteness, the symmetry, and the linearity axioms.

Definition 1.9 (Euclidean n-space). *The linear space* $\mathbb{R}^n$ *with the Euclidean dot product (1.8) is called the Euclidean n-space and is denoted by* E^n.

The dot product above allows defining the *Euclidean norm* as

$$\|\mathbf{x}\| = (\mathbf{x} \bullet \mathbf{x})^{1/2} = \left(\sum_{i=1}^{n} x_i^2\right)^{1/2},$$

which is denoted by $\|.\|$. The distance of two vectors is defined as

$$d(\mathbf{x}, \mathbf{y}) = \|\mathbf{x} - \mathbf{y}\|.$$

Remark 1.5 *If other dot products instead of (1.8) are used, other norms and distances can be obtained.*

If $\mathbf{x}$ and $\mathbf{y}$ are two nonzero vectors, the *angle* θ between $\mathbf{x}$ and $\mathbf{y}$ is defined as (see Fig. 1.5)

$$\cos\theta = \frac{\mathbf{x} \bullet \mathbf{y}}{\|\mathbf{x}\|\|\mathbf{y}\|}; \quad 0 \leq \theta \leq \pi.$$

Definition 1.10 (Orthogonal vectors). *The vectors* $\mathbf{x}$ *and* $\mathbf{y}$ *are said to be orthogonal if and only if* $\mathbf{x} \bullet \mathbf{y} = 0$.

A basis $\{\mathbf{v}_1, \ldots, \mathbf{v}_n\}$ of E^n is called *orthonormal* iff $\mathbf{v}_i \bullet \mathbf{v}_j = \delta_{ij}$, where δ_{ij} are the *Kronecker deltas*:

$$\delta_{ij} = \begin{cases} 0, & \text{if } i \neq j, \\ 1, & \text{if } i = j. \end{cases}$$

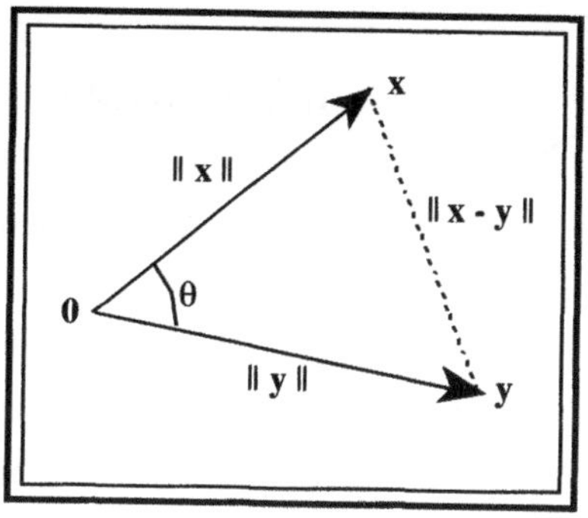

Fig. 1.5 Graphical illustration of the angle between two vectors.

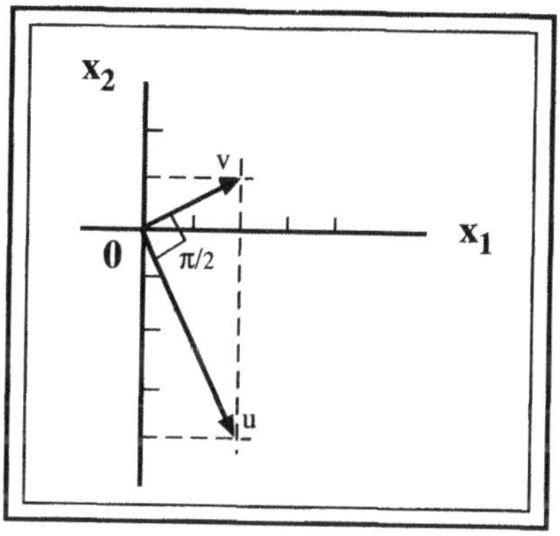

Fig. 1.6 Graphical illustration of two orthogonal vectors.

The canonical basis $\{\mathbf{e}_1, \ldots, \mathbf{e}_n\}$ is an orthonormal basis of E^n.

Remark 1.6 *Note that in* $\mathbb{R}^n$ *with the Euclidean dot product, if two non-null vectors* $\mathbf{u}$ *and* $\mathbf{v}$ *are orthogonal, then the angle* θ *between* $\mathbf{u}$ *and* $\mathbf{v}$ *is* $\theta = \pi/2$, *since*

$$\cos\theta = \frac{\mathbf{u} \bullet \mathbf{v}}{\|\mathbf{u}\|\|\mathbf{v}\|} = 0.$$

Example 1.11 (Two orthogonal vectors). Consider the Euclidean dot product in $\mathbb{R}^2$. Let $\mathbf{u} = (2, -4)^T$ and $\mathbf{v} = (2, 1)^T$ be two vectors in $\mathbb{R}^2$. Since

$$\mathbf{u} \bullet \mathbf{v} = 2 \times 2 + (-4) \times 1 = 4 - 4 = 0,$$

$\mathbf{u}$ and $\mathbf{v}$ are orthogonal, as can be seen in Fig. 1.6, where the angle θ between both vectors is $\pi/2$.

■

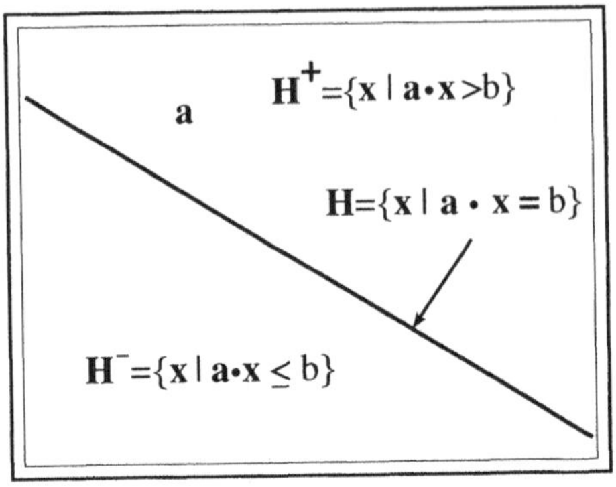

Fig. 1.7 Illustration of a halfspace in E^2.

1.3.1 Hyperplanes and Halfspaces

Definition 1.11 (Hyperplane). *A hyperplane in E^n is a set of the form*

$$H \equiv \{\mathbf{x} \in E^n | \mathbf{a} \bullet \mathbf{x} = b\}, \tag{1.9}$$

where $\mathbf{a} \in E^n$, $\mathbf{a} \neq \mathbf{0}$, *and* $b \in \mathbb{R}$.

Hyperplanes generalize the concept of lines in E^2 or planes in E^3.

The constant b in (1.9) can be eliminated if a particular $\mathbf{x}_0 \in H$ is known:

$$H \equiv \{\mathbf{x} \in E^n | \mathbf{a} \bullet (\mathbf{x} - \mathbf{x}_0) = 0\}. \tag{1.10}$$

In this case, we have

$$H \equiv \mathbf{x}_0 + \mathbf{B}_\rho,$$

where $\mathbf{B}_\rho \equiv \{\mathbf{x} \in E^n | \mathbf{a} \bullet \mathbf{x} = 0\}$ is a linear subspace of E^n of dimension $n-1$.

A hyperplane divides the space E^n into two halfspaces (see Fig. 1.7):

$$H^- \equiv \{\mathbf{x} \in E^n | \mathbf{a} \bullet \mathbf{x} \leq b\}, \tag{1.11}$$

$$H^+ \equiv \{\mathbf{x} \in E^n | \mathbf{a} \bullet \mathbf{x} > b\}. \tag{1.12}$$

The hyperplane (1.11) is a closed halfspace and the hyperplane (1.12) is an open halfspace.

1.4 ORTHOGONAL SETS AND DECOMPOSITIONS

Definition 1.12 (Orthogonal subspaces). *Two linear subspaces $\mathcal{U}$ and $\mathcal{V}$ are said to be orthogonal iff*

$$\mathbf{u} \bullet \mathbf{v} = 0; \ \forall \mathbf{u} \in \mathcal{U}, \forall \mathbf{v} \in \mathcal{V}.$$

If $\{\mathbf{u}_1, \ldots, \mathbf{u}_k\}$ and $\{\mathbf{v}_1, \ldots, \mathbf{v}_p\}$ are two sets of generators of $\mathcal{U}$ and $\mathcal{V}$, respectively, then the subspaces are orthogonal if and only if

$$\mathbf{u}_i \bullet \mathbf{v}_j = 0; \ i = 1, \ldots, k; \ j = 1, \ldots, p.$$

Example 1.12 (Orthogonal subspaces). Consider the subspaces $\mathcal{U} \equiv \{\mathbf{x} \in E^2 \mid x_1 + x_2 = 0\}$ and $\mathcal{V} \equiv \{\mathbf{x} \in E^2 \mid x_1 - x_2 = 0\}$. If $\mathbf{u} \in \mathcal{U}$ and $\mathbf{v} \in \mathcal{V}$, $\mathbf{u} = (a, -a)^T$ and $\mathbf{v} = (b, b)^T$ with $a, b \in \mathbb{R}$; then

$$\mathbf{u} \bullet \mathbf{v} = ab + (-a)b = 0.$$

Therefore, $\mathcal{U}$ and $\mathcal{V}$ are orthogonal. ∎

Definition 1.13 (Linear subspace orthogonal to a vector). *Let $\mathbf{u} \in E^n$, the linear subspace orthogonal to $\mathbf{u}$ is*

$$\mathbf{u}^\perp \equiv \{\mathbf{v} \in E^n \mid \mathbf{u} \bullet \mathbf{v} = 0\},$$

that is, the set of all vectors in E^n, which are orthogonal to $\mathbf{u}$.

Remark 1.7 *The set $\mathbf{u}^\perp$ is a linear subspace of $\mathbb{R}^n$.*

Example 1.13 (Linear subspace orthogonal to a vector). The linear subspace orthogonal to $\mathbf{u} = (1, -1, 3, 2)^T \in E^4$ is

$$\mathbf{u}^\perp \equiv \left\{\mathbf{x} \in \mathbb{R}^4 \mid x_1 - x_2 + 3x_3 + 2x_4 = 0\right\}.$$

∎

Definition 1.14 (Linear subspace orthogonal to a linear subspace). *Let $\mathcal{U}$ be a linear subspace in E^n. Its orthogonal subspace is*

$$\mathcal{U}^\perp \equiv \{\mathbf{v} \in E^n \mid \mathbf{u} \bullet \mathbf{v} = 0; \ \forall \mathbf{u} \in \mathcal{U}\},$$

that is, the set of all vectors in E^n, which are orthogonal to every vector in $\mathcal{U}$.

It can be seen that if $\{\mathbf{u}_1, \ldots, \mathbf{u}_k\}$ is a set of generators of $\mathcal{U}$ then

$$\mathcal{U}^\perp \equiv \bigcap_{i=1}^{k} \mathbf{u}_i^\perp.$$

Definition 1.15 (Direct sum). *We say that a linear subspace $\mathcal{V}$ is a direct sum of two subspaces $\mathcal{U}$ and $\mathcal{W}$, and we shall denote it as $\mathcal{V} = \mathcal{U} \oplus \mathcal{W}$, iff*

(i) $\forall \mathbf{v} \in \mathcal{V}$ there exists $\mathbf{u} \in \mathcal{U}, \mathbf{w} \in \mathcal{W}$ such that $\mathbf{v} = \mathbf{u} + \mathbf{w}$,

(ii) $\mathcal{U} \cap \mathcal{W} = \{0\}$.

Definition 1.16 (Orthogonal decomposition of a linear subspace with respect to a vector). *Let $\mathcal{V} \subset E^n$ be a linear subspace and $\mathbf{u} \in E^n$. The orthogonal decomposition of $\mathcal{V}$ with respect to $\mathbf{u}$ is*

$$\mathcal{V} \equiv \mathbf{u}_{\mathcal{V}}^{\perp} \oplus \mathcal{W},$$

where $\mathbf{u}_{\mathcal{V}}^{\perp} \equiv \mathcal{V} \cap \mathbf{u}^{\perp}$ is the linear subspace orthogonal to $\mathbf{u}$ in $\mathcal{V}$ and $\mathcal{W}$ is its $\mathcal{V}$-complement.

If in Definition 1.16 we consider $\mathcal{U}$ a linear subspace instead of a vector $\mathbf{u}$, we have the definition of orthogonal decomposition with respect to a linear subspace.

1.5 MATRICES

Definition 1.17 (Matrix). *A matrix is a rectangular array of elements of a field.*

We say that a matrix is of order $m \times n$ if it has m rows and n columns. Matrices of real numbers are an important tool used in this book. We use a boldfaced uppercase letter to denote a matrix and the corresponding lowercase letter with two subindices to refer to its elements. With this we have

$$\mathbf{A} = \begin{pmatrix} a_{11} & a_{12} & \cdots & a_{1n} \\ a_{21} & a_{22} & \cdots & a_{2n} \\ \cdots & \cdots & \cdots & \cdots \\ a_{m1} & a_{m2} & \cdots & a_{mn} \end{pmatrix}.$$

When $m = n$ we say that $\mathbf{A}$ is a *square matrix.*

The columns of $\mathbf{A}$ are denoted by $\mathbf{a}_1, \ldots, \mathbf{a}_n \in \mathbb{R}^m$ and its rows by $\mathbf{a}_1^T, \ldots, \mathbf{a}_m^T \in \mathbb{R}^n$; then we have

$$\mathbf{A} = (\mathbf{a}_1, \ldots, \mathbf{a}_n) = \begin{pmatrix} \mathbf{a}_1^T \\ \vdots \\ \mathbf{a}_m^T \end{pmatrix}.$$

Let $\mathcal{M}_{m\times n}(\mathbb{R})$ be the set of matrices of order $m \times n$ over $\mathbb{R}$. We can define two operations that give $\mathcal{M}_{m\times n}(\mathbb{R})$ a linear space structure.

Definition 1.18 (Sum of matrices). *The sum of two matrices $\mathbf{A} \in \mathcal{M}_{m\times n}(\mathbb{R})$ and $\mathbf{B} \in \mathcal{M}_{m\times n}(\mathbb{R})$ is another matrix $\mathbf{C} = \mathbf{A} + \mathbf{B} \in \mathcal{M}_{m\times n}(\mathbb{R})$, whose (i, j)-element is $c_{ij} = a_{ij} + b_{ij}$.*

Remark 1.8 *Note that only two matrices of the same dimensions can be added up, and that the resulting matrix has the same dimensions as the summands.*

Definition 1.19 (Multiplication by a scalar). *The multiplication of a matrix* $\mathbf{A} \in \mathcal{M}_{m\times n}(\mathbb{R})$ *by a scalar* $\alpha \in \mathbb{R}$ *is another matrix* $\mathbf{C} = \alpha\mathbf{A} \in \mathcal{M}_{m\times n}(\mathbb{R})$ *whose* (i, j)*-element is* $c_{ij} = \alpha a_{ij}$.

Remark 1.9 *Note that when multiplying a matrix by a scalar, the resulting matrix has the same dimensions as the initial matrix.*

Definition 1.20 (Product of two matrices). *The product of two matrices* $\mathbf{A} \in \mathcal{M}_{m\times n}(\mathbb{R})$ *and* $\mathbf{B} \in \mathcal{M}_{n\times p}(\mathbb{R})$ *is another matrix* $\mathbf{C} = \mathbf{AB} \in \mathcal{M}_{m\times p}(\mathbb{R})$ *whose* (i, j)*-element is*

$$c_{ij} = \sum_{k=1}^{n} a_{ik} b_{kj}. \tag{1.13}$$

Remark 1.10 *Matrix multiplication is associative and distributive with respect to addition but is not commutative.*

Definition 1.21 (Identity matrix). *The identity matrix* $\mathbf{I}_n$ *is a square matrix whose* (i, j)*-elements are* 0 *if* $i \neq j$ *and* 1 *if* $i = j$.

The identity matrix satisfies

$$\begin{aligned} \mathbf{A}\mathbf{I}_n &= \mathbf{A}, \forall \mathbf{A} \in \mathcal{M}_{m\times n}(\mathbb{R}), \\ \mathbf{I}_n\mathbf{A} &= \mathbf{A}, \forall \mathbf{A} \in \mathcal{M}_{n\times m}(\mathbb{R}). \end{aligned}$$

Definition 1.22 (Transpose matrix). *Let* $\mathbf{A} \in \mathcal{M}_{m\times n}(\mathbb{R})$ *be a matrix of real numbers. Its transpose* $\mathbf{A}^T$ *is a matrix of order* $n \times m$ *formed by interchanging the rows and columns of* $\mathbf{A}$.

Example 1.14 (Transpose matrix). To obtain the transpose of the matrix

$$\mathbf{A} = \begin{pmatrix} -1 & 2 & 0 \\ 0 & 1 & 1 \\ -2 & 0 & 1 \end{pmatrix},$$

we interchange rows and columns and get

$$\mathbf{A}^T = \begin{pmatrix} -1 & 0 & -2 \\ 2 & 1 & 0 \\ 0 & 1 & 1 \end{pmatrix}.$$

■

Of course, $(\mathbf{A}^T)^T = \mathbf{A}$ and also $(\mathbf{A} + \mathbf{B})^T = \mathbf{A}^T + \mathbf{B}^T$, $(\mathbf{AB})^T = \mathbf{B}^T\mathbf{A}^T$.

1.5.1 Partitioned Matrices

A submatrix of any matrix results from removing columns or rows or both from the matrix. A given matrix can be partitioned into submatrices or blocks by inserting cuts between rows or columns. For example, the matrix

$$\mathbf{A} = \begin{pmatrix} a_{11} & a_{12} & a_{13} & a_{14} & a_{15} \\ a_{21} & a_{22} & a_{23} & a_{24} & a_{25} \\ a_{31} & a_{32} & a_{33} & a_{34} & a_{35} \\ a_{41} & a_{42} & a_{43} & a_{44} & a_{45} \\ a_{51} & a_{52} & a_{53} & a_{54} & a_{55} \end{pmatrix}$$

can be partitioned into four submatrices as follows:

$$\mathbf{A} = \left(\begin{array}{c|c} \mathbf{A}_{11} & \mathbf{A}_{12} \\ \hline \mathbf{A}_{21} & \mathbf{A}_{22} \end{array}\right) = \left(\begin{array}{ccc|cc} a_{11} & a_{12} & a_{13} & a_{14} & a_{15} \\ a_{21} & a_{22} & a_{23} & a_{24} & a_{25} \\ a_{31} & a_{32} & a_{33} & a_{34} & a_{35} \\ \hline a_{41} & a_{42} & a_{43} & a_{44} & a_{45} \\ a_{51} & a_{52} & a_{53} & a_{54} & a_{55} \end{array}\right),$$

where

$$\mathbf{A}_{11} = \begin{pmatrix} a_{11} & a_{12} & a_{13} \\ a_{21} & a_{22} & a_{23} \\ a_{31} & a_{32} & a_{33} \end{pmatrix}, \quad \mathbf{A}_{12} = \begin{pmatrix} a_{14} & a_{15} \\ a_{24} & a_{25} \\ a_{34} & a_{35} \end{pmatrix},$$

$$\mathbf{A}_{21} = \begin{pmatrix} a_{41} & a_{42} & a_{43} \\ a_{51} & a_{52} & a_{53} \end{pmatrix}, \quad \mathbf{A}_{22} = \begin{pmatrix} a_{44} & a_{45} \\ a_{54} & a_{55} \end{pmatrix}.$$

Sums and multiplications of partitioned matrices can be performed in terms of correctly selected blocks.

We use partitioned matrices throughout the book. In many situations, it is more efficient working with blocks than with elements.

1.5.2 Inverse, Determinant, and Rank of a Matrix

Definition 1.23 (Nonsingular matrix). *A square matrix* $\mathbf{A}$ *is said to be nonsingular if there exists a matrix* $\mathbf{A}^{-1}$ *such that*

$$\mathbf{A}\mathbf{A}^{-1} = \mathbf{A}^{-1}\mathbf{A} = \mathbf{I}.$$

The matrix $\mathbf{A}^{-1}$ *is known as the inverse matrix of* $\mathbf{A}$. *A square matrix with no inverse is said to be singular. A square matrix with an inverse is called regular.*

Remark 1.11 *The inverse matrix* $\mathbf{A}^{-1}$ *is unique and satisfies*

$$\left(\mathbf{A}^{-1}\right)^{-1} = \mathbf{A} \quad \textit{and} \quad \left(\mathbf{A}^{-1}\right)^{T} = \left(\mathbf{A}^{T}\right)^{-1}.$$

When $\mathbf{A}$ *and* $\mathbf{B}$ *are both nonsingular of order* n*, then so is* $\mathbf{AB}$ *and*

$$(\mathbf{AB})^{-1} = \mathbf{B}^{-1}\mathbf{A}^{-1}.$$

Definition 1.24 (Determinant). *The determinant of a square matrix* $\mathbf{A}$ *is defined as*

$$|\mathbf{A}| = \sum_{i=1}^{n} a_{i1}A_{i1}, \tag{1.14}$$

where A_{i1} *is the cofactor defined as* $(-1)^{i+1}$ *times the determinant of the matrix resulting after removing the* i*-th row and the first column of* $\mathbf{A}$*. The determinant of a matrix* 1×1 *is the element itself.*

Remark 1.12 *Alternative definitions are*

$$|\mathbf{A}| = \sum_{i=1}^{n} a_{ik}A_{ik} = \sum_{i=1}^{n} a_{ki}A_{ki}, \tag{1.15}$$

where A_{ij} *is* $(-1)^{i+j}$ *times the determinant of the matrix resulting after removing the* i*-th row and the* j*-th column of* $\mathbf{A}$*. Both definitions of the determinat are recursive.*

The determinant satisfies some interesting properties, such as:

- A square matrix $\mathbf{A}$ is nonsingular iff $|\mathbf{A}| \neq 0$.
- $|\mathbf{A}^{-1}| = |\mathbf{A}|^{-1}$.
- $|\mathbf{A}^T| = |\mathbf{A}|$.
- $|\mathbf{AB}| = |\mathbf{A}||\mathbf{B}|$.
- The vectors $\mathbf{a}_1, \ldots, \mathbf{a}_n \in \mathbb{R}^n$ are linearly independent iff

$$|(\mathbf{a}_1, \ldots, \mathbf{a}_n)| \neq 0.$$

Definition 1.25 (Rank). *The rank of* $\mathbf{A} \in \mathcal{M}_{m \times n}(\mathbb{R})$ *is defined as:*

(i) The maximum number of linearly independent rows of $\mathbf{A}$*,*

(ii) The maximum number of linearly independent columns of $\mathbf{A}$*, or*

(iii) The order of the largest submatrix of $\mathbf{A}$ *having nonzero determinant.*

All of the above definitions are equivalent.

1.6 SYSTEMS OF LINEAR EQUATIONS

Let

$$\begin{array}{ccccccccc} a_{11}x_1 & + & a_{12}x_2 & + & \cdots & + & a_{1n}x_n & = & b_1 \\ a_{21}x_1 & + & a_{22}x_2 & + & \cdots & + & a_{2n}x_n & = & b_2 \\ \cdots & & \cdots & & \cdots & & \cdots & & \cdots \\ a_{m1}x_1 & + & a_{m2}x_2 & + & \cdots & + & a_{mn}x_n & = & b_m \end{array} \tag{1.16}$$

be a linear system of m equations with n unknowns. Let

$$\mathbf{A} = \begin{pmatrix} a_{11} & \cdots & a_{1n} \\ \cdots & \cdots & \cdots \\ a_{m1} & \cdots & a_{mn} \end{pmatrix}, \quad \mathbf{x} = \begin{pmatrix} x_1 \\ \vdots \\ x_n \end{pmatrix}, \text{ and } \mathbf{b} = \begin{pmatrix} b_1 \\ \vdots \\ b_m \end{pmatrix}.$$

Then the system in (1.16) can be written in matrix form as

$$\mathbf{Ax} = \mathbf{b}. \tag{1.17}$$

The matrix $\mathbf{A}$ is called the *coefficient matrix*, and $\mathbf{b}$, the *vector of independent terms*.

The following theorems establish some conditions for the existence and uniqueness of solution of system (1.17).

Theorem 1.1 (Existence of solution). *The system (1.17) has solution iff*

$$rank(\mathbf{A}) = rank(\mathbf{A} : \mathbf{b}),$$

where $(\mathbf{A} : \mathbf{b})$ *is the matrix with the columns of* $\mathbf{A}$ *and* $\mathbf{b}$.

Theorem 1.2 (Uniqueness of solution). *The system (1.17) has a unique solution iff*

$$rank(\mathbf{A}) = rank(\mathbf{A} : \mathbf{b}) = n.$$

When $\mathbf{b} = \mathbf{0}$ the system (1.17) is said to be *homogeneous*. There is an important relationship between the solution of a linear system and the solution of its associated homogeneous system.

Theorem 1.3 (General solution). *The general solution of the complete system (1.17) is*

$$\mathbf{x}_0 + \mathbf{x}_h, \tag{1.18}$$

where $\mathbf{x}_0$ *is a particular solution of (1.17) and* $\mathbf{x}_h$ *is the general solution of its associated homogeneous system.*

Remark 1.13 *The set of solutions of a homogeneous system is a linear subspace of* $\mathbb{R}^n$*, so the general solution of (1.17) can be expressed as*

$$\mathbf{x}_0 + \mathbf{B}_\rho,$$

where $\mathbf{B}_\rho$ *is the subspace generated by* $\{\mathbf{b}_1, \ldots, \mathbf{b}_k\}$.

Exercises

1.1 Let $\mathcal{M}_n(\mathbb{R})$ be the set of square matrices of dimension n with coefficients in $\mathbb{R}$. Prove that it is a linear space considering the usual sum of matrices and the product of a matrix by a scalar.

1.2 Consider the following vectors in $\mathbb{R}^3$:

$$\mathbf{v}_1 = (a, 0, 2)^T,$$

$$\mathbf{v}_2 = (1, b, 3)^T,$$

$$\mathbf{v}_3 = (2, 1, c)^T,$$

where $a, b, c \in \mathbb{R}$. What are the values of a, b, and c that make $\mathbf{v}_1, \mathbf{v}_2$, and $\mathbf{v}_3$ orthogonal vectors?

1.3 A factory makes three types of screws (large, medium, and small) using two different machines. The first machine makes 50 large, 70 medium, and 80 small screws per hour. The second machine makes 90 large, 50 medium, and 80 small screws per hour. Taking into account that when one machine works the other one cannot be used, calculate the time that each machine has to work in order to make 80 large, 55 medium, and 80 small screws.

1.4 If $\mathbf{A}$ and $\mathbf{B}$ are square matrices of order n, find examples of $\mathbf{A}$ and $\mathbf{B}$, where:

(a) $\mathbf{AB} \neq \mathbf{BA}$.

(b) $(\mathbf{A} + \mathbf{B})^2 \neq \mathbf{A}^2 + \mathbf{B}^2 + 2\mathbf{AB}$.

(c) $(\mathbf{A} + \mathbf{B})(\mathbf{A} - \mathbf{B}) \neq \mathbf{A}^2 - \mathbf{B}^2$.

1.5 A matrix $\mathbf{A} \in \mathcal{M}_n(\mathbb{R})$ is said to be *orthogonal* if $\mathbf{AA}^T = \mathbf{I}_n$.

(a) Calculate the value of λ that makes the following matrix orthogonal:

$$\mathbf{A} = \begin{pmatrix} \lambda & 0 & \lambda \\ \lambda & 0 & -\lambda \\ 0 & 1 & 0 \end{pmatrix}.$$

(b) Show that the rows of an orthogonal matrix in $\mathcal{M}_n(\mathbb{R})$ form an orthogonal basis of $\mathbb{R}^n$.

(c) Show that the following matrix is orthogonal:

$$\begin{pmatrix} \cos\alpha & -\sin\alpha & 0 \\ \sin\alpha & \cos\alpha & 0 \\ 0 & 0 & 1 \end{pmatrix}.$$

1.6 A square matrix $\mathbf{A}$ is said to be *idempotent* if $\mathbf{A}^2 = \mathbf{A}$.

(a) Show that the following matrix is idempotent:

$$\begin{pmatrix} 2 & -2 & -4 \\ -1 & 3 & 4 \\ 1 & -2 & -3 \end{pmatrix}.$$

(b) Show that if $\mathbf{AB} = \mathbf{A}$ and $\mathbf{BA} = \mathbf{B}$, then $\mathbf{A}$ and $\mathbf{B}$ are both idempotent.

(c) Show that if $\mathbf{A}$ is idempotent, then $\mathbf{A}^n = \mathbf{A}$ for all positive integers n.

1.7 Compute the determinant of the Vandermonde matrix:

$$\begin{vmatrix} 1 & x_1 & x_1^2 & \cdots & x_1^{n-1} \\ 1 & x_2 & x_2^2 & \cdots & x_2^{n-1} \\ \cdots & \cdots & \cdots & \cdots & \cdots \\ 1 & x_n & x_n^2 & \cdots & x_n^{n-1} \end{vmatrix}.$$

1.8 Let

$$\mathbf{A} = \begin{pmatrix} 0 & \alpha & 1 \\ 2 & 1 & \alpha \\ 0 & 1 & \alpha \end{pmatrix}.$$

(a) Calculate α such that $rank(\mathbf{A}) = 2$.

(b) Calculate α such that $rank(\mathbf{A}) = 3$.

1.9 Prove that the function $T : \mathbb{R}^3 \times \mathbb{R}^3 \longrightarrow \mathbb{R}$ defined by

$$T(\mathbf{x}, \mathbf{y}) = (x_1 - x_2)y_1 + (-x_1 + 3x_2)y_3 + x_3y_3,$$

where $\mathbf{x} = (x_1, x_2, x_3)$ and $\mathbf{y} = (y_1, y_2, y_3)$, is a dot product in $\mathbb{R}^3$.

1.10 Consider the system

$$\begin{array}{rrrcr} x & +2y & & = & -1 \\ 2x & & +z & = & -2 \\ -x & +2y & +z & = & 0 \end{array}$$

Find its solution by calculating the inverse of the coefficient matrix.

2

Orthogonal Sets

2.1 INTRODUCTION AND MOTIVATION

Orthogonal sets are used frequently in many areas of mathematics. They allow solving problems in an elegant, direct, and easily tractable manner. In addition, their geometric interpretations facilitate the understanding of processes and algorithms.

This chapter is dedicated to orthogonal sets. It describes some necessary concepts and key methods that are systematically used in the remaining chapters of the book.

In Section 2.2 we introduce the pivoting transformation, which transforms a linear subspace into itself, a very important property that needs to be analyzed to fully understand what the pivoting process leads to. This transformation gives a method for obtaining the orthogonal decomposition of a linear subspace V^0 with respect to a given vector $\mathbf{u}$. This means that V^0 is decomposed as a direct sum of two linear subspaces $\mathcal{W}$ and $\mathcal{V}$, such that $\mathcal{W}$ is orthogonal to $\mathbf{u}$ and $\mathcal{V}$ is its V^0-complement. Next, this method is extended to a linear subspace $\mathcal{U}$, by the sequential application of the previous method to all of its generators.

In Section 2.3 an algorithm for obtaining the set orthogonal to a linear subspace, in another given linear subspace, and its complement is presented.

Finally, in the last section, a Mathematica module for solving these important problems is given. Both the algorithm and the module are illustrated by several examples.

2.2 ORTHOGONAL DECOMPOSITIONS

This section presents a method for obtaining orthogonal decompositions and their fundamentals. The main tool consists of the so called *pivoting transformation*, which transforms a set of vectors $\mathbf{V}^j \equiv \{\mathbf{v}_1^j, \ldots, \mathbf{v}_n^j\}$ into another set of vectors $\mathbf{V}^{j+1} \equiv \{\mathbf{v}_1^{j+1}, \ldots, \mathbf{v}_n^{j+1}\}$ by

$$\mathbf{v}_k^{j+1} = \begin{cases} \mathbf{v}_k^j / t_j^j; & if\ k = j, \\ \mathbf{v}_k^j - \dfrac{t_k^j}{t_j^j}\mathbf{v}_j^j; & if\ k \neq j, \end{cases} \tag{2.1}$$

where $t_j^j \neq 0$ and $t_k^j; k \neq j$ are arbitrary real numbers. This transformation is used in well-known methods, such as the Gauss elimination method.

A close look at (2.1) shows that the pivoting transformation consists of:

1. *A normalization step*, in which the pivot vector is divided by t_j^j, and
2. *A pivoting process*, in which we subtract from each vector $\mathbf{v}_k$ the new pivot vector multiplied by the corresponding value t_k^j.

This transformation can be formulated in matrix form as follows. Define the matrices

$$\mathbf{V}^j = \left(\mathbf{v}_1^j, \ldots, \mathbf{v}_n^j\right) \text{ and } \mathbf{M}_j = \left(\mathbf{e}_1, \ldots, \mathbf{e}_{j-1}, \mathbf{t}_j, \mathbf{e}_{j+1}, \ldots, \mathbf{e}_n\right)^T, \tag{2.2}$$

where $\mathbf{e}_i$ is the ith column of the identity matrix and $\mathbf{t}_j = (t_1^j, \ldots, t_n^j)^T$. Since $t_j^j \neq 0$ the matrix $\mathbf{M}_j$ is invertible and from (2.1) we get

$$\mathbf{V}^{j+1} = \mathbf{V}^j \mathbf{M}_j^{-1}, \tag{2.3}$$

which is the matrix formulation of (2.1). It can be proved that $\mathbf{M}_j^{-1}$ is the identity matrix with its jth row replaced by

$$\mathbf{t}_j^* = \frac{1}{t_j^j}\left(-t_1^j, \ldots, -t_{j-1}^j, 1, -t_{j+1}^j, \ldots, -t_n^j\right)^T;$$

that is,

$$\mathbf{M}_j^{-1} = \left(\mathbf{e}_1, \ldots, \mathbf{e}_{j-1}, \mathbf{t}_j^*, \mathbf{e}_{j+1}, \ldots, \mathbf{e}_n\right)^T. \tag{2.4}$$

The following lemma shows that given a matrix $\mathbf{V}$, its columns (or rows) can be transformed using the pivoting transformation without changing the linear subspace they generate. The lemma also prepares the notation for the process to be used later. More specifically, since in the orthogonalization process, to be explained, the columns of a matrix are to be transformed sequentially, we use a superindex to indicate the step number.

Lemma 2.1 *Let $\mathcal{L}(\mathbf{V}^j) = \mathcal{L}\{\mathbf{v}_1^j, \ldots, \mathbf{v}_n^j\}$ be the linear subspace generated or spanned by the set of vectors $\{\mathbf{v}_1^j, \ldots, \mathbf{v}_n^j\}$. Consider the pivoting transformation (2.1) and let $\mathcal{L}(\mathbf{V}^{j+1}) = \mathcal{L}\{\mathbf{v}_1^{j+1}, \ldots, \mathbf{v}_n^{j+1}\}$; then $\mathcal{L}(\mathbf{V}^j) = \mathcal{L}(\mathbf{V}^{j+1})$.*

Proof: $\mathcal{L}(\mathbf{V}^{j+1}) \subseteq \mathcal{L}(\mathbf{V}^j)$, because the $\mathcal{L}(\mathbf{V}^{j+1})$ generators belong to $\mathcal{L}(\mathbf{V}^j)$. But we also have $\mathcal{L}(\mathbf{V}^j) \subseteq \mathcal{L}(\mathbf{V}^{j+1})$, because we can write

$$\begin{aligned}
&\alpha_1 \mathbf{v}_1^j + \ldots + \alpha_n \mathbf{v}_n^j \\
&= \alpha_1 \left(\mathbf{v}_1^j - \frac{t_1^j}{t_j^j} \mathbf{v}_j^j \right) + \cdots + \left(\frac{\alpha_1 t_1^j + \cdots + \alpha_n t_n^j}{t_j^j} \right) \mathbf{v}_j^j + \cdots + \alpha_n \left(\mathbf{v}_n^j - \frac{t_n^j}{t_j^j} \mathbf{v}_j^j \right) \\
&= \alpha_1 \mathbf{v}_1^{j+1} + \cdots + \left(\alpha_1 t_1^j + \cdots + \alpha_n t_n^j \right) \mathbf{v}_j^{j+1} + \cdots + \alpha_n \mathbf{v}_n^{j+1},
\end{aligned} \tag{2.5}$$

which shows that any vector from $\mathcal{L}(\mathbf{V}^j)$ can be written as a linear combination of the $\mathcal{L}(\mathbf{V}^{j+1})$ generators.

An easier alternative proof comes from (2.3), because $\mathbf{V}^{j+1}$, the matrix whose columns are the generators of $\mathcal{L}(\mathbf{V}^{j+1})$, is obtained from $\mathbf{V}^j$ after postmultiplying it by a nonsingular matrix. ■

Remark 2.1 *The important implication of the lemma is that if $t_j^j \neq 0$ this transformation does not alter the linear subspace spanned or generated by the columns of a given matrix $\mathbf{V}$. This important property will be used throughout the book.*

Example 2.1 (Pivoting transformation). Consider the matrix

$$\mathbf{V}^1 = \begin{pmatrix} 1 & 2 & -1 \\ 0 & -1 & 3 \\ -1 & 1 & 0 \\ 3 & 0 & 4 \end{pmatrix}, \tag{2.6}$$

the linear subspace spanned by its columns $\mathcal{L}(\mathbf{V}^1) = \mathcal{L}\{\mathbf{v}_1^1, \mathbf{v}_2^1, \mathbf{v}_3^1\}$, where

$$\begin{aligned}
\mathbf{v}_1^1 &= (1, 0, -1, 3)^T, \\
\mathbf{v}_2^1 &= (2, -1, 1, 0)^T, \\
\mathbf{v}_3^1 &= (-1, 3, 0, 4)^T,
\end{aligned}$$

and the following real numbers: $t_1^1 = 2$, $t_2^1 = -1$, and $t_3^1 = 3$. Then, after using the transformation (2.1), we get the subspace $\mathcal{L}(\mathbf{V}^2) = \mathcal{L}\{\mathbf{v}_1^2, \mathbf{v}_2^2, \mathbf{v}_3^2\}$, where

$$\begin{aligned}
\mathbf{v}_1^2 &= (1, 0, -1, 3)^T/2 = (1/2, 0, -1/2, 3/2)^T, \\
\mathbf{v}_2^2 &= (2, -1, 1, 0)^T - \frac{-1}{2}(1, 0, -1, 3)^T = (5/2, -1, 1/2, 3/2)^T, \\
\mathbf{v}_3^2 &= (-1, 3, 0, 4)^T - \frac{3}{2}(1, 0, -1, 3)^T = (-5/2, 3, 3/2, -1/2)^T.
\end{aligned}$$

Obviously, both subspaces are the same, because the generators of $\mathcal{L}(\mathbf{V}^1)$ belong to $\mathcal{L}(\mathbf{V}^2)$ and vice versa.

Note that the initial matrix $\mathbf{V}^1$ is transformed to

$$\mathbf{V}^2 = \mathbf{V}^1\mathbf{M}_1^{-1} = \begin{pmatrix} 1 & 2 & -1 \\ 0 & -1 & 3 \\ -1 & 1 & 0 \\ 3 & 0 & 4 \end{pmatrix} \begin{pmatrix} 1/2 & 1/2 & -3/2 \\ 0 & 1 & 0 \\ 0 & 0 & 1 \end{pmatrix}$$

$$= \begin{pmatrix} 1/2 & 5/2 & -5/2 \\ 0 & -1 & 3 \\ -1/2 & 1/2 & 3/2 \\ 3/2 & 3/2 & -1/2 \end{pmatrix}.$$

■

2.2.1 Orthogonal Decomposition with Respect to a Vector

In the following theorem the procedure for obtaining the orthogonal decomposition of a linear subspace $\mathcal{L}(\mathbf{V}^j)$ with respect to a given vector is presented. In fact, we show that the pivoting process of Lemma 2.1, with an adequate selection of $t_k^j; k = 1, \ldots, n$, leads to this decomposition.

Theorem 2.1 (Orthogonal decomposition with respect to a given vector). *Assume now a vector $\mathbf{u}_j \neq \mathbf{0}$ and let $t_k^j = \mathbf{u}_j \bullet \mathbf{v}_k^j; k = 1, \ldots, n$ be the dot products of $\mathbf{u}_j$ and $\mathbf{v}_k^j$. If $t_j^j \neq 0$, then*

$$\mathbf{u}_j \bullet \mathbf{v}_k^{j+1} = \delta_{jk}, \tag{2.7}$$

where the $\mathbf{v}_1^{j+1}, \ldots, \mathbf{v}_n^{j+1}$ vectors are those in (2.1), and δ_{jk} are the Kronecker deltas ($\delta_{ij} = 1$, if $i = j$; and $\delta_{ij} = 0$, otherwise).

In addition, the orthogonal set to $\mathbf{u}_j$ in $\mathcal{L}(\mathbf{V}^j)$ is

$$\mathcal{L}(\mathbf{V}_j)^\perp \equiv \{\mathbf{v} \in \mathcal{L}(\mathbf{V}^j) | \mathbf{u}_j \bullet \mathbf{v} = 0\} \equiv \mathcal{L}\left(\mathbf{v}_1^{j+1}, \ldots, \mathbf{v}_{j-1}^{j+1}, \mathbf{v}_{j+1}^{j+1}, \ldots, \mathbf{v}_n^{j+1}\right),$$

and its complement is $\mathcal{L}(\mathbf{v}_j^{j+1})$.

In other words, the transformation (2.1) leads to the set orthogonal to $\mathbf{u}_j$ and its complement.

Proof: It is clear from (2.1) that $\mathbf{u}_j \bullet \mathbf{v}_j^{j+1} = 1$ and that $\mathbf{u}_j$ is orthogonal to the set of vectors $\{\mathbf{v}_1^{j+1}, \ldots, \mathbf{v}_{j-1}^{j+1}, \mathbf{v}_{j+1}^{j+1}, \ldots, \mathbf{v}_n^{j+1}\}$ because we have

$$\mathbf{u}_j \bullet \mathbf{v}_k^{j+1} = \mathbf{u}_j \bullet \left(\mathbf{v}_k^j - \frac{t_k^j}{t_j^j}\mathbf{v}_j^j\right) = t_k^j - \frac{t_k^j t_j^j}{t_j^j} = 0;\ k \neq j. \tag{2.8}$$

Since $\mathbf{u}_j$ is not the null vector, it is clear that $dim(\mathcal{L}(\mathbf{V}_j)^\perp) = d - 1$, where $d = dim(\mathcal{L}(\mathbf{V}^j))$. On the other hand, because of Lemma 2.1 we have

$$d - 1 \leq dim\left[\mathcal{L}\left(\mathbf{v}_1^{j+1}, \ldots, \mathbf{v}_{j-1}^{j+1}, \mathbf{v}_{j+1}^{j+1}, \ldots, \mathbf{v}_n^{j+1}\right)\right] \leq d.$$

Thus,

$$\mathcal{L}(V_j)^{\perp} \equiv \mathcal{L}\left(\mathbf{v}_1^{j+1}, \ldots, \mathbf{v}_{j-1}^{j+1}, \mathbf{v}_{j+1}^{j+1}, \ldots, \mathbf{v}_n^{j+1}\right)$$

and

$$\mathcal{L}(V_j) \equiv \mathcal{L}\left(\mathbf{v}_j^{j+1}\right) \oplus \mathcal{L}\left(\mathbf{v}_1^{j+1}, \ldots, \mathbf{v}_{j-1}^{j+1}, \mathbf{v}_{j+1}^{j+1}, \ldots, \mathbf{v}_n^{j+1}\right).$$

■

Remark 2.2 *Note that Theorem 2.1 allows obtaining the orthogonal set* $\mathbf{V}_j^{\perp}$ *to a given vector* $\mathbf{u}_j$ *in any case. If* $t_j^j = 0$*, we can reorder the* $\mathbf{v}$ *vectors until we satisfy the condition* $t_j^j \neq 0$ *or we find that* $t_j^j = 0, \forall j = 1, \ldots, n$*, in which case the orthogonal set to* $\mathbf{u}_j$ *in* $\mathcal{L}(\mathbf{V}^j)$ *is all* $\mathcal{L}(\mathbf{V}^j)$.

Example 2.2 (Orthogonal decomposition of a subspace with respect to a vector). The orthogonal decomposition of subspace $\mathcal{L}(\mathbf{V}^1)$ in Example 2.1 with respect to the vector $\mathbf{u} = (3/4, 5/4, -5/4, 0)^T$ can be obtained using Theorem 2.1. To this aim, we first calculate the dot products of $\mathbf{u}$ and the generators of $\mathcal{L}(\mathbf{V}^1)$:

$$\begin{aligned} t_1^1 &= \mathbf{u} \bullet \mathbf{v}_1^1 = 2, \\ t_2^1 &= \mathbf{u} \bullet \mathbf{v}_2^1 = -1, \\ t_3^1 &= \mathbf{u} \bullet \mathbf{v}_3^1 = 3. \end{aligned}$$

Next, using the transformation (2.1), we obtain the subspace $\mathcal{L}(\mathbf{V}^2)$ in Example 2.1 and then we conclude that the orthogonal subspace of $\mathcal{L}(\mathbf{V}^1)$ with respect to the vector $\mathbf{u}$ is

$$\mathcal{L}(\mathbf{V}_1)^{\perp} = \mathcal{L}\{(5/2, -1, 1/2, 3/2)^T, (-5/2, 3, 3/2, -1/2)^T\},$$

and its complement is $\mathcal{L}\{(1/2, 0, -1/2, 3/2)^T\}$.

The above procedure can be organized in a tabular form (see Table 2.1). We start with the matrix $\mathbf{V}^1$ and the vector $\mathbf{u}$, in the leftmost column. Next, we calculate the dot products, which are written in the last row of the table. Choosing the first non-null dot product, we obtain the pivot column (boldfaced in the table). Finally, we perform the pivoting process. The last two columns in the output part of Table 2.1 are the generators of the orthogonal subspace and the first column is the generator of the complement subspace (this has been italicized).

■

Example 2.3 (Orthogonal decomposition of $\mathbb{R}^3$ with respect to a vector). Consider the vector $\mathbf{u} = (1, -1, 3)^T$ in $\mathbb{R}^3$. With the algorithm above we get the orthogonal decomposition of $\mathbb{R}^3$ with respect to $\mathbf{u}$ (see Table 2.2):

$$\mathbb{R}^3 = \mathcal{L}\{\mathbf{v}_1\} \oplus \mathcal{L}\{\mathbf{v}_2, \mathbf{v}_3\},$$

Table 2.1 Iteration for obtaining $\mathcal{L}(\mathbf{V}_1)^{\perp}$.

	Iteration 1			**Output**		
3/4	**1**	2	−1	*1/2*	5/2	−5/2
5/4	**0**	−1	3	*0*	−1	3
−5/4	**−1**	1	0	*−1/2*	1/2	3/2
0	**3**	0	4	*3/2*	3/2	−1/2
	2	−1	3			

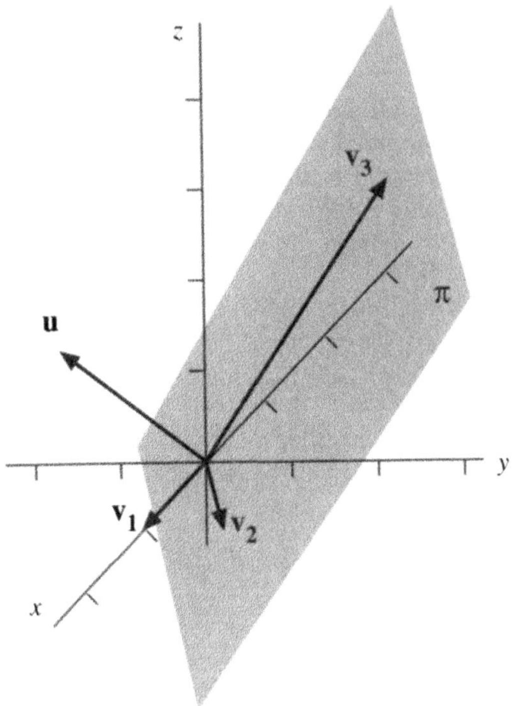

Fig. 2.1 Illustration of the orthogonal decomposition of $\mathbb{R}^3$ with respect to the vector $\mathbf{u} = (1, -1, 3)^T$.

where $\mathbf{v}_1 = (1, 0, 0)^T$, $\mathbf{v}_2 = (1, 1, 0)^T$, and $\mathbf{v}_3 = (-3, 0, 1)^T$. Figure 2.1 shows these vectors and the set orthogonal to $\mathbf{u}$.

As a more interesting alternative, we can use the algorithm again to obtain a decomposition with mutually orthogonal vectors as follows:

- Calculate the orthogonal set of $\mathbf{u}_{\mathbf{v}}^{\perp}$ in $\mathbb{R}^3$ as shown in Table 2.3. Observe that we are only interested in calculating the orthogonal subspace. Thus, we write only the corresponding part of the tabular representation.

Table 2.2 Iteration for obtaining the orthogonal decomposition of $\mathbb{R}^3$ with respect to the vector $\mathbf{u} = (1, -1, 3)^T$.

Iteration 1				Output		
1	**1**	0	0	*1*	1	-3
-1	**0**	1	0	*0*	1	0
3	**0**	0	1	*0*	0	1
	1	-1	3			

Table 2.3 Set orthogonal to $\mathbf{u}_{\mathbf{v}}^{\perp}$ in $\mathbb{R}^3$.

Iteration 1				Iteration 2			Output
1	**1**	0	0	-3	**-1**	0	1/3
1	**0**	1	0	0	**1**	0	-1/3
0	**0**	0	1	1	**0**	1	1
	1	1	0		**3**	1	

- Now consider the first resulting generator of $\mathbf{u}_{\mathbf{v}}^{\perp}$ and calculate the orthogonal subspace of this vector in the linear subspace $\mathbf{u}_{\mathbf{v}}^{\perp}$ (see Table 2.4). The resulting vector together with the first lead to an orthogonal basis for $\mathbf{u}_{\mathbf{v}}^{\perp}$.

Thus, we get

$$\mathcal{V} \equiv \mathbf{u}_{\mathbf{v}}^{\perp} \oplus \mathcal{W}, \tag{2.9}$$

where

$$\begin{aligned} \mathbf{u}_{\mathbf{v}}^{\perp} &\equiv \mathcal{L}\{(1,1,0)^T, (-3/2, 3/2, 1)^T\} \\ \mathcal{W} &\equiv \mathcal{L}\{(1/3, -1/3, 1)^T\}, \end{aligned}$$

whose generators form an orthogonal basis of $\mathcal{V}$. It can be seen that this process is the Gram–Schmidt method. ∎

Table 2.4 Set orthogonal to vector $(1, 1, 0)^T$ in $\mathbf{u}_{\mathbf{v}}^{\perp}$.

Iteration 1			Output
1	**1**	- 3	-3/2
1	**1**	0	3/2
0	**0**	1	1
	2	-3	

2.2.2 Orthogonal Decomposition with Respect to a Linear Subspace

Next, we extend the orthogonalization procedure to a linear subspace.

Lemma 2.2 *If we sequentially apply the transformation in Lemma 2.1 to a set of linearly independent vectors $\{\mathbf{u}_1, \ldots, \mathbf{u}_j\}$, the orthogonalization and normalization properties in (2.7) are kept. In other words, we have*

$$\mathbf{u}_r \bullet \mathbf{v}_k^{j+1} = \delta_{rk}; \; \forall r \le j; \; \forall j, \tag{2.10}$$

where δ_{rk} are the Kronecker deltas.

Proof: We prove this by induction over j.
Step 1: The lemma is true for $j = 1$, because from (2.7) we have

$$\mathbf{u}_1 \bullet \mathbf{v}_k^2 = \delta_{1k}.$$

Step j: We assume that the lemma is true for j, that is,

$$\mathbf{u}_r \bullet \mathbf{v}_k^{j+1} = \delta_{rk}; \; \forall r \le j; \; \forall j. \tag{2.11}$$

Step j+1: We prove that it is true for $j+1$. In fact, we can write

$$\mathbf{u}_r \bullet \mathbf{v}_k^{j+2} = \begin{cases} \delta_{rk}, & \text{if} \quad r = j+1, \\ \mathbf{u}_r \bullet \dfrac{\mathbf{v}_k^{j+1}}{t_{j+1}^{j+1}} = \dfrac{\delta_{rk}}{t_{j+1}^{j+1}} = 0 = \delta_{rk}, & \text{if} \quad r < k = j+1, \\ \mathbf{u}_r \bullet \left(\mathbf{v}_k^{j+1} - \dfrac{t_k^{j+1}}{t_{j+1}^{j+1}} \mathbf{v}_{j+1}^{j+1} \right) & \\ = \delta_{rk} - \dfrac{t_k^{j+1}}{t_{j+1}^{j+1}} \delta_{rj+1} = \delta_{rk}, & \text{if} \quad r < j+1 \neq k, \end{cases} \tag{2.12}$$

where in the first case we have used (2.7) and in the second and third cases we have used (2.1) and (2.7). ∎

Theorem 2.2 (Orthogonal decomposition with respect to a given linear subspace). *Assume now the linear subspace $\mathcal{L}\{\mathbf{u}_1, \ldots, \mathbf{u}_n\}$. We can sequentially use Theorem 2.1 to obtain the orthogonal set to $\mathcal{L}\{\mathbf{u}_1, \ldots, \mathbf{u}_n\}$ in a given subspace $\mathcal{L}(\mathbf{V}^0)$. Let t_i^j be the dot product of $\mathbf{u}_j$ and $\mathbf{v}_i^j$. Then assuming, without loss of generality, that $t_q^{q-1} \neq 0$, we obtain*

$$\begin{aligned} \mathcal{L}(\mathbf{V}^{q-1}) &\equiv \mathcal{L}\left(\mathbf{v}_1^{q-1} - \frac{t_1^{q-1}}{t_q^{q-1}} \mathbf{v}_q^{q-1}, \ldots, \mathbf{v}_q^{q-1}, \ldots, \mathbf{v}_n^{q-1} - \frac{t_n^{q-1}}{t_q^{q-1}} \mathbf{v}_q^{q-1} \right) \\ &\equiv \mathcal{L}\left(\mathbf{v}_1^q, \ldots, \mathbf{v}_n^q\right) \equiv \mathcal{L}(\mathbf{V}^q), \end{aligned}$$

and

$$\mathcal{L}(\mathbf{V}_q)^\perp \equiv \{\mathbf{v} \in \mathbf{V}^0 | \mathbf{u}_1 \bullet \mathbf{v} = 0, \ldots, \mathbf{u}_q \bullet \mathbf{v} = 0\} \equiv \mathcal{L}\left(\mathbf{v}_{q+1}^q, \ldots, \mathbf{v}_n^q\right),$$

Table 2.5 Iterations for obtaining the orthogonal vector to the linear subspace $\mathcal{P}$.

Iteration 1				**Iteration 2**				**Output**		
3/2	**1**	0	0	1	*2/3*	**-2/3**	0	*0*	*1*	-1
1	**0**	1	0	0	*0*	**1**	0	*1*	*-3/2*	3/2
0	**0**	0	1	1	*0*	**0**	1	*0*	*0*	1
	3/2	1	0		*2/3*	**−2/3**	1			

where $\mathcal{L}(\mathbf{V}^q)$ and $\mathcal{L}(\mathbf{V}^q)^\perp$ are the transformed linear subspace and the orthogonal set associated with the q-th step. In addition, we have

$$\mathbf{u}_1 \bullet \mathbf{v}_1^q = 1, \mathbf{u}_1 \bullet \mathbf{v}_i^q = 0, \forall i \neq 1, \ldots, \mathbf{u}_q \bullet \mathbf{v}_q^q = 1, \mathbf{u}_q \bullet \mathbf{v}_i^q = 0, \forall i \neq q.$$

The proof can easily be obtained using Lemma 2.2.

Example 2.4 (Vector orthogonal to a linear subspace). The vector $\mathbf{u}$ orthogonal to the linear subspace $\mathcal{P} = \mathcal{L}\{\mathbf{v}_1, \mathbf{v}_2\}$, where

$$\begin{aligned} \mathbf{v}_1 &= (3/2, 1, 0)^T, \\ \mathbf{v}_2 &= (1, 0, 1)^T, \end{aligned}$$

can be obtained by the previous method (see Table 2.5).

The result is illustrated in Fig. 2.2. The orthogonal vector is $\mathbf{u} = (-2, 3, 2)^T$. ∎

2.3 THE ORTHOGONALIZATION MODULE

In this section we describe an algorithm for obtaining orthogonal decompositions. This algorithm is based on the theory developed in the previous section.

Assume the linear subspace $\mathcal{L}(\mathbf{U}) \equiv \mathcal{L}\{\mathbf{u}_1, \ldots, \mathbf{u}_n\}$. We can obtain the orthogonal subspace in a given subspace $\mathcal{L}(\mathbf{V}^0)$ to $\mathcal{L}(\mathbf{U})$ as follows:

The matrices $\mathbf{V}^j$ for $j = 1, \ldots, n$ are obtained by means of

$$\mathbf{V}^{j+1} = \mathbf{V}^j \mathbf{M}_j^{-1};\ j = 0, \ldots, n-1, \tag{2.13}$$

where $\mathbf{M}_j$ is defined in (2.2) with $\mathbf{t}_j = \mathbf{u}_j \mathbf{V}^j$. Then, assuming non-null dot products in the pivot columns, we have

$$\begin{aligned} \mathbf{u}_k^T \mathbf{V}^j &= \mathbf{u}_k^T \mathbf{V}^k \mathbf{M}_k^{-1} \mathbf{M}_{k+1}^{-1} \cdots \mathbf{M}_j^{-1} = \\ &= \mathbf{t}_k^T \mathbf{M}_k^{-1} \mathbf{M}_{k+1}^{-1} \cdots \mathbf{M}_j^{-1} = \\ &= \mathbf{e}_k^T \mathbf{M}_{k+1}^{-1} \cdots \mathbf{M}_j^{-1} = \mathbf{e}_k^T;\ \forall k \leq j. \end{aligned} \tag{2.14}$$

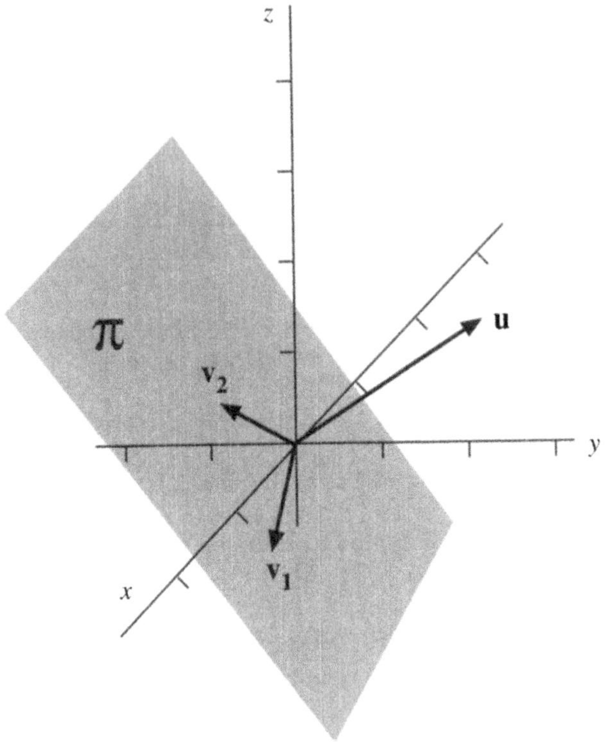

Fig. 2.2 Orthogonal vector to the linear subspace given in Example 2.4.

Let $\{\mathbf{w}_1, \ldots, \mathbf{w}_s\}$ be the set of columns of matrix $\mathbf{V}^n$; then

$$\mathcal{L}(\mathbf{V}^0) \equiv \mathcal{L}(\mathbf{U})^{\perp} \oplus \mathcal{L}\{\mathbf{w}_1, \ldots, \mathbf{w}_n\},$$

where

$$\mathcal{L}(\mathbf{U})^{\perp} \equiv \mathcal{L}\{\mathbf{w}_{n+1}, \ldots, \mathbf{w}_s\}.$$

In general, Theorem 2.2 allows building the following algorithm.

Algorithm 2.1 Orthogonal set to a linear subspace and its complement.

- **Input:** Two linear subspaces $\mathcal{L}(\mathbf{V}^0) \equiv \mathcal{L}(\mathbf{v}_1, \ldots, \mathbf{v}_s) \subseteq \mathbb{R}^n$ and $\mathcal{L}(\mathbf{U}) \equiv \mathcal{L}(\mathbf{u}_1, \ldots, \mathbf{u}_m) \subseteq \mathbb{R}^n$.
- **Output:** The orthogonal linear subspace $\mathcal{L}(\mathbf{W}_2)$ to $\mathcal{L}(\mathbf{U})$ in $\mathcal{L}(\mathbf{V}^0)$ and its complement $\mathcal{L}(\mathbf{W}_1)$.

Step 1: Set $\mathbf{W} = \mathbf{V}^0$ (the matrix with $\mathbf{v}_j; j = 1, \ldots, s$ as columns).

Step 2: Let $i = 1$ and $\ell = 0$.

Step 3: Calculate the dot products $t_j^\ell = \mathbf{u}_i \bullet \mathbf{w}_j; j = 1, \ldots, s$.

Step 4: For $j = \ell$ to s locate the pivot column r_ℓ as the first column not orthogonal to $\mathbf{u}_i$, that is, $t_{r_\ell}^\ell \neq 0$. If there is no such a column go to Step 7. Otherwise, continue with Step 5.

Step 5: Increase ℓ in one unit, divide the r_ℓ column by $t_{r_\ell}^\ell$, and if $r_\ell \neq \ell$ switch columns ℓ and r_ℓ and associated dot products $t_{r_\ell}^\ell$ and t_ℓ^ℓ.

Step 6: For $j = 1$ to s and $j \neq r_\ell$ do the following: If $t_j^\ell \neq 0$ do $w_{kj} = w_{kj} - t_j^\ell * w_{ki}$ for $k = 1, \ldots, n$.

Step 7: If $i = m$ go to Step 8. Otherwise, increase i in one unit and go to Step 3.

Step 8: Return $\mathcal{L}(\mathbf{W}_2) \equiv \mathcal{L}(\mathbf{w}_\ell, \ldots, \mathbf{w}_s)$ as the orthogonal subspace of $\mathcal{L}(\mathbf{U})$ in $\mathcal{L}(\mathbf{V}^0)$ and $\mathcal{L}(\mathbf{W}_1) \equiv \mathcal{L}(\mathbf{w}_1, \ldots, \mathbf{w}_{\ell-1})$ as its complement.

The process described in Algorithm 2.1 can be organized in a tabular form as in the previous examples. A detailed description is given in the following example.

Example 2.5 (Orthogonal decomposition). Consider the linear subspace

$$\mathcal{L}(\mathbf{U}) \equiv \mathcal{L}\left\{(2, 0, -1, 3, 1)^T, (1, 0, 0, 1, -2)^T, (-1, 1, 2, -2, 0)^T\right\}$$

of $\mathcal{L}(\mathbf{V}^0) \equiv \mathbb{R}^5$.

First, to obtain the orthogonal decomposition of $\mathcal{L}(\mathbf{V}^0)$ with respect to $\mathcal{L}(\mathbf{U})$, we construct the initial table (see Iteration 1 in Table 2.6). The first column of this table is the first generator of $\mathcal{L}(\mathbf{U})$ and the generators of the subspace to be decomposed are in the other columns. The last row contains the inner products of the vector in the first column by the corresponding column vectors.

Next, the first non-null element in the last row is identified and the corresponding column is selected as the pivot column, which is boldfaced in Iteration 1.

Finally, it is necessary to perform the pivoting process and to update the first column and the last row of the table with the next generator of $\mathcal{L}(\mathbf{U})$ and the new inner products. Then, we get the second table (see Iteration 2 in Table 2.6). In order to select the pivot column, we have to look for the first non-null element in the last row starting with its second element because we are in the second iteration. Then, the selected column is the third one, and before performing the pivoting process interchange of second and third columns must be done.

Table 2.6 Iterations for obtaining the orthogonal decomposition of $\mathcal{L}(\mathbf{V}^0)$ with respect to $\mathcal{L}(\mathbf{U})$. Pivot columns are boldfaced.

Iteration 1						Iteration 2					
2	**1**	0	0	0	0	1	*1/2*	0	**1/2**	-3/2	-1/2
0	**0**	1	0	0	0	0	*0*	1	**0**	0	0
-1	**0**	0	1	0	0	0	*0*	0	**1**	0	0
3	**0**	0	0	1	0	1	*0*	0	**0**	1	0
1	**0**	0	0	0	1	-2	*0*	0	**0**	0	1
	2	0	-1	3	1		*1/2*	0	**1/2**	-1/2	-5/2
Modified second table						**Iteration 3**					
1	*1/2*	**1/2**	0	-3/2	-1/2	-1	*0*	*1*	**0**	-1	2
0	*0*	**0**	1	0	0	1	*0*	*0*	**1**	0	0
0	*0*	**1**	0	0	0	2	*-1*	*2*	**0**	1	5
1	*0*	**0**	0	1	0	-2	*0*	*0*	**0**	1	0
-2	*0*	**0**	0	0	1	0	*0*	*0*	**0**	0	1
	1/2	**1/2**	0	-1/2	-5/2		*-2*	*3*	**1**	1	8

Output				
0	*1*	*0*	-1	2
2	*-3*	*1*	-1	-8
-1	*2*	*0*	1	5
0	*0*	*0*	1	0
0	*0*	*0*	0	1

We repeat the pivoting process, incorporate the last generators of $\mathcal{L}(\mathbf{U})$ and obtain the new dot products. We select the pivot column, starting at column three and get this column as the pivot. Next, we repeat the normalization and pivoting processes and, finally, we get the Output in Table 2.6.

In the last table, the first three vectors are the generators of the complement subspace and the last two are the generators of the orthogonal subspace. Italicized columns are used in all iterations to refer to the complementary subspace.

Thus, the orthogonal decomposition becomes

$$\mathbb{R}^5 \equiv \mathcal{L}\left\{(0,2,-1,0,0)^T,(1,-3,2,0,0)^T,(0,1,0,0,0)^T\right\} \oplus \mathcal{L}(\mathbf{U})^{\perp},$$

where

$$\mathcal{L}(\mathbf{U})^{\perp} \equiv \mathcal{L}\left\{(-1,-1,1,1,0)^T,(2,-8,5,0,1)^T\right\}.$$

■

Example 2.6 (Orthogonal decomposition of a subspace with respect to the vector). In this example we consider the linear subspace

$$\mathcal{V} \equiv \left\{(x_1,x_2,x_3,x_4)^T \in \mathbb{R}^4 \mid x_1+x_2+x_3+x_4=0\right\}$$

Table 2.7 Orthogonal decomposition.

Iteration 1				Output		
1	**1**	1	1	*-1*	2	3
2	**-1**	0	0	*1*	-1	-2
0	**0**	-1	0	*0*	-1	0
-1	**0**	0	-1	*0*	0	-1
	-1	1	2			

and the vector $\mathbf{u} = (1, 2, 0, -1)^T$, and we obtain the orthogonal decomposition of subspace $\mathcal{V}$ with respect to the vector $\mathbf{u}$.

First, we consider a set of generators of $\mathcal{V}$,

$$\mathcal{V} \equiv \mathcal{L}\{(1, -1, 0, 0)^T, (1, 0, -1, 0)^T, (1, 0, 0, -1)^T\},$$

and apply the algorithm, as shown in Table 2.7. We get

$$\mathcal{V} \equiv \mathbf{u}_{\mathbf{v}}^{\perp} \oplus \mathcal{W},$$

where

$$\begin{aligned} \mathbf{u}_{\mathbf{v}}^{\perp} &\equiv \mathcal{L}\{(2, -1, -1, 0)^T, (3, -2, 0, -1)^T\}, \\ \mathcal{W} &\equiv \mathcal{L}\{(-1, 1, 0, 0)^T\}. \end{aligned}$$

As a more interesting alternative, we can use the algorithm again to obtain a decomposition with mutually orthogonal vectors. First, we obtain the set in $\mathcal{V}$ orthogonal to $\mathbf{u}_{\mathbf{v}}^{\perp}$, as shown in Table 2.8 and get $\mathcal{L}\{(1, 3, -1, -3)\}$. Next, we obtain the set in $\mathbf{u}_{\mathbf{v}}^{\perp}$ orthogonal to the vector $(2, -1, -1, 0)$, and get $\mathcal{L}\{(1, -2, 4, -3)\}$ (see Table 2.9). Thus, we finally get

$$\mathcal{V} \equiv \mathbf{u}_{\mathbf{v}}^{\perp} \oplus \mathcal{W}_1,$$

where

$$\begin{aligned} \mathbf{u}_{\mathbf{v}}^{\perp} &\equiv \mathcal{L}\{(2, -1, -1, 0)^T, (1, -2, 4, -3)^T\}, \\ \mathcal{W}_1 &\equiv \mathcal{L}\{(1, 3, -1, -3)^T\}. \end{aligned}$$

However, now $\mathbf{u}_{\mathbf{v}}^{\perp}$ is orthogonal to $\mathcal{W}_1$, and the generators in $\mathbf{u}_{\mathbf{v}}^{\perp}$ are orthogonal. It can be seen that this process is the Gram–Schmidt method.

■

Example 2.7 (Orthogonal decomposition of $\mathbb{R}^5$ with respect to the linear subspace). The orthogonal decomposition of $\mathbb{R}^5$ with respect to the linear subspace $\mathcal{L}\left\{(1, 2, 0, 1, 0)^T, (2, 1, -1, 0, 0)^T, (0, 2, 1, 0, 1)^T\right\}$ can be

Table 2.8 Set orthogonal to $\mathbf{u}_{\mathcal{V}}^{\perp}$ in $\mathcal{V}$.

Iteration 1				Iteration 2			Output
2	**1**	1	1	3	**0**	1/3	1/3
-1	**−1**	0	0	-2	**1**	2/3	1
-1	**0**	-1	0	0	**−1**	0	-1/3
0	**0**	0	-1	-1	**0**	-1	-1
	3	3	2		**−2**	2/3	

Table 2.9 Obtaining the set in $\mathbf{u}_{\mathcal{V}}^{\perp}$ orthogonal to the vector $(2, -1, -1, 0)^T$.

Iteration 1			Output
2	**2**	3	1/3
-1	**−1**	-2	-2/3
-1	**−1**	0	4/3
0	**0**	-1	-1
	6	8	

obtained using Theorem 2.2. The solution is the matrix

$$\begin{pmatrix} -\mathit{3} & \mathit{2} & \mathit{2} & 3 & -2 \\ \mathit{2} & -\mathit{1} & -\mathit{1} & -2 & 1 \\ -\mathit{4} & \mathit{2} & \mathit{3} & 4 & -3 \\ \mathit{0} & \mathit{0} & \mathit{0} & 1 & 0 \\ \mathit{0} & \mathit{0} & \mathit{0} & 0 & 1 \end{pmatrix},$$

which shows that the orthogonal decomposition of $\mathbb{R}^5$ associated with the above linear subspace is

$$\mathcal{L}\left\{(-3, 2, -4, 0, 0)^T, (2, -1, 2, 0, 0)^T, (2, -1, 3, 0, 0)^T\right\}$$
$$\oplus \mathcal{L}\left\{(3, -2, 4, 1, 0)^T, (-2, 1, -3, 0, 1)^T\right\}.$$

■

2.4 MATHEMATICA PROGRAM

The Mathematica implementation of a module for obtaining the orthogonal set to a given linear subspace in another linear subspace given its generators is shown in Program 2.1. The function `Orthogonal` defined in this program has several optional parameters to calculate determinants, inverse matrices, and ranks and to solve linear systems, which will be described in the following chapters.

Program 2.1 (Module for obtaining the set orthogonal to a linear subspace.)

```
Options[Orthogonal]={Space->Automatic,
                     Complement->False,
                     SingularMatrixQ->False,
                     Determinant->False,
                     SystemofEquations->False};

Orthogonal[Generator_?MatrixQ,opts___]:=
Module[
  {m2,m1,m,n,k,i,j,kk,T,aux,col,B1,smq,detopt,output,
   se,mess=0,Gen,orthog={},complopt,compl={},d=1},
  n=Length[Generator];
  If[n==0,Print["Empty Generator"];Return[{}]];
  m=Length[Generator[[1]]];
  k=1;
  {B1,complopt,smq,detopt,se}=
     {Space,Complement,SingularMatrixQ,Determinant,
           SystemofEquations}/.{opts}/.Options[Orthogonal];
  If[TrueQ[B1==Automatic],B1=IdentityMatrix[m];
     m1=m,m1=Length[B1]];
  Gen=Generator;
  If[se,m2=m1-1,m2=m1];
  Do[col=0;T=B1.Gen[[i]];
     Do[If[,TrueQ[T[[j]]==0],col=j;Break[]],{j,k,m2}];
     If[col,=0,If[col,=k,d*=-1;
        aux=B1[[col]];B1[[col]]=B1[[k]];B1[[k]]=aux;
        aux=T[[k]];T[[k]]=T[[col]];T[[col]]=aux];
     If[detopt,d*=T[[k]]];B1[[k]]=Map[(#/T[[k]])&,B1[[k]]];
        Do[If[((j,=k)&&(,TrueQ[T[[j]]==0])),
           B1[[j]]-=T[[j]]*B1[[k]]],{j,1,m1}];
     If[k<=m2,k++,Break[]],d=0;If[smq,Print["Singular matrix"]];
       If[se && (T[[m1]],=0),Print["Incompatible System"];
          mess=1;Break[]]],
   {i,1,Length[Generator]}];
   If[mess==1,Return[{}]];
 Do[If[i<k,AppendTo[compl,B1[[i]]],
           AppendTo[orthog,B1[[i]]]],{i,1,m1}];
 If[Length[compl]>0,Transpose[compl],compl];
 If[Length[orthog]>0,Transpose[orthog],orthog];
 If[(,complopt)&&(,detopt),Return[orthog],output={orthog}];
 If[complopt,output=Append[output,compl]];
```

```
  If[detopt,output=Append[output,d]];
  Return[output]
]
```

The numerical and symbolic power of the proposed method is illustrated by means of the following examples.

Example 2.8 (Mathematica program). In this example we use the Mathematica program for obtaining the orthogonal decomposition of Example 2.3. The following command returns a set of generators of the orthogonal subspace to the vector $\mathbf{u} = (1, -1, 3)^T$ in $\mathbb{R}^3$:

```
In[1]:= Orthogonal[{{1,-1,3}}]
Out[1]= {{1, 1, 0}, {-3, 0, 1}}
```

The `Orthogonal` command admits the optional parameter `Complement`. If its value is `True` the command returns the generators of the orthogonal and the complement subspaces.

```
In[2]:= Orthogonal[{{1,-1,3}},Complement->True]
Out[2]= {{{1, 1, 0}, {-3, 0, 1}}, {{1, 0, 0}}}
```

■

Example 2.9 (Orthogonal decomposition with Mathematica). The orthogonal decomposition of Example 2.5 can be calculated with the following command:

```
In[3]:= Orthogonal[{{2,0,-1,3,1},{1,0,0,1,-2},{-1,1,2,-2,0}},
                    Complement->True]
Out[3]= {{{-1, -1, 1, 1, 0}, {2, -8, 5, 0, 1}},
          {{0, 2, -1, 0, 0}, {1, -3, 2, 0, 0}, {0, 1, 0, 0, 0}}}
```

■

Example 2.10 (Orthogonal decomposition starting with a given linear subspace). If the linear subspace to be decomposed is not E^n, it has to be indicated with the optional parameter `Space`, as can be seen in the following command, which allows obtaining the decomposition in Example 2.6

```
In[4]:= Orthogonal[{{1,2,0,-1}},
                    Space->{{1,-1,0,0},{1,0,-1,0},{1,0,0,-1}},
                    Complement->True]
Out[4]= {{{2, -1, -1, 0}, {3, -2, 0, -1}}, {{-1, 1, 0, 0}}}
```

■

Example 2.11 (Orthogonal decomposition of $\mathbb{R}^4$ with respect to the linear subspace). The orthogonal decomposition of $\mathbb{R}^4$ with respect to the linear subspace $\mathcal{L}\{(1,2,0,1),(1,1,0,a),(1,2,1,b)\}$, where a and b are symbolic values, can be obtained using Algorithm 2.1 or the above Mathematica program, using the command

```
In[5]:= Orthogonal[{{1,2,0,1},{1,1,0,a},{1,2,1,b}},
                   Complement->True]
Out[5]= {{{-1 - 2 (-1 + a), -1 + a, 1 - b, 1}},
         {{-1, 1, -1, 0}, {2, -1, 0, 0}, {0, 0, 1, 0}}}
```

This shows that the orthogonal decomposition of $\mathbb{R}^4$ associated with the above linear subspace is

$$\mathcal{L}\{(1-2a,-1+a,1-b,1)^T\}+\mathcal{L}\{(-1,1,-1,0)^T,(2,-1,0,0)^T,(0,0,1,0)^T\}.$$

■

Exercises

2.1 Let $\mathcal{U} \equiv \mathcal{L}\{(1,2,3,0,2)^T,(2,-1,0,1,0)^T,(0,-2,1,3,0)^T\} \subseteq \mathbb{R}^5$ be a linear subspace, and $t_1 = 1/2, t_2 = 2/3$, and $t_3 = 1/2$ be three non-null real numbers. Apply the pivoting process defined in Lemma 2.1 to the generators of $\mathcal{U}$, and prove that the resulting vectors are also generators of $\mathcal{U}$.

2.2 Let $\mathcal{V} \equiv \mathcal{L}\{(2,1,0,-1)^T,(1,0,2,-1)^T,(1,1,0,1)^T\} \subseteq \mathbb{R}^4$ be a linear subspace, and suppose that after applying Lemma 2.1 the following vectors are obtained:

$$\begin{aligned}\mathbf{v}_1 &= (1,1/2,0,-1/2)^T,\\ \mathbf{v}_2 &= (1/2,-1/4,2,-3/4)^T,\\ \mathbf{v}_3 &= (0,1/2,0,3/2)^T.\end{aligned}$$

Calculate the t_1, t_2, and t_3 values used in the process.

2.3 Obtain the linear subspace orthogonal to the vector $(3,2,1,4)^T$ in the linear subspace $\mathcal{L}\{(1,2,-1,0)^T,(2,3,4,1)^T\}$.

2.4 Show that the orthogonal decomposition of $\mathbb{R}^5$ with respect to the linear subspace $\mathcal{L}\{(1,2,0,1,0)^T,(2,1,-1,0,0)^T,(0,2,1,0,1)^T\}$ is

$$\mathcal{L}\{(-3,2,-4,0,0)^T,(2,-1,2,0,0)^T,(2,-1,3,0,0)^T\}$$

$$\oplus\mathcal{L}\{(3,-2,4,1,0)^T,(-2,1,-3,0,1)^T\}$$

2.5 Consider the vector $\mathbf{u} = (1,1,1)^T \in \mathbb{R}^3$.

(a) Calculate the orthogonal decomposition of $\mathbb{R}^3$ with respect to $\mathbf{u}$.

(b) Give the equation of the plane orthogonal to $\mathbf{u}$ and draw it.

2.6 Find the orthogonal vector to the linear subspace

$$\mathcal{N} \equiv \mathcal{L}\{(1,0,2)^T, (-1,0,1)^T\} \subseteq \mathbb{R}^3.$$

2.7 Obtain the orthogonal decomposition of $\mathbb{R}^5$ with respect to

$$\mathcal{V} \equiv \mathcal{L}\{(a,1,2,a,0)^T, (1,a,2,a,0)^T\} \subseteq \mathbb{R}^5,$$

where a is a parameter, using as the initial basis of $\mathbb{R}^5$ the following vectors:

$$\begin{array}{l}\{(1,0,1,2,1)^T, (0,-1,2,0,-2)^T, (2,0,1,-1,-1)^T,\\ (0,-1,2,1,1)^T, (2,-1,-2,0,0)^T\}.\end{array}$$

2.8 Calculate the orthogonal decomposition of $\mathbb{R}^4$ with respect to

$$\mathcal{V} \equiv \mathcal{L}\{(a,b,2,a)^T, (0,a,1,2)^T, (a,0,b,1)^T\} \subseteq \mathbb{R}^4,$$

where $a, b \in \mathbb{R}$.

2.9 Find the orthogonal decomposition of $\mathbb{R}^3$ with respect to the plane

$$\Pi \equiv \{(x,y,z)^T | 2x + y - 3z = 0\}.$$

2.10 Find the orthogonal decomposition of $\mathbb{R}^4$ with respect to

$$\mathcal{U} \equiv \left\{(x,y,z,t)^T \in \mathbb{R}^4 | x + y + t = 0;\ z + t = 0\right\}.$$

2.11 Use Lemma 2.1 to obtain the dimension of the subspace

$$\mathcal{L}\{(2,-1,4,3)^T, (-1,0,2,1)^T, (1,-1,6,4)^T, (3,0,-1,0)^T\}.$$

Hint: Perform the pivoting process to obtain an equivalent set of generators such that the associated row matrix is *upper triangular* (all the elements below the main diagonal are zero). The dimension of the subspace is the number of non–zero elements in the main diagonal.

3

Matrix Calculations Using Orthogonal Sets

3.1 INTRODUCTION

Usual techniques in matrix algebra have their root in the concept of linear equality and operate exclusively with the matrix elements, rows, and columns in its *primal sense.* For a complete treatment of the problem of inverting matrices by standard methods the reader is referred, for example, to Hadi (1996) or Golub and Van Loan (1989). The techniques to be developed in this book are based on *dual, polar, or orthogonal concepts* (see Farkas (1902)), which are associated with linear inequalities and belong to the polarity and the alternative theories. As we shall see, the dual (orthogonal) approach, in addition to being useful, opens the field of applications of linear equalities with no increased complexity.

In this chapter we show how orthogonal sets described in Chapter 2 can be applied to solve matrix problems.

In Section 3.2, a powerful method for inverting a matrix is presented. The orthogonal decomposition view facilitates the understanding of the whole orthogonalization process and also allows using it for calculating the rank and determinant of the initial matrix, as shown in Sections 3.3 and 3.4, respectively. In Section 3.5 an algorithm for obtaining the rank and calculating the determinant of a matrix and, if it is nonsingular, obtaining its inverse is given. In Section 3.6 the method's complexity, which is shown to coincide with the complexity of the Gauss elimination procedure, is calculated. In Section 3.7 we show how to update inverses and determinants with little extra computational effort when changes of a few rows and/or columns of the original matrix are made.

Most existing methods for calculating the inverse of a matrix or its determinant require restarting the whole process from scratch when a single element or several elements of the original matrix are changed. Thus, the huge amount of computational effort for these calculations is lost if changes are made to the original matrix. However, there are some alternatives, as the Sherman-Morrison and the Woodbury methods (see Gill, Golub, Murray, and Saunders (1974), Press, Teukolsky, Vetterling, and Flannery (1985) (pages 73–77), and Burden and Faires (1985)), that allow calculating the new inverse, with a reduced number of multiplications and additions, when some special types of modifications are performed on the initial matrix. These methods have arisen in matrix theory. The proposed method is directly applicable to this dynamic updating process.

In Section 3.8 the proposed method is shown to be appropriate for calculating the symbolic inverse or the determinant of a matrix, that is, for dealing with matrices with symbols or parameters as elements instead of numbers. In fact, it is shown that this problem can easily be solved using numerical procedures, which are greatly simplified by this inversion method. In Section 3.9 the method is extended to block matrices. In Section 3.10 an algorithm for updating the inverse of a matrix after adding and/or removing rows and/or columns is given.

In summary, in this chapter we give a method for inverting matrices, calculating determinants, and obtaining the rank of a matrix, which is computationally efficient and allows dynamic updating of the inverses and determinants when rows of the initial matrix are changed. In addition, we show that the method is especially suitable for symbolic computations.

The method is defined for column manipulation of matrices, but it is also valid for row manipulation. Finally, the method is shown to be applicable to block–defined matrices with the extra advantage of a feasible and obvious parallel implementation.

3.2 INVERTING A MATRIX

In this section a method for calculating the inverse of a matrix is given. Suppose we are interested in finding the inverse of the square matrix $\mathbf{A} = (\mathbf{a}_1, \ldots, \mathbf{a}_n)^T$. Assume that Algorithm 2.1 is applied to the rows of $\mathbf{A}$ using a nonsingular initial matrix $\mathbf{V}^0$. Then the matrices $\mathbf{V}^j$ for $j = 1, \ldots, n$ are obtained, using the transformations

$$\mathbf{V}^{j+1} = \mathbf{V}^j \mathbf{M}_j^{-1}; \; j = 0, \ldots, n-1, \tag{3.1}$$

where $\mathbf{M}_j$ is defined in (2.2) with $\mathbf{t}_j^T = \mathbf{a}_j^T \mathbf{V}^j$. Then it satisfies

$$\begin{aligned} \mathbf{a}_j^T \mathbf{V}^n &= \mathbf{a}_j^T \mathbf{V}^j \mathbf{M}_j^{-1} \mathbf{M}_{j+1}^{-1} \cdots \mathbf{M}_n^{-1} \\ &= \mathbf{t}_j^T \mathbf{M}_j^{-1} \mathbf{M}_{j+1}^{-1} \cdots \mathbf{M}_n^{-1} \\ &= \mathbf{e}_j^T \mathbf{M}_{j+1}^{-1} \cdots \mathbf{M}_n^{-1} = \mathbf{e}_j^T. \end{aligned} \tag{3.2}$$

This proves that $\mathbf{A}^{-1} = \mathbf{V}^n$; that is, the inverse of $\mathbf{A}$ is the matrix whose columns are in the final tableau obtained using Algorithm 2.1.

The above process allows obtaining the orthogonal decomposition of the linear space $\mathbb{R}^n$ with respect to any row of $\mathbf{A}$. Let $\{\mathbf{v}_1, \ldots, \mathbf{v}_n\}$ be the columns of $\mathbf{V}^n$. Then

$$\mathbb{R}^n \equiv \mathcal{L}(\mathbf{v}_1, \ldots, \mathbf{v}_{i-1}, \mathbf{v}_{i+1}, \ldots, \mathbf{v}_n) \oplus \mathcal{L}(\mathbf{v}_i)$$

is the orthogonal decomposition of $\mathbb{R}^n$ with respect to $\mathbf{a}_i$. Thus, we have

$$\mathbf{a}_i \bullet \mathbf{v}_j = \delta_{ij},$$

with δ_{ij} being the Kronecker deltas.

Example 3.1 (Numerical example). Consider the matrix

$$\mathbf{A} = \begin{pmatrix} 1 & 2 & 0 & 1 & 0 \\ 2 & 1 & -1 & 0 & 0 \\ 0 & 2 & 1 & 0 & 1 \\ 0 & 0 & 0 & 1 & 2 \\ 0 & -1 & 0 & 2 & 1 \end{pmatrix}. \tag{3.3}$$

Table 3.1 shows the iterations for inverting the above matrix using Algorithm 2.1. The resulting inverse matrix is obtained in the last iteration as

$$\mathbf{A}^{-1} = \begin{pmatrix} -1 & 1 & 1 & -1 & 1 \\ 3/4 & -3/8 & -3/8 & 1/2 & -5/8 \\ -5/4 & 5/8 & 13/8 & -3/2 & 11/8 \\ 1/2 & -1/4 & -1/4 & 0 & 1/4 \\ 1/4 & 1/8 & 1/8 & 1/2 & -1/8 \end{pmatrix}. \tag{3.4}$$

■

3.3 THE RANK OF A MATRIX

As we have seen, the orthogonal decomposition view facilitates the understanding of the whole orthogonalization process and also allows us to use it for calculating the inverse of a matrix. Here we shall see that it also allows one to determine the rank of a matrix.

In an n-dimensional linear space, the rank of a matrix $\mathbf{U}$ coincides with the dimension of the complement of the linear subspace generated by its rows with respect to $\mathbb{R}^n$, or with n minus the dimension of its orthogonal subspace. Thus, if during the orthogonalization process we start with a nonsingular matrix as columns, we can find a pivot in all iterations, and then the corresponding matrix is full rank. Otherwise, the rank is equal to the number of pivot columns we can find.

Table 3.1 Iterations for inverting the matrix in Example 3.1. Pivot columns are boldfaced.

Iteration 1						**Iteration 2**					
1	**1**	0	0	0	0	2	1	**−2**	0	−1	0
2	**0**	1	0	0	0	1	0	**1**	0	0	0
0	**0**	0	1	0	0	−1	0	**0**	1	0	0
1	**0**	0	0	1	0	0	0	**0**	0	1	0
0	**0**	0	0	0	1	0	0	**0**	0	0	1
	1	2	0	1	0		2	**−3**	−1	−2	0
Iteration 3						**Iteration 4**					
0	−1/3	2/3	**2/3**	1/3	0	0	−3	2	2	**3**	−2
2	2/3	−1/3	**−1/3**	−2/3	0	0	2	−1	−1	**−2**	1
1	0	0	**1**	0	0	0	−4	2	3	**4**	−3
0	0	0	**0**	1	0	1	0	0	0	**1**	0
1	0	0	**0**	0	1	2	0	0	0	**0**	1
	4/3	−2/3	**1/3**	−4/3	1		0	0	0	**1**	2
Iteration 5						**Output**					
0	−3	2	2	3	**−8**		−1	1	1	−1	1
−1	2	−1	−1	−2	**5**		3/4	−3/8	−3/8	1/2	−5/8
0	−4	2	3	4	**−11**		−5/4	5/8	13/8	−3/2	11/8
2	0	0	0	1	**−2**		1/2	−1/4	−1/4	0	1/4
1	0	0	0	0	**1**		−1/4	1/8	1/8	1/2	−1/8
	−2	1	1	4	**−8**						

Example 3.2 (Rank of a matrix). Assume that we are interested in calculating the rank of the matrix

$$\mathbf{A} = \begin{pmatrix} 1 & 0 & 1 & 1 & 1 \\ 0 & 1 & 1 & 0 & 1 \\ 1 & 1 & 2 & 1 & 2 \\ 1 & 1 & 0 & 0 & 0 \\ 1 & -1 & 0 & 1 & 0 \end{pmatrix}.$$

In Table 3.2 we show the iterations for obtaining its rank. We can see that the rank of $\mathbf{A}$ is 3, since the third and fifth iterations have no pivots.

■

3.4 CALCULATING THE DETERMINANT OF A MATRIX

The following theorem shows that the determinant of a matrix can be calculated by means of Algorithm 2.1.

Table 3.2 Iterations for calculating the rank of the matrix in Example 3.2. Pivot columns are boldfaced.

Iteration 1						Iteration 2					
1	**1**	0	0	0	0	0	*1*	**0**	−1	−1	−1
0	**0**	1	0	0	0	1	*0*	**1**	0	0	0
1	**0**	0	1	0	0	1	*0*	**0**	1	0	0
1	**0**	0	0	1	0	0	*0*	**0**	0	1	0
1	**0**	0	0	0	1	1	*0*	**0**	0	0	1
	1	0	1	1	1		*0*	**1**	1	0	1

Iteration 3						Iteration 4					
1	*1*	*0*	−1	−1	−1	1	*1*	*0*	**−1**	−1	−1
1	*0*	*1*	−1	0	−1	1	*0*	*1*	**−1**	0	−1
2	*0*	*0*	1	0	0	0	*0*	*0*	**1**	0	0
1	*0*	*0*	0	1	0	0	*0*	*0*	**0**	1	0
2	*0*	*0*	0	0	1	0	*0*	*0*	**0**	0	1
	1	*1*	0	0	0		*1*	*1*	**−2**	−1	−2

Iteration 5					
1	*1/2*	*−1/2*	*1/2*	−1/2	0
−1	*−1/2*	*1/2*	*1/2*	1/2	0
0	*1/2*	*1/2*	*−1/2*	−1/2	−1
1	*0*	*0*	*0*	1	0
0	*0*	*0*	*0*	0	1
	1	*−1*	*1*	0	0

Theorem 3.1 (Determinant of a matrix). *The determinant of a matrix* **A** *can be calculated by Algorithm 2.1 by multiplying the normalizing constants* $t^{\ell}_{r_\ell}; l = 1, \ldots, n$ *used in Step 5 and* $(-1)^p$, *where* p *is the number of interchanges of columns that have occurred when executing the algorithm.*

Proof: Assume that we start in Step 1 with an identity matrix, $\mathbf{W} = \mathbf{I}_n$, which has a determinant of one. In the inverting process we transform this matrix using two different transformations: the pivoting step, which does not alter the determinant value of the matrix, and the normalization step, which divides its determinant by $t^{\ell}_{r_\ell}$ (see (2.1)). In addition, we multiply it by -1 each time we switch columns. Since $|\mathbf{A}^{-1}| = |\mathbf{A}|^{-1}$, we have

$$|\mathbf{A}| = \prod_{i=1}^{n} (-1)^p t^i_{r_i}. \tag{3.5}$$

■

Example 3.3 (Determinant of a matrix). The determinant of the matrix in Example 3.1 is obtained by multiplying the normalizing constants, that is, the last values in the boldfaced columns in Table 3.1. Thus, we have

$$1 \times (-3) \times (1/3) \times 1 \times (-8) = 8.$$

■

Remark 3.1 *If instead of starting with the identity matrix* $\mathbf{I}_n$, *we start with a nonsingular matrix* $\mathbf{B}$ *with determinant* $|\mathbf{B}|$, *Expression (3.5) becomes*

$$|\mathbf{A}| = |\mathbf{B}|^{-1} \prod_{i=1}^{n} (-1)^p t^i_{r_i}. \tag{3.6}$$

Example 3.4 (Determinant of a matrix starting with an arbitrary nonsingular matrix). Here we show how to obtain the determinant of matrix $\mathbf{A}$ in Example 3.1 if we start with the nonsingular matrix

$$\mathbf{B} = \begin{pmatrix} 1 & 0 & 0 & 0 & 0 \\ 0 & 1 & 0 & 1 & 0 \\ 0 & 0 & 1 & 0 & 0 \\ 0 & -1 & 0 & 0 & 0 \\ 0 & 0 & 0 & 0 & 2 \end{pmatrix},$$

with determinant $|\mathbf{B}| = 2$. Now in the first iteration, we use matrix $\mathbf{B}$ instead of the unit matrix, as shown in Table 3.3, and we continue the process keeping the values of the pivots. Then the determinant of $\mathbf{A}$ according to (3.6) becomes

$$|\mathbf{A}| = \frac{1}{2}[1 \times (-1) \times (-1) \times (-1) \times (-16)] = 8.$$

■

3.5 ALGORITHM FOR MATRIX CALCULATIONS

The previous theorems allow calculating the inverse of a matrix and its determinant and rank by means of the following algorithm.

Algorithm 3.1 Inverse of a matrix, determinant, and rank.

- **Input:** A square matrix $\mathbf{A}$ of dimension n.
- **Output:** The inverse $\mathbf{A}^{-1}$ and its determinant, or a message indicating that the matrix is singular, its rank, and a set of linearly independent rows spanning the same linear subspace as that of the input matrix $\mathbf{A}$.

Step 1: Let $\mathbf{B} = \mathbf{I}_n$, where $\mathbf{I}_n$ is the identity matrix of dimension n.

Step 2: Let $i = 1$, det $= 1$, and $Basis = \emptyset$.

Table 3.3 Process for obtaining the pivot values t^i_{ri}.

Iteration 1						Iteration 2					
1	**1**	0	0	0	0	2	1	**−1**	0	−2	0
2	**0**	1	0	1	0	1	0	**1**	0	1	0
0	**0**	0	1	0	0	−1	0	**0**	1	0	0
1	**0**	−1	0	0	0	0	0	**−1**	0	0	0
0	**0**	0	0	0	2	0	0	**0**	0	0	2
	1	1	0	2	0		2	**−1**	−1	−3	0
Iteration 3						**Iteration 4**					
0	−1	1	**1**	1	0	0	3	−1	−1	**−3**	2
2	2	−1	**−1**	−2	0	0	−2	1	1	**2**	−2
1	0	0	**1**	0	0	0	4	−2	−1	**−4**	2
0	−2	1	**1**	3	0	1	2	−1	−1	**−1**	2
1	0	0	**0**	0	2	2	0	0	0	**0**	2
	4	−2	**−1**	−4	2		2	−1	−1	**−1**	6
Iteration 5						**Output**					
0	−3	2	2	3	**−16**		−1	1	1	−1	1
−1	2	−1	−1	−2	**10**		3/4	−3/8	−3/8	1/2	−5/8
0	−4	2	3	4	**−22**		−5/4	5/8	13/8	−3/2	11/8
2	0	0	0	1	**−4**		1/2	−1/4	−1/4	0	1/4
1	0	0	0	0	**2**		−1/4	1/8	1/8	1/2	−1/8
	−2	1	1	4	**−16**						

Step 3: Calculate the scalar products $t^i_j; j = 1, \ldots, n$, of the ith row by the columns of $\mathbf{B}$.

Step 4: For $j = i$ to n locate the pivot column r_i as the first column not orthogonal to the ith row, that is, $t^i_{r_i} \neq 0$. If there is no such a column print the message "Singular Matrix", make det = 0, and go to Step 7.

Step 5: Append the ith row to *Basis*, divide the r_ith column by $t^i_{r_i}$, let $det = det * t^i_{r_i}$, and switch the ith and r_i columns and dot products if $i \neq r_i$.

Step 6: For $j = 1$ to n do the following: If $t^i_j \neq 0$ and $j \neq i$ let $b_{kj} = b_{kj} - t^i_j * b_{ki}$.

Step 7: If $i = n$ go to Step 8. Otherwise, increase i by one and go to Step 3.

Step 8: Return $\mathbf{B}$ as the inverse of $\mathbf{A}$, *det* as its determinant, the number of elements in *Basis* as the rank, and *Basis* as the basis.

In Section 3.11 we give some Mathematica programs to invert matrices, calculate determinants, and obtain the rank of a matrix using this algorithm.

Table 3.4 Process for calculating the rank of a matrix starting from a nonidentity matrix.

Iteration 1					Iteration 2				
0	**1**	0	0	1	1	*1/2*	**−1/2**	−1/2	1
1	**2**	1	1	0	0	*1*	**0**	0	0
0	**1**	1	1	0	0	*1/2*	**1/2**	1/2	0
0	**2**	0	−1	0	1	*1*	**−1**	−2	0
	2	1	1	0		*3/2*	**−3/2**	−5/2	1
Iteration 3					**Output**				
2	*0*	*1/3*	**1/3**	2/3	3	*0*	*0*	*1/3*	1/3
1	*1*	*0*	**0**	0	1	*1*	*0*	*0*	0
−1	*1*	*−1/3*	**−1/3**	1/3	−1	1	*0*	*−1/3*	2/3
0	*0*	*2/3*	**−1/3**	−2/3	1	*0*	*1*	*−1/3*	−1/3
	0	*1*	**1**	1		*0*	*1*	*1*	0

Remark 3.2 *Even though, for simplicity, we start Algorithm 3.1 with an identity matrix,* $\mathbf{I}_n$*, we can start with any nonsingular matrix* $\mathbf{B}$ *and the inverse of its determinant* $|\mathbf{B}|^{-1}$ *to obtain the same result.*

Example 3.5 (Rank of a matrix starting from an arbitrary nonsingular matrix). In this example we illustrate how to obtain the rank of a matrix $\mathbf{A}$ starting from a nonsingular matrix $\mathbf{B} \neq \mathbf{I}$. Consider the matrices

$$\mathbf{A} = \begin{pmatrix} 0 & 1 & 0 & 0 \\ 1 & 0 & 0 & 1 \\ 2 & 1 & -1 & 0 \\ 3 & 1 & -1 & 1 \end{pmatrix} \quad \text{and} \quad \mathbf{B} = \begin{pmatrix} 1 & 0 & 0 & 1 \\ 2 & 1 & 1 & 0 \\ 1 & 1 & 1 & 0 \\ 2 & 0 & -1 & 0 \end{pmatrix}.$$

Then we start with the matrix $\mathbf{B}$ as shown in Table 3.4 and conclude that $rank(\mathbf{A}) = 3$.

■

Remark 3.3 *In Algorithm 3.5 we start with the identity matrix, and we manipulate it by columns (or rows) until we get the inverse matrix* $\mathbf{A}^{-1}$*. This proves that the proposed method is the dual of the Gauss elimination procedure, which starts with* $\mathbf{A}$ *and manipulates its columns (or rows) until it becomes the identity matrix.*

Remark 3.4 *Note that at the end of Step 6 for any* i *value the resulting* $\mathbf{B}$ *matrix is the inverse of the matrix resulting from replacing the first* i *rows in the identity matrix by the first rows of the matrix to be inverted. Similarly, if instead of starting with* $\mathbf{B} = \mathbf{I}_n$*, we start with a matrix* $\mathbf{C}$*, the resulting* $\mathbf{B}$ *matrix is the inverse of the matrix resulting from replacing the first* i *rows in matrix* $\mathbf{C}$.

Table 3.5 Inverses of row-modified matrices.

	Iteration 1			$\mathbf{B}_1^{-1}$		
0	**–1/3**	2/3	1/3	–1/2	1/2	0
1	**2/3**	–1/3	–2/3	1	0	0
0	**0**	0	1	0	0	1
	2/3	–1/3	–2/3			
	Iteration 2			$\mathbf{B}_2^{-1}$		
–1	–1/2	**1/2**	0	0	–1	2
0	1	**0**	0	1	0	0
2	0	**0**	1	0	0	1
	1/2	**1/2**	2			
	Iteration 3			$\mathbf{B}_3^{-1}$		
1	0	–1	**2**	–1	0	1
1	1	0	**0**	1	0	0
0	0	0	**1**	–1/2	1/2	1/2
	1	–1	**2**			

Example 3.6 (Inverse of a row-modified matrix). Let

$$\mathbf{A} = \begin{pmatrix} 1 & 2 & 1 \\ 2 & 1 & 0 \\ 0 & 0 & 1 \end{pmatrix}$$

and its inverse

$$\mathbf{A}^{-1} = \begin{pmatrix} -1/3 & 2/3 & 1/3 \\ 2/3 & -1/3 & -2/3 \\ 0 & 0 & 1 \end{pmatrix}.$$

Consider the vectors

$$\begin{aligned} \mathbf{a}_1 &= (0,1,0)^T, \\ \mathbf{a}_2 &= (-1,0,2)^T, \\ \mathbf{a}_3 &= (1,1,0)^T. \end{aligned}$$

As we can see in Table 3.5 the resulting matrices $\mathbf{B}_1, \mathbf{B}_2$, and $\mathbf{B}_3$ of the pivoting process are the inverses of matrix $\mathbf{A}$ with the rows sequentially modified by $\mathbf{a}_1^T, \mathbf{a}_2^T$, and $\mathbf{a}_3^T$, respectively. Thus, we have

$$\mathbf{B}_1 = \mathbf{A}_1^{-1} = \begin{pmatrix} 0 & 1 & 0 \\ 2 & 1 & 0 \\ 0 & 0 & 1 \end{pmatrix}^{-1} = \begin{pmatrix} -1/2 & 1/2 & 0 \\ 1 & 0 & 0 \\ 0 & 0 & 1 \end{pmatrix},$$

$$\mathbf{B}_2 = \mathbf{A}_2^{-1} = \begin{pmatrix} 0 & 1 & 0 \\ -1 & 0 & 2 \\ 0 & 0 & 1 \end{pmatrix}^{-1} = \begin{pmatrix} 0 & -1 & 2 \\ 1 & 0 & 0 \\ 0 & 0 & 1 \end{pmatrix},$$

and

$$\mathbf{B}_3 = \mathbf{A}_3^{-1} = \begin{pmatrix} 0 & 1 & 0 \\ -1 & 0 & 2 \\ 1 & 1 & 0 \end{pmatrix}^{-1} = \begin{pmatrix} -1 & 0 & 1 \\ 1 & 0 & 0 \\ -1/2 & 1/2 & 1/2 \end{pmatrix}.$$

■

3.6 COMPLEXITY

In this section we derive the complexity of the proposed inversion method and compare it with the Gauss elimination procedure. In Table 3.6 we give the required number of products and sums in the iterations. Thus, we have

$$Total\ number\ of\ products\ = n^3,$$

and

$$Total\ number\ of\ sums\ = n^3 - 2n^2 + n.$$

Similarly, in Table 3.7 we give the required number of products and sums in the iterations for the Gauss elimination method for inverting a matrix. Thus, we have

$$Total\ number\ of\ products\ = n^3,$$

and

$$Total\ number\ of\ sums\ = n^3 - 2n^2 + n,$$

which shows that the complexities of the two algorithms are exactly the same.

If we are interested in using this algorithm for symbolic calculations we need to avoid normalizations (divisions). Thus, we need to:

1. Remove the normalization step and modify the pivoting process by replacing (2.1) by

$$\mathbf{v}_k^{j+1} = \begin{cases} \mathbf{v}_k^j; & \text{if } k = j \\ t_j^j \mathbf{v}_k^j - t_k^j \mathbf{v}_j^j; & \text{if } k \neq j \end{cases} \tag{3.7}$$

2. Normalize all columns at the end by

$$\mathbf{v}_k^{n+1} = \frac{\mathbf{v}_k^n}{(\mathbf{u}_k \bullet \mathbf{v}_k^n)}; k = 1, \ldots, n. \tag{3.8}$$

The complexity associated with these modifications, as illustrated in Table 3.8, is

$$Total\ number\ of\ products\ = \frac{3n^3 + n^2 - 2n}{2},$$

and

$$Total\ number\ of\ sums\ = n^3 - n^2.$$

Table 3.6 Required number of products and sums for the proposed inversion method.

Products or divisions					
Phase	Iter. 1	Iter. 2	...	Iter. n	Total
Dot Products	0	n	...	$n(n-1)$	$(n^3-n^2)/2$
Normalization	1	2	...	n	$(n^2+n)/2$
Pivoting	$n-1$	$2(n-1)$	...	$n(n-1)$	$(n^3-n)/2$
Sums					
Phase	Iter. 1	Iter. 2	...	Iter. n	Total
Dot Products	0	$n-1$	...	$(n-1)^2$	$(n^3-2n^2+n)/2$
Pivoting	0	$n-1$	...	$(n-1)^2$	$(n^3-2n^2+n)/2$

Table 3.7 Required number of products and sums for the Gauss elimination method for inverting a matrix.

Products or divisions					
Phase	Iter. 1	Iter. 2	...	Iter. n	Total
Normalizations	n	n	...	n	n^2
Pivoting	$n(n-1)$	$n(n-1)$	...	$n(n-1)$	$n^2(n-1)$
Sums					
Pivoting	$(n-1)^2$	$(n-1)^2$	...	$(n-1)^2$	n^3-2n^2+n

3.7 INVERSES AND DETERMINANTS OF ROW-MODIFIED MATRICES

One of the main advantages of the proposed method is that, given the inverse of a matrix, we can obtain the inverse of another matrix with s modified rows

Table 3.8 Required number of products and sums for the symbolically oriented proposed inversion method.

Products or divisions					
Phase	Iter. 1	Iter. 2	Iter. n	Iter. $n+1$	Total
Dot Products	0	n	$n(n-1)$	0	$\frac{(n^3-n^2)}{2}$
Normalization	0	0	0	n^2	n^2
Pivoting	$2(n-1)$	$4(n-1)$	$2n(n-1)$	0	(n^3-n)
Sums					
Phase	Iter. 1	Iter. 2	Iter. n	Iter. $n+1$	Total
Dot Products	0	$n-1$	$(n-1)^2$	0	$\frac{(n^3-2n^2+n)}{2}$
Pivoting	0	$n-1$	$(n-1)^2$	0	$\frac{(n^3-2n^2+n)}{2}$
Normalization	0	0	0	$n(n-1)$	$n(n-1)$

in s additional iterations and we do not need to restart the process from the very beginning.

This possibility has many interesting applications. Assume, for example, an engineer dealing with the design of a structure. The standard process consists of a trial and error process in which different candidate structures are sequentially modified as a result of previous calculations. This implies solving a system of thousands of equations or inverting a sequentially changing stiffness matrix where in each iteration only some of its rows and columns are modified. The same is applicable to finite element methods, where sometimes we change a few elements and start solving the resulting system of equations again.

The following theorem gives the inverse of a matrix as a function of the inverse of another identical matrix except for one row.

Theorem 3.2 (Inverse of a row-modified matrix). *Let* $\mathbf{A}$ *be a nonsingular matrix. Assume that* $\mathbf{C}$ *is another matrix, which is obtained from* $\mathbf{A}$ *by modifying the s-th row, that is,*

$$a_{ij} = c_{ij}, \; \textit{if } i \neq s, j = 1, \ldots, n. \tag{3.9}$$

Let t_k be the dot product of the s-th row (the modified row) of $\mathbf{C}$ *by the k-th column of* $\mathbf{A}^{-1}$, *that is,*

$$t_k = \sum_{i=1}^{n} c_{si} a^*_{ik}, \tag{3.10}$$

where a^*_{ij} is the ij element of A^{-1}. Then the inverse $\mathbf{C}^{-1}$ *of* $\mathbf{C}$ *has elements*

$$c^*_{jk} = \begin{cases} a^*_{jk} - \dfrac{t_k}{t_s} a^*_{js}, & \text{if } k \neq s, \\ \dfrac{a^*_{jk}}{t_s}, & \text{if } k = s. \end{cases} \tag{3.11}$$

Proof: Let us prove that the product of the $\mathbf{C}$ matrix and the matrix in (3.11) is the identity matrix. We have

$$\sum_{j=1}^{n} c_{ij} c^*_{jk} = \begin{cases} \sum_{j=1} a_{ij} a^*_{jk} - \dfrac{t_k}{t_s} \sum_{j=1}^{n} a_{ij} a^*_{js} = \delta_{ik}, & \text{if } i \neq s \text{ and } k \neq s, \\ \sum_{j=1} c_{sj} a^*_{jk} - \dfrac{t_k}{t_s} \sum_{j=1}^{n} c_{sj} a^*_{js} = 0, & \text{if } i = s \text{ and } k \neq s, \\ \sum_{j=1}^{n} a_{ij} a^*_{jk}/t_s = 0, & \text{if } i \neq s \text{ and } k = s, \\ \sum_{j=1}^{n} c_{sj} a^*_{jk}/t_s = 1, & \text{if } i = s \text{ and } k = s, \end{cases} \tag{3.12}$$

where δ_{ik}, as before, are the Kronecker deltas. Expression (3.12) shows that we get the identity matrix. ■

The idea of the method is simple. If we know the inverse of a matrix, we know a basis of the orthogonal sets associated with any of the rows in the original matrix. Thus, modifying a row requires only determining the orthogonal set associated with the modified row, which is obtained using, as the starting basis, the columns of the old inverse matrix, which has $n-1$ vectors orthogonal to each of the common rows, and using transformations that do not alter the orthogonal properties of the column vectors or the normalization properties. Thus, only one more iteration is required for each modified row.

The following theorem gives the determinant of the modified matrix as a function of the old determinant.

Theorem 3.3 (Determinant of a row-modified matrix). *Let* $\mathbf{A}$ *and* $\mathbf{C}$ *be the matrices in Theorem 3.2. Then the determinant of* $\mathbf{C}$ *is*

$$|\mathbf{C}| = |\mathbf{A}| t_s = |\mathbf{A}| \sum_{j=1}^{n} c_{sj} a^*_{js}; \tag{3.13}$$

that is, the new determinant is the old one times the dot product of the new row by its corresponding column in $\mathbf{A}^{-1}$.

Table 3.9 Additional iteration for inverting the matrix in Example 3.1 after changing the third row. (Pivot column is boldfaced).

Extra iteration					
1	-1	1	**1**	-1	1
0	3/4	-3/8	**-3/8**	1/2	-5/8
1	-5/4	5/8	**13/8**	-3/2	11/8
1	1/2	-1/4	**-1/4**	0	1/4
1	-1/4	1/8	**1/8**	1/2	-1/8
	-2	3/2	**5/2**	-2	5/2
Output					
	-1/5	2/5	2/5	-1/5	0
	9/20	-3/20	-3/20	1/5	-1/4
	1/20	-7/20	13/20	-1/5	-1/4
	3/10	-1/10	-1/10	-1/5	1/2
	-3/20	1/20	1/20	3/5	-1/4

Proof: The proof is obvious from Algorithm 3.1 and Theorem 3.2. ■

The following example illustrates the method.

Example 3.7 (Modifying a row in a matrix). Assume now that we modify a single row (the third row), which is boldfaced below, in matrix (3.3), that is,

$$\mathbf{A} = \begin{pmatrix} 1 & 2 & 0 & 1 & 0 \\ 2 & 1 & -1 & 0 & 0 \\ \mathbf{1} & \mathbf{0} & \mathbf{1} & \mathbf{1} & \mathbf{1} \\ 0 & 0 & 0 & 1 & 2 \\ 0 & -1 & 0 & 2 & 1 \end{pmatrix}. \tag{3.14}$$

The above algorithm allows us to obtain the new inverse in a single iteration by starting from the previous inverse. The resulting process is shown in Table 3.9. Thus, we obtain the corresponding modified inverse

$$\mathbf{A}^{-1} = \begin{pmatrix} -1/5 & 2/5 & 2/5 & -1/5 & 0 \\ 9/20 & -3/20 & -3/20 & 1/5 & -1/4 \\ 1/20 & -7/20 & 13/20 & -1/5 & -1/4 \\ 3/10 & -1/10 & -1/10 & -1/5 & 1/2 \\ -3/20 & 1/20 & 1/20 & 3/5 & -1/4 \end{pmatrix}. \tag{3.15}$$

According to (3.13) the new determinant is the old one with 8 times the normalizing constant, that is, $8 \times 5/2 = 20$. ■

Remark 3.5 *The above methods can also be applied to columns instead of rows. To this aim, instead of premultiplying the initial modified matrix by the inverse, we postmultiply the inverse by the initial modified matrix.*

Table 3.10 Updating the inverse matrix after changing a column.

Extra iteration						**Output**				
1	0	0	0	1						
-1	1	1	-1	1	0	-1	1	1	-1	1
3/4	**-3/8**	**-3/8**	**1/2**	**-5/8**	**1/8**	6	-3	-3	4	-5
-5/4	5/8	13/8	-3/2	11/8	1/8	-2	1	2	-2	2
1/2	-1/4	-1/4	0	1/4	3/4	-4	2	2	-3	4
-1/4	1/8	1/8	1/2	-1/8	-3/8	2	-1	-1	2	-2

Example 3.8 (Inverse of a column-modified matrix). Assume now that in matrix (3.3) we modify a single column (boldfaced) and get

$$\mathbf{A}^* = \begin{pmatrix} 1 & \mathbf{1} & 0 & 1 & 0 \\ 2 & \mathbf{0} & -1 & 0 & 0 \\ 0 & \mathbf{0} & 1 & 0 & 1 \\ 0 & \mathbf{0} & 0 & 1 & 2 \\ 0 & \mathbf{1} & 0 & 2 & 1 \end{pmatrix}.$$

The resulting process is shown in Table 3.10, where now the first row of the first table is the modified column of $\mathbf{A}$, the starting matrix is its inverse (see (3.4)), the dot products are in the sixth column, and the second table is the inverse matrix of $\mathbf{A}^*$.

■

3.8 INVERSES OF SYMBOLIC MATRICES

In this section we show how to obtain the inverses and determinants of symbolic matrices. Assume that we have a matrix $\mathbf{C}$, the sth row of which is symbolic; that is, it includes symbols or parameters. Suppose also that there exists at least a set of numeric values for those parameters leading to a nonsingular matrix $\mathbf{A}$ whose inverse $\mathbf{A}^{-1}$ has elements a^*_{ij}. Then we can apply the results of Theorem 3.2 to obtain the inverse of $\mathbf{C}$. In fact, we can write

$$\begin{aligned} c^*_{jk} &= \frac{a^*_{jk}\left(\sum\limits_{\ell=1}^{n} c_{s\ell}a^*_{\ell s}\right) - a^*_{js}\left(\sum\limits_{\ell=1}^{n} c_{s\ell}a^*_{\ell k}\right)}{\sum\limits_{\ell=1}^{n} c_{s\ell}a^*_{\ell s}} \\ &= \frac{\sum\limits_{\ell=1}^{n} c_{s\ell}\left(a^*_{\ell s}a^*_{jk} - a^*_{\ell k}a^*_{js}\right)}{\sum\limits_{\ell=1}^{n} c_{s\ell}a^*_{\ell s}}, \quad \text{if } k \neq s, \end{aligned} \tag{3.16}$$

and

$$c_{jk}^* = \frac{a_{jk}^*}{\sum_{\ell=1}^{n} c_{k\ell} a_{\ell k}^*}, \text{ if } k = s. \tag{3.17}$$

This gives the elements of $\mathbf{C}^{-1}$, which are rational functions in the parameters $\{c_{s\ell}; \ell = 1, \ldots, n\}$. Thus, the coefficients of $c_{s\ell}$ in the numerator and denominator of c_{jk}^* are $a_{\ell s}^* a_{jk}^* - a_{\ell k}^* a_{js}^*$ and $a_{\ell s}^*$, respectively.

From (3.13) we also have

$$|\mathbf{C}| = |\mathbf{A}| \sum_{\ell=1}^{n} c_{s\ell} a_{\ell s}^*, \tag{3.18}$$

which shows that the determinant is a linear function of the parameters.

Example 3.9 (Inverse of a symbolic matrix). Assume that we are interested in calculating the inverse of the matrix

$$\mathbf{A} = \begin{pmatrix} 1 & 2 & 0 & 1 & 0 \\ 2 & 1 & -1 & 0 & 0 \\ a & 0 & b & 1 & c \\ 0 & 0 & 0 & 1 & 2 \\ 0 & -1 & 0 & 2 & 1 \end{pmatrix}. \tag{3.19}$$

Observe that this matrix can be obtained from matrix (3.3) in Example 3.1 by modifying a single row (the third row). The inverse of (3.3) is obtained in Table 3.1, so we can calculate the inverse of $\mathbf{A}$ as function of parameters a, b, and c using (3.16) and (3.17). Thus, $\mathbf{A}^{-1}$ becomes

$$\frac{1}{k} \begin{pmatrix} -2-3b+c & 8b & 8 & 2-b-5c & -4+2b+2c \\ 3a+6b & -3b & -3 & -1+a+2b+2c & 2-2a-4b-c \\ -4+3a+2c & 2-8a-c & 13 & 3+a-8c & -6-2a+3c \\ 2a+4b & -2b & -2 & -2a-3b+c & 4a+6b \\ -a-2b & b & 1 & -1+5a+8b & -2a-3b \end{pmatrix},$$

where $k = 8a + 13b + c - 2$. ■

3.9 EXTENSIONS TO PARTITIONED MATRICES

The methods described in previous sections can be extended to partitioned matrices (see Fig. 3.1) as follows:

1. The dot products are now replaced by

$$\mathbf{C}_{ij} = \sum_{k=1}^{p} \mathbf{A}_{ik} * \mathbf{B}_{kj}, \tag{3.20}$$

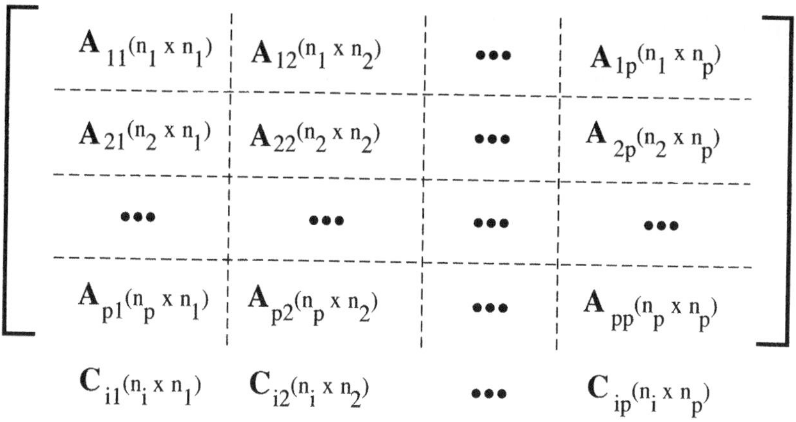

Fig. 3.1 Illustration of a partitioned matrix, its dimensions, and generalized dot products.

where $*$ is the matrix product.

2. The normalization becomes

$$\mathbf{B}_{ji} = \mathbf{B}_{ji} * \mathbf{C}_{ii}^{-1}; j = 1, \ldots, p. \tag{3.21}$$

3. The pivoting process becomes

$$\mathbf{B}_{jk} = \mathbf{B}_{jk} - \mathbf{B}_{ji} * \mathbf{C}_{ik}; \forall k \neq i; j = 1, \ldots, p. \tag{3.22}$$

Algorithm 3.1 for partitioned matrices transforms to Algorithm 3.2.

Algorithm 3.2 Inverse of a block matrix.

- **Input:** A square matrix $\mathbf{A}$ of dimension n divided into blocks.
- **Output:** The inverse $\mathbf{A}^{-1}$ in block form or a message indicating that the matrix is singular.

Step 1: Let $\mathbf{B} = \mathbf{I}_n$, where $\mathbf{I}_n$ is the identity matrix of dimension n and divide it into the same blocks as the input matrix.

Step 2: Let $i = 1$.

Step 3: Obtain the matrix $\mathbf{C}_{ii}$ using (3.20).

Step 4: If $\mathbf{C}_{ii}$ is nonsingular, go to Step 5. Otherwise, interchange the necessary columns in matrices $\mathbf{B}_{ki}; k = 1, \ldots, p$ and matrices $\mathbf{B}_{k\ell}; \ell > i; k = 1, \ldots, p$ for making $\mathbf{C}_{ii}$ nonsingular. If this is not possible print the message "Singular Matrix".

Step 5: Perform the normalization step in (3.21).

Step 6: Perform the pivoting process in (3.22).

Step 7: If $i = p$, go to Step 8. Otherwise, increase i by one and go to Step 3.

Step 8: Return the blocks of $\mathbf{B}$ as the inverse of $\mathbf{A}$.

Remark 3.6 *The block matrix approach can be interpreted as a modification of the method described previously, in which instead of working with the rows of the initial matrix one by one, we work with $n_1, \ldots, n_p$ rows in each step.*

Example 3.10 (Inverse of a partitioned matrix). Consider the partitioned matrix

$$\left(\begin{array}{cc|ccc|c} 1 & 1 & 1 & 2 & -1 & 0 \\ 0 & 1 & 1 & 1 & -1 & 1 \\ \hline 1 & 2 & -1 & -1 & -1 & 0 \\ 0 & 1 & 2 & 1 & -1 & -1 \\ 1 & 1 & 1 & 0 & 0 & 0 \\ \hline 2 & 0 & 0 & 0 & 1 & 1 \end{array}\right). \tag{3.23}$$

In Table 3.11 we show the iterations associated with Algorithm 3.2. Thus, the inverse of (3.23) is

$$\left(\begin{array}{rr|rrr|r} -1/2 & 0 & 1/2 & 3/2 & -2 & 3/2 \\ 7/6 & -1/3 & -1/2 & -5/2 & 11/3 & -13/6 \\ \hline -2/3 & 1/3 & 0 & 1 & -2/3 & 2/3 \\ 7/6 & -1/3 & -1/2 & -3/2 & 5/3 & -7/6 \\ 4/3 & -2/3 & -1 & -3 & 13/3 & -7/3 \\ \hline -1/3 & 2/3 & 0 & 0 & -1/3 & 1/3 \end{array}\right). \tag{3.24}$$

Assume now that we change some elements in the initial matrix. Then, we can calculate the new inverse in only one iteration. In fact, we can interchange rows and columns in such a manner that all the modified elements appear in the first (last) rows and columns. After this process, we have modified elements only in a square matrix $\mathbf{A}_{11}$ ($\mathbf{A}_{pp}$) and we just need to apply one iteration. Of course, in computer implementations one does not need to interchange rows and columns, but only be aware of the modified elements. ■

Remark 3.7 *The resulting complexity of the block version of the method is exactly the same as the simple method (i.e., n^3 products). However, it has important advantages since it is suitable for a parallel implementation.*

Table 3.11 Iterations for inverting the matrix in Example 3.15. Pivot block columns are boldfaced.

Iteration 1

1	0	**1**	**0**	0	0	0	0
1	1	**0**	**1**	0	0	0	0
1	1	**0**	**0**	1	0	0	0
2	1	**0**	**0**	0	1	0	0
-1	-1	**0**	**0**	0	0	1	0
0	1	**0**	**0**	0	0	0	1
		1	**1**	1	2	-1	0
		0	**1**	1	1	-1	1

Iteration 2

1	0	1	*1*	*-1*	**0**	**-1**	**0**	1
2	1	1	*0*	*1*	**-1**	**-1**	**1**	-1
-1	2	1	*0*	*0*	**1**	**0**	**0**	0
-1	1	0	*0*	*0*	**0**	**1**	**0**	0
-1	-1	0	*0*	*0*	**0**	**0**	**1**	0
0	-1	0	*0*	*0*	**0**	**0**	**0**	1
			1	*1*	**-3**	**-4**	**1**	-1
			0	*1*	**1**	**0**	**0**	-2
			1	*0*	**0**	**-2**	**1**	0

Iteration 3

2	*1*	*-3*	*1/2*	*3/2*	*-1/2*	**9/2**
0	*-1*	*4*	*-1/2*	*-5/2*	*3/2*	**-13/2**
0	*0*	*-1*	*0*	*1*	*0*	**2**
0	*0*	*2*	*-1/2*	*-3/2*	*1/2*	**-7/2**
1	*-1*	*4*	*-1*	*-3*	*2*	**-7**
1	*0*	*0*	*0*	*0*	*0*	**1**
	1	*-2*	*0*	*0*	*1*	**3**

Output

-1/2	*0*	*1/2*	*3/2*	*-2*	*3/2*
7/6	*-1/3*	*-1/2*	*-5/2*	*11/3*	*-13/6*
-2/3	*1/3*	*0*	*1*	*-2/3*	*2/3*
7/6	*-1/3*	*-1/2*	*-3/2*	*5/3*	*-7/6*
4/3	*-2/3*	*-1*	*-3*	*13/3*	*-7/3*
-1/3	*2/3*	*0*	*0*	*-1/3*	*1/3*

3.10 INVERSES OF MODIFIED MATRICES

In this section we solve the problem of obtaining the inverses of matrices resulting from other matrices by adding or removing the same number of rows and columns.

3.10.1 Inverses of Matrices After Adding Rows and Columns

Theorem 3.4 (Inverse of a partitioned matrix). *Let* $\mathbf{A}$ *be a regular matrix of dimension* $n \times n$ *with inverse* $\mathbf{A}^{-1}$. *Assume that we have block-partitioned* $\mathbf{A}$ *as follows:*

$$\mathbf{A} = \left(\begin{array}{c|c} \mathbf{A}_{11} & \mathbf{A}_{12} \\ \hline \mathbf{A}_{21} & \mathbf{A}_{22} \end{array} \right), \tag{3.25}$$

where $\mathbf{A}_{11}, \mathbf{A}_{12}, \mathbf{A}_{21}$, *and* $\mathbf{A}_{22}$ *are the block matrices of dimensions* $r \times r$, $r \times (n-r)$,$(n-r) \times r$, *and* $(n-r) \times (n-r)$, *respectively. Then its inverse is*

$$\mathbf{A}^{-1} = \left(\begin{array}{c|c} \mathbf{B}^{-1} & -\mathbf{B}^{-1}\mathbf{A}_{12}\mathbf{A}_{22}^{-1} \\ \hline -\mathbf{A}_{22}^{-1}\mathbf{A}_{21}\mathbf{B}^{-1} & \mathbf{A}_{22}^{-1}\left(\mathbf{I}_{(n-r)} + \mathbf{A}_{21}\mathbf{B}^{-1}\mathbf{A}_{12}\mathbf{A}_{22}^{-1}\right) \end{array} \right), \tag{3.26}$$

where

$$\mathbf{B} = \mathbf{A}_{11} - \mathbf{A}_{12}\mathbf{A}_{22}^{-1}\mathbf{A}_{21}. \tag{3.27}$$

Proof: In this initial proof we simply show that $\mathbf{A}\mathbf{A}^{-1} = \mathbf{I}_n$. Using the block multiplication of matrices (3.25) and (3.26), considering (3.27), and after some simplifications, we get

$$\mathbf{A}\mathbf{A}^{-1} = \left(\begin{array}{c|c} \mathbf{I}_r & \mathbf{0}_{r \times (n-r)} \\ \hline \mathbf{0}_{(n-r) \times r} & \mathbf{I}_{(n-r)} \end{array} \right) = \mathbf{I}_n, \tag{3.28}$$

which proves the theorem. ∎

Alternative proof: In this proof we use the pivoting transformation for block matrices to obtain the inverse (3.26).

In the first step we use the second block row and use the second block column as pivot column. In the second step we use the first block row and use the first block column as pivot column. Finally, in the third step, we obtain the desired inverse (see Table 3.12). ∎

Remark 3.8 *Note that Theorem 3.4 allows calculating the inverse of a matrix after adding* n *rows and* n *columns based on the inverse of the initial matrix* $\mathbf{A}_{22}$. *The added rows and columns are defined by* $\mathbf{A}_{11}, \mathbf{A}_{12}$, *and* $\mathbf{A}_{21}$. *Note*

Table 3.12 Steps for inverting matrix (3.25) using the pivoting transformation for block matrices.

	First step			**Second step**	
$\mathbf{A}_{21}$	$\mathbf{I}_r$	$\mathbf{0}_{r\times(n-r)}$	$\mathbf{A}_{11}$	$\mathbf{I}_p$	$\mathbf{0}_{r\times(n-r)}$
$\mathbf{A}_{22}$	$\mathbf{0}_{(n-r)\times r}$	$\mathbf{I}_{(n-r)}$	$\mathbf{A}_{12}$	$-\mathbf{A}_{22}^{-1}\mathbf{A}_{21}$	$\mathbf{A}_{22}^{-1}$
	$\mathbf{A}_{21}$	$\mathbf{A}_{22}$		$\mathbf{A}_{11}-\mathbf{A}_{12}\mathbf{A}_{22}^{-1}\mathbf{A}_{21}$	$\mathbf{A}_{12}\mathbf{A}_{22}^{-1}$

Third step	
$\mathbf{B}^{-1}$	$-\mathbf{B}^{-1}\mathbf{A}_{12}\mathbf{A}_{22}^{-1}$
$-\mathbf{A}_{22}^{-1}\mathbf{A}_{11}\mathbf{B}^{-1}$	$\mathbf{A}_{22}^{-1}+\mathbf{A}_{22}^{-1}\mathbf{A}_{21}\mathbf{B}^{-1}\mathbf{A}_{12}\mathbf{A}_{22}^{-1}$

that the only condition for the added rows and columns to lead to a nonsingular matrix is that the **B** *matrix be invertible.*

Example 3.11 (Inverse of a partitioned matrix). Consider the following partitioned matrix:

$$\mathbf{A} = \left(\begin{array}{cc|cc} 1 & 2 & 1 & 1 \\ 0 & 1 & 3 & 1 \\ \hline 2 & 1 & 0 & 1 \\ 1 & 1 & 1 & 1 \end{array}\right), \tag{3.29}$$

where

$$\mathbf{A}_{11} = \begin{pmatrix} 1 & 2 \\ 0 & 1 \end{pmatrix}; \mathbf{A}_{12} = \begin{pmatrix} 1 & 1 \\ 3 & 1 \end{pmatrix}; \mathbf{A}_{21} = \begin{pmatrix} 2 & 1 \\ 0 & 1 \end{pmatrix}; \mathbf{A}_{22} = \begin{pmatrix} 0 & 1 \\ 1 & 1 \end{pmatrix}.$$

We are interested in calculating $\mathbf{A}^{-1}$. Since

$$\mathbf{A}_{22}^{-1} = \begin{pmatrix} -1 & 1 \\ 1 & 0 \end{pmatrix},$$

from (3.27) we obtain

$$\begin{aligned} \mathbf{B} &= \mathbf{A}_{11} - \mathbf{A}_{12}\mathbf{A}_{22}^{-1}\mathbf{A}_{21} \\ &= \begin{pmatrix} 1 & 2 \\ 0 & 1 \end{pmatrix} - \begin{pmatrix} 1 & 1 \\ 3 & 1 \end{pmatrix}\begin{pmatrix} -1 & 1 \\ 1 & 0 \end{pmatrix}\begin{pmatrix} 2 & 1 \\ 1 & 1 \end{pmatrix} = \begin{pmatrix} 0 & 1 \\ 1 & 0 \end{pmatrix}, \end{aligned}$$

whose determinant is -1. Thus, since Remark 3.8 guarantees that matrix $\mathbf{A}$ has an inverse, we can use Theorem 3.4 for obtaining $\mathbf{A}^{-1}$, and we calculate (see (3.26))

$$
\begin{aligned}
\mathbf{B}^{-1} &= \begin{pmatrix} 0 & 1 \\ 1 & 0 \end{pmatrix} \\
-\mathbf{B}^{-1}\mathbf{A}_{12}\mathbf{A}_{22}^{-1} &= \begin{pmatrix} 2 & -3 \\ 0 & -1 \end{pmatrix} \\
-\mathbf{A}_{22}^{-1}\mathbf{A}_{21}\mathbf{B}^{-1} &= \begin{pmatrix} 0 & 1 \\ -1 & -2 \end{pmatrix} \\
\mathbf{A}_{22}^{-1}(\mathbf{I}+\mathbf{A}_{21}\mathbf{B}^{-1}\mathbf{A}_{12}\mathbf{A}_{22}^{-1}) &= \begin{pmatrix} 1 & -2 \\ -3 & 7 \end{pmatrix}.
\end{aligned}
\tag{3.30}
$$

Consequently, we finally get

$$
\mathbf{A}^{-1} = \begin{pmatrix} 0 & 1 & 2 & -3 \\ 1 & 0 & 0 & -1 \\ 0 & 1 & 1 & -2 \\ -1 & -2 & -3 & 7 \end{pmatrix}.
$$

■

3.10.2 Inverses of Matrices After Removing Rows and Columns

Theorem 3.5 (Inverse of a matrix after removing n rows and n columns). *Let* $\mathbf{A}$ *be the nonsingular matrix in (3.25) whose inverse* $\mathbf{A}^{-1}$ *is partitioned as follows:*

$$
\mathbf{A}^{-1} = \left(\begin{array}{c|c} \mathbf{A}^{11} & \mathbf{A}^{12} \\ \hline \mathbf{A}^{21} & \mathbf{A}^{22} \end{array} \right). \tag{3.31}
$$

Then we have

$$
\mathbf{A}_{22}^{-1} = \mathbf{A}^{22} - \mathbf{A}^{21}{\mathbf{A}^{11}}^{-1}\mathbf{A}^{12}. \tag{3.32}
$$

Proof: In fact, we can modify the matrix (3.25) to

$$
\mathbf{B} = \left(\begin{array}{c|c} \mathbf{I}_r & \mathbf{0}_{r\times(n-r)} \\ \hline \mathbf{A}_{21} & \mathbf{A}_{22} \end{array} \right), \tag{3.33}
$$

whose inverse is

$$
\mathbf{B}^{-1} = \left(\begin{array}{c|c} \mathbf{I}_r & \mathbf{0}_{r\times(n-r)} \\ \hline -\mathbf{A}_{22}^{-1}\mathbf{A}_{21} & \mathbf{A}_{22}^{-1} \end{array} \right). \tag{3.34}
$$

Table 3.13 Obtaining the inverse of matrix (3.33).

$\mathbf{I}_r$	$\mathbf{A}^{11}$	$\mathbf{A}^{12}$	$\mathbf{I}_r$	$\mathbf{0}_{r\times(n-r)}$
$\mathbf{0}_{(n-r)\times r}$	$\mathbf{A}^{21}$	$\mathbf{A}^{22}$	$\mathbf{A}^{21}{\mathbf{A}^{11}}^{-1}$	$\mathbf{A}^{22}-\mathbf{A}^{21}{\mathbf{A}^{11}}^{-1}\mathbf{A}^{12}$
	$\mathbf{A}^{11}$	$\mathbf{A}^{12}$		

By incorporating the rows associated with $(\ \mathbf{I}\ \mid\ \mathbf{0}\ )$ and using the pivoting process, starting with the matrix in (3.31) (see Table 3.13), we obtain the new inverse matrix

$$\mathbf{B}^{-1}=\left(\begin{array}{c|c}\mathbf{I}_r & \mathbf{0}_{r\times(n-r)}\\ \hline \mathbf{A}^{21}{\mathbf{A}^{11}}^{-1} & \mathbf{A}^{22}-\mathbf{A}^{21}{\mathbf{A}^{11}}^{-1}\mathbf{A}^{12}\end{array}\right). \tag{3.35}$$

Now, since the inverse is unique, we identify the lower right blocks in (3.34) and (3.35) and get (3.32).

Alternative proof: Because of the uniqueness of the inverse matrix, we can identify the blocks in (3.26) and (3.31) to get

$$\mathbf{A}^{12} = -\mathbf{A}^{11}\mathbf{A}_{12}\mathbf{A}_{22}^{-1}, \tag{3.36}$$

$$\mathbf{A}^{21} = -\mathbf{A}_{22}^{-1}\mathbf{A}_{21}\mathbf{A}^{11}, \tag{3.37}$$

$$\mathbf{A}^{22} = \mathbf{A}_{22}^{-1}\left(\mathbf{I}+\mathbf{A}_{21}\mathbf{A}^{11}\mathbf{A}_{12}\mathbf{A}_{22}^{-1}\right), \tag{3.38}$$

and then we can write

$$\mathbf{A}^{22}=\mathbf{A}_{22}^{-1}+\mathbf{A}_{22}^{-1}\mathbf{A}_{21}\mathbf{A}^{11}\mathbf{A}_{12}\mathbf{A}_{22}^{-1}=\mathbf{A}_{22}^{-1}+\mathbf{A}^{21}{\mathbf{A}^{11}}^{-1}\mathbf{A}^{12}, \tag{3.39}$$

from which we get (3.32). ∎

Remark 3.9 *Note that again the only condition for the removed rows and columns to lead to a nonsingular matrix is that the* $\mathbf{A}^{11}$ *matrix is invertible.*

The set $\mathcal{C}$ of pairs (i,j) of rows and columns that can be removed from a regular matrix $\mathbf{A}$ for it to continue to be nonsingular is

$$\mathcal{C}=\{(i,j)|a_{ji}^{-1}\neq 0\},$$

where a_{ji}^{-1} is an element of $\mathbf{A}^{-1}$.

Note that $a_{ji}^{-1}=Adj_{ij}/|\mathbf{A}|$, where $|\mathbf{A}|$ is the determinant of $\mathbf{A}$, and Adj_{ij} is its adjoint determinant, that is, the determinant of the matrix resulting after removing the ith row and the jth column times $(-1)^{i+j}$.

Example 3.12 (Rows and columns that can be removed to keep a matrix nonsingular). In this example we determine the set $\mathcal{C}$ of pairs (i, j) of rows and columns that can be removed from the regular matrix

$$\mathbf{A} = \begin{pmatrix} 1 & 1 & 0 & 0 \\ 0 & 0 & -1 & 0 \\ 0 & 1 & 0 & 0 \\ 1 & -1 & 0 & -1 \end{pmatrix}, \tag{3.40}$$

for it to continue to be nonsingular. First, we calculate the inverse:

$$\mathbf{A}^{-1} = \begin{pmatrix} 1 & 0 & -1 & 0 \\ 0 & 0 & 1 & 0 \\ 0 & -1 & 0 & 0 \\ 1 & 0 & -2 & -1 \end{pmatrix}, \tag{3.41}$$

and then

$$\mathcal{C} = \{(1,1), (3,1), (3,2), (2,3), (1,4), (3,4), (4,4)\}. \tag{3.42}$$

■

Example 3.13 (Inverse of a matrix after removing a row and a column). Consider the matrix

$$\mathbf{A} = \begin{pmatrix} 1 & 1 & 0 & 0 \\ 0 & 0 & -1 & 0 \\ 0 & 1 & 0 & 0 \\ 1 & -1 & 0 & -1 \end{pmatrix}, \tag{3.43}$$

with inverse

$$\mathbf{A}^{-1} = \begin{pmatrix} 1 & 0 & -1 & 0 \\ 0 & 0 & 1 & 0 \\ 0 & -1 & 0 & 0 \\ 1 & 0 & -2 & -1 \end{pmatrix}, \tag{3.44}$$

and assume that we want to calculate the inverse of the resulting matrix after removing the third row ($i = 3$) and the second column ($j = 4$):

$$\mathbf{A} = \begin{pmatrix} 1 & 1 & 0 & \mathbf{0} \\ 0 & 0 & -1 & \mathbf{0} \\ \mathbf{0} & \mathbf{1} & \mathbf{0} & \mathbf{0} \\ 1 & -1 & 0 & \mathbf{-1} \end{pmatrix}. \tag{3.45}$$

The first step consists of moving the selected rows and columns until the matrix to be inverted is in the lower right block. This can be achieved by multiplying $\mathbf{A}$ by the following permutation matrices: $\mathbf{P}_{12}, \mathbf{P}_{23}, \mathbf{P}_{34}, \mathbf{P}_{23}$, and $\mathbf{P}_{12}$, where $\mathbf{P}_{ij}$ is obtained from $\mathbf{I}_4$ after permuting its ith row and jth column.

Table 3.14 Updating the inverse matrix

Extra iteration					Output			
1	**−2**	1	0	−1	1	0	0	0
0	**−1**	1	0	0	1/2	1/2	0	1/2
0	**1**	0	0	0	−1/2	1/2	0	−1/2
0	**0**	0	−1	0	0	0	−1	0
	−2	1	0	−1				

Thus, we obtain the permuted matrix

$$\mathbf{A}^* = \mathbf{P}_{12}\mathbf{P}_{23}\mathbf{A}\mathbf{P}_{34}\mathbf{P}_{23}\mathbf{P}_{12} = \begin{pmatrix} 0 & 0 & 1 & 0 \\ 0 & 1 & 1 & 0 \\ 0 & 0 & 0 & -1 \\ -1 & 1 & -1 & 0 \end{pmatrix}. \tag{3.46}$$

Taking into account that $\mathbf{P}_{ij}^{-1} = \mathbf{P}_{ij}^{\mathsf{T}}$ we calculate

$$(\mathbf{A}^*)^{-1} = \begin{pmatrix} -2 & 1 & 0 & -2 \\ -1 & 1 & 0 & 0 \\ 1 & 0 & 0 & 0 \\ 0 & 0 & -1 & 0 \end{pmatrix}. \tag{3.47}$$

If we remove the first column and the first row of the final matrix in Table 3.14, we get

$$\begin{pmatrix} 1 & 1 & 0 \\ 0 & 0 & -1 \\ 1 & -1 & 0 \end{pmatrix}^{-1} = \mathbf{A}^{22} - \mathbf{A}^{21}{\mathbf{A}^{11}}^{-1}\mathbf{A}^{12} = \begin{pmatrix} 1/2 & 0 & 1/2 \\ 1/2 & 0 & -1/2 \\ 0 & -1 & 0 \end{pmatrix}. \tag{3.48}$$

■

Example 3.14 (Inverse of a matrix after adding a row and a column). Consider the matrix

$$\mathbf{A} = \begin{pmatrix} 1 & 1 & 0 \\ 0 & 0 & -1 \\ 1 & -1 & 0 \end{pmatrix}, \tag{3.49}$$

whose inverse is (3.48). Suppose we are interested in calculating the inverse of the matrix

$$\mathbf{B} = \begin{pmatrix} 1 & 1 & 0 & \mathbf{0} \\ 0 & 0 & -1 & \mathbf{0} \\ \mathbf{0} & \mathbf{1} & \mathbf{0} & \mathbf{0} \\ 1 & -1 & 0 & \mathbf{-1} \end{pmatrix}, \tag{3.50}$$

Table 3.15 Updating the inverse matrix (phase I).

Phase I				
1	1	0	0	0
0	0	1/2	0	1/2
1	0	1/2	0	–1/2
0	0	0	–1	0
	1	1/2	0	–1/2

Output of phase I			
1	–1/2	0	1/2
0	1/2	0	1/2
0	1/2	0	–1/2
0	0	–1	0

Table 3.16 Updating the inverse matrix (phase II).

Phase II				
0	1	0	0	0
0	–1/2	1/2	1/2	0
0	0	0	0	–1
–1	1/2	1/2	–1/2	0
	–1/2	–1/2	1/2	0

Output of phase II			
–2	–1	1	0
1	1	0	0
0	0	0	–1
–1	0	0	0

obtained after inserting the boldfaced row and column in matrix **A**.

First, we permute rows and columns in order to get the inserted row and column in the first position. Then we have

$$\mathbf{D} = \begin{pmatrix} \mathbf{0} & \mathbf{0} & \mathbf{1} & \mathbf{0} \\ \mathbf{0} & 1 & 1 & 0 \\ \mathbf{0} & 0 & 0 & -1 \\ \mathbf{-1} & 1 & -1 & 0 \end{pmatrix}. \tag{3.51}$$

Next, we apply the algorithm to the matrix

$$\begin{pmatrix} \mathbf{I} & \mathbf{0} \\ \mathbf{0} & \mathbf{A}^{-1} \end{pmatrix}$$

as shown in Tables 3.15 and 3.16.

Finally, transposing the output of phase II and undoing the permutations of rows and columns, we get

$$\mathbf{B}^{-1} = \begin{pmatrix} 1 & 0 & -1 & 0 \\ 0 & 0 & 1 & 0 \\ 0 & -1 & 0 & 0 \\ 1 & 0 & -2 & -1 \end{pmatrix}. \tag{3.52}$$

■

3.11 MATHEMATICA PROGRAMS

The following Mathematica programs use the Orthogonal Module in Chapter 2 to invert matrices, calculate determinants, and obtain the rank of a given matrix.

Program 3.1 (Modules for obtaining the rank, determinant, and inverse of a matrix).

```
(* Inverse of a square matrix *)
InverseMatrix[m_?MatrixQ]:=
  If[Length[m],=Length[m[[1]]],Print["Non square matrix"],
      Transpose[Part[Orthogonal[m,Complement->True,
                                     SingularMatrixQ->True],2]]];

(* Determinant of a square matrix *)
Determinant[m_?MatrixQ]:=
  If[Length[m],=Length[m[[1]]],Print["Non square matrix"],
      Part[Orthogonal[m,Determinant->True],2]];

(* Rank of a matrix *)
MatrixRank[m_?MatrixQ]:=
  Length[m[[1]]]-Length[Orthogonal[m]]
```

The following examples illustrate the use of these programs.

Example 3.15 (Inverting a matrix). The matrix

$$\mathbf{A} = \begin{pmatrix} 1 & 4 & 0 \\ 2 & 1 & 1 \\ -2 & 1 & 3 \end{pmatrix} \tag{3.53}$$

can be inverted using the command

```
In[]:= InverseMatrix[{{1,4,0},{2,1,1},{-2,1,3}}]
             1   2   2      4    1   1       2   3  7
Out[]= {{-(--),-,-(--)},{--,-(--),--},{-(--),--,--}}
            15   5   15     15   10  30      15  10 30
```

Thus, we obtain

$$\mathbf{A}^{-1} = \begin{pmatrix} -1/15 & 2/15 & -2/15 \\ 4/15 & -1/10 & 1/30 \\ -2/15 & 3/10 & 7/30 \end{pmatrix}. \tag{3.54}$$

■

Example 3.16 (Determinant of a matrix). The determinant of the matrix in Example 3.15 can be calculated by means of the command

```
In[]:= Determinant[{{1,4,0},{2,1,1},{-2,1,3}}]
Out[]= -30
```

■

Example 3.17 (Rank of a matrix). We can use the function `MatrixRank` in Program 3.1 for calculating the rank of a matrix, as the following example shows.

```
In[]:= MatrixRank[{{-1,-5,9,-10},{1,2,3,-2},{0,0,4,-2},
                   {1,3,-1,2}}]
Out[]= 3
```

■

Program 3.2 (Module for obtaining the inverse of a row-modified matrix).

```
(* Inverse of a row modified matrix *)

ModifiedInverse[InitialInverse_,Row_,RowNumber_]:=
  Module[{T,MM,n,j},
  MM=Transpose[InitialInverse];
  n=Length[MM];
  T=MM.Row;
  If[TrueQ[T[[RowNumber]]==0],Print["Singular matrix"];
  Return[{0}],
  MM[[RowNumber]]=Map[(#/T[[RowNumber]])&,MM[[RowNumber]]];
  Do[If[((j!=RowNumber)&&(!TrueQ[T[[j]]==0])),
    MM[[j]]-=T[[j]]*MM[[RowNumber]]],
  {j,1,n}];
  Return[Transpose[MM]]
  ]
```

The following example illustrates the use of this program.

Example 3.18 (Inverting a row-modified matrix). We can use the Mathematica command `ModifiedInverse` for solving Example 3.7.

```
In[]:=
A={{1,2,0,1,0},{2,1,-1,0,0},{1,0,1,1,1},{0,0,0,1,2},
{0,-1,0,2,1}};

A1=InverseMatrix[A]

Out[]=
      1    2  2    1          9     3       3    1    1
{{-(-), -, -, -(-), 0}, {--, -(--), -(--), -, -(-)},
      5    5  5    5         20    20      20    5    4
```

```
  1     7     13    1     1     3     1      1     1  1
{--, -(--), --, -(-), -(-)}, {--, -(--), -(--), -(-), -},
 20    20    20    5     4     10    10     10    5  2

   3    1   1  3   1
{-(--), --, --, -, -(-)}}
   20   20  20 5   4
In[]:= ModifiedInverse[A1,{1,0,1,1,1},3]

Out[]=
   1   2  2   1        9     3      3   1   1
{{-(-), -, -, -(-), 0}, {--, -(--), -(--), -, -(-)},
   5   5  5   5        20    20     20  5   4

  1     7     13    1     1     3     1      1     1  1
{--, -(--), --, -(-), -(-)}, {--, -(--), -(--), -(-), -},
 20    20    20    5     4     10    10     10    5  2

   3    1   1  3   1
{-(--), --, --, -, -(-)}}
   20   20  20 5   4
```

■

Exercises

3.1 Calculate the determinant and the inverse of the following matrix using Algorithm 2.1:

$$\mathbf{A} = \begin{pmatrix} -1 & 1 & 0 & -3 & 1 \\ 1 & 0 & -1 & 2 & -2 \\ -1 & -2 & 1 & -1 & -2 \\ -2 & -1 & -1 & 1 & -2 \\ 1 & 3 & -2 & 0 & 2 \end{pmatrix}.$$

3.2 Obtain the determinant of the matrix:

$$\mathbf{P} = \begin{pmatrix} 1 & 1 & 2 & -2 \\ 2 & -2 & 0 & 1 \\ 3 & -1 & 2 & -1 \\ 1 & 0 & -2 & -1 \end{pmatrix}.$$

3.3 Use the following nonsingular matrix $\mathbf{B}$ whose determinant is 4 as an initial matrix for obtaining the determinant of matrix $\mathbf{P}$ in Exercise 3.2:

$$\mathbf{B} = \begin{pmatrix} 1 & 0 & 2 & -1 \\ 0 & -2 & -2 & 1 \\ 1 & -1 & 2 & -1 \\ 2 & 0 & -2 & -1 \end{pmatrix}.$$

3.4 Let $\mathbf{A}$ be the matrix

$$\mathbf{A} = \begin{pmatrix} a & a & 2 & a \\ 2 & a & 1 & a \\ 1 & a & a & 0 \\ a & 0 & 1 & 1 \end{pmatrix},$$

where a is a parameter.

(a) How does the rank of $\mathbf{A}$ depend on the parameter a?

(b) Calculate its determinant.

(c) Obtain its inverse.

3.5 Jimmy is nervous because today he has a test for obtaining a job in the local administration. One of the questions consists of calculating the inverse of the following matrix:

$$\mathbf{A} = \begin{pmatrix} 0 & 1 & -3 & 1 & -2 \\ 1 & -3 & 2 & -2 & 1 \\ 1 & 0 & -2 & 1 & -2 \\ -2 & 0 & 1 & 2 & 0 \\ 1 & 3 & -2 & 1 & 2 \end{pmatrix}.$$

Unfortunately, Jimmy makes an error in copying the data, so he is really calculating the inverse of the matrix

$$\mathbf{A}' = \begin{pmatrix} 0 & 1 & -3 & \mathbf{-2} & -2 \\ 1 & -3 & 2 & \mathbf{1} & 1 \\ 1 & 0 & -2 & 1 & -2 \\ -2 & 0 & 1 & 2 & 0 \\ 1 & 3 & -2 & 1 & 2 \end{pmatrix}.$$

When the test time is close to finish and $\mathbf{A}'^{-1}$ has been calculated, he discovers his mistake. Help Jimmy to obtain the right inverse with a few calculations using his wrong inverse.

3.6 Consider the matrix $\mathbf{A}$ and its inverse $\mathbf{A}^{-1}$

$$\mathbf{A} = \begin{pmatrix} 0 & 2 & 3 & -2 & 1 \\ 1 & 2 & -2 & 1 & 0 \\ 1 & 0 & -2 & -1 & 1 \\ -1 & -1 & -2 & 1 & 1 \\ 1 & 0 & 1 & 1 & -2 \end{pmatrix},$$

$$\mathbf{A}^{-1} = \begin{pmatrix} 17/20 & -1/5 & 19/20 & 13/10 & 31/20 \\ -1/5 & 2/5 & -2/5 & -3/5 & -3/5 \\ 3/5 & -1/5 & 1/5 & 4/5 & 4/5 \\ 3/4 & 0 & 1/4 & 3/2 & 5/4 \\ 11/10 & -1/5 & 7/10 & 9/5 & 13/10 \end{pmatrix}.$$

Now, calculate the inverse of $\mathbf{A}^*$, resulting from $\mathbf{A}$ by adding the boldfaced row and column, based on the inverse of the initial matrix, where

$$\mathbf{A}^* = \begin{pmatrix} 0 & 2 & \mathbf{0} & 3 & -2 & 1 \\ 1 & 2 & \mathbf{1} & -2 & 1 & 0 \\ 1 & 0 & \mathbf{-1} & -2 & -1 & 1 \\ \mathbf{2} & \mathbf{0} & \mathbf{1} & \mathbf{-1} & \mathbf{0} & \mathbf{1} \\ -1 & -1 & \mathbf{0} & -2 & 1 & 1 \\ 1 & 0 & \mathbf{1} & 1 & 1 & -2 \end{pmatrix}.$$

3.7 Given the matrix

$$\mathbf{A} = \begin{pmatrix} 1 & 0 & -1 & 1 \\ a & 2 & b & 0 \\ 2 & 1 & 2 & 1 \\ 0 & -1 & -1 & 0 \end{pmatrix}$$

(a) Show that if $-2 - 2a + b = 0$, $\mathbf{A}$ is a singular matrix.

(b) Show that if a and b do not satisfy the above condition, the inverse matrix $\mathbf{A}^{-1}$ of $\mathbf{A}$ is

$$\begin{pmatrix} \frac{2-b}{-2-2a+b} & \frac{-2}{-2-2a+b} & \frac{-2+b}{-2-2a+b} & \frac{-6+b}{-2-2a+b} \\ -\frac{a}{-2-2a+b} & -\frac{1}{-2-2a+b} & \frac{a}{-2-2a+b} & \frac{3a-b}{-2-2a+b} \\ \frac{a}{-2-2a+b} & \frac{1}{-2-2a+b} & -\frac{a}{-2-2a+b} & \frac{2-a}{-2-2a+b} \\ \frac{-4-a+2b}{-2-2a+b} & \frac{3}{-2-2a+b} & \frac{2-a-b}{-2-2a+b} & \frac{8-a-b}{-2-2a+b} \end{pmatrix}.$$

3.8 Given the set of vectors

$$W = \{(1,0,0,1), (2,1,-1,0), (0,1,2,-1), (3,3,0,-2)\}$$

(a) Using orthogonal sets, check whether or not they are linearly independent.

(b) Determine the rank of the matrix having those vectors as rows or columns.

(c) Show that $\mathcal{L}\{(-1, \frac{5}{3}, -\frac{1}{3}, 1)\}$ is the linear subspace orthogonal to W.

3.9 Given the following matrix $\mathbf{A}$ and its inverse $\mathbf{A}^{-1}$

$$\mathbf{A} = \begin{pmatrix} 1 & 0 & -1 & 1 \\ 0 & 0 & 1 & 2 \\ 1 & -1 & 1 & 0 \\ 2 & 0 & 0 & 0 \end{pmatrix}; \ \mathbf{A}^{-1} = \begin{pmatrix} 0 & 0 & 0 & 1/2 \\ -2/3 & 1/3 & -1 & 5/6 \\ -2/3 & 1/3 & 0 & 1/3 \\ 1/3 & 1/3 & 0 & -1/6 \end{pmatrix}.$$

(a) Give a list of all pairs of the form ($row, column$) such that eliminating the dcorresponding row and column in $\mathbf{A}^{-1}$ a regular matrix is obtained.

(b) Starting from the inverse $\mathbf{A}^{-1}$, obtain the inverse of the matrix

$$\mathbf{B} = \begin{pmatrix} 1 & 0 & -1 & 1 \\ 0 & 0 & 1 & 2 \\ 2 & 1 & 1 & 0 \\ 2 & 0 & 0 & 0 \end{pmatrix}$$

3.10 Show that the determinant of the matrix

$$\mathbf{A} = \begin{pmatrix} 1 & 0 & 2 & -1 & 0 \\ 0 & 1 & 0 & 0 & 0 \\ 0 & 0 & 1 & 0 & 0 \\ 0 & a & 0 & b & c \\ 0 & 0 & 0 & 0 & 1 \end{pmatrix}$$

is b. **Hint:** Note that **A** is the unit matrix with two modified rows.

3.11 Taking into account that the inverse of the matrix

$$\mathbf{B} = \begin{pmatrix} 1 & 0 & 2 \\ 1 & 1 & 1 \\ 0 & 0 & 1 \end{pmatrix}$$

is

$$\mathbf{B}^{-1} \begin{pmatrix} 1 & 0 & -2 \\ -1 & 1 & 1 \\ 0 & 0 & 1 \end{pmatrix}$$

and that **A** comes from **B** by adding rows and columns, show that the inverse of the matrix

$$\mathbf{A} = \begin{pmatrix} 1 & 0 & 2 & -1 & 1 \\ 0 & 1 & 0 & -1 & 0 \\ 0 & 0 & 1 & 0 & 2 \\ 0 & 0 & 1 & 1 & 1 \\ 0 & 1 & 0 & 0 & 1 \end{pmatrix}$$

is

$$\mathbf{A}^{1} = \begin{pmatrix} 1 & -2 & -1 & -1 & 2 \\ 0 & 1/2 & -1/2 & 1/2 & 1/2 \\ 0 & 1 & 0 & 1 & -1 \\ 0 & -1/2 & -1/2 & 1/2 & 1/2 \\ 0 & -1/2 & 1/2 & -1/2 & 1/2 \end{pmatrix}$$

4 More Applications of Orthogonal Sets

In this chapter we include more applications of the orthogonal sets technique described in Chapter 2 to the solution of problems in linear algebra. In Section 4.1 we show how to obtain the intersection of two linear subspaces. In Section 4.2 we obtain the reciprocal image of a linear subspace by a linear transformation and, in Section 4.3, the conditions for the images of two linear subspaces by two linear transformations to coincide.

4.1 INTERSECTION OF TWO LINEAR SUBSPACES

Theorem 2.2 allows us to obtain the intersection of two linear subspaces S_1 and S_2 by noting that

$$S_1 \cap S_2 \equiv S_1 \cap (S_2^{\perp})^{\perp} \equiv S_2 \cap (S_1^{\perp})^{\perp}. \tag{4.1}$$

In fact, we can obtain first $S_2^{\perp}$, the orthogonal to S_2, by letting $\mathcal{L}(\mathbf{V}^0) \equiv \mathbb{R}^n$ in Theorem 2.2 and then find the orthogonal to $S_2^{\perp}$ in S_1, using S_1 as $\mathcal{L}(\mathbf{V}^0)$. Alternatively, we can obtain first $S_1^{\perp}$, the orthogonal to S_1, by letting $\mathcal{L}(\mathbf{V}^0) \equiv R^n$ in Theorem 2.2 and then find the orthogonal to $S_1^{\perp}$ in S_2, using S_2 as $\mathcal{L}(\mathbf{V}^0)$.

Example 4.1 (Intersection of moving planes). Consider two moving planes in $\mathbb{R}^3$:

$$\Pi_1(t) \equiv \mathcal{L}\{(0,0,1)^T, (t, t+1, 0)^T\} \quad \text{and} \quad \Pi_2(t) \equiv \mathcal{L}\{(1,0,0)^T, (0, t, 1-t)^T\},$$

where the time parameter $t \in [0,1]$. Figure 4.1 shows the initial ($t = 0$), the intermediate ($0 < t < 1$), and the final ($t = 1$) positions of the planes. The

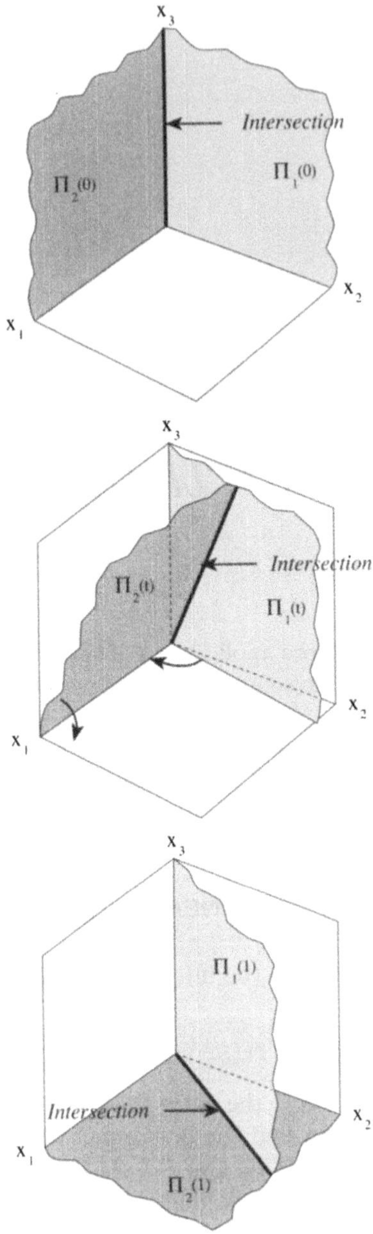

Fig. 4.1 Intersection of two moving planes.

intersection of $\Pi_1(t)$ and $\Pi_2(t)$ is always a linear subspace generated by a unique vector, that is, a line passing through the origin.

Table 4.1 Orthogonal subspace to $\Pi_1(t)$.

Iteration 1				Iteration 2				Output		
0	1	0	**0**	t	*0*	**0**	1	*0*	*0*	1
0	0	1	**0**	t+1	*0*	**1**	0	*0*	*1/(t + 1)*	–t/(t + 1)
1	0	0	**1**	0	*1*	**0**	0	*1*	*0*	0
	0	0	**1**		*0*	**t + 1**	t			

Table 4.2 Orthogonal subspace to $\Pi_1(t)^\perp$ in $\Pi_2(t)$.

Iteration 1			Output	
1	**1**	0	*1*	$t^2/(t+1)$
–t/(t+1)	**0**	t	*0*	t
0	**0**	$1-t$	*0*	$1-t$
	1	$-t^2/(t+1)$		

In order to calculate this intersection as a function of time t, we proceed as follows:

1. We first obtain the orthogonal set $\Pi_1(t)^\perp$, by letting $\mathcal{L}(\mathbf{V}^0) \equiv \mathbb{R}^3$ in Theorem 2.2 (see Table 4.1):

$$\Pi_1(t)^\perp \equiv \mathcal{L}\left\{\left(1, -\frac{t}{t+1}, 0\right)^T\right\}.$$

2. Finally, we obtain the set orthogonal to $\Pi_1(t)^\perp$ in $\Pi_2(t)$, using $\Pi_2(t)$ as $\mathcal{L}(\mathbf{V}^0)$ (see Table 4.2):

$$\Pi_1(t) \cap \Pi_2(t) \equiv \mathcal{L}\left\{\left(\frac{t^2}{t+1}, t, 1-t\right)^T\right\}.$$

As can be seen, initially ($t = 0$) the intersection is the linear subspace generated by the vector $(0, 0, 1)^T$, and at the end ($t = 1$) it is the linear subspace generated by $(1/2, 1, 0)^T$.

■

Example 4.2 (Intersection of moving subspaces). Consider the linear subspaces $S_1 \equiv \mathcal{L}\{\mathbf{v}_1, \mathbf{v}_2, \mathbf{v}_3, \mathbf{v}_4\}$ and $S_2^\perp \equiv \mathcal{L}\{\mathbf{v}_1, \mathbf{v}_4\}$, where

$$\begin{array}{lcl}
\mathbf{v}_1 &=& (\;1, \quad \sin 2t, \quad -1, \quad \cos t\;)^T, \\
\mathbf{v}_2 &=& (\;\cos t, \quad 1, \quad \sin 2t, \quad -1\;)^T, \\
\mathbf{v}_3 &=& (\;-1, \quad \cos t, \quad 1, \quad \sin 2t\;)^T, \\
\mathbf{v}_4 &=& (\;\sin 2t, \quad -1, \quad \cos t, \quad 1\;)^T,
\end{array}$$

and assume that we wish to:

1. Determine the intersection $Q_1 \equiv S_1 \cap S_2$ for all values of the time parameter $0 \le t \le 2\pi$.
2. Find the t-values for which we have $Q_1 \equiv Q_2$, where $Q_2 \equiv S_1 \cap S_3$ and $S_3^{\perp} \equiv \mathcal{L}\{\mathbf{v}_1, \mathbf{v}_2\}$.

Then we have:

1. By definition we can write

$$Q_1 \equiv \{\mathbf{v} \in S_1 | \mathbf{v} \in S_2\}; \mathbf{v} \in S_2 \Leftrightarrow \ \mathbf{v} \bullet \mathbf{v}_1 = 0 \text{ and } \mathbf{v} \bullet \mathbf{v}_4 = 0.$$

Using the procedure in Theorem 2.2 and starting with $\mathbf{v}_1$, we get

$$\begin{aligned} p &= \mathbf{v}_1 \bullet \mathbf{v}_1 = \sin^2 2t + \cos^2 t + 2, \\ q &= \mathbf{v}_1 \bullet \mathbf{v}_3 = 2\sin 2t \cos t - 2, \\ \mathbf{v}_1 &\bullet \mathbf{v}_2 = \mathbf{v}_1 \bullet \mathbf{v}_4 = 0. \end{aligned} \tag{4.2}$$

Since $p \neq 0; \forall t$, using the orthogonalization procedure in Theorem 2.2, we obtain

$$\{\mathbf{v} \in S_1 | \mathbf{v} \bullet \mathbf{v}_1 = 0\} \equiv \mathcal{L}\{\mathbf{u}_1 = p\mathbf{v}_3 - q\mathbf{v}_1, \mathbf{v}_2, \mathbf{v}_4)\},$$

and proceeding with $\mathbf{v}_4$ and taking into account that

$$\mathbf{v}_4 \bullet \mathbf{v}_2 = q; \mathbf{v}_4 \bullet \mathbf{v}_4 = p; \mathbf{v}_4 \bullet \mathbf{u}_1 = 0,$$

we have

$$Q_1 \equiv \mathcal{L}\{p\mathbf{v}_3 - q\mathbf{v}_1, p\mathbf{v}_2 - q\mathbf{v}_4\} \equiv \mathcal{L}\{\mathbf{u}_1, \mathbf{u}_2\}.$$

2. By a similar process we get

$$Q_2 \equiv \mathcal{L}\{p\mathbf{v}_3 - q\mathbf{v}_1, p\mathbf{v}_4 - q\mathbf{v}_2\} \equiv \mathcal{L}\{\mathbf{u}_1, \mathbf{u}_3\}.$$

Since Q_2 is orthogonal to $\mathbf{v}_2$ and $Q_1 = Q_2$, then Q_1 is orthogonal to $\mathbf{v}_2$ and, in particular, $\mathbf{u}_2$ is orthogonal to $\mathbf{v}_2$. Similarly, since Q_1 is orthogonal to $\mathbf{v}_4$ and $Q_1 = Q_2$, then Q_2 is orthogonal to $\mathbf{v}_4$ and, in particular, $\mathbf{u}_3$ is orthogonal to $\mathbf{v}_4$; that is,

$$\begin{aligned} &Q_1 \equiv Q_2 \Rightarrow \mathbf{u}_2 \bullet \mathbf{v}_2 = 0, \mathbf{u}_3 \bullet \mathbf{v}_4 = 0 \Rightarrow p^2 - q^2 = 0 \\ &\Rightarrow p = -q \Rightarrow \cos t(2\sin t + 1) = 0 \Rightarrow t \in A \equiv \left\{\frac{\pi}{2}, \frac{3\pi}{2}, \frac{7\pi}{6}, \frac{11\pi}{6}\right\} \end{aligned}$$

and, conversely,

$$p = -q \Rightarrow \mathbf{u}_2 = \mathbf{u}_3 \Rightarrow Q_1 \equiv Q_2.$$

Thus, we get

$$Q_1 \equiv Q_2 \Leftrightarrow t \in A \equiv \left\{\frac{\pi}{2}, \frac{3\pi}{2}, \frac{7\pi}{6}, \frac{11\pi}{6}\right\}.$$

■

Table 4.3 Process for obtaining the sets $\mathcal{L}^{\perp}\left\{(1,-1,2)^T\right\}$ and $\mathcal{L}^{\perp}\left\{(1,-1,2)^T,(2,3,4)^T\right\}$ in Example 4.3.

Iteration 1				**Iteration 2**			**Output**
1	**1**	0	0	2	**1**	–2	–2
–1	**0**	1	0	3	**1**	0	0
2	**0**	0	1	4	**0**	1	1
	1	–1	2		**5**	0	

4.2 RECIPROCALS IMAGES IN LINEAR TRANSFORMATIONS

The above methods are also useful for obtaining the reciprocal image of a linear subspace by a linear transformation. Assume that we are interested in obtaining the set of vectors $\mathbf{x}$ such that $\mathbf{Ax}$ belongs to $\mathcal{L}\{\mathbf{s}_1,\ldots,\mathbf{s}_r\}$. Since we can write

$$\begin{aligned}\mathbf{Ax}\in\mathcal{L}\{\mathbf{s}_1,\ldots,\mathbf{s}_r\} \quad &\Leftrightarrow \quad \left[\mathcal{L}^{\perp}\{\mathbf{s}_1,\ldots,\mathbf{s}_r\}\right]^T\mathbf{Ax}=\mathbf{0}\\ &\Leftrightarrow \quad \mathbf{x}\in\left[\mathcal{L}^{\perp}\{\mathbf{s}_1,\ldots,\mathbf{s}_r\}^T\mathbf{A}\right]^{\perp},\end{aligned} \tag{4.3}$$

we can use Theorem 2.2 to solve this problem. Thus, we proceed as follows:

1. We obtain the orthogonal subspace to $\mathcal{L}\{\mathbf{s}_1,\ldots,\mathbf{s}_r\}$.

2. We multiply matrix $\left[\mathcal{L}^{\perp}\{\mathbf{s}_1,\ldots,\mathbf{s}_r\}\right]^T$ (the matrix whose rows are the generators of $\mathcal{L}^{\perp}\{\mathbf{s}_1,\ldots,\mathbf{s}_r\}$) and $\mathbf{A}$.

3. We obtain the orthogonal subspace to the set generated by the rows of the resulting matrix.

Example 4.3 (Reciprocal image of the linear subspace by a linear transformation). Consider the linear transformation with associated matrix

$$\mathbf{A}=\begin{pmatrix}3&-1&1\\2&1&-1\\-1&-1&-1\end{pmatrix}. \tag{4.4}$$

The reciprocal image of the linear subspace $\mathcal{L}\left\{(1,-1,2)^T\right\}$ by the transformation of matrix $\mathbf{A}$ is the linear subspace $\mathcal{L}\left\{(0,3,1)^T\right\}$.

In Table 4.3 (Steps 1 and 2) we show how to obtain the orthogonal set to $\mathcal{L}\left\{(1,-1,2)^T\right\}$, which becomes

$$\mathcal{L}^{\perp}\left\{(1,-1,2)^T\right\}=\mathcal{L}\left\{(1,1,0)^T,(-2,0,1)^T\right\}. \tag{4.5}$$

Table 4.4 Process for obtaining the orthogonal set to $\mathcal{L}\{(-7,1,-3)^T\}$ and $\mathcal{L}\{(5,0,0)^T,(-7,1,-3)^T\}$.

Iteration 1				Iteration 2			Output
-7	**1**	0	0	5	**1/7**	-3/7	0
1	**0**	1	0	0	**1**	0	3
-3	**0**	0	1	0	**0**	1	1
	-7	1	-3		**5/7**	-15/7	

Next, we calculate

$$\begin{aligned}\mathcal{L}\left\{(1,1,0)^T,(-2,0,1)^T\right\}^T\mathbf{A} &\Rightarrow \begin{pmatrix}1 & 1 & 0\\ -2 & 0 & 1\end{pmatrix}\begin{pmatrix}3 & -1 & 1\\ 2 & 1 & -1\\ -1 & -1 & -1\end{pmatrix}\\ &= \begin{pmatrix}5 & 0 & 0\\ -7 & 1 & -3\end{pmatrix},\end{aligned} \tag{4.6}$$

and finally, we obtain $\mathcal{L}\left\{(0,3,1)^T\right\}$ as the orthogonal to

$$\mathcal{L}\left\{(-7,1,-3)^T,(5,0,0)^T\right\},$$

as shown in Table 4.4. Thus, we obtain the above solution.

Similarly, the reciprocal of the linear subspace $\mathcal{L}\left\{(1,-1,2)^T,(2,3,4)^T\right\}$ by the transformation of matrix $\mathbf{A}$ is the linear subspace

$$\mathcal{L}\left\{(1/7,1,0)^T,(-3/7,0,1)^T\right\}.$$

In Table 4.3 we show that the orthogonal set to $\mathcal{L}\left\{(1,-1,2)^T,(2,3,4)^T\right\}$ is

$$\mathcal{L}^{\perp}\left\{(1,-1,2)^T,(2,3,4)^T\right\} = \mathcal{L}\left\{(-2,0,1)^T\right\}. \tag{4.7}$$

Next, we calculate

$$\begin{aligned}\mathcal{L}\left\{(-2,0,1)^T\right\}^T\mathbf{A} &\Rightarrow \begin{pmatrix}-2 & 0 & 1\end{pmatrix}\begin{pmatrix}3 & -1 & 1\\ 2 & 1 & -1\\ -1 & -1 & -1\end{pmatrix}\\ &= \begin{pmatrix}-7 & 1 & -3\end{pmatrix},\end{aligned} \tag{4.8}$$

and finally, in Table 4.4 (Steps 1 and 2), we show that the orthogonal to $\mathcal{L}\left\{(-7,1,-3)^T\right\}$ is $\mathcal{L}\left\{(1/7,1,0)^T,(-3/7,0,1)^T\right\}$. ■

In Section 4.4 we give a Mathematica program to obtain the reciprocal image of a linear subspace by a linear transformation.

4.3 OTHER APPLICATIONS

Next, we show how to obtain the conditions under which the images of two linear transformations coincide. Consider two linear transformations defined on the same linear subspace and with associated matrices **A** and **B**. Suppose we wish to find the linear subspaces associated with $\mathbf{x}_a$ and $\mathbf{x}_b$ such that $\mathbf{A}\mathbf{x}_a = \mathbf{B}\mathbf{x}_b$; that is, their images by **A** and **B**, respectively, coincide. Then we have

$$(\mathbf{A} : -\mathbf{B}) \begin{pmatrix} \mathbf{x}_a \\ \mathbf{x}_b \end{pmatrix} = 0 \Rightarrow \begin{pmatrix} \mathbf{x}_a \\ \mathbf{x}_b \end{pmatrix} \in (\mathbf{A} : -\mathbf{B})^{\perp}. \tag{4.9}$$

Example 4.4 (Conditions under which the images of two linear transformations coincide). Let

$$\mathbf{A} = \begin{pmatrix} 1 & -1 \\ 2 & -2 \end{pmatrix}; \ \mathbf{B} = \begin{pmatrix} 1 & 1 \\ 2 & 0 \end{pmatrix}. \tag{4.10}$$

Then we have

$$(\mathbf{A} : -\mathbf{B}) = \begin{pmatrix} 1 & -1 & -1 & -1 \\ 2 & -2 & -2 & 0 \end{pmatrix} \tag{4.11}$$

and

$$(\mathbf{A} : -\mathbf{B})^{\perp} = \begin{pmatrix} 1 & 1 \\ 1 & 0 \\ 0 & 1 \\ 0 & 0 \end{pmatrix}, \tag{4.12}$$

which implies

$$\mathbf{x}_a = (\rho_1 + \rho_2, \rho_1)^T \text{ and } \mathbf{x}_b = (\rho_2, 0)^T. \tag{4.13}$$

■

The previous results are also applicable to solving the following important problems:

- Determine matrices **C** and **D** such that **AC** = **BD**.
- Determine whether or not two matrices **A** and **B** are equivalent, that is, whether or not a matrix **C** exists such that **AC** = **CB** and is the most general matrix satisfying this condition.

Example 4.5 (Matrices such that AC = BD). Consider again the matrices **A** and **B** in Example 4.4. Then, from the results above, we obtain that the most general **C** and **D** matrices of dimension $p \times p$ such that **AC** = **BD** holds are

$$\mathbf{C} = \begin{pmatrix} \rho_1 + \rho_2 & \rho_3 + \rho_4 & \cdots & \rho_{2p-1} + \rho_{2p} \\ \rho_1 & \rho_3 & \cdots & \rho_{2p-1} \end{pmatrix}; \ \mathbf{D} = \begin{pmatrix} \rho_2 & \rho_4 & \cdots & \rho_{2p} \\ 0 & 0 & \cdots & 0 \end{pmatrix}; \tag{4.14}$$

since $\mathbf{A}\mathbf{c}_i = \mathbf{B}\mathbf{d}_i$, and (4.13) implies that $\mathbf{c}_i = (\rho_{2i-1} + \rho_{2i}, \rho_{2i-1})^T$ and $\mathbf{d}_i = (\rho_{2i}, 0)^T$ for $i = 1, \ldots, p$. ∎

Example 4.6 (Equivalent matrices). In linear algebra, two matrices $\mathbf{A}$ and $\mathbf{B}$ are said to be equivalent if a nonsingular $\mathbf{C}$ matrix exists such that

$$\mathbf{AC} = \mathbf{CB}. \tag{4.15}$$

Consider the following two matrices

$$\mathbf{A} = \begin{pmatrix} -2 & 4 & -4 \\ 3 & 0 & 4 \\ 2 & -1 & 3 \end{pmatrix}, \quad \mathbf{B} = \begin{pmatrix} 2 & 1 & 1 \\ 1 & 0 & 2 \\ 1 & 1 & -1 \end{pmatrix}. \tag{4.16}$$

If the above mentioned matrix $\mathbf{C}$ exists, then (4.15) can be written as

$$\left(\begin{array}{ccc|ccc|ccc} -2 & 0 & 0 & 4 & 0 & 0 & -4 & 0 & 0 \\ 0 & -2 & 0 & 0 & 4 & 0 & 0 & -4 & 0 \\ 0 & 0 & -2 & 0 & 0 & 4 & 0 & 0 & -4 \\ \hline 3 & 0 & 0 & 0 & 0 & 0 & 4 & 0 & 0 \\ 0 & 3 & 0 & 0 & 0 & 0 & 0 & 4 & 0 \\ 0 & 0 & 3 & 0 & 0 & 0 & 0 & 0 & 4 \\ \hline 2 & 0 & 0 & -1 & 0 & 0 & 3 & 0 & 0 \\ 0 & 2 & 0 & 0 & -1 & 0 & 0 & 3 & 0 \\ 0 & 0 & 2 & 0 & 0 & -1 & 0 & 0 & 3 \end{array}\right) \left(\begin{array}{c} c_{11} \\ c_{12} \\ c_{13} \\ \hline c_{21} \\ c_{22} \\ c_{23} \\ \hline c_{31} \\ c_{32} \\ c_{33} \end{array}\right)$$

$$= \left(\begin{array}{ccc|ccc|ccc} 2 & 1 & 1 & 0 & 0 & 0 & 0 & 0 & 0 \\ 1 & 0 & 1 & 0 & 0 & 0 & 0 & 0 & 0 \\ 1 & 2 & -1 & 0 & 0 & 0 & 0 & 0 & 0 \\ \hline 0 & 0 & 0 & 2 & 1 & 1 & 0 & 0 & 0 \\ 0 & 0 & 0 & 1 & 0 & 1 & 0 & 0 & 0 \\ 0 & 0 & 0 & 1 & 2 & -1 & 0 & 0 & 0 \\ \hline 0 & 0 & 0 & 0 & 0 & 0 & 2 & 1 & 1 \\ 0 & 0 & 0 & 0 & 0 & 0 & 1 & 0 & 1 \\ 0 & 0 & 0 & 0 & 0 & 0 & 1 & 2 & -1 \end{array}\right) \left(\begin{array}{c} c_{11} \\ c_{12} \\ c_{13} \\ \hline c_{21} \\ c_{22} \\ c_{23} \\ \hline c_{31} \\ c_{32} \\ c_{33} \end{array}\right), \tag{4.17}$$

where c_{ij} are the elements of $\mathbf{C}$ and we have used the block representation to illustrate the relation of the new matrices to $\mathbf{A}$ and $\mathbf{B}$.

Equation (4.17) is equivalent to

$$\left(\begin{array}{ccc|ccc|ccc} -4 & -1 & -1 & 4 & 0 & 0 & -4 & 0 & 0 \\ -1 & -2 & -1 & 0 & 4 & 0 & 0 & -4 & 0 \\ -1 & -2 & -1 & 0 & 0 & 4 & 0 & 0 & -4 \\ \hline 3 & 0 & 0 & -2 & -1 & -1 & 4 & 0 & 0 \\ 0 & 3 & 0 & -1 & 0 & -1 & 0 & 4 & 0 \\ 0 & 0 & 3 & -1 & -2 & 1 & 0 & 0 & 4 \\ \hline 2 & 0 & 0 & -1 & 0 & 0 & 1 & -1 & -1 \\ 0 & 2 & 0 & 0 & -1 & 0 & -1 & 3 & -1 \\ 0 & 0 & 2 & 0 & 0 & -1 & -1 & -2 & 4 \end{array}\right) \left(\begin{array}{c} c_{11} \\ c_{12} \\ c_{13} \\ \hline c_{21} \\ c_{22} \\ c_{23} \\ \hline c_{31} \\ c_{32} \\ c_{33} \end{array}\right) = \mathbf{0}. \tag{4.18}$$

The orthogonal set to the linear subspace generated by the rows of the square matrix in (4.18), that is, the solution of (4.18), is the linear subspace generated by the vectors:

$$\left(\frac{4}{9}, \frac{8}{9}, \frac{8}{9}, \frac{17}{9}, \frac{7}{9}, \frac{7}{9}, 1, 0, 0\right)^T,$$

$$(1, -1, 1, 1, 1, 0, 0, 1, 0)^T,$$

$$\left(\frac{7}{9}, \frac{5}{9}, -\frac{13}{9}, \frac{5}{9}, \frac{1}{9}, \frac{10}{9}, 0, 0, 1\right)^T,$$

which implies that the most general matrix $\mathbf{C}$ satisfying $\mathbf{AC} = \mathbf{CB}$ is

$$\mathbf{C} = \begin{pmatrix} \frac{4\rho_1}{9} + \rho_2 + \frac{7\rho_3}{9} & \frac{8\rho_1}{9} - \rho_2 + \frac{5\rho_3}{9} & \frac{8\rho_1}{9} + \rho_2 - \frac{13\rho_3}{9} \\ \frac{17\rho_1}{9} + \rho_2 + \frac{5\rho_3}{9} & \frac{7\rho_1}{9} + \rho_2 + \frac{\rho_3}{9} & \frac{7\rho_1}{9} + \frac{10\rho_3}{9} \\ \rho_1 & \rho_2 & \rho_3 \end{pmatrix}, \tag{4.19}$$

where ρ_1, ρ_2, and ρ_3 are arbitrary real constants. ∎

4.4 MATHEMATICA PROGRAMS

The following Mathematica programs use the Orthogonal Module in Chapter 2 to obtain the intersection of two linear subspaces and the reciprocal image of a linear subspace by a linear transformation.

Program 4.1 (Module for obtaining the intersection of two linear subspaces).

```
(* Intersection of two linear spaces *)
IntersectionSpace[A_,B_]:=Module[{Ao,Bo},
  Ao=Orthogonal[A];
  If[Ao,={},Return[Orthogonal[Ao,Space->B]]];
  Bo=Orthogonal[B];
  If[Bo,={},Return[Orthogonal[Bo,Space->A]],
                    Return[IdentityMatrix[Length[B[[1]]]]]]
]
```

Example 4.7 (Intersection of moving planes). Example 4.1 can be solved using the following Mathematica command

```
In[]:=
IntersectionSpace[{{0,0,1},{t,t+1,0}},
                  {{1,0,0},{0,t,1-t}}]
Out[]=
    2
   t
{{-----, t, 1 - t}}
  1 + t
```

■

The following Mathematica program uses the Orthogonal Module in Chapter 2 to obtain the reciprocal image of a linear subspace by a linear transformation.

Program 4.2 (Module for obtaining the reciprocal image of a linear subspace by a linear transformation).

```
(* Reciprocal image of a linear subspace by a linear transformation *)
ReciprocalImage[Space_,Transformation_]:=Module[{},
 Orthogonal[Orthogonal[Space].Transformation]
]
```

Example 4.8 (Reciprocal image of a linear transformation). The problem described in Example 4.3 can be solved using the following Mathematica command:

```
In[]:=
  ReciprocalImage[{{1,-1,2}},{{3,-1,1},{2,1,-1},
                  {-1,-1,-1}}]
Out[]=
  {{0, 3, 1}}
```

■

Exercises

4.1 Obtain the intersection of planes Π_1 and Π_2 in $\mathbb{R}^3$, where

$$\Pi_1 \equiv \{(x_1, x_2, x_3)^T | -x_1 + 2x_2 + x_3 = 0\},$$

and

$$\Pi_2 \equiv \{(x_1, x_2, x_3)^T | 3x_1 - x_2 + x_3 = 0\}.$$

4.2 Consider two moving planes in $\mathbb{R}^3$:

$$\begin{array}{rcl} \Pi_1(t) & \equiv & \mathcal{L}\{(t, 2t, 1)^T, (1-t, -2t, 0)^T\}, \\ \Pi_2(t) & \equiv & \mathcal{L}\{(0, 1, t)^T, (1, 0, -t)^T\}, \end{array}$$

where the time parameter $t \in [0, 1]$.

(a) Find the intersection of both planes as a function of t.

(b) When does this intersection become the line passing through the origin and having direction $(1, 4, 2)^T$?

(c) What happens when $t = 1/2$?

4.3 Let $\mathcal{U}$ be the linear subspace $\mathcal{U} \equiv \mathcal{L}\{(-1, 0, 2, 0)^T, (0, 2, 1, 1)^T\}$. Find the reciprocal image of $\mathcal{U}$ and $\mathcal{U}^\perp$ with respect to the linear transformation with associated matrix

$$\mathbf{T} = \begin{pmatrix} 1 & 0 & 1 & 2 \\ 0 & 1 & 2 & 1 \\ 2 & 1 & 1 & 0 \\ 1 & 1 & 0 & 0 \end{pmatrix}.$$

4.4 Two mathematicians are studying a linear transformation, but with different associated matrices. While the first one uses matrix $\mathbf{A}$, the second one uses matrix $\mathbf{B}$, where

$$\mathbf{A} = \begin{pmatrix} 1 & -1 & 2 \\ 0 & 1 & -2 \\ 1 & -2 & 1 \\ 1 & 1 & 0 \end{pmatrix}; \quad \mathbf{B} = \begin{pmatrix} -2 & -1 & 1 \\ 0 & -1 & -1 \\ 1 & 1 & 0 \\ -2 & 1 & 0 \end{pmatrix}.$$

Find the linear subspaces associated with $\mathbf{x}_a$ and $\mathbf{x}_b$ such that $\mathbf{A}\mathbf{x}_a = \mathbf{B}\mathbf{x}_b$, that is, their images by $\mathbf{A}$ and $\mathbf{B}$, respectively, coincide.

4.5 Obtain the intersection of the following subspaces:

$$\begin{array}{rcl} \mathcal{U} & = & \mathcal{L}\{(1, 1, -1, 2), (0, 2, 1, 0), (-1, 0, 0, -2)\}, \\ \mathcal{V} & = & \mathcal{L}\{(0, 0, 1, -1), (-1, 0, -2, -1)\}. \end{array}$$

4.6 Calculate the reciprocal image of the plane

$$\Pi = \{(x, y, z) \mid x + y + z = 0\},$$

by the linear transformation with associated matrix

$$\mathbf{C} = \begin{pmatrix} 1 & 0 & 1 \\ 1 & 2 & -1 \\ 0 & 1 & 2 \end{pmatrix}.$$

4.7 Obtain the set of matrices **C**, **D** such that **AC** = **BD**, where

$$\mathbf{A} = \begin{pmatrix} 2 & -1 & 0 \\ 1 & 0 & 1 \\ 0 & -2 & 0 \end{pmatrix}; \quad \mathbf{B} = \begin{pmatrix} 1 & 0 & -1 \\ 1 & 0 & 2 \\ 0 & -1 & 0 \end{pmatrix}.$$

4.8 Find the general matrix **C** satisfying the constraint **AC** = **CB**, with

$$\mathbf{A} = \begin{pmatrix} 1 & -2 & 0 & -2 \\ 2 & 0 & -1 & 0 \\ 1 & -1 & -2 & 1 \\ 0 & 1 & 2 & 1 \end{pmatrix}; \quad \mathbf{B} = \begin{pmatrix} 1 & 13/3 & 2 & 2 \\ 1 & 7 & -1 & 10 \\ -2 & -15 & -1 & -17 \\ 0 & -19/3 & 0 & -7 \end{pmatrix}.$$

5

Orthogonal Sets and Systems of Linear Equations

5.1 INTRODUCTION

This chapter is devoted to systems of linear equations. When dealing with systems of linear equations, the most common problems are:

- Determining whether or not a system of linear equations is compatible, that is, it admits at least one solution.
- Determining whether or not a system of linear equations has a unique solution.
- Finding the general solution, when it exists.
- Checking the equivalence of two systems of equations, in the sense of having the same solutions.
- Solving the system in some selected variables.
- Given the solution of an initial linear system of equations, finding the solution of a new system of equations resulting after:
  - one or several equations are modified,
  - one or several equations and/or variables are removed, or
  - one or several equations and/or variables are added.

We start in Section 5.2 by analyzing the compatibility problem. We show how orthogonal sets can be used to solve these problems in an elegant manner. In Section 5.3 we discuss the problem of solving systems of linear equations, using orthogonal sets and the orthogonalization techniques described in Chapter 2. We also discuss the uniqueness problem. In Section 5.4 the complexity of the proposed method is calculated and compared with the Gauss elimination procedure.

In Section 5.5 we apply the method to analyze the equivalence of two systems of equations. In Section 5.6 we show how the proposed method can be applied for solving the system of equations in some selected variables.

Most of existing methods for solving systems of equations require restarting the whole process from the very beginning when a single coefficient or several coefficients of the original system of equations are changed. Thus, the huge amount of computational effort for these calculations is lost if changes are made to the original system. However, there are some alternatives, such as the Sherman-Morrison and the Woodbury methods (see Gill, Golub, Murray, and Saunders (1974), Press, Teukolsky, Vetterling, and Flannery (pages 73–77) (1985), and Burden and Faires (1985)), that allow calculating the new solution, with a small number of additional multiplications and additions, when some special types of modification are performed on the initial system. The proposed method is directly applicable to this dynamically updating process. In fact, we show, in Section 5.7, how to update solutions with little extra computational effort when some equations and/or variables are added or removed from the original system of equations. Finally, in Section 5.8, we give some applications.

In summary, in this chapter, we show that the methods described in Chapter 2 are very useful for solving the most important problems of systems of equations, including the analysis of compatibility, finding a solution, or obtaining a solution for some selected variables. We also give a method for updating the solution of a system of linear equations when equations and/or unknowns are added to or removed from the system. The method starts by solving the initial system of equations and keeping information of the orthogonal space and its complement of the linear space generated by the rows of the matrix associated with the linear system. In addition, information is stored on the pivot columns associated with each equation and on the basic independent equations on which the linearly dependent equations depend. With this information, the solution of any modified system can be updated in each of the four cases above. Several examples of applications illustrate the methods.

5.2 COMPATIBILITY OF A SYSTEM OF LINEAR EQUATIONS

In this section we show how to apply the orthogonal methods to analyze the compatibility of a given system of equations. Consider the system of linear equations

$$u_{i1}x_1 + \cdots + u_{in}x_n = u_{i,n+1}; \; i = 1, \ldots, m, \tag{5.1}$$

where $u_{i1}, \dots, u_{i,n+1};\ i = 1, \dots, m$ are real numbers and $x_1, \dots, x_n$ are the unknowns. System (5.1) can also be written as

$$\mathbf{u}_1 x_1 + \cdots + \mathbf{u}_n x_n = \mathbf{u}_{n+1} \tag{5.2}$$

or

$$\mathbf{U}\mathbf{x} = \mathbf{u}_{n+1}, \tag{5.3}$$

where $\mathbf{u}_j = (u_{1j}, \dots, u_{mj})^T; j = 1, \dots, n+1$ are column vectors, $\mathbf{x} = (x_1, \dots, x_n)^T$ is the column of unknowns, and $\mathbf{U} = (\mathbf{u}_1, \dots, \mathbf{u}_n)$.

If system (5.1) has a solution, then a certain $\mathbf{x}^*$ exists such that $\mathbf{U}\mathbf{x}^* = \mathbf{u}_{n+1}$; that is,

$$\mathbf{U}\mathbf{x} = \mathbf{u}_{n+1} \text{ has a solution } \Leftrightarrow \mathbf{U}\mathbf{x}^* = \mathbf{u}_{n+1} \Leftrightarrow \mathbf{u}_{n+1} \in \mathcal{L}\{\mathbf{u}_1, \dots, \mathbf{u}_n\}, \tag{5.4}$$

which, in its primal conception, allows us to write

$$\mathbf{u}_{n+1} \in \mathcal{L}\{\mathbf{u}_1, \dots, \mathbf{u}_n\} \Leftrightarrow rank(\mathbf{U}) = rank(\mathbf{U} : \mathbf{u}_{n+1}). \tag{5.5}$$

This expression together with (5.4) leads to the well-known Rouche's Theorem.

However, the dual interpretation we are dealing with in this book allows us to go further. In fact, if $\mathcal{L}(\mathbf{w}_1, \dots, \mathbf{w}_p)$ is the orthogonal of $\mathcal{L}\{\mathbf{u}_1, \dots, \mathbf{u}_n\}$ in $\mathbb{R}^m$, that is, if

$$\mathcal{L}\{\mathbf{w}_1, \dots, \mathbf{w}_p\} \equiv \left\{\mathbf{w} \in \mathbb{R}^m | \mathbf{U}^T\mathbf{w} = \mathbf{0}\right\}, \tag{5.6}$$

we also have

$$\mathbf{u}_{n+1} \in \mathcal{L}\{\mathbf{u}_1, \dots, \mathbf{u}_n\} \Leftrightarrow \mathbf{W}^T\mathbf{u}_{n+1} = \mathbf{0}, \tag{5.7}$$

where $\mathbf{W} = (\mathbf{w}_1, \dots, \mathbf{w}_p)$, and finally,

$$\boxed{\mathbf{U}\mathbf{x} = \mathbf{u}_{n+1} \text{ has a solution } \Leftrightarrow \mathbf{W}^T\mathbf{u}_{n+1} = \mathbf{0}.} \tag{5.8}$$

Thus, analyzing the compatibility of the system of equations (5.1), (5.2), or (5.3) reduces to finding the space $\mathcal{L}(\mathbf{U})^\perp \equiv \mathcal{L}\{\mathbf{w}_1, \dots, \mathbf{w}_p\}$ orthogonal to $\mathcal{L}\{\mathbf{u}_1, \dots, \mathbf{u}_n\}$ and checking whether or not $\mathbf{W}^T\mathbf{u}_{n+1} = \mathbf{0}$.

The computational process arising from (5.8) has a complexity which coincides exactly with that for the Gauss elimination procedure. However, it has one important advantage: $\mathbf{W}$ is independent of $\mathbf{u}_{n+1}$ and so we can analyze the compatibility of any symbolic vector $\mathbf{u}_{n+1}$ without extra computations.

Algorithm 2.1 allows determination of the compatibility of a system of linear equations. However, since we do not need the complement of $\mathcal{L}(\mathbf{U})^\perp$, it can be simplified as follows:

Algorithm 5.1 Compatibility of a system of linear equations.

- **Input:** Coefficients $u_{ij}; i = 1, \ldots, m; j = 1, \ldots, n+1$ defining a system of m linear equations in n unknowns.

- **Output:** Compatibility conditions for the system to have solution.

Step 1: Let $\mathbf{W} = \mathbf{I}_n$, where $\mathbf{I}_n$ is the identity matrix of dimension n.

Step 2: Let $i = 1$ and $k = n$.

Step 3: Calculate the dot products $t^i_j; j = 1, \ldots, k$, of the ith row by the columns of $\mathbf{W}$.

Step 4: For $j = 1$ to k locate the pivot column r_i as the first column not orthogonal to the ith row, that is, $t^i_{r_i} \neq 0$. If there is no such a column go to Step 8.

Step 5: Divide the r_i column by $t^i_{r_i}$ and switch columns and dot products of the first and r_i columns if $r_i \neq 1$.

Step 6: For $j = 2$ to k if $t^i_j \neq 0$, let $W_{kj} = W_{kj} - t^i_j * W_{ki}$.

Step 7: Remove the first column from $\mathbf{W}$ and let $k = k - 1$.

Step 8: If $i = m$ go to Step 9. Otherwise, increase i by one and go to Step 3.

Step 9: Return $\mathbf{W}^T \mathbf{u}_{n+1} = 0$ as the compatibility conditions.

Remark 5.1 *This algorithm is similar to Algorithm 2.1 described in Chapter 2 for performing orthogonal decompositions. The main difference is that, in this case, we don't need to calculate the complement space, so we can remove the pivot column after each iteration. Consequently, we save computational effort. This difference is reflected in the tabular form of the algorithm.*

The following examples illustrate the use of Algorithm 5.1.

Example 5.1 (Compatibility of a linear system of equations). Suppose that we are interested in determining the conditions under which the system of equations

$$\begin{array}{cccccccc} x_1 & +2x_2 & & +x_4 & -4x_5 & = & a \\ 2x_1 & +x_2 & -x_3 & & -x_5 & = & b \\ x_1 & +2x_2 & +x_3 & & +x_5 & = & c \\ 5x_1 & +2x_2 & -3x_3 & +x_4 & +x_5 & = & d \\ x_1 & +x_2 & -x_3 & +x_4 & -5x_5 & = & e \end{array} \tag{5.9}$$

is compatible. Then, using Algorithm 5.1, we get (see Table 5.1)

$$\mathbf{W} \equiv \mathcal{L}\left\{(-1, -1/3, 2/3, 0, 1)^T\right\}, \tag{5.10}$$

Table 5.1 Iterations for determining the compatibility of the system of equations in Example 5.1.

Iteration 1						Iteration 2				
1	**1**	0	0	0	0	2	**–2**	–1	–5	–1
2	**0**	1	0	0	0	1	**1**	0	0	0
1	**0**	0	1	0	0	2	**0**	1	0	0
5	**0**	0	0	1	0	2	**0**	0	1	0
1	**0**	0	0	0	1	1	**0**	0	0	1
	1	2	1	5	1		**–3**	0	–8	–1

Iteration 3				Iteration 4			Output	
0	**–1**	1/3	–1/3	1	**0**	–1	–4	–1
–1	**0**	–8/3	–1/3	0	**–8/3**	–1/3	–1	–1/3
1	**1**	0	0	0	**1/3**	2/3	1	2/3
–3	**0**	1	0	1	**1**	0	1	0
–1	**0**	0	1	1	**0**	1	–5	1
	1	–1/3	–2/3		**1**	0		0

which implies the following compatibility condition:

$$\mathbf{w}_1 \bullet (a,b,c,d,e)^T = 0 \Rightarrow 3a + b - 2c - 3e = 0. \tag{5.11}$$

■

Example 5.2 (Compatibility of a linear system of equations). Let us analyze the compatibility of the following system of equations, which depends on the parameters a and b:

$$\begin{array}{rrrrrcl} 2x_1 & +x_2 & +3x_3 & +x_4 & & = & a \\ -x_1 & +x_2 & & -2x_4 & +x_5 & = & a-2b \\ 3x_1 & +x_2 & +4x_3 & +2x_4 & & = & b \\ -2x_1 & -x_2 & -3x_3 & -x_4 & -2x_5 & = & -1 \\ x_1 & -2x_2 & -x_3 & +3x_4 & +x_5 & = & b \end{array} \tag{5.12}$$

Using Algorithm 5.1, we get (see Table 5.2)

$$\mathbf{W} \equiv \mathcal{L}\left\{(-7,2,6,1,0)^T, (11,-1,-8,0,1)^T\right\}, \tag{5.13}$$

which implies the following compatibility condition:

$$\begin{pmatrix} -7 & 2 & 6 & 1 & 0 \\ 11 & -1 & -8 & 0 & 1 \end{pmatrix} \begin{pmatrix} a \\ a-2b \\ b \\ -1 \\ b \end{pmatrix} = \begin{pmatrix} -1-5a+2b \\ 10a-5b \end{pmatrix} = \begin{pmatrix} 0 \\ 0 \end{pmatrix}. \tag{5.14}$$

Then the system is compatible if and only if $a = -1$ and $b = -2$. ■

Table 5.2 Iterations for determining the compatibility of the system of equations in Example 5.2.

Iteration 1						Iteration 2				
2	**1**	0	0	0	0	1	**1/2**	–1/2	1	–1/2
–1	**0**	1	0	0	0	1	**1**	0	0	0
3	**0**	0	1	0	0	1	**0**	1	0	0
–2	**0**	0	0	1	0	–1	**0**	0	1	0
1	**0**	0	0	0	1	–2	**0**	0	0	1
	2	–1	3	–2	1		**3/2**	–1/2	0	–5/2

Iteration 3				Iteration 4			
3	–4/3	1	1/3	1	–4/3	1	1/3
0	1/3	0	5/3	–2	1/3	0	5/3
4	1	0	0	2	1	0	0
–3	0	1	0	–1	0	1	0
–1	0	0	1	3	0	0	1
	0	0	0		0	0	0

Iteration 5				Output	
0	**–4/3**	1	1/3	–7	11
1	**1/3**	0	5/3	2	–1
0	**1**	0	0	6	–8
–2	**0**	1	0	1	0
1	**0**	0	1	0	1
	1/3	–2	8/3		

5.3 SOLVING A SYSTEM OF LINEAR EQUATIONS

In this section we analyze how to solve the system of equations (5.1) using orthogonal sets. To this aim we incorporate a new artificial variable x_{n+1} and analyze the following equivalent system of equations

$$\begin{aligned} u_{i1}x_1 + \cdots + u_{in}x_n - u_{i,n+1}x_{n+1} &= 0; \quad i = 1, \ldots, m \\ x_{n+1} &= 1 \end{aligned} \tag{5.15}$$

or

$$\begin{aligned} \mathbf{u}_i^T \bullet \mathbf{x} &= 0; \quad i = 1, \ldots, m \\ x_{n+1} &= 1 \end{aligned} \tag{5.16}$$

where now $\mathbf{x}$ includes $n+1$ components, and $\mathbf{u}_i^T$ is the ith row of $\mathbf{U}$.

Equation (5.16) shows that the solution of (5.1) is the subset of the orthogonal space to $\mathcal{L}\left\{\mathbf{u}_1^T, \ldots, \mathbf{u}_m^T\right\}$, such that $x_{n+1} = 1$. Thus, we obtain the solution of (5.1) as follows.

Let $\mathbf{X}^0 \equiv \mathcal{L}\left\{\mathbf{w}_1^0, \ldots, \mathbf{w}_{n+1}^0\right\} \equiv \mathbb{R}^{n+1}$ with $\mathbf{w}_j^0 = \mathbf{e}_j$; $j = 1, \ldots, n+1$, where $\mathbf{e}_j$ is the vector with all zeroes except for the jth component, which is one. The solution of (5.16) is the set $\mathbf{X}^m \cap \{\mathbf{x} \in \mathbb{R}^{n+1} | x_{n+1} = 1\}$, where

$$\mathbf{X}^m \equiv \{\mathbf{x} \in \mathbf{X}^0 | \mathbf{u}_i^T \bullet \mathbf{x} = 0, i = 1, \ldots, m\} \equiv \mathcal{L}\{\mathbf{w}_1^m, \ldots, \mathbf{w}_{n_m}^m\}, \tag{5.17}$$

which we obtain sequentially in m steps and for the pth step we have

$$\mathbf{X}^p \equiv \{\mathbf{x} \in \mathbf{X}^{p-1} | \mathbf{u}_p^T \bullet \mathbf{x} = 0\}, \mathbf{X}^{p-1} \equiv \mathcal{L}\left\{\mathbf{w}_1^{p-1}, \ldots, \mathbf{w}_{n_{p-1}}^{p-1}\right\}. \tag{5.18}$$

In each step of the computational process we must deal with one of the following situations:

1. *The new equation $\mathbf{u}_p^T \bullet \mathbf{x} = 0$ is a linear combination of the previous equations:* This occurs when

$$\mathbf{u}_p^T \bullet \mathbf{w}_i^{p-1} = 0, i = 1, \ldots, n_{p-1}. \tag{5.19}$$

   In this case $\mathbf{X}^p = \mathbf{X}^{p-1}$ and the process continues.

2. *The system of equations is incompatible:* This occurs when

$$\mathbf{u}_p^T \bullet \mathbf{w}_i^{p-1} = 0, \forall i = 1, \ldots, n_{p-1} - 1 \wedge \mathbf{u}_p^T \bullet \mathbf{w}_{n_{p-1}}^{p-1} \neq 0. \tag{5.20}$$

   In this case $x_{n+1} = 0 \neq 1$ and the system is incompatible.

3. *The new equation is not a linear combination of the previous equations but is compatible with them:*

$$\mathbf{u}_p^T \bullet \mathbf{w}_i^{p-1} = t_i^p \neq 0, \text{ for some } i \neq n_{p-1}. \tag{5.21}$$

   In this case we have

$$\mathbf{w}_j^{p+1} = \mathbf{w}_j^p - \frac{t_j^p}{t_i^p} \mathbf{w}_i^p, \forall j \neq i. \tag{5.22}$$

   We suppress vector $\mathbf{w}_i^p$ and continue.

After this sequential process we have $\mathbf{X}^m \equiv \mathcal{L}\{\mathbf{w}_1^m, \ldots, \mathbf{w}_{n_m}^m\}$. Since the last generator $\mathbf{w}_{n_m}^m$ is the only one having a non-negative last component, of value 1 (see Examples 5.3 and 5.4), after using condition $x_{n+1} = 1$, the solution of the initial system becomes

$$\hat{\mathbf{w}}_{n_m}^m + \mathcal{L}\{\hat{\mathbf{w}}_1^m, \ldots, \hat{\mathbf{w}}_{n_m-1}^m\}, \tag{5.23}$$

where $\hat{\mathbf{w}}$ is the vector obtained from $\mathbf{w}$ by removing its last component.

When $\mathcal{L}(\hat{\mathbf{w}}_1^m, \ldots, \hat{\mathbf{w}}_{n_m-1}^m)$ degenerates to the empty set, we have uniqueness of solution.

Consequently, we obtain the following algorithm.

Algorithm 5.2 Solving a system of linear equations.

- **Input:** A system of m linear equations in n unknowns in the form (5.15), that is, $\mathbf{U}$ of dimensions $m \times (n+1)$.
- **Output:** The general solution of the system of equations or a message indicating that no solution exists.

Step 1: Let $\mathbf{W} = \mathbf{I}_{n+1}$, where $\mathbf{I}_{n+1}$ is the identity matrix of dimension $n+1$.

Step 2: Let $i = 1$ and $\ell = 0$.

Step 3: Calculate the dot products $t_i^j = \mathbf{u}_i^T \bullet \mathbf{w}_j; j = 1, \ldots, n+1$, of the ith row of $\mathbf{U}$ by the columns of $\mathbf{W}$.

Step 4: For $j = \ell + 1$ to n locate the pivot column r_ℓ as the first column not orthogonal to the ith row, that is, $t_i^{r_\ell} \neq 0$. If there is no such column, go to Step 7 if $t_i^{n+1} = 0$; otherwise, print "Incompatible System of Equations" and quit.

Step 5: Let $\ell = \ell + 1$ and divide the r_ℓ column by $t_i^{r_\ell}$. If $r_\ell \neq \ell$, switch columns ℓ and r_ℓ, and associated dot products.

Step 6: For $j = 1$ to $n+1$ and $j \neq r_\ell$, if $t_i^j \neq 0$ let $w_{kj} = w_{kj} - t_i^j * w_{k\ell}$, for $k = 1, \ldots, n+1$.

Step 7: If $i = m$ go to Step 8. Otherwise, increase i by one and go to Step 3.

Step 8: Return $\hat{\mathbf{w}}_{n+1} + \mathcal{L}\{\hat{\mathbf{w}}_{\ell+1}, \ldots, \hat{\mathbf{w}}_n\}$ as the general solution of the system, where $\hat{\mathbf{w}}_j$ is the jth column of $\mathbf{W}$ without the last element.

Thus, the solution of the initial system becomes

$$\hat{\mathbf{w}}_{n+1} + \mathcal{L}\{\hat{\mathbf{w}}_{\ell+1}, \ldots, \hat{\mathbf{w}}_n\}, \tag{5.24}$$

where $\hat{\mathbf{w}}$ is the vector obtained from $\mathbf{w}$ by removing its last component and

$$\mathbb{R}^n \equiv \mathcal{L}\{\hat{\mathbf{w}}_1, \ldots, \hat{\mathbf{w}}_n\}. \tag{5.25}$$

Remark 5.2 *Note that we can start at Step 1 with* $\mathbf{W} = \mathbf{A}$, *where* $\mathbf{A}$ *is any nonsingular matrix. However, for the sake of simplicity, we have used* $\mathbf{W} = \mathbf{I}_{n+1}$.

Remark 5.3 *Note also that the linear space* $\mathcal{L}\{\mathbf{w}_{\ell+1}, \ldots, \mathbf{w}_n\}$ *is the orthogonal set to* $\mathcal{L}\{\mathbf{u}_1^T, \ldots, \mathbf{u}_m^T\}$ *and* $\mathcal{L}\{\mathbf{w}_1, \ldots, \mathbf{w}_\ell\}$ *its complement. Actually, this last space is not necessary to solve the linear system of equations, so we can remove the pivot column after each iteration as we did in Algorithm 5.1. The following examples illustrate this difference: the first one performs the process with all columns, but the second one removes the pivot columns.*

Example 5.3 (Solving a system of linear equations). To solve the system of equations

$$\begin{array}{rrrrcr} x_1 & -x_2 & +2x_3 & -2x_4 & = & 3 \\ 3x_1 & -2x_2 & & +x_4 & = & -1 \\ x_1 & & -2x_3 & +2x_4 & = & 1 \end{array} \tag{5.26}$$

we incorporate a new artificial variable x_5 and we consider the equivalent system of equations

$$\begin{array}{rrrrrcr} x_1 & -x_2 & +2x_3 & -2x_4 & -3x_5 & = & 0 \\ 3x_1 & -2x_2 & & +x_4 & +x_5 & = & 0 \\ x_1 & & -2x_3 & +2x_4 & -x_5 & = & 0 \\ & & & & x_5 & = & 1 \end{array} \tag{5.27}$$

Then, using Algorithm 5.2, we get the general solution (see Table 5.3)

$$\begin{pmatrix} x_1 \\ x_2 \\ x_3 \\ x_4 \end{pmatrix} = \begin{pmatrix} 9 \\ 14 \\ 4 \\ 0 \end{pmatrix} + \rho \begin{pmatrix} 1 \\ 2 \\ 3/2 \\ 1 \end{pmatrix}, \tag{5.28}$$

where ρ is an arbitrary real constant. This solution is obtained from expression (5.23) and

$$\begin{array}{l} \mathbf{X}^0 \equiv \mathcal{L}\{(1,0,0,0,0)^T,(0,1,0,0,0)^T,(0,0,1,0,0)^T,(0,0,0,1,0)^T,(0,0,0,0,1)^T\}, \\ \mathbf{X}^1 \equiv \mathcal{L}\left\{(1,1,0,0,0)^T,(-2,0,1,0,0)^T,(2,0,0,1,0)^T,(3,0,0,0,1)^T\right\}, \\ \mathbf{X}^2 \equiv \mathcal{L}\left\{(4,6,1,0,0)^T,(-5,-7,0,1,0)^T,(-7,-10,0,0,1)^T\right\}, \\ \mathbf{X}^3 \equiv \mathcal{L}\left\{(1,2,3/2,1,0)^T,(9,14,4,0,1)^T\right\}. \end{array}$$

■

Example 5.4 (Solving a system of linear equations). To solve the system of equations

$$\begin{array}{rrrrcr} x_1 & +2x_2 & & +x_4 & = & 0 \\ 2x_1 & +x_2 & -x_3 & & = & 0 \\ & +2x_2 & +x_3 & & = & -1 \end{array} \tag{5.29}$$

we incorporate a new artificial variable x_5 and we consider the equivalent system of equations

$$\begin{array}{rrrrrcr} x_1 & +2x_2 & & +x_4 & & = & 0 \\ 2x_1 & +x_2 & -x_3 & & & = & 0 \\ & +2x_2 & +x_3 & & +x_5 & = & 0 \\ & & & & x_5 & = & 1 \end{array} \tag{5.30}$$

Table 5.3 Iterations for solving the system of equations in Example 5.3. The vectors $\mathbf{u}_i^T; i = 1, 2, 3$ are shown in the first column, the dot products $t_i^j = \mathbf{u}_i^T \bullet \mathbf{w}_j$ are shown in the last row, and the $\mathbf{w}_j$ vectors, in the remaining columns and rows. Pivot columns are boldfaced.

Iteration 1						**Iteration 2**					
1	**1**	0	0	0	0	3	*1*	**1**	−2	2	3
−1	**0**	1	0	0	0	−2	*0*	**1**	0	0	0
2	**0**	0	1	0	0	0	*0*	**0**	1	0	0
−2	**0**	0	0	1	0	1	*0*	**0**	0	1	0
−3	**0**	0	0	0	1	1	*0*	**0**	0	0	1
	1	−1	2	−2	−3		*3*	**1**	−6	7	0
Iteration 3						**Output**					
1	*−2*	*1*	**4**	−5	−7		*2*	*−1*	*2*	1	9
0	*−3*	*1*	**6**	−7	−10		*3*	*−2*	*3*	2	14
−2	*0*	*0*	**1**	0	0		*1*	*−1/2*	*1/2*	3/2	4
2	*0*	*0*	**0**	1	0		*0*	*0*	*0*	1	0
−1	*0*	*0*	**0**	0	1		*0*	*0*	*0*	0	1
	−2	*1*	**2**	−3	−8						

Then, using Algorithm 5.2 and removing the pivot columns after each iteration, we get (see Table 5.4)

$$\mathbf{X}^m = \mathbf{X}^3 \equiv (-2, 1, -3, 0)^T \oplus \mathcal{L}\left\{(3, -2, 4, 1)^T\right\}. \tag{5.31}$$

In other words, the general solution of the system of equations (5.29) is

$$\begin{pmatrix} x_1 \\ x_2 \\ x_3 \\ x_4 \end{pmatrix} = \begin{pmatrix} -2 \\ 1 \\ -3 \\ 0 \end{pmatrix} + \rho \begin{pmatrix} 3 \\ -2 \\ 4 \\ 1 \end{pmatrix}, \tag{5.32}$$

where ρ is an arbitrary real constant. ■

Example 5.5 (Solving a system of linear equations). To solve the system of equations

$$\begin{array}{rrrrrcr} x_1 & +2x_2 & & +x_4 & -x_5 & = & 4 \\ 2x_1 & +x_2 & -x_3 & & +3x_5 & = & 0 \\ -x_1 & & +2x_3 & +x_4 & & = & -1 \end{array} \tag{5.33}$$

we incorporate a new artificial variable x_6 and we consider the equivalent system of equations:

$$\begin{array}{rrrrrrcr} x_1 & +2x_2 & & +x_4 & -x_5 & -4x_6 & = & 0 \\ 2x_1 & +x_2 & -x_3 & & +3x_5 & & = & 0 \\ -x_1 & & +2x_3 & +x_4 & & +x_6 & = & 0 \\ & & & & & x_6 & = & 1 \end{array} \tag{5.34}$$

Table 5.4 Iterations for solving the system of equations in Example 5.4 (Pivot columns are bolfaced).

Iteration 1						Iteration 2				
1	**1**	0	0	0	0	2	**–2**	0	–1	0
2	**0**	1	0	0	0	1	**1**	0	0	0
0	**0**	0	1	0	0	–1	**0**	1	0	0
1	**0**	0	0	1	0	0	**0**	0	1	0
0	**0**	0	0	0	1	0	**0**	0	0	1
	1	2	0	1	0		**–3**	–1	–2	0

Iteration 3				Iteration 4	
0	**2/3**	1/3	0	3	–2
2	**–1/3**	–2/3	0	–2	1
1	**1**	0	0	4	–3
0	**0**	1	0	1	0
1	**0**	0	1	0	1
	1/3	–4/3	1		

Then, using Algorithm 5.2, we get

$$\mathbf{X} \equiv (-\frac{5}{2}, \frac{13}{4}, -\frac{7}{4}, 0, 0)^T \oplus \mathcal{L}\left\{(0, -\frac{1}{2}, -\frac{1}{2}, 1, 0)^T, (-\frac{7}{2}, \frac{9}{4}, -\frac{7}{4}, 0, 1)^T\right\}. \tag{5.35}$$

In other words, the general solution of the system of equations (5.33) is

$$\begin{pmatrix} x_1 \\ x_2 \\ x_3 \\ x_4 \\ x_5 \end{pmatrix} = \begin{pmatrix} -5/2 \\ 13/4 \\ -7/4 \\ 0 \\ 0 \end{pmatrix} + \rho_1 \begin{pmatrix} 0 \\ -1/2 \\ -1/2 \\ 1 \\ 0 \end{pmatrix} + \rho_2 \begin{pmatrix} -7/2 \\ 9/4 \\ -7/4 \\ 0 \\ 1 \end{pmatrix}, \tag{5.36}$$

where ρ_1 and ρ_2 are arbitrary real constants. ■

In Section 5.9 we give Mathematica programs to solve a system of linear equations using Algorithms 5.1 and 5.2.

5.4 COMPLEXITY

In this section we derive the complexity of the proposed method for solving linear systems of equations.

In Table 5.5 we give the required number of products and sums in the different steps. Thus, we have

$$Total\ number\ of\ products = \frac{m(1 + 3m - 4m^2 + 6mn)}{6}$$

Table 5.5 Required number of products and sums for the proposed orthogonalization method.

Products or divisions		
Phase	Iter. k $(1 \leq k \leq m)$	Total
Dot Products	$(k-1)(n-k+1)$	$(m-1)m(3n-2m+1)/6$
Normalization	k	$(m^2+m)/2$
Pivoting	$k(n-k)$	$m(m+1)(3n-2m-1)/6$
Sums		
Phase	Iter. k	Total
Dot Products	$(k-1)(n-k+1)$	$(m-1)m(3n-2m+1)/6$
Pivoting	$(k-1)(n-k)$	$(12-4m-6m^2-2m^3-6n+3mn+3m^2n)/6$

and

$$Total\ number\ of\ sums = \frac{12-5m-3m^2-4m^3-6n+6m^2n}{6},$$

which implies a total number of flops of

$$\frac{6-2m-4m^3-3n+6m^2n}{3}.$$

If we are interested in using this algorithm for symbolic calculations we need to avoid normalizations (divisions). Thus, we need to remove the normalization step and modify the pivoting process by replacing (2.1) by

$$\mathbf{w}_k^{j+1} = \begin{cases} \mathbf{w}_k^j; & \text{if } k = j, \\ t_j^j \mathbf{w}_k^j - t_k^j \mathbf{w}_j^j; & \text{if } k \neq j. \end{cases} \quad (5.37)$$

5.5 CHECKING SYSTEMS EQUIVALENCE

In this section we show how to use the above methods to determine whether two systems of equations lead to the same solution. We need the following lemma.

Lemma 5.1 *Let* $\mathbf{A}$ *be a matrix containing as rows the generators of a linear space* $\mathcal{A}$*. We denote by* $\mathbf{A}^{\perp}$ *the matrix whose columns are a set of generators of the orthogonal space to* $\mathcal{A}$*, and* $\mathcal{L}\{\mathbf{A}^{\perp}\}$ *the orthogonal space. Then, the orthogonals* $\mathcal{L}\{\mathbf{A}^{\perp}\}$ *and* $\mathcal{L}\{\mathbf{B}^{\perp}\}$ *associated with two matrices* $\mathbf{A}$ *and* $\mathbf{B}$ *coincide if and only if* $\mathbf{A}\mathbf{B}^{\perp} = \mathbf{0}$ *and* $\mathbf{B}\mathbf{A}^{\perp} = \mathbf{0}$.

Proof: If the two orthogonals coincide, columns of $\mathbf{B}^\perp$ belong to $\mathcal{L}\{\mathbf{A}^\perp\}$ and then $\mathbf{AB}^\perp = \mathbf{0}$. Similarly, we have $\mathbf{BA}^\perp = \mathbf{0}$.

Conversely, if $\mathbf{AB}^\perp = \mathbf{0}$ and $\mathbf{BA}^\perp = \mathbf{0}$ we have

$$\left.\begin{array}{l}\mathbf{AB}^\perp = \mathbf{0} \Rightarrow \mathcal{L}\{\mathbf{B}^\perp\} \subseteq \mathcal{L}\{\mathbf{A}^\perp\}\\ \mathbf{BA}^\perp = \mathbf{0} \Rightarrow \mathcal{L}\{\mathbf{A}^\perp\} \subseteq \mathcal{L}\{\mathbf{B}^\perp\}\end{array}\right\} \Rightarrow \mathcal{L}\{\mathbf{A}^\perp\} \equiv \mathcal{L}\{\mathbf{B}^\perp\}. \quad (5.38)$$

■

Assume the two systems of equations $\mathbf{Ax} = \mathbf{0}$ and $\mathbf{By} = \mathbf{0}$ are equivalent, where we have assumed that the artificial variables have already been incorporated into the systems. The solution of the system $\mathbf{Ax} = \mathbf{0}$ is the linear space generated by the columns of $\mathbf{A}^\perp$ and the solution of the system $\mathbf{By} = \mathbf{0}$ is the linear space generated by the columns of $\mathbf{B}^\perp$. If both solutions coincide we must have $\mathbf{AB}^\perp = \mathbf{0}$ and $\mathbf{BA}^\perp = \mathbf{0}$, as the following theorem shows.

Theorem 5.1 *Two systems of linear equations $\mathbf{Ax} = \mathbf{0}$ and $\mathbf{Bx} = \mathbf{0}$ are equivalent if and only if $\mathbf{AB}^\perp = \mathbf{0}$ and $\mathbf{BA}^\perp = \mathbf{0}$.*

Proof: The proof is indeed obvious using Lemma 5.1. Note that $\mathcal{L}\{\mathbf{A}^\perp\} = \{\mathbf{x}|\mathbf{Ax} = \mathbf{0}\}$ is the solution of $\mathbf{Ax} = \mathbf{0}$ and $\mathcal{L}\{\mathbf{B}^\perp\} = \{\mathbf{x}|\mathbf{Bx} = \mathbf{0}\}$ is the solution of $\mathbf{Bx} = \mathbf{0}$. ■

Example 5.6 (Equivalent systems of equations). Consider the two systems of equations

$$\begin{array}{rrrcr} 3x_1 & +2x_2 & +x_3 & = & 6\\ x_1 & -x_2 & -x_3 & = & -1 \end{array} \quad (5.39)$$

and

$$\begin{array}{rrrcr} 4x_1 & +x_2 & & = & 5\\ 2x_1 & +3x_2 & +2x_3 & = & 7 \end{array} \quad (5.40)$$

which, after adding the artificial variable $x_4 = 1$, are equivalent to the two systems

$$\begin{array}{rrrrcr} 3x_1 & +2x_2 & +x_3 & -6x_4 & = & 0\\ x_1 & -x_2 & -x_3 & +x_4 & = & 0\\ & & & x_4 & = & 1 \end{array} \quad (5.41)$$

and

$$\begin{array}{rrrrcr} 4x_1 & +x_2 & & -5x_4 & = & 0\\ 2x_1 & +3x_2 & +2x_3 & -7x_4 & = & 0\\ & & & x_4 & = & 1 \end{array} \quad (5.42)$$

To analyze whether or not the two equations in (5.41) and (5.42) are equivalent we obtain the matrices

$$\mathbf{A} = \begin{pmatrix} 3 & 2 & 1 & -6\\ 1 & -1 & -1 & 1 \end{pmatrix} \quad (5.43)$$

and

$$\mathbf{B} = \begin{pmatrix} 4 & 1 & 0 & -5\\ 2 & 3 & 2 & -7 \end{pmatrix}, \quad (5.44)$$

and their associated orthogonals:

$$\mathbf{A}^{\perp} = \mathbf{B}^{\perp} = \begin{pmatrix} 1/5 & 4/5 \\ -4/5 & 9/5 \\ 1 & 0 \\ 0 & 1 \end{pmatrix}. \tag{5.45}$$

Finally, we obtain

$$\mathbf{AB}^{\perp} = \begin{pmatrix} 3 & 2 & 1 & -6 \\ 1 & -1 & -1 & 1 \end{pmatrix} \begin{pmatrix} 1/5 & 4/5 \\ -4/5 & 9/5 \\ 1 & 0 \\ 0 & 1 \end{pmatrix} = \begin{pmatrix} 0 & 0 \\ 0 & 0 \end{pmatrix} \tag{5.46}$$

and

$$\mathbf{BA}^{\perp} = \begin{pmatrix} 4 & 1 & 0 & -5 \\ 2 & 3 & 2 & -7 \end{pmatrix} \begin{pmatrix} 1/5 & 4/5 \\ -4/5 & 9/5 \\ 1 & 0 \\ 0 & 1 \end{pmatrix} = \begin{pmatrix} 0 & 0 \\ 0 & 0 \end{pmatrix}, \tag{5.47}$$

which indicate that the two systems are equivalent. ■

Example 5.7 (Equivalent systems of equations). Consider the two systems of equations

$$\begin{array}{rrrcl} x_1 & -x_2 & +2x_3 & = & 3 \\ 3x_1 & +2x_2 & -4x_3 & = & 6 \end{array} \tag{5.48}$$

and

$$\begin{array}{rrrcl} 2x_1 & -7x_2 & +14x_3 & = & 9 \\ 2x_1 & +3x_2 & -6x_3 & = & 3 \end{array} \tag{5.49}$$

which are equivalent to

$$\begin{array}{rrrrcl} x_1 & -x_2 & +2x_3 & -3x_4 & = & 0 \\ 3x_1 & +2x_2 & -4x_3 & -6x_4 & = & 0 \\ & & & x_4 & = & 1 \end{array} \tag{5.50}$$

and

$$\begin{array}{rrrrcl} 2x_1 & -7x_2 & +14x_3 & -9x_4 & = & 0 \\ 2x_1 & +3x_2 & -6x_3 & -3x_4 & = & 0 \\ & & & x_4 & = & 1 \end{array} \tag{5.51}$$

The corresponding matrices are

$$\mathbf{A} = \begin{pmatrix} 1 & -1 & 2 & -3 \\ 3 & 2 & -4 & -6 \end{pmatrix} \tag{5.52}$$

and

$$\mathbf{B} = \begin{pmatrix} 2 & -7 & 14 & -9 \\ 2 & 3 & -6 & -3 \end{pmatrix}, \tag{5.53}$$

and their associated orthogonals are

$$\mathbf{A}^{\perp} = \mathbf{B}^{\perp} = \begin{pmatrix} 0 & 12/5 \\ 2 & -3/5 \\ 1 & 0 \\ 0 & 1 \end{pmatrix}. \tag{5.54}$$

Thus, we have

$$\mathbf{AB}^{\perp} = \begin{pmatrix} 1 & -1 & 2 & -3 \\ 3 & 2 & -4 & -6 \end{pmatrix} \begin{pmatrix} 0 & 12/5 \\ 2 & -3/5 \\ 1 & 0 \\ 0 & 1 \end{pmatrix} = \begin{pmatrix} 0 & 0 \\ 0 & 0 \end{pmatrix} \tag{5.55}$$

and

$$\mathbf{BA}^{\perp} = \begin{pmatrix} 2 & -7 & 14 & -9 \\ 2 & 3 & -6 & -3 \end{pmatrix} \begin{pmatrix} 0 & 12/5 \\ 2 & -3/5 \\ 1 & 0 \\ 0 & 1 \end{pmatrix} = \begin{pmatrix} 0 & 0 \\ 0 & 0 \end{pmatrix}. \tag{5.56}$$

Hence, the two systems are equivalent. ■

In Section 5.9 a Mathematica program is given to check equivalence of two systems of equations.

5.6 SOLVING A SYSTEM IN SOME SELECTED VARIABLES

In some cases we are interested in knowing the solution of some selected unknowns of a system of equations. In this section we show that the above method allows solving this problem with an important reduction in the complexity.

Consider again the system in (5.15)

$$\begin{array}{rcl} u_{i1}x_1 + \cdots + u_{in}x_n - u_{i,n+1}x_{n+1} & = & 0; \quad i = 1, \ldots, m \\ x_{n+1} & = & 1 \end{array} \tag{5.57}$$

and assume without loss of generality that we are interested in the first r variables $x_1, \ldots, x_r$. Then system (5.57) can also be written as

$$\begin{array}{rcl} u_{i1}x_1 + \cdots + u_{ir}x_r - u_{i,n+1}x_{n+1} & = & -u_{ir+1}x_1 + \cdots - u_{in}x_n; \; \forall i \\ x_{n+1} & = & 1 \end{array} \tag{5.58}$$

or

$$\mathbf{U}_a\mathbf{x}_a = -\mathbf{U}_b\mathbf{x}_b; \; x_{n+1} = 1, \tag{5.59}$$

where $\mathbf{x}_a = (x_1, \ldots, x_r, x_{n+1})^T$, $\mathbf{x}_b = (x_{r+1}, \ldots, x_n)^T$, and $\mathbf{U}_a$ and $\mathbf{U}_b$ are their corresponding matrices.

Expression (5.59) shows that $\mathbf{x}_a$ is the reciprocal image of the linear space generated by the columns of $\mathbf{U}_b$ under the transformation associated with the matrix $\mathbf{U}_a$; thus, we have the case described in Section 4.2. Therefore, the methods described there are applicable.

Example 5.8 (Solving a linear system in selected variables). Consider the system of equations

$$\begin{array}{rrrrrrrcr} -x_1 & +x_2 & +x_3 & +2x_4 & -3x_5 & +x_6 & -x_7 & = & -2 \\ x_1 & -x_2 & & -x_4 & +x_5 & & +x_7 & = & -1 \\ & & +x_3 & & -x_5 & +x_6 & +2x_7 & = & -1 \\ & & -x_3 & -x_4 & & -x_6 & -3x_7 & = & -1 \end{array} \tag{5.60}$$

and suppose that we are interested only in variables x_1, x_2, and x_3. Using the artificial variable x_8 we can write system (5.60) as

$$\begin{array}{rrrrrrrrcr} -x_1 & +x_2 & +x_3 & +2x_4 & -3x_5 & +x_6 & -x_7 & +2x_8 & = & 0 \\ x_1 & -x_2 & & -x_4 & +x_5 & & +x_7 & +x_8 & = & 0 \\ & & +x_3 & & -x_5 & +x_6 & +2x_7 & +x_8 & = & 0 \\ & & -x_3 & +x_4 & & -x_6 & -3x_7 & +x_8 & = & 0 \\ & & & & & & & x_8 & = & 1 \end{array} \tag{5.61}$$

and selecting the variables of interest x_1, x_2, and x_3 and the artificial variable x_8 we have

$$\begin{array}{rrrrcrrrr} -x_1 & +x_2 & +x_3 & +2x_8 & = & -2x_4 & +3x_5 & -x_6 & +x_7 \\ x_1 & -x_2 & & +x_8 & = & +x_4 & -x_5 & & -x_7 \\ & & +x_3 & +x_8 & = & & +x_5 & -x_6 & -2x_7 \\ & & -x_3 & +x_8 & = & -x_4 & & +x_6 & +3x_7 \\ & & & x_8 & = & 1 & & & \end{array} \tag{5.62}$$

To obtain the solution for x_1, x_2, x_3, and x_8 we first obtain the orthogonal of the space $\mathcal{L}\{(-2,1,0,-1)^T, (3,-1,1,0)^T, (-1,0,-1,1)^T, (1,-1,-2,3)^T\}$, generated by the columns on the right-hand side of (5.62), as indicated in Table 5.6. This set is $\mathcal{L}\{(0,1,1,1)^T\}$.

Next, we calculate

$$\begin{pmatrix} 0 & 1 & 1 & 1 \end{pmatrix} \begin{pmatrix} -1 & 1 & 1 & 2 \\ 1 & -1 & 0 & 1 \\ 0 & 0 & 1 & 1 \\ 0 & 0 & -1 & 1 \end{pmatrix} = \begin{pmatrix} 1 & -1 & 0 & 3 \end{pmatrix}, \tag{5.63}$$

and finally, we obtain the orthogonal set to $\mathcal{L}\{(1,-1,0,3)^T\}$, as shown in Table 5.7. This set is $\mathcal{L}\{(1,1,0,0)^T, (0,0,1,0)^T, (-3,0,0,1)^T\}$.

Thus, the solution of system (5.62) is

$$\begin{pmatrix} x_1 \\ x_2 \\ x_3 \\ x_8 \end{pmatrix} = a \begin{pmatrix} 1 \\ 1 \\ 0 \\ 0 \end{pmatrix} + b \begin{pmatrix} 0 \\ 0 \\ 1 \\ 0 \end{pmatrix} + c \begin{pmatrix} -3 \\ 0 \\ 0 \\ 1 \end{pmatrix}. \tag{5.64}$$

Table 5.6 Obtaining the Example 5.8 orthogonal set to $\mathcal{L}\left\{(-2,1,0,-1)^T,(3,-1,1,0)^T,\ (-1,0,-1,1)^T,(1,-1,-2,3)^T\right\}$.

Iteration 1					Iteration 2			
-2	**1**	0	0	0	3	**1/2**	0	-1/2
1	**0**	1	0	0	-1	**1**	0	0
0	**0**	0	1	0	1	**0**	1	0
-1	**0**	0	0	1	0	**0**	0	1
	-2	1	0	-1		**1/2**	1	-3/2

Iteration 3			Iteration 4			Output
-1	-1	1	1	**-1**	1	0
0	-2	3	-1	**-2**	3	1
-1	1	0	-2	**1**	0	1
1	0	1	3	**0**	1	1
	0	0		**-1**	1	

Table 5.7 Obtaining the orthogonal set to $\mathcal{L}\{(1,-1,0,3)^T\}$ in Example 5.8.

Iteration 1					Output		
1	**1**	0	0	0	1	0	-3
-1	**0**	1	0	0	1	0	0
0	**0**	0	1	0	0	1	0
3	**0**	0	0	1	0	0	1
	1	-1	0	3			

Since $x_8 = 1$, then $c = 1$ and the solution of the selected variables becomes

$$\begin{pmatrix} x_1 \\ x_2 \\ x_3 \end{pmatrix} = \begin{pmatrix} -3 \\ 0 \\ 0 \end{pmatrix} + a \begin{pmatrix} 1 \\ 1 \\ 0 \end{pmatrix} + b \begin{pmatrix} 0 \\ 0 \\ 1 \end{pmatrix} = \begin{pmatrix} a-3 \\ a \\ b \end{pmatrix}. \tag{5.65}$$

■

5.7 MODIFYING SYSTEMS OF EQUATIONS

In this section we discuss how to find the solution of a system of equations resulting from some modifications of a given system of equations whose solution has previously been calculated. We show how the proposed methods allow important savings in computations. We consider the cases of adding or removing equations or unknowns.

Table 5.8 Adding one equation.

Extra iteration					
1	*2*	*-1*	*2*	**1**	9
3	*3*	*-2*	*3*	**2**	14
2	*1*	*-1/2*	*1/2*	**3/2**	4
-5	*0*	*0*	*0*	**1**	0
-4	*0*	*0*	*0*	**0**	1
	13	*-8*	*12*	**5**	55

Output				
-3/5	*3/5*	*-2/5*	*1/5*	-2
-11/5	*6/5*	*-9/5*	*2/5*	-8
-29/10	*19/10*	*-31/10*	*3/10*	-25/2
-13/5	*8/5*	*-12/5*	*1/5*	-11
0	*0*	*0*	*0*	1

5.7.1 Adding Equations

If k equations are added to system (5.1) we only need repeating Steps 3 to 6 in Algorithm 5.2 k more times to introduce the new equations and update the solution. Thus, the procedure for adding new equations is straight forward.

Example 5.9 (Adding new equations to a linear system of equations). Suppose we add the new equation

$$x_1 + 3x_2 + 2x_3 - 5x_4 = 4 \tag{5.66}$$

to system (5.26). To update the solution we must perform an additional iteration of the algorithm. Table 5.8 shows this additional iteration, starting with the output in Table 5.3. From it, we can see that the solution of the new system is

$$\begin{pmatrix} x_1 \\ x_2 \\ x_3 \\ x_4 \end{pmatrix} = \begin{pmatrix} -2 \\ -8 \\ -25/2 \\ -11 \end{pmatrix}. \tag{5.67}$$

■

Example 5.10 (Adding new equations to a linear system of equations). Consider the following system of equations

$$\begin{array}{cccccc} 2x_1 & -x_2 & +x_3 & +3x_4 & & = & 4 \\ & x_2 & -x_3 & -2x_4 & +3x_5 & = & 3 \\ 2x_1 & & +x_3 & & -x_5 & = & -2 \end{array} \tag{5.68}$$

whose general solution is

$$\begin{pmatrix} x_1 \\ x_2 \\ x_3 \\ x_4 \\ x_5 \end{pmatrix} = \begin{pmatrix} 7/2 \\ -6 \\ -9 \\ 0 \\ 0 \end{pmatrix} + \rho_1 \begin{pmatrix} -1/2 \\ 3 \\ 1 \\ 1 \\ 0 \end{pmatrix} + \rho_2 \begin{pmatrix} -3/2 \\ 1 \\ 4 \\ 0 \\ 1 \end{pmatrix}, \tag{5.69}$$

as shown in Table 5.9.

Table 5.9 Solving the system.

			Iteration 1			
2	**1**	0	0	0	0	0
–1	**0**	1	0	0	0	0
1	**0**	0	1	0	0	0
3	**0**	0	0	1	0	0
0	**0**	0	0	0	1	0
–4	**0**	0	0	0	0	1
	2	–1	1	3	0	–4
			Iteration 2			
0	*1/2*	**1/2**	–1/2	–3/2	0	2
1	*0*	**1**	0	0	0	0
–1	*0*	**0**	1	0	0	0
–2	*0*	**0**	0	1	0	0
3	*0*	**0**	0	0	1	0
–3	*0*	**0**	0	0	0	1
	0	**1**	–1	–2	3	–3
			Iteration 3			
2	*1/2*	*1/2*	**0**	–1/2	–3/2	7/2
0	*0*	*1*	**1**	2	–3	3
1	*0*	*0*	**1**	0	0	0
0	*0*	*0*	**0**	1	0	0
–1	*0*	*0*	**0**	0	1	0
2	*0*	*0*	**0**	0	0	1
	1	*1*	**1**	–1	–4	9
			Output			
	1/2	*1/2*	*0*	–1/2	–3/2	7/2
	–1	*0*	*1*	3	1	–6
	–1	*–1*	*1*	1	4	–9
	0	*0*	*0*	1	0	0
	0	*0*	*0*	0	1	0
	0	*0*	*0*	0	0	1

Table 5.10 Adding a redundant equation.

Iteration 4						
0	*1/2*	*1/2*	*0*	−1/2	−3/2	7/2
1	*-1*	*0*	*1*	3	1	-6
−2	*−1*	*−1*	*1*	1	4	−9
−1	*0*	*0*	*0*	1	0	0
7	*0*	*0*	*0*	0	1	0
−12	*0*	*0*	*0*	0	0	1
	1	*2*	*−1*	0	0	0

Now, suppose that we add a new equation

$$x_2 - 2x_3 - x_4 + 7x_5 = 12 \tag{5.70}$$

in system (5.68). Then we need one step more of the algorithm to obtain the solution of the modified system. As we can see in Table 5.10, it's impossible to select a new pivot column because the inner products associated with the last three columns are null. This means that the added equation is redundant. In fact, it is the first equation plus two times the second, minus the third, as indicated by the first three inner products in the table. Then the solution of the new system is again (5.69).

Finally, if we introduce a new nonredundant equation

$$\frac{2}{3}x_2 - x_3 + 2x_5 = 0 \tag{5.71}$$

in system (5.68), we need to perform an additional iteration (see Table 5.11) in order to obtain the solution of the new system:

$$\begin{pmatrix} x_1 \\ x_2 \\ x_3 \\ x_4 \\ x_5 \end{pmatrix} = \begin{pmatrix} 6 \\ -21 \\ -14 \\ -5 \\ 0 \end{pmatrix} + \rho_1 \begin{pmatrix} -13/6 \\ 5 \\ 16/3 \\ 4/3 \\ 1 \end{pmatrix}. \tag{5.72}$$

■

5.7.2 Removing Unknowns

Assume that in the system of equations (5.1) we eliminate the k unknowns $x_{r_1}, \ldots, x_{r_k}$; that is, we want to solve the new system of equations resulting from making $u_{ir_j} = 0; i = 1, \ldots, m; j = 1, \ldots, k$. In this case, the new solution is obtained by adding the k additional equations $x_{r_j} = 0; j = 1, \ldots, k$. In other words, this case reduces to a particular case of the previous situation (adding equations).

Table 5.11 Adding a nonredundant equation.

			Iteration 5			
0	*1/2*	*1/2*	*0*	**−1/2**	−3/2	7/2
2/3	*−1*	*0*	*1*	**3**	1	−6
−1	*−1*	*−1*	*1*	**1**	4	−9
0	*0*	*0*	*0*	**1**	0	0
2	*0*	*0*	*0*	**0**	1	0
0	*0*	*0*	*0*	**0**	0	1
	1/3	*1*	*−1/3*	**1**	−4/3	5
			Output			
	2/3	*1*	*−1/6*	*−1/2*	−13/6	6
	−2	*−3*	*2*	*3*	5	−21
	−4/3	*−2*	*4/3*	*1*	16/3	−14
	−1/3	*−1*	*1/3*	*1*	4/3	−5
	0	*0*	*0*	*0*	1	0
	0	*0*	*0*	*0*	0	1

Example 5.11 (Removing equations from a linear system of equations). Assume now that we remove the unknown x_4 from the system of equations (5.26); that is, we consider the new system of equations

$$\begin{array}{rrrcr} x_1 & -x_2 & +2x_3 & = & 3 \\ 3x_1 & -2x_2 & & = & -1 \\ x_1 & & -2x_3 & = & 1 \end{array} \tag{5.73}$$

This is equivalent to adding the new equation $x_4 = 0$. Thus, we introduce this equation and obtain the results in Table 5.12, showing that the new solution is

$$\begin{pmatrix} x_1 \\ x_2 \\ x_3 \end{pmatrix} = \begin{pmatrix} 9 \\ 14 \\ 4 \end{pmatrix}. \tag{5.74}$$

If we also remove the unknown x_3, we get the system

$$\begin{array}{rrcr} x_1 & -x_2 & = & 3 \\ 3x_1 & -2x_2 & = & -1 \\ x_1 & & = & 1 \end{array} \tag{5.75}$$

Thus, we introduce the equation $x_3 = 0$ and find that there are no more columns to choose as pivot column. Since, in addition, we have $\mathbf{u}_5^T \bullet \mathbf{w}_5 = 4 \neq 0$, where $\mathbf{u}_5^T = (0, 0, 1, 0, 0)$ is the vector associated with the new equation $x_3 = 0$, the new system is incompatible. ■

Table 5.12 Steps corresponding to the cases of removing the unknowns x_4 and x_3 from the system of equations (5.26).

Iteration 4						Iteration 5					
0	2	-1	2	**1**	9	0	2	-1	2	1	9
0	3	-2	3	**2**	14	0	3	-2	3	2	14
0	1	-1/2	1/2	**3/2**	4	1	1	-1/2	1/2	3/2	4
1	0	0	0	**1**	0	0	0	0	0	1	0
0	0	0	0	**0**	1	0	0	0	0	0	1
	0	0	0	**1**	0		1	-1/2	1/2	3/2	4

Table 5.13 Solving the system (5.76).

Iteration 1						Iteration 2					
1	**1**	0	0	0	0	2	1	**-2**	-1	1	1
2	**0**	1	0	0	0	-1	0	**1**	0	0	0
1	**0**	0	1	0	0	2	0	**0**	1	0	0
-1	**0**	0	0	1	0	1	0	**0**	0	1	0
-1	**0**	0	0	0	1	3	0	**0**	0	0	1
	1	2	1	-1	-1		2	**-5**	0	3	5
Iteration 3						**Output**					
3	1/5	2/5	**-1**	-1/5	-1		3/40	11/40	1/8	-1/5	-1/2
1	2/5	-1/5	**0**	3/5	1		2/5	-1/5	0	3/5	1
-5	0	0	**1**	0	0		1/8	1/8	-1/8	0	-1/2
0	0	0	**0**	1	0		0	0	0	1	0
-2	0	0	**0**	0	1		0	0	0	0	1
	1	1	**-8**	0	-4						

Example 5.12 (Removing unknowns from a linear system of equations). Consider the following system of equations

$$\begin{array}{rrrrcr} x_1 & +2x_2 & +x_3 & -x_4 & = & 1 \\ 2x_1 & -x_2 & +2x_3 & +x_4 & = & -3 \\ 3x_1 & +x_2 & -5x_3 & & = & 2 \end{array} \tag{5.76}$$

whose general solution is

$$\begin{pmatrix} x_1 \\ x_2 \\ x_3 \\ x_4 \end{pmatrix} = \begin{pmatrix} -1/2 \\ 1 \\ -1/2 \\ 0 \end{pmatrix} + \rho_1 \begin{pmatrix} -1/5 \\ 3/5 \\ 0 \\ 1 \end{pmatrix}, \tag{5.77}$$

as shown in Table 5.13.

Suppose that we remove the unknown x_3 from system (5.76). This is equivalent to adding the new equation $x_3 = 0$ and we get Table 5.14. The

Table 5.14 Removing an unknown from the system (5.76).

Iteration 4					
0	*3/40*	*11/40*	*1/8*	*-1/5*	-1/2
0	*2/5*	*-1/5*	*0*	*3/5*	1
1	*1/8*	*1/8*	*-1/8*	*0*	-1/2
0	*0*	*0*	*0*	*1*	0
0	*0*	*0*	*0*	*0*	1
	1/8	*1/8*	*-1/8*	*0*	-1/2

Table 5.15 Removing an unknown from the system (5.76).

Fourth iteration					
0	3/40	11/40	1/8	**−1/5**	−1/2
0	2/5	−1/5	0	**3/5**	1
0	1/8	1/8	−1/8	**0**	−1/2
1	0	0	0	**1**	0
0	0	0	0	**0**	1
	0	0	0	**1**	0
Output					
	3/40	11/40	1/8	**−1/5**	−1/2
	2/5	−1/5	0	**3/5**	1
	1/8	1/8	−1/8	**0**	−1/2
	0	0	0	**1**	0
	0	0	0	**0**	1

resulting system is incompatible because the fourth column, the only one that can be selected as the pivot column, has an associated inner product equal to zero and the last inner product is non-null.

However, if we remove the unknown x_4 from (5.76) instead of x_3, the resulting system has the following solution:

$$\begin{pmatrix} x_1 \\ x_2 \\ x_3 \end{pmatrix} = \begin{pmatrix} -1/2 \\ 1 \\ -1/2 \end{pmatrix}, \tag{5.78}$$

as shown in Table 5.15.

■

5.7.3 Removing Equations

Contrary to previous cases, an efficient treatment of the elimination of one equation or several equations requires some extra information, which has been ignored in Algorithm 5.2. To avoid this, we need to modify Algorithm 5.2 and keep account of the pivot column corresponding to each equation. We consider two essentially different cases: removing one equation and removing more than one equation.

5.7.3.1 Removing One Equation This is the simpler case to solve. We analyze the following two alternative cases:

1. **Redundant equations have previously been removed.** Note that this can easily be done, since redundant equations are detected in Step 4 of Algorithm 5.2. In this case, obtaining the new solution is simple: If the removed equation has associated pivot column $\mathbf{w}^p$ and the solution of the initial system is

$$\hat{\mathbf{w}}_{n+1} + \mathcal{L}\left\{\hat{\mathbf{w}}_{\ell+1}, \ldots, \hat{\mathbf{w}}_n\right\}, \tag{5.79}$$

then the solution of the new system obtained by removing that equation from the original system becomes

$$\hat{\mathbf{w}}_{n+1} + \mathcal{L}\left\{\hat{\mathbf{w}}_{\ell+1}, \ldots, \hat{\mathbf{w}}_n, \hat{\mathbf{w}}_p\right\}. \tag{5.80}$$

2. **Redundant equations remain in the system.** This is the usual case, since when redundant equations exist, we have several choices to get a nonredundant system of equations, and perhaps we want to keep all of them in order to be free in the future to choose which equations to keep and which to remove. Then we also have two cases:

   (i) The equation to be removed has associated pivot column $\mathbf{w}_p$ and all other equations are orthogonal to $\mathbf{w}_p$. In this case, the new solution is again (5.80).

   (ii) Otherwise, the solution of the new system remains the same.

Example 5.13 (Updating solutions of linear systems of equations). Suppose we wish to solve the nonredundant system of equations

$$\begin{array}{rrrrcr} x & +2y & & +u & = & 0 \\ 2x & +y & -z & & = & 0 \\ & +2y & +z & & = & -1 \end{array} \tag{5.81}$$

The solution is given in Table 5.16. Suppose further that we wish to remove one or two equations from the system (5.81). Then their corresponding solutions can immediately be obtained from the columns of Table 5.16 and (5.80), as shown in Table 5.17.

Table 5.16 Iterations for solving the system of equations in Example 5.13 but keeping and pivoting all columns.

Iteration 1						Iteration 2					
1	**1**	0	0	0	0	2	*1*	**−2**	0	−1	0
2	**0**	1	0	0	0	1	*0*	**1**	0	0	0
0	**0**	0	1	0	0	−1	*0*	**0**	1	0	0
1	**0**	0	0	1	0	0	*0*	**0**	0	1	0
0	**0**	0	0	0	1	0	*0*	**0**	0	0	1
	1	2	0	1	0		*2*	**−3**	−1	−2	0
Iteration 3						**Output**					
0	*−1/3*	*2/3*	**2/3**	1/3	0		*−3*	*2*	*2*	3	−2
2	*2/3*	*−1/3*	**−1/3**	−2/3	0		*2*	*−1*	*−1*	−2	1
1	*0*	*0*	**1**	0	0		*−4*	*2*	*3*	4	−3
0	*0*	*0*	**0**	1	0		*0*	*0*	*0*	1	0
1	*0*	*0*	**0**	0	1		*0*	*0*	*0*	0	1
	4/3	*−2/3*	**1/3**	−4/3	1						

■

Example 5.14 (Updating solutions of linear systems of equations). Consider the following system of equations

$$\begin{array}{rrrrcr} 2x_1 & -3x_2 & +x_3 & -x_4 & = & 2 \\ x_1 & +2x_2 & & -2x_4 & = & 1 \\ x_1 & -5x_2 & +x_3 & +x_4 & = & 1 \\ & 2x_2 & -2x_3 & +x_4 & = & 3 \\ x_1 & -2x_2 & +4x_3 & -4x_4 & = & -5 \\ & x_2 & -x_3 & & = & 0 \end{array} \tag{5.82}$$

whose solution is

$$\begin{pmatrix} x_1 \\ x_2 \\ x_3 \\ x_4 \end{pmatrix} = \begin{pmatrix} 4 \\ 3/2 \\ 3/2 \\ 3 \end{pmatrix}, \tag{5.83}$$

as shown in Table 5.18. Table 5.18 shows that the third and fifth equations of system (5.82) are redundant, so if we remove them from the system the solution (5.83) is maintained. But if we remove some of the nonredundant equations we have to update the solution as follows:

- If we remove the first, second, or fourth equations of system (5.82) the solution remains the same because the redundant equations depend on them.
  - The pivot column of the first equation is $\mathbf{w}_1$ and $\mathbf{w}_1 \bullet \mathbf{u}_3^T = 1$, so the third equation depends on the first one.

Table 5.17 Solutions of different systems of equations obtained by removing equations from system (5.81), shown in the upper left corner of this table.

System	Solution $\begin{pmatrix} x \\ y \\ z \\ u \end{pmatrix}$
$\begin{aligned} x + 2y + u &= 0 \\ 2x + y - z &= 0 \\ +2y + z &= -1 \end{aligned}$	$\begin{pmatrix} -2 \\ 1 \\ -3 \\ 0 \end{pmatrix} + \rho_1 \begin{pmatrix} 3 \\ -2 \\ 4 \\ 1 \end{pmatrix}$
$\begin{aligned} x + 2y + u &= 0 \\ 2x + y - z &= 0 \end{aligned}$	$\begin{pmatrix} -2 \\ 1 \\ -3 \\ 0 \end{pmatrix} + \rho_1 \begin{pmatrix} 3 \\ -2 \\ 4 \\ 1 \end{pmatrix} + \rho_2 \begin{pmatrix} 2 \\ -1 \\ 3 \\ 0 \end{pmatrix}$
$\begin{aligned} x + 2y + u &= 0 \\ 2y + z &= -1 \end{aligned}$	$\begin{pmatrix} -2 \\ 1 \\ -3 \\ 0 \end{pmatrix} + \rho_1 \begin{pmatrix} 3 \\ -2 \\ 4 \\ 1 \end{pmatrix} + \rho_2 \begin{pmatrix} 2 \\ -1 \\ 2 \\ 0 \end{pmatrix}$
$\begin{aligned} 2x + y - z &= 0 \\ 2y + z &= -1 \end{aligned}$	$\begin{pmatrix} -2 \\ 1 \\ -3 \\ 0 \end{pmatrix} + \rho_1 \begin{pmatrix} 3 \\ -2 \\ 4 \\ 1 \end{pmatrix} + \rho_2 \begin{pmatrix} -3 \\ 2 \\ -4 \\ 0 \end{pmatrix}$
$x + 2y + u = 0$	$\begin{pmatrix} -2 \\ 1 \\ -3 \\ 0 \end{pmatrix} + \rho_1 \begin{pmatrix} 3 \\ -2 \\ 4 \\ 1 \end{pmatrix} + \rho_2 \begin{pmatrix} 2 \\ -1 \\ 3 \\ 0 \end{pmatrix} + \rho_3 \begin{pmatrix} 2 \\ -1 \\ 2 \\ 0 \end{pmatrix}$
$2x + y - z = 0$	$\begin{pmatrix} -2 \\ 1 \\ -3 \\ 0 \end{pmatrix} + \rho_1 \begin{pmatrix} 3 \\ -2 \\ 4 \\ 1 \end{pmatrix} + \rho_2 \begin{pmatrix} 2 \\ -1 \\ 3 \\ 0 \end{pmatrix} + \rho_3 \begin{pmatrix} -3 \\ 2 \\ -4 \\ 0 \end{pmatrix}$
$2y + z = -1$	$\begin{pmatrix} -2 \\ 1 \\ -3 \\ 0 \end{pmatrix} + \rho_1 \begin{pmatrix} 3 \\ -2 \\ 4 \\ 1 \end{pmatrix} + \rho_2 \begin{pmatrix} -3 \\ 2 \\ -4 \\ 0 \end{pmatrix} + \rho_3 \begin{pmatrix} 2 \\ -1 \\ 2 \\ 0 \end{pmatrix}$

- The pivot column of the second equation is $\mathbf{w}_2$ and $\mathbf{w}_2 \bullet \mathbf{u}_3^T = -1$, $\mathbf{w}_2 \bullet \mathbf{u}_5^T = 1$, so the third and fifth equations depend on the second one.
- The pivot column of the fourth equation is $\mathbf{w}_3$ and $\mathbf{w}_3 \bullet \mathbf{u}_5^T = -2$, so the fifth equation depends on the fourth one.

- If in system (5.82) we remove the sixth equation, which has associated pivot column $\mathbf{w}_4$, the general solution is

$$\begin{pmatrix} x_1 \\ x_2 \\ x_3 \\ x_4 \end{pmatrix} = \begin{pmatrix} 4 \\ 3/2 \\ 3/2 \\ 3 \end{pmatrix} + \rho_1 \begin{pmatrix} -5/3 \\ -7/6 \\ -13/6 \\ -2 \end{pmatrix}, \tag{5.84}$$

because all other equations are orthogonal to $\mathbf{w}_4$.

■

5.7.3.2 Removing More Than One Equation The above method allows removing a single equation $\mathbf{u}_k^T \bullet \mathbf{x} = 0$ because it has the following information:

1. The column vectors $\mathbf{w}_j; j = 1, \ldots, n+1$ of the final complete tableau.
2. The pivot column corresponding to each equation.

However, it allows iterating the process only with nonredundant systems (see Example 5.13). When redundant equations remain in the system the above information has to be updated. In other words, after removing equation $\mathbf{u}_k^T \bullet \mathbf{x} = 0$, some or all $\mathbf{w}_j; j = 1, \ldots, n+1$ and the pivot equations can change. Thus, an updating process becomes necessary for the method to be able to solve the case of removing more than one equation from the original system.

To this aim, we also need to know the list of equations that are not orthogonal to each $\mathbf{w}_j; j = 1, \ldots, n$. Thus, we define:

1. A function $\phi(j)$, which associates each column of $\mathbf{w}_j$ with its corresponding pivot equation or no equation at all (equation 0) when such an equation does not exist:

$$\phi(j) = \begin{cases} i & \text{if } \mathbf{u}_i^T \bullet \mathbf{w}_j = 1;\ \mathbf{u}_i^T \bullet \mathbf{w}_k = 0;\ \forall k \neq j, \\ 0, & \text{otherwise}, \end{cases} \quad j = 1, \ldots, n. \tag{5.85}$$

The set of equations $\{\mathbf{u}_k^T \bullet \mathbf{x} = 0 | k = \phi(j); j = 1, \ldots, n; k \neq 0\}$ is called the set of *basic equations*.

It is worthwhile mentioning that, since each pivot column corresponds to a single equation, we have

$$\phi(\ell) = k \Rightarrow \phi(j) \neq k, \forall j \neq \ell. \tag{5.86}$$

Table 5.18 Solving the system (5.82).

			Iteration 1		
2	**1**	0	0	0	0
-3	**0**	1	0	0	0
1	**0**	0	1	0	0
-1	**0**	0	0	1	0
-2	**0**	0	0	0	1
	2	-3	1	-1	-2
			Iteration 2		
1	*1/2*	**3/2**	-1/2	1/2	1
2	*0*	**1**	0	0	0
0	*0*	**0**	1	0	0
-2	*0*	**0**	0	1	0
-1	*0*	**0**	0	0	1
	1/2	**7/2**	-1/2	-3/2	0
			Iteration 3		
1	*2/7*	*3/7*	-2/7	8/7	1
-5	*-1/7*	*2/7*	1/7	3/7	0
1	*0*	*0*	1	0	0
1	*0*	*0*	0	1	0
-1	*0*	*0*	0	0	1
	1	*-1*	0	0	0
			Iteration 4		
0	*2/7*	*3/7*	**-2/7**	8/7	1
2	*-1/7*	*2/7*	**1/7**	3/7	0
-2	*0*	*0*	**1**	0	0
1	*0*	*0*	**0**	1	0
-3	*0*	*0*	**0**	0	1
	-2/7	*4/7*	**-12/7**	13/7	-3
			Iteration 5		
1	*1/3*	*1/3*	*1/6*	5/6	3/2
-2	*-1/6*	*1/3*	*-1/12*	7/12	-1/4
4	*-1/6*	*1/3*	*-7/12*	13/12	-7/4
-4	*0*	*0*	*0*	1	0
5	*0*	*0*	*0*	0	1
	0	*1*	*-2*	0	0

Table 5.18 (Continued).

Iteration 6					
0	*1/3*	*1/3*	*1/6*	**5/6**	3/2
1	*-1/6*	*1/3*	*-1/12*	**7/12**	-1/4
-1	*-1/6*	*1/3*	*-7/12*	**13/12**	-7/4
0	*0*	*0*	*0*	**1**	0
0	*0*	*0*	*0*	**0**	1
	0	*0*	*1/2*	**-1/2**	3/2
Output					
	1/3	*1/3*	*1*	*-5/3*	4
	-1/6	*1/3*	*1/2*	*-7/6*	3/2
	-1/6	*1/3*	*1/2*	*-13/6*	3/2
	0	*0*	*1*	*-2*	3
	0	*0*	*0*	*0*	1

2. A function $\psi(j)$, which associates each column j with the list of non-basic equations that are not orthogonal to $\mathbf{w}_j$; that is,

$$\psi(j) = \{k \mid t_k^j = \mathbf{u}_k^T \bullet \mathbf{w}_j \neq 0 \wedge \phi(j) \neq k\}; j = 1, \ldots, n. \tag{5.87}$$

   Note that this list contains the numbers of equations that are linear combinations of basic equations.

The solution of the system of equations can then be written as

$$\hat{\mathbf{w}}_{n+1} + \mathcal{L}\left(\{\hat{\mathbf{w}}_j | \phi(j) = 0; j = 1, \ldots, n\}\right). \tag{5.88}$$

The following lemma shows how to update $(\phi(j), \psi(j)); j = 1, \ldots, n$ and $\mathbf{w}_j; j = 1, \ldots, n+1$ when we remove equation $\mathbf{u}_k^T \bullet \mathbf{x} = 0$. An asterisk is used to refer to updated functions.

Lemma 5.2 *Let the ϕ and ψ functions be as defined in (5.85) and (5.87), respectively. If we remove equation $\mathbf{u}_k^T \bullet \mathbf{x} = 0$, the vectors $\mathbf{w}_j; j = 1, \ldots, n+1$ and the $(\phi(j), \psi(j)); j = 1, \ldots, n$ functions must be updated as follows:*

1. **If $\mathbf{u}_k^T \bullet \mathbf{x} = 0$ is a nonredundant equation,** *that is, if*

$$\exists \ell \in \{1, \ldots, n\} | \phi(\ell) = k \wedge \psi(\ell) = \emptyset,$$

the updating expressions are

$$\phi^*(j) = \begin{cases} 0, & if\ j = \ell, \\ \phi(j), & otherwise, \end{cases} \quad j = 1, \ldots, n, \tag{5.89}$$

$$\psi^*(j) = \psi(j);\ j = 1, \ldots, n, \tag{5.90}$$

$$\mathbf{w}_j^* = \mathbf{w}_j;\ j = 1, \ldots, n+1. \tag{5.91}$$

2. **If $\mathbf{u}_k^T \bullet \mathbf{x} = 0$ is an initially nonredundant equation that becomes redundant because other equations linearly depend on it,** *that is, if $\exists \ell \in \{1,\ldots,n\} | \phi(\ell) = k \wedge \psi(\ell) \neq \emptyset$, we select one r in $\psi(\ell)$ and the updating expressions become*

$$\phi^*(j) = \begin{cases} r, & \text{if } j = \ell, \\ \phi(j), & \text{otherwise}, \end{cases} \quad j = 1,\ldots,n, \tag{5.92}$$

$$\psi^*(j) = \begin{cases} A_{\ell j}, & \text{if } r \in \psi(j) \wedge j \neq \ell, \\ \psi(j) - \{r\}, & \text{if } j = \ell, \\ \psi(j), & \text{otherwise}, \end{cases} \quad j = 1,\ldots,n, \tag{5.93}$$

$$\mathbf{w}_j^* = \begin{cases} \mathbf{w}_j - t_r^j \mathbf{w}_\ell / t_r^\ell, & \text{if } r \in \psi(j) \wedge j \neq \ell, \\ \mathbf{w}_j / t_r^\ell, & \text{if } j = \ell, \\ \mathbf{w}_j, & \text{otherwise}, \end{cases} \quad j = 1,\ldots,n+1, \tag{5.94}$$

where

$$A_{\ell j} = (\psi(\ell) - \psi(j)) \cup (\psi(j) - \psi(\ell)) \cup \{v \in \psi(\ell) \cap \psi(j) | t_v^j t_r^\ell \neq t_r^j t_v^\ell\}. \tag{5.95}$$

3. **If $\mathbf{u}_k^T \bullet \mathbf{x} = 0$ is a redundant equation,** *that is, if $\forall \ell \in \{1,\ldots,n\}$, $\phi(\ell) \neq k$, the updating expressions are*

$$\phi^*(j) = \phi(j); \; j = 1,\ldots,n, \tag{5.96}$$

$$\psi^*(j) = \psi(j) - \{k\}; \; j = 1,\ldots,n, \tag{5.97}$$

$$\mathbf{w}_j^* = \mathbf{w}_j; \; j = 1,\ldots,n+1. \tag{5.98}$$

The proof of this lemma is given in the Appendix to this chapter.

Based on Lemma 5.2, we can obtain the following algorithm.

Algorithm 5.3 Dynamically updating the solution of a system of equations.

- **Input:**
  1. The column vectors $\mathbf{w}_j; j = 1,\ldots,n+1$ of the final tableau for the initial system of equations.
  2. The equations $\phi(j)$ corresponding to the columns of $\mathbf{w}_j; j = 1,\ldots,n$.
  3. The lists $\psi(j); j = 1,\ldots,n$ of redundant equations that are not orthogonal to each $\mathbf{w}_j; j = 1,\ldots,n$.
  4. An equation $\mathbf{u}_k^T \bullet \mathbf{x} = 0$ to be removed.
- **Output:** The updated values of the first three of the above items.

Step 1: Determine whether or not $\mathbf{u}_k^T \bullet \mathbf{x} = 0$ is a basic equation, that is, if there exists $\ell \in \{1,\ldots,n\}$ such that $\phi(\ell) = k$.

Table 5.19 Initial tableau with the ϕ and ψ functions.

Iteration 1					
$\psi(j)$	3	3 5	5		
$\phi(j)$	1	2	4	6	
$\mathbf{w}_j$	$\mathbf{w}_1$	$\mathbf{w}_2$	$\mathbf{w}_3$	$\mathbf{w}_4$	$\mathbf{w}_5$
	1/3 –1/6 –1/6 0 0	1/3 1/3 1/3 0 0	1 1/2 1/2 1 0	–5/3 –7/6 -13/6 –2 0	4 3/2 3/2 3 1

Step 2: If such an ℓ does not exist, remove k from all $\psi(j); j = 1, \ldots, n$ and go to Step 5. Otherwise, continue with Step 3.

Step 3: If $\psi(\ell) = \emptyset$, let $\phi(\ell) = 0$ and go to Step 5. Otherwise, continue with Step 4.

Step 4: Take one $r \in \psi(\ell)$ and

1. Let $\phi(\ell) = r$ and $\mathbf{w}_\ell = \mathbf{w}_\ell / t_r^\ell$.
2. For all $j \neq \ell$ and $\psi(j)$ containing r, let $\psi(j) = A_{\ell j}$, with $A_{\ell j}$ defined in (5.95), and $\mathbf{w}_j = \mathbf{w}_j - t_r^j \mathbf{w}_\ell$.

Step 5: Return $\mathbf{w}_j; j = 1, \ldots, n+1$ and $(\phi(j), \psi(j)); j = 1, \ldots, n$.

Note that Algorithm 5.3 allows us to obtain the new solution of the resulting system of equations using (5.88). The following examples illustrate Algorithm 5.3.

Example 5.15 (Removing more than one equation). In this example we consider system (5.82) and we obtain its general solution after removing the second, fifth, and sixth equations. The basic equations of this system are equations numbers 1, 2, 4, and 6; the rest, that is, equations 3 and 5, are nonbasic. In Table 5.19 we can see the ϕ and ψ functions together the $\mathbf{w}_j$ vectors obtained using the Algorithm 5.2 over the system (5.82).

1. **We remove equation 6**: This is a nonredundant equation (case 1 in Lemma 5.2), so we get Table 5.20.

2. **We remove equation 2**: This is a nonredundant equation but other equations linearly depend on it (case 2 in Lemma 5.2). First, we update

Table 5.20 Tableau with the ϕ and ψ functions after removing equation 6.

Iteration 2					
$\psi(j)$	3	3 5	5		
$\phi(j)$	1	2	4	0	
$\mathbf{w}_j$	$\mathbf{w}_1$	$\mathbf{w}_2$	$\mathbf{w}_3$	$\mathbf{w}_4$	$\mathbf{w}_5$
	1/3	1/3	1	–5/3	4
	–1/6	1/3	1/2	–7/6	3/2
	–1/6	1/3	1/2	–13/6	3/2
	0	0	1	–2	3
	0	0	0	0	1

the ϕ and ψ functions using the updating expressions (5.92) and (5.93), where $l = 2$ and $r = 3 \in \psi(2)$:

$$\begin{array}{ll} \phi^*(1) = 1 & \psi^*(1) = A_{21} = \{5\} \\ \phi^*(2) = 3 & \psi^*(2) = \psi(2) - \{3\} = \{5\} \\ \phi^*(3) = 4 & \psi^*(3) = \psi(3) = \{5\} \\ \phi^*(4) = 0 & \psi^*(4) = \psi(4) = \{\} \end{array}$$

Finally, expression (5.94) allows us to obtain

$$\begin{aligned} \mathbf{w}_1^* &= \mathbf{w}_1 - t_3^1 \mathbf{w}_2 / t_3^2 = \mathbf{w}_1 + \mathbf{w}_2, \\ \mathbf{w}_2^* &= \mathbf{w}_2 / t_3^2 = -\mathbf{w}_2, \\ \mathbf{w}_j^* &= \mathbf{w}_j, \quad j = 3, 4, 5. \end{aligned}$$

Thus, we get Table 5.21.

3. **We remove equation 5**: This is a redundant equation (case 3 in Lemma 5.2). Expressions (5.96), (5.97), and (5.98) are evaluated in Table 5.22.

   Thus, the general solution of system (5.82) after removing the second, fifth, and sixth equations is

$$\begin{pmatrix} x_1 \\ x_2 \\ x_3 \\ x_4 \end{pmatrix} = \begin{pmatrix} 4 \\ 3/2 \\ 3/2 \\ 3 \end{pmatrix} + \rho_1 \begin{pmatrix} -5/3 \\ -7/6 \\ -13/6 \\ -2 \end{pmatrix}.$$

■

Table 5.21 Tableau with the ϕ and ψ functions after removing equation 2.

Iteration 3					
$\psi(j)$	5	5	5		
$\phi(j)$	1	3	4	0	
$\mathbf{w}_j$	$\mathbf{w}_1$	$\mathbf{w}_2$	$\mathbf{w}_3$	$\mathbf{w}_4$	$\mathbf{w}_5$
	2/3	-1/3	1	-5/3	4
	1/6	-1/3	1/2	-7/6	3/2
	1/6	-1/3	1/2	-13/6	3/2
	0	0	1	-2	3
	0	0	0	0	1

Table 5.22 Tableau with the ϕ and ψ functions after removing equation 5.

Iteration 4					
$\psi(j)$					
$\phi(j)$	1	3	4	0	
$\mathbf{w}_j$	$\mathbf{w}_1$	$\mathbf{w}_2$	$\mathbf{w}_3$	$\mathbf{w}_4$	$\mathbf{w}_5$
	2/3	-1/3	1	-5/3	4
	1/6	-1/3	1/2	-7/6	3/2
	1/6	-1/3	1/2	-13/6	3/2
	0	0	1	-2	3
	0	0	0	0	1

Example 5.16 (Removing more than one equation). Consider the following system of equations, where we have already included the artificial variable $x_7 = 1$:

$$\begin{pmatrix} 1 & -2 & -1 & 0 & 1 & -1 & 1 \\ -1 & 1 & 1 & 1 & 0 & 2 & -3 \\ 2 & 0 & 2 & 0 & -1 & 1 & 2 \\ 0 & -1 & 0 & 1 & 1 & 1 & -2 \\ 4 & -3 & 0 & -1 & 0 & -2 & 6 \\ 0 & 2 & 1 & -2 & 2 & 0 & -1 \\ 1 & 0 & 1 & 0 & -1 & 1 & 1 \\ 2 & -2 & 1 & 2 & -3 & 1 & 3 \\ 1 & 3 & 4 & -1 & 1 & 3 & -2 \end{pmatrix} \begin{pmatrix} x_1 \\ x_2 \\ x_3 \\ x_4 \\ x_5 \\ x_6 \\ x_7 \end{pmatrix} = \mathbf{0}. \tag{5.99}$$

Using Algorithm 5.2 and the definitions of the ϕ and ψ functions, we obtain the $\mathbf{w}$-tableau and the new ϕ and ψ functions in Table 5.23. The ψ sets indicate that equation 4 is a linear combination of equations 1 and 2; equation 5 is a linear combination of equations $1, 2$, and 3; equation 8 is a linear combination of equations 3 and 6; equation 9 is a linear combination of equations $2, 3$, and 6; and all other equations are independent.

From Equation (5.88) we know that the solution of (5.99) is

$$\begin{pmatrix} x_1 \\ x_2 \\ x_3 \\ x_4 \\ x_5 \\ x_6 \end{pmatrix} = \begin{pmatrix} -6 \\ -5 \\ 5 \\ -3 \\ 0 \\ 0 \end{pmatrix} + \rho \begin{pmatrix} 6 \\ 6 \\ -6 \\ 4 \\ 1 \\ 1 \end{pmatrix}. \tag{5.100}$$

Next, we consider the following cases:

1. **We remove equation 7**: Then we have $\phi(5) = 7$ and $\psi(5) = \emptyset$ and we are in case 1. Thus, we make $\phi^*(5) = 0$ and the new solution becomes

$$\begin{pmatrix} x_1 \\ x_2 \\ x_3 \\ x_4 \\ x_5 \\ x_6 \end{pmatrix} = \begin{pmatrix} -6 \\ -5 \\ 5 \\ -3 \\ 0 \\ 0 \end{pmatrix} + \rho_1 \begin{pmatrix} 6 \\ 6 \\ -6 \\ 4 \\ 1 \\ 1 \end{pmatrix} + \rho_2 \begin{pmatrix} -9 \\ -19/2 \\ 8 \\ -15/2 \\ -2 \\ 0 \end{pmatrix}. \tag{5.101}$$

2. **We remove equation 4**: Then there is no ℓ such that $\phi(\ell) = 4$ and we are in case 3. Thus, we make $\psi^*(1) = \{5\}$, $\psi^*(2) = \{5, 9\}$ and the new solution is again (5.100).

3. **We sequentially remove equations 3, 6, 5, and 8**:

   (a) **We remove equation 3**: We have $\ell = 3$ since $\phi(3) = 3$ and $\psi(3) = \{5, 8, 9\}$ and we are in case 2. Thus, we take, for example, $r = 5$ and let $\phi^*(3) = 5$. Next, we calculate $A_{31} = \{4, 8, 9\}$, $A_{32} = \{4, 8, 9\}$, $A_{33} = \{8, 9\}$, and $A_{34} = \{8, 9\}$, we update the ψ, we normalize $\mathbf{w}_3$, which in this case is not necessary since $t_5^3 = 1$, and we perform the pivoting to columns 1 and 2 (see Table 5.24).

   (b) **We remove equation 6**: We have $\ell = 4$ since $\phi(4) = 6$ and $\psi(4) = \{8, 9\}$ and we are in case 2. Thus, we take, for example, $r = 8$ and let $\phi^*(4) = 8$. Next, we calculate $A_{41} = \{4, 9\}$, $A_{42} = \{4, 9\}$, $A_{43} = \{9\}$, and $A_{44} = \{9\}$, we update the ψ, we normalize $\mathbf{w}_4$, which in this case is not necessary since $t_8^4 = 1$, and we perform the pivoting to columns $1, 2$, and 3 (see Table 5.24).

   (c) **We remove equation 5**: We have $\ell = 3$ since $\phi(3) = 5$ and $\psi(3) = \{9\}$ and we are in case 2. Thus, we take $r = 9$ and let

Table 5.23 Initial tableau with the ϕ and ψ functions.

Iteration 1							
$\psi(j)$		9	9				
	5	5	8	9			
	4	4	5	8			
$\phi(j)$	1	2	3	6	7	0	
$\mathbf{w}_j$	$\mathbf{w}_1$	$\mathbf{w}_2$	$\mathbf{w}_3$	$\mathbf{w}_4$	$\mathbf{w}_5$	$\mathbf{w}_6$	$\mathbf{w}_7$
	–2	–2	5	–1	–9	6	–6
	–5/2	–2	5	–1	–19/2	6	–5
	2	2	–4	1	8	–6	5
	–3/2	–1	4	–1	–15/2	4	–3
	0	0	1	0	–2	1	0
	0	0	0	0	0	1	0
	0	0	0	0	0	0	1

$\phi^*(3) = 9$. Next, we calculate $A_{31} = \{4\}$, $A_{32} = \{4\}$, and $A_{34} = \emptyset$, we update the ψ, we normalize $\mathbf{w}_3$, and we perform the pivoting to columns 1, 2, and 4 (see Table 5.25).

(d) **We remove equation 8**: We have $\ell = 4$ since $\phi(4) = 8$ and $\psi(4) = \emptyset$ and we are in case 1. Thus, we let $\phi(4) = 0$.

Thus, according to (5.88), the solution of the new system is

$$\begin{pmatrix} x_1 \\ x_2 \\ x_3 \\ x_4 \\ x_5 \\ x_6 \end{pmatrix} = \begin{pmatrix} -6 \\ -5 \\ 5 \\ -3 \\ 0 \\ 0 \end{pmatrix} + \rho_1 \begin{pmatrix} 6 \\ 6 \\ -6 \\ 4 \\ 1 \\ 1 \end{pmatrix} + \rho_2 \begin{pmatrix} 3 \\ 3 \\ -5/2 \\ 5/2 \\ 1/2 \\ 0 \end{pmatrix}. \tag{5.102}$$

■

5.7.4 Adding One Unknown

Addition of a new variable forces us to work with $\mathbb{R}^{n+2}$ instead of $\mathbb{R}^{n+1}$. Let $\mathbf{e}_{n+2}$ be the vector with all zero components but the last one, which is one. Since

$$\mathbb{R}^{n+1} \equiv \mathcal{L}\{\mathbf{w}_1, \ldots, \mathbf{w}_{n+1}\}, \tag{5.103}$$

we have

$$\mathbb{R}^{n+2} \equiv \mathcal{L}\left\{\begin{pmatrix} \mathbf{w}_1 \\ 0 \end{pmatrix}, \ldots, \begin{pmatrix} \mathbf{w}_{n+1} \\ 0 \end{pmatrix}, \mathbf{e}_{n+2}\right\}. \tag{5.104}$$

Table 5.24 Illustration of how the tableau and the ϕ and ψ functions are updated after removing equations 3 and 6.

Iteration 2							
$\psi(j)$	9 8 4	9 8 4	9 8	9 8			
$\phi(j)$	1	2	5	6	7	0	
$\mathbf{w}_j$	$\mathbf{w}_1$	$\mathbf{w}_2$	$\mathbf{w}_3$	$\mathbf{w}_4$	$\mathbf{w}_5$	$\mathbf{w}_6$	$\mathbf{w}_7$
	–7	3	5	–1	–9	6	–6
	–15/2	3	5	–1	–19/2	6	–5
	6	–2	–4	1	8	–6	5
	–11/2	3	4	–1	–15/2	4	–3
	–1	1	1	0	–2	1	0
	0	0	0	0	0	1	0
	0	0	0	0	0	0	1
Iteration 3							
$\psi(j)$	9 4	9 4	9	9			
$\phi(j)$	1	2	5	8	7	0	
$\mathbf{w}_j$	$\mathbf{w}_1$	$\mathbf{w}_2$	$\mathbf{w}_3$	$\mathbf{w}_4$	$\mathbf{w}_5$	$\mathbf{w}_6$	$\mathbf{w}_7$
	–6	2	4	1	–9	6	–6
	–13/2	2	4	1	–19/2	6	–5
	5	–1	–3	–1	8	–6	5
	–9/2	2	3	1	–15/2	4	–3
	–1	1	1	0	–2	1	0
	0	0	0	0	0	1	0
	0	0	0	0	0	0	1

Table 5.25 Illustration of how the tableau and the ϕ and ψ functions are updated after removing equations 5 and 8.

Iteration 4							
$\psi(j)$	4	4					
$\phi(j)$	1	2	9	8	7	0	
$\mathbf{w}_j$	$\mathbf{w}_1$	$\mathbf{w}_2$	$\mathbf{w}_3$	$\mathbf{w}_4$	$\mathbf{w}_5$	$\mathbf{w}_6$	$\mathbf{w}_7$
	–2	–4	2	3	–9	6	–6
	–5/2	–4	2	3	–19/2	6	–5
	2	7/2	–3/2	–5/2	8	–6	5
	–3/2	–5/2	3/2	5/2	–15/2	4	–3
	0	–1/2	1/2	1/2	–2	1	0
	0	0	0	0	0	1	0
	0	0	0	0	0	0	1
Iteration 5							
$\psi(j)$	4	4					
$\phi(j)$	1	2	9	0	7	0	
$\mathbf{w}_j$	$\mathbf{w}_1$	$\mathbf{w}_2$	$\mathbf{w}_3$	$\mathbf{w}_4$	$\mathbf{w}_5$	$\mathbf{w}_6$	$\mathbf{w}_7$
	–2	–4	2	3	–9	6	–6
	–5/2	–4	2	3	–19/2	6	–5
	2	7/2	–3/2	–5/2	8	–6	5
	–3/2	–5/2	3/2	5/2	–15/2	4	–3
	0	–1/2	1/2	1/2	–2	1	0
	0	0	0	0	0	1	0
	0	0	0	0	0	0	1

Suppose now that we apply Algorithm 5.2 and start in Step 1 with the matrix

$$\mathbf{W} = \begin{pmatrix} \mathbf{w}_1 & \dots & \mathbf{w}_{n+1} & \mathbf{0} \\ 0 & \dots & 0 & 1 \end{pmatrix}, \tag{5.105}$$

instead of matrix $\mathbf{I}_{n+1}$. Suppose also that we operate first with the basic equations. Since the last components of the first $n+1$ columns are zero, the new scalar products we get for the first $n+1$ columns are the same as the scalar products we obtain at the end of the process of solving the system for $n+1$ variables. Thus, we need no normalization or pivoting for the first $n+1$ columns. So, the only pivoting operation corresponds to the $(n+2)$th column. Consequently, all the the first $n+1$ columns remain unchanged and the $(n+2)$th becomes

$$\mathbf{w}_{n+2} = \mathbf{e}_{n+2} - \sum_{i \in I} u_{i,n+2} \begin{pmatrix} \mathbf{w}_i \\ 0 \end{pmatrix}, \tag{5.106}$$

where $u_{i,n+2}; i = 1, \dots, m$ are the coefficients associated with the new unknown, I is the set of pivot columns and $\mathbf{w}_i$ is the pivot column associated with the ith equation.

Next, we look for one nonbasic equation such that $\mathbf{u}_k^T \bullet \mathbf{w}_{n+2} \neq 0$. If this equation exists, we perform the standard pivoting and the new solution becomes

$$\begin{pmatrix} \hat{\mathbf{w}}_{n+1} \\ 0 \end{pmatrix} + \mathcal{L}\left\{ \begin{pmatrix} \hat{\mathbf{w}}_{\ell+1} \\ 0 \end{pmatrix}, \dots, \begin{pmatrix} \hat{\mathbf{w}}_n \\ 0 \end{pmatrix} \right\}. \tag{5.107}$$

Otherwise, the new solution is

$$\begin{pmatrix} \hat{\mathbf{w}}_{n+1} \\ 0 \end{pmatrix} + \mathcal{L}\left\{ \begin{pmatrix} \hat{\mathbf{w}}_{\ell+1} \\ 0 \end{pmatrix}, \dots, \begin{pmatrix} \hat{\mathbf{w}}_n \\ 0 \end{pmatrix}, \hat{\mathbf{w}}_{n+2} \right\}. \tag{5.108}$$

Remark 5.4 *Note that since the initial system of equations is assumed to be compatible, the new system is always compatible, because $x_{n+2} = 0$ leads to the old set of solutions.*

Example 5.17 (Adding one unknown). Consider the system of equations

$$\begin{array}{rrcl} x_1 & +x_2 & = & 2 \\ 2x_1 & -x_2 & = & 1 \\ 3x_1 & & = & 3 \end{array} \tag{5.109}$$

which, using the artificial variable $x_3 = 1$, can be written as

$$\begin{array}{rrrcl} x_1 & +x_2 & -2x_3 & = & 0 \\ 2x_1 & -x_2 & -x_3 & = & 0 \\ 3x_1 & & -3x_3 & = & 0 \\ & & x_3 & = & 1 \end{array} \tag{5.110}$$

Table 5.26 Tableau with the ϕ and ψ functions.

	Output		
$\psi(j)$	3	3	
$\phi(j)$	1	2	
$\mathbf{w}_j$	$\mathbf{w}_1$	$\mathbf{w}_2$	$\mathbf{w}_3$
	1/3 2/3 0	1/3 -1/3 0	1 1 1

Operating with Algorithm 5.2 (see Table 5.26) we get

$$\mathbf{W} \equiv \mathbb{R}^{n+1} \equiv \mathbb{R}^3 \equiv \mathcal{L}\left\{\begin{pmatrix} 1/3 \\ 2/3 \\ 0 \end{pmatrix}, \begin{pmatrix} 1/3 \\ -1/3 \\ 0 \end{pmatrix}, \begin{pmatrix} 1 \\ 1 \\ 1 \end{pmatrix}\right\}, \tag{5.111}$$

where the first two column vectors are the pivot columns associated with the first two equations in (5.110). Thus, the solution of (5.109) is

$$\begin{pmatrix} x_1 \\ x_2 \end{pmatrix} = \begin{pmatrix} 1 \\ 1 \end{pmatrix}. \tag{5.112}$$

Suppose now that we add a new unknown x_4 to the above system and get

$$\begin{array}{rrrrcl} x_1 & +x_2 & -2x_3 & +x_4 & = & 0 \\ 2x_1 & -x_2 & -x_3 & +x_4 & = & 0 \\ 3x_1 & & -3x_3 & +x_4 & = & 0 \end{array} \tag{5.113}$$

In order to update the solution we consider the vectors

$$\begin{pmatrix} \mathbf{w}_1 \\ 0 \end{pmatrix} = \begin{pmatrix} 1/3 \\ 2/3 \\ 0 \\ 0 \end{pmatrix}, \quad \begin{pmatrix} \mathbf{w}_2 \\ 0 \end{pmatrix} = \begin{pmatrix} 1/3 \\ -1/3 \\ 0 \\ 0 \end{pmatrix}, \quad \begin{pmatrix} \mathbf{w}_3 \\ 0 \end{pmatrix} = \begin{pmatrix} 1 \\ 1 \\ 1 \\ 0 \end{pmatrix},$$

and using expression (5.106), we have

$$\mathbf{w}_4 = \begin{pmatrix} 0 \\ 0 \\ 0 \\ 1 \end{pmatrix} - \begin{pmatrix} 1/3 \\ 2/3 \\ 0 \\ 0 \end{pmatrix} - \begin{pmatrix} 1/3 \\ -1/3 \\ 0 \\ 0 \end{pmatrix} = \begin{pmatrix} -2/3 \\ -1/3 \\ 0 \\ 1 \end{pmatrix}.$$

Thus, using the starting matrix

$$\mathbf{W}^* \equiv \begin{pmatrix} 1/3 & 1/3 & 1 & -2/3 \\ 2/3 & -1/3 & 1 & -1/3 \\ 0 & 0 & 1 & 0 \\ 0 & 0 & 0 & 1 \end{pmatrix}, \tag{5.114}$$

we find that $\mathbf{u}_3^T \bullet \mathbf{w}_4 = -1$, and then we perform the pivoting for the last column to get

$$\mathbf{W}^* \equiv \mathbb{R}^{n+2} \equiv \mathbb{R}^4 \equiv \mathcal{L}\left\{\left(\begin{array}{c} -1/3 \\ 1/3 \\ 0 \\ 1 \end{array}\right), \left(\begin{array}{c} -1/3 \\ -2/3 \\ 0 \\ 1 \end{array}\right), \left(\begin{array}{c} 1 \\ 1 \\ 1 \\ 0 \end{array}\right), \left(\begin{array}{c} 2/3 \\ 1/3 \\ 0 \\ -1 \end{array}\right)\right\}; \tag{5.115}$$

hence the new solution becomes

$$\left(\begin{array}{c} x_1 \\ x_2 \\ x_4 \end{array}\right) = \left(\begin{array}{c} 1 \\ 1 \\ 0 \end{array}\right). \tag{5.116}$$

Alternatively, suppose that we consider the system

$$\begin{array}{lllll} x_1 & +x_2 & -2x_3 & +x_4 & = 0 \\ 2x_1 & -x_2 & -x_3 & +x_4 & = 0 \\ 3x_1 & & -3x_3 & +2x_4 & = 0 \end{array} \tag{5.117}$$

We find that $\mathbf{u}_3^T \bullet \mathbf{w}_4 = 0$, so no extra pivot is necessary with the last equation. Then, considering matrix (5.114), the new solution becomes

$$\left(\begin{array}{c} x_1 \\ x_2 \\ x_4 \end{array}\right) = \left(\begin{array}{c} 1 \\ 1 \\ 0 \end{array}\right) + \rho \left(\begin{array}{c} -2/3 \\ -1/3 \\ 1 \end{array}\right), \tag{5.118}$$

for any $\rho \in \mathbb{R}$. ■

5.8 APPLICATIONS

In this section we illustrate the methods developed in the previous sections using four interesting applications:

1. Electric circuits.
2. Dynamics of particle motion.
3. Cubic splines.
4. Least squares methods.

5.8.1 Electric Circuits

Electric circuits or networks are built by interconnecting circuit elements. In this example we consider only batteries and resistors. Batteries connected in series have composite terminal voltages that depend on the manner in which

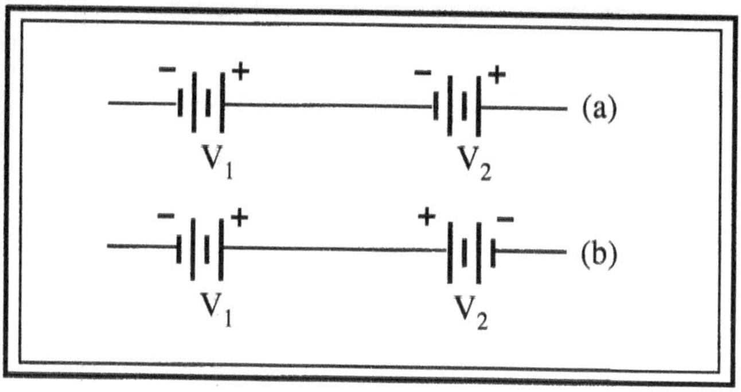

Fig. 5.1 Two different ways of connecting the same batteries in series.

the battery polarities are arranged. As an example, in Fig. 5.1(a) the terminal voltage is $V_1 + V_2$, while in Fig. 5.1(b) the terminal voltage is $V_1 - V_2$.

The problems associated with all electric circuits involving batteries and resistors can be solved using two important laws:

1. Ohm's law: When a resistor is connected to a battery, the relationship between the voltage V and the electric current of intensity I is

$$V = RI,$$

where R is the resistance of the resistor.

2. Kirchhoff's voltage law: The algebraic sum of voltages around a closed loop must equal zero. Note that some of these voltages may be positive and some may be negative.

Example 5.18 (Electric circuit). Consider the closed circuit in Fig. 5.2. According to the above laws, the currents in the three loops can be calculated by solving the following system of equations:

$$\begin{pmatrix} R_1+R_2+R_3+R_4 & -R_4 & 0 \\ -R_4 & R_4+R_5 & -R_5 \\ 0 & -R_5 & R_5+R_6 \end{pmatrix} \begin{pmatrix} I_1 \\ I_2 \\ I_3 \end{pmatrix} = \begin{pmatrix} -V_1+V_2 \\ V_3 \\ -V_2+V_4 \end{pmatrix}. \tag{5.119}$$

This system is always compatible, as can be seen using Algorithm 5.1. If now we remove resistors R_4 and R_6, system (5.119) becomes

$$\begin{pmatrix} R_1+R_2+R_3 & 0 & 0 \\ 0 & R_5 & -R_5 \\ 0 & -R_5 & R_5 \end{pmatrix} \begin{pmatrix} I_1 \\ I_2 \\ I_3 \end{pmatrix} = \begin{pmatrix} -V_1+V_2 \\ V_3 \\ -V_2+V_4 \end{pmatrix}. \tag{5.120}$$

Using Algorithm 5.1, the compatibility condition for system (5.120) becomes

$$-V_2 + V_3 + V_4 = 0.$$

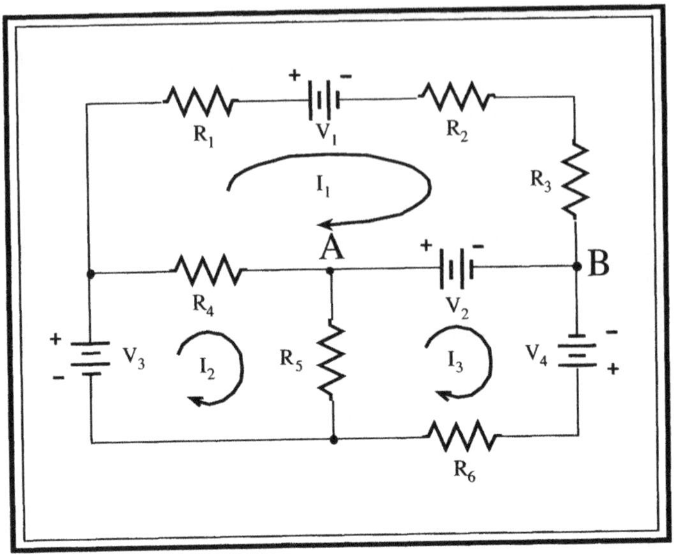

Fig. 5.2 Closed circuit with four batteries and six resistors.

Table 5.27 Steps for obtaining the electric currents corresponding to the circuit in Fig. 5.2.

	Iteration 1					**Iteration 2**			
7	**1**	0	0	0	-3	*1/7*	**3/7**	0	5/7
-3	**0**	1	0	0	9	*0*	**1**	0	0
0	**0**	0	1	0	-6	*0*	**0**	1	0
-5	**0**	0	0	1	-5	*0*	**0**	0	1
	7	-3	0	5		*-3/7*	**72/7**	-6	-50/7
	Iteration 3					**Output**			
0	*0.667*	*0.055*	**0.333**	1.11		*0.21*	*0.16*	*0.14*	0.76
-6	*0.055*	*0.12*	**0.777**	0.92		*0.16*	*0.38*	*0.333*	0.11
7	*0*	*0*	**1**	0		*0.14*	*0.333*	*0.42*	-1.04
8	*0*	*0*	**0**	1		*0*	*0*	*0*	1
	-0.51	*-0.77*	**2.333**	2.44					

This condition establishes that the sum of V_3 and V_4 voltages is equal to the V_2 voltage. This condition becomes apparent in Fig. 5.2 when we consider the potential difference between points A and B, and $R_4 = R_6 = 0$.

Table 5.28 Iterations for obtaining the currents of circuit in Fig. 5.2.

	Extra iteration				Output			
9	**0.21**	*0.16*	*0.14*	0.76	*0.14*	*0.11*	*0.10*	0.53
–3	**0.16**	*0.38*	*0.33*	0.11	*0.11*	*0.34*	*0.30*	–0.06
0	**0.14**	*0.33*	*0.42*	–1.04	*0.09*	*0.30*	*0.39*	–1.20
–5	**0.00**	*0.00*	*0.00*	1.00	0.00	*0.00*	*0.00*	*1.00*
	1.41	*0.30*	*0.26*	1.50				

Assume now that we are interested in solving the circuit associated with the following values for resistances and voltages:

$$\begin{array}{ll} R_1 = 2\Omega, & V_1 = 5 \text{ V}, \\ R_2 = 1.5\Omega, & V_2 = 10 \text{ V}, \\ R_3 = 0.5\Omega, & V_3 = 5 \text{ V}, \\ R_4 = 3\Omega, & V_4 = 2 \text{ V}, \\ R_5 = 6\Omega, & \\ R_6 = 1\Omega. & \end{array}$$

Then, the system to be solved is

$$\begin{array}{rcrrr} 5 & = & 7I_1 & -3I_2 & \\ 5 & = & -3I_1 & +9I_2 & -6I_3 \\ -8 & = & & -6I_2 & +7I_3 \end{array} \tag{5.121}$$

As can be seen in Table 5.27 the values of the corresponding currents are

$$\begin{array}{rcr} I_1 & = & 0.76 \text{ mA}, \\ I_2 & = & 0.11 \text{ mA}, \\ I_3 & = & -1.04 \text{ mA}. \end{array}$$

We observe that I_3 is negative. This means that the current I_3 circulates in the opposite direction.

Assume now that we replace $R_1 = 2\Omega$ by $R_1 = 4\Omega$. Then the system of equations (5.121) transforms to

$$\begin{array}{rcrrr} 5 & = & 9I_1 & -3I_2 & \\ 5 & = & -3I_1 & +9I_2 & -6I_3 \\ -8 & = & & -6I_2 & +7I_3 \end{array} \tag{5.122}$$

Using the process for modifying rows, the new solution can be obtained as shown in Table 5.28.

Thus, the new solution becomes

$$\begin{array}{rcr} I_1 & = & 0.53 \text{ mA}, \\ I_2 & = & -0.06 \text{ mA}, \\ I_3 & = & -1.2 \text{ mA}. \end{array}$$

Note that not only the current intensities have changed but one of their signs. ∎

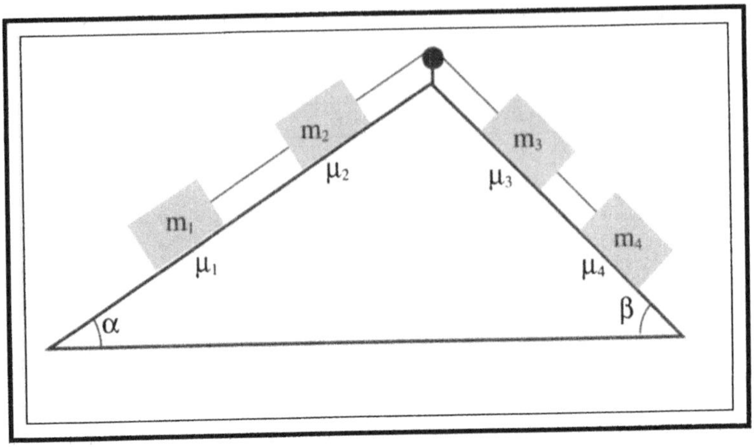

Fig. 5.3 Example of a mechanical system with four masses.

5.8.2 Dynamics of Particle Motion

Example 5.19 (Mechanical system). Consider a mechanical system composed of four masses m_1, m_2, m_3, and m_4 lying on two declining planes, connected by a string passing round a pulley at the top (see Fig. 5.3). Let μ_i be the friction coefficient associated with mass m_i, $i = 1, 2, 3, 4$.

According to the Newton's second law, the algebraic sum of forces acting on a body is the product of its mass by its acceleration. Fig. 5.4 shows the forces acting on each of the four bodies in the mechanical system, where F_i is the weight component in the movement direction, T_i is the string tension, and R_i is the friction force. Note that

$$F_i = \begin{cases} m_i g \sin\alpha, & \text{if } i = 1, 2, \\ m_i g \sin\beta, & \text{if } i = 3, 4, \end{cases}$$
$$R_i = \begin{cases} \mu_i m_i g \cos\alpha, & \text{if } i = 1, 2, \\ \mu_i m_i g \cos\beta, & \text{if } i = 3, 4. \end{cases}$$

Suppose that the system moves from left to right. The string tensions and the acceleration can be obtained by solving the following system of equations:

$$\begin{aligned} T_1 - F_1 - R_1 &= m_1 a \\ T_2 - T_1 - F_2 - R_2 &= m_2 a \\ T_3 + F_3 - T_2 - R_3 &= m_3 a \\ F_4 - T_3 - R_4 &= m_4 a \end{aligned} \qquad (5.123)$$

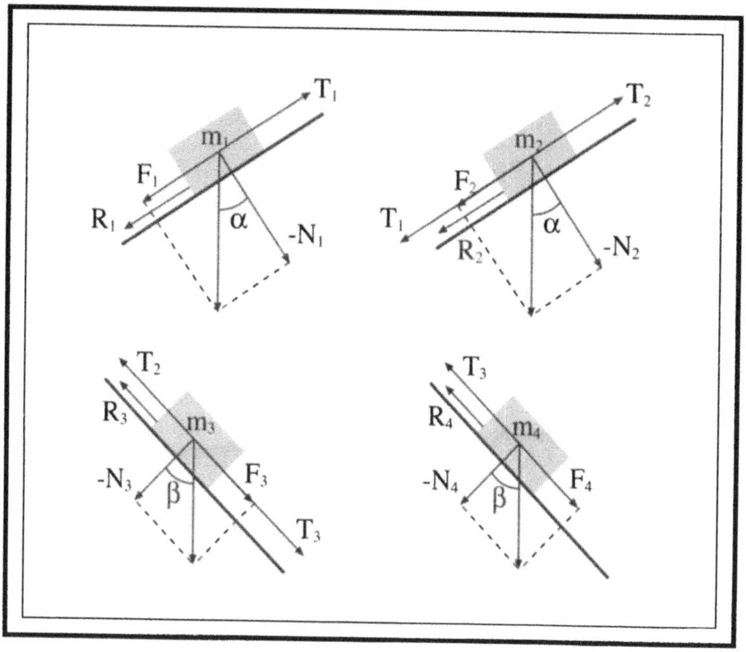

Fig. 5.4 Forces acting on each of the four masses.

or in its equivalent matrix form

$$\begin{pmatrix} -1 & 0 & 0 & m_1 \\ 1 & -1 & 0 & m_2 \\ 0 & 1 & -1 & m_3 \\ 0 & 0 & 1 & m_4 \end{pmatrix} \begin{pmatrix} T_1 \\ T_2 \\ T_3 \\ a \end{pmatrix} = \begin{pmatrix} -m_1 g(\sin\alpha + \mu_1 \cos\alpha) \\ -m_2 g(\sin\alpha + \mu_2 \cos\alpha) \\ m_3 g(\sin\beta - \mu_3 \cos\beta) \\ m_4 g(\sin\beta - \mu_4 \cos\beta) \end{pmatrix}$$
$$T_1, T_2, T_3, a \geq 0. \tag{5.124}$$

On the other hand, if the movement is from right to left, the system of equations (5.123) becomes

$$\begin{array}{lcl} F_1 - T_1 - R_1 & = & m_1 a \\ T_1 - T_2 + F_2 - R_2 & = & m_2 a \\ T_2 - F_3 - T_3 - R_3 & = & m_3 a \\ T_3 - F_4 - R_4 & = & m_4 a \end{array} \tag{5.125}$$

■

5.8.3 Cubic Splines

In this section we analyze an application of a linear system of equations to computer aided design (CAD). We use the methods described in this chapter to work with cubic splines. Consider n points $P_i = (x_i, y_i)$, $i = 1, \ldots, n$ such that $x_i < x_{i+1}$. A cubic spline is a piecewise polynomial curve, that is, a curve $s(x)$ passing through them defined by $n-1$ polynomial functions of the form

$$p_i(x) = a_i + b_i x + c_i x^2 + d_i x^3; \;\; \forall x \in [x_i, x_{i+1}],\; i = 1, \ldots, n-1 \qquad (5.126)$$

and such that

$$\begin{array}{rcll} p_{i-1}(x_i) & = & p_i(x_i), & i = 2, \ldots, n-1, \qquad (5.127)\\ p'_{i-1}(x_i) & = & p'_i(x_i), & i = 2, \ldots, n-1, \qquad (5.128)\\ p''_{i-1}(x_i) & = & p''_i(x_i), & i = 2, \ldots, n-1. \qquad (5.129) \end{array}$$

To obtain a cubic spline the $4n-4$ parameters in (5.126) have to be calculated, using the $4n-6$ equations: conditions (5.127), (5.128), and (5.129), together with the interpolation conditions:

$$p_i(x_i) = y_i, \;\; i = 1, \ldots, n-1 \;\; \text{and} \;\; p_{n-1}(x_n) = y_n. \qquad (5.130)$$

Then we have only two degrees of freedom. Thus, to obtain a unique solution, two extra conditions are required, as, for example, giving the values $p'_1(x_1) = z_1$ and $p'_{n-1}(x_n) = z_n$ or $p''_1(x_1) = p''_{n-1}(x_n) = 0$. In this last case, the spline is said to be natural.

Example 5.20 (Splines). Suppose that we want to find the natural spline passing through the points

$$P_1 = (1, 2), P_2 = (2, 1/2), P_3 = (4, 3), P_4 = (6, 2).$$

Then we have to solve the system of equations $\mathbf{Ax} = \mathbf{b}$, where

$$\mathbf{A} = \begin{pmatrix} 1 & -1 & 0 & 2 & -2 & 0 & 4 & -4 & 0 & 8 & -8 & 0 \\ 0 & 1 & -1 & 0 & 4 & -4 & 0 & 16 & -16 & 0 & 64 & -64 \\ 0 & 0 & 0 & 1 & -1 & 0 & 4 & -4 & 0 & 12 & -12 & 0 \\ 0 & 0 & 0 & 0 & 1 & -1 & 0 & 8 & -8 & 0 & 48 & -48 \\ 0 & 0 & 0 & 0 & 0 & 0 & 2 & -2 & 0 & 12 & -12 & 0 \\ 0 & 0 & 0 & 0 & 0 & 0 & 0 & 2 & -2 & 0 & 24 & -24 \\ 1 & 0 & 0 & 1 & 0 & 0 & 1 & 0 & 0 & 1 & 0 & 0 \\ 0 & 1 & 0 & 0 & 2 & 0 & 0 & 4 & 0 & 0 & 8 & 0 \\ 0 & 0 & 1 & 0 & 0 & 4 & 0 & 0 & 16 & 0 & 0 & 64 \\ 0 & 0 & 1 & 0 & 0 & 6 & 0 & 0 & 36 & 0 & 0 & 216 \\ 0 & 0 & 0 & 0 & 0 & 0 & 2 & 0 & 0 & 6 & 0 & 0 \\ 0 & 0 & 0 & 0 & 0 & 0 & 0 & 0 & 2 & 0 & 0 & 36 \end{pmatrix},$$

$$\begin{array}{rcl} \mathbf{x} & = & (a_1, a_2, a_3, b_1, b_2, b_3, c_1, c_2, c_3, d_1, d_2, d_3)^T, \\ \mathbf{b} & = & (0, 0, 0, 0, 0, 0, 2, 1/2, 3, 2, 0, 0)^T. \end{array}$$

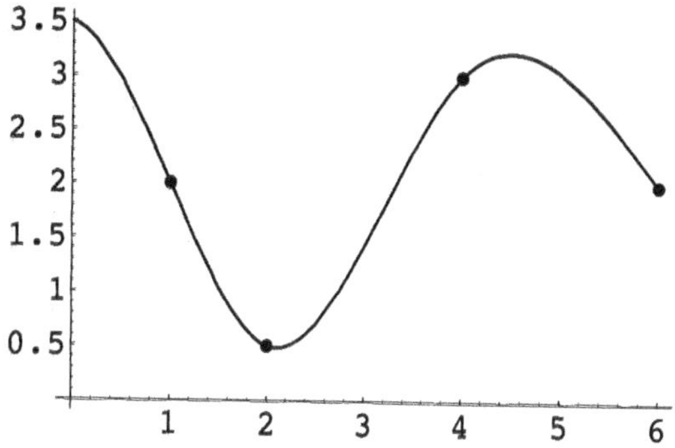

Fig. 5.5 Example of a cubic spline.

The unique solution of the above system is

$$
\begin{aligned}
a_1 &= 3.5, & a_2 &= 11.9, & a_3 &= -29.9,\\
b_1 &= -0.34, & b_2 &= -12.95, & b_3 &= 18.4,\\
c_1 &= -1.74, & c_2 &= 4.57, & c_3 &= -3.27,\\
d_1 &= 0.58, & d_2 &= -0.47, & d_3 &= 0.182.
\end{aligned}
$$

Fig. 5.5 shows the given points together with the cubic spline passing through them.

■

The method used above to calculate a cubic spline can be greatly improved as follows. Consider the linear space

$$S_3 = \{s(x) \in C^{(2)}[x_1, x_n] \mid s(x)|_{[x_i, x_{i+1}]} \in \mathcal{P}_3(\mathbb{R}), i = 1, \ldots, n-1\}, \quad (5.131)$$

where $C^{(2)}[x_1, x_n]$ is the set of real functions that are continuous and have first and second continuous derivatives in the interval $[x_1, x_n]$, that is, the space of functions defined on $[x_1, x_n]$ by $n-1$ polynomial segments of degree 3 and with continuity up to the second derivative. It can be shown that

$$\{1, x, x^2, x^3, (x-x_2)_+^3, \ldots, (x-x_{n-1})_+^3\}, \quad (5.132)$$

where the subindex + indicates that the positive part is a basis of S_3. Then a cubic spline is determined by $n+2$ coefficients:

$$s(x) = a + bx + cx^2 + dx^3 + e_2(x-x_2)_+^3 + \cdots + e_{n-1}(x-x_{n-1})_+^3. \quad (5.133)$$

These coefficients can be calculated by solving the following system of equations:

$$\begin{pmatrix} 1 & x_1 & x_1^2 & x_1^3 & 0 & \cdots & 0 \\ 1 & x_2 & x_2^2 & x_2^3 & 0 & \cdots & 0 \\ 1 & x_3 & x_3^2 & x_3^3 & (x_3-x_2)^3 & \cdots & 0 \\ \cdots & \cdots & \cdots & \cdots & \cdots & \cdots & \cdots \\ 1 & x_n & x_n^2 & x_n^3 & (x_n-x_2)^3 & \cdots & (x_n-x_{n-1})^3 \\ 0 & 0 & 2 & 6x_1 & 0 & \cdots & 0 \\ 0 & 0 & 2 & 6x_n & 6(x_n-x_2) & \cdots & 6(x_n-x_{n-1}) \end{pmatrix} \mathbf{z} = \begin{pmatrix} y_1 \\ y_2 \\ \vdots \\ y_n \\ 0 \\ 0 \end{pmatrix}, \tag{5.134}$$

where

$$\mathbf{z} = (a, b, c, d, e_2, \ldots, e_{n-1})^T.$$

The two last conditions in system (5.134) imply $s''(x_1) = s''(x_n) = 0$.

As can be shown, when a single P_i is modified, a single equation in system (5.134) has to be modified and the method described in this chapter for updating the solutions can be applied.

Example 5.21 (Cubic natural spline). Consider again the points in Example 5.20. The cubic natural spline passing through them can be calculated by solving the following system of equations:

$$\begin{pmatrix} 1 & 1 & 1 & 1 & 0 & 0 \\ 1 & 2 & 4 & 8 & 0 & 0 \\ 1 & 4 & 16 & 64 & 8 & 0 \\ 1 & 6 & 36 & 216 & 64 & 8 \\ 0 & 0 & 2 & 6 & 0 & 0 \\ 0 & 0 & 2 & 36 & 24 & 12 \end{pmatrix} \begin{pmatrix} a \\ b \\ c \\ d \\ e_2 \\ e_3 \end{pmatrix} = \begin{pmatrix} 2 \\ 0.5 \\ 3 \\ 2 \\ 0 \\ 0 \end{pmatrix}. \tag{5.135}$$

The solution of this system is (see Table 5.29):

$$\begin{array}{llllll} a &=& 3.5, & b &=& -0.34, & c &=& -1.74, \\ d &=& 0.58, & e_2 &=& -1.05, & e_3 &=& 0.65. \end{array}$$

■

5.8.4 Least Squares Methods

It is common to measure pairs of values of variables, which are associated in some way, and then try to fit some curve to these points. For example, if we have n points (α_i, β_i), $i = 1, \ldots, n$, a convenient curve might be a polynomial expression of the form

$$p(x) = a_0 + a_1 x + a_2 x^2 + \cdots + a_m x^m. \tag{5.136}$$

If $n = m + 1$ and the α's are all different, then there is a unique curve that passes through all the points. It can be found by solving the following linear

Table 5.29 Obtaining the solution of system (5.135).

Solution						
2.000	-1.000	0.000	0.000	1.000	0.000	3.500
-0.640	0.410	0.270	-0.045	-1.890	0.030	-0.340
-0.550	0.890	-0.410	0.068	1.090	-0.045	-1.740
0.180	-0.300	0.136	-0.023	-0.170	0.015	0.580
-0.295	0.510	-0.284	0.068	0.216	-0.045	-1.050
0.136	-0.284	0.227	-0.079	-0.022	0.136	0.653
0.000	0.000	0.000	0.000	0.000	0.000	1.000

system of equations:

$$\begin{pmatrix} 1 & \alpha_1 & \alpha_1^2 & \dots & \alpha_1^n \\ 1 & \alpha_2 & \alpha_2^2 & \dots & \alpha_2^n \\ \dots & \dots & \dots & \dots & \dots \\ 1 & \alpha_n & \alpha_n^2 & \dots & \alpha_n^n \end{pmatrix} \begin{pmatrix} a_0 \\ a_1 \\ \dots \\ a_m \end{pmatrix} = \begin{pmatrix} \beta_1 \\ \beta_2 \\ \dots \\ \beta_n \end{pmatrix}. \tag{5.137}$$

However, if $m < n$, then, in general, not all the given points can lie on the curve; that is, the system (5.137) is incompatible. In this case we can find the "best" curve fitting the given points using the least squares method. If we assume that the α_i are all different and we define the error e_i, at each point, as the difference between the $p(\alpha_i)$ and β_i values, we can choose a_i, $i = 0, 1, \dots, m$, so that the total squared error is as small as possible. Then we must minimize the function

$$S(a_0, a_1, \dots, a_m) = \sum_{i=1}^{n} (p(\alpha_i) - \beta_i)^2. \tag{5.138}$$

For S to be a minimum, each of its partial derivatives must be zero, that is,

$$\begin{aligned} \frac{\partial S(a_0, a_1, \dots, a_m)}{\partial a_0} &= \sum_{i=1}^{n} 2(p(\alpha_i) - \beta_i) = 0, \\ \frac{\partial S(a_0, a_1, \dots, a_m)}{\partial a_1} &= \sum_{i=1}^{n} 2(p(\alpha_i) - \beta_i)\alpha_i = 0, \\ \dots\dots\dots\dots\dots & \quad \dots\dots\dots\dots\dots\dots \\ \frac{\partial S(a_0, a_1, \dots, a_m)}{\partial a_m} &= \sum_{i=1}^{n} 2(p(\alpha_i) - \beta_i)\alpha_i^m = 0. \end{aligned} \tag{5.139}$$

The system above can be written in the form

Table 5.30 Data to be fitted.

i	α_i	β_i
1	0.5	-1.95
2	1.0	-0.90
3	1.5	0.59
4	2.0	-0.93
5	2.5	-0.30
6	3.0	-2.42
7	3.5	-1.89
8	4.0	0.38

$$\begin{pmatrix} \sum\alpha_i^0 & \sum\alpha_i^1 & \dots & \sum\alpha_i^m \\ \sum\alpha_i^1 & \sum\alpha_i^2 & \dots & \sum\alpha_i^{m+1} \\ \dots & \dots & \dots & \dots \\ \sum\alpha_i^m & \sum\alpha_i^{m+1} & \dots & \sum\alpha_i^{m+m} \end{pmatrix} \begin{pmatrix} a_0 \\ a_1 \\ \dots \\ a_m \end{pmatrix} = \begin{pmatrix} \sum\beta_i \\ \sum\alpha_i\beta_i \\ \dots \\ \sum\alpha_i^m\beta_i \end{pmatrix}. \tag{5.140}$$

If we use the notation

$$\mathbf{A}^T = \begin{pmatrix} 1 & 1 & \dots & 1 \\ \alpha_1 & \alpha_2 & \dots & \alpha_n \\ \dots & \dots & \dots & \dots \\ \alpha_1^m & \alpha_2^m & \dots & \alpha_n^m \end{pmatrix}, \quad \mathbf{a} = \begin{pmatrix} a_0 \\ a_1 \\ \dots \\ a_m \end{pmatrix}, \quad \mathbf{b} = \begin{pmatrix} \beta_1 \\ \beta_2 \\ \dots \\ \beta_n \end{pmatrix},$$

system (5.139) can be written as

$$\mathbf{A}^T\mathbf{A}\mathbf{a} = \mathbf{A}^T\mathbf{b}. \tag{5.141}$$

If $\alpha_i \neq \alpha_j$ for all i, j, $rank(\mathbf{A}) = rank(\mathbf{A}^T\mathbf{A}) = m < n$ and (5.141) has a unique solution.

Example 5.22 (Least squares method). Suppose that we want to fit the points in Table 5.30 using the third degree polynomial

$$p(x) = a_0 + a_1x + a_2x^2 + a_3x^3.$$

According to (5.141), we have to solve the system

$$\begin{pmatrix} 8.00 & 18.00 & 51.00 & 162.00 \\ 18.00 & 51.00 & 162.00 & 548.25 \\ 51.00 & 162.00 & 548.25 & 1930.50 \\ 162.00 & 548.25 & 1930.50 & 6983.81 \end{pmatrix} \begin{pmatrix} a_0 \\ a_1 \\ a_2 \\ a_3 \end{pmatrix} = \begin{pmatrix} -7.4262 \\ -15.9646 \\ -44.5275 \\ -133.3850 \end{pmatrix}.$$

Its unique solution is

$$\begin{pmatrix} a_0 \\ a_1 \\ a_2 \\ a_3 \end{pmatrix} = \begin{pmatrix} -6.332800 \\ 10.816500 \\ -5.564370 \\ 0.816803 \end{pmatrix}.$$

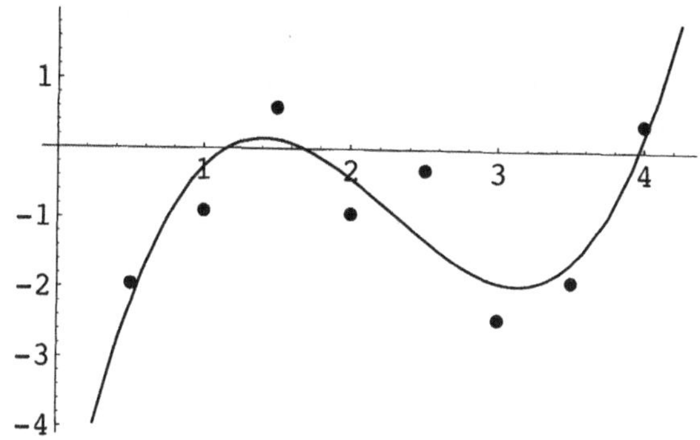

Fig. 5.6 Least squares function.

Fig. 5.6 shows the given points together with the least-squares resulting third degree polynomial.

■

5.9 MATHEMATICA PROGRAMS

The following Mathematica program uses the Orthogonal Module in Chapter 2 to analyze the compatibility of a system of linear equations.

Program 5.1 (Analyzing the compatibility of a linear system)

```
(* Compatibility of a linear system of equations *)
CompatibleQ[Coefficients_,IndependentTerm_]:=
  Module[{aux},
    aux=Orthogonal[Transpose[Coefficients]];
    If[aux={},aux1=aux.IndependentTerm;aux1.aux1==0,
             Return[True]]
  ]
```

Example 5.23 (Compatibility with Mathematica). To check if the linear system of equations in Example 5.1 is compatible for $(a, b, c, d, e) = (1, 0, 0, 1, 2)$, we use the Mathematica command

```
In[]:= CompatibleQ[{{1,2,0,1,-4},{2,1,-1,0,-1},{1,2,1,0,1},
                   {5,2,-3,1,1},{1,1,-1,1,-5}},{1,0,0,1,2}]
Out[]= False
```

■

The following Mathematica program uses the Orthogonal Module in Chapter 2 to solve a system of linear equations.

Program 5.2 (Solving a linear system of equations)

```
(* Solving a system of linear equations *)
SolveSystem[Coefficients_,IndependentTerm_]:=Module[{A,B},
  A=Transpose[Append[Transpose[Coefficients],-IndependentTerm]];
  B=Orthogonal[A,SystemofEquations->True];
  If[B,={},Transpose[Drop[Transpose[B],-1]],B]
]
```

Example 5.24 (Solving a system of linear equations with Mathematica). The solution of Example 5.5 can be obtained with the following Mathematica command:

```
In[]:= SolveSystem[{{1,2,0,1,-1},{2,1,-1,0,3},{-1,0,2,1,0}},
                   {4,0,-1}]

                1    1           7  9   7           5  13   7
Out[]= {{0,-(-),-(-),1,0},{-(-),-,-(-),0,1},{-(-),--,-(-),0,0}}
                2    2           2  4   4           2  4    4
```

■

The following Mathematica program uses the Orthogonal Module in Chapter 2 to check whether or not two systems of linear equations are equivalent.

Program 5.3 (Equivalence of two systems of equations)

```
(* Equivalence of two linear systems of equations *)
EquivalentQ[Coefficients1_,IndependTerm1_,
           Coefficients2_,IndependTerm2_]:=Module[{A,B,aux},
  A=Transpose[Append[Transpose[Coefficients1],-IndependTerm1]];
  B=Transpose[Append[Transpose[Coefficients2],-IndependTerm2]];
  aux=Flatten[{Orthogonal[B].Transpose[A],
               Orthogonal[A].Transpose[B]}];
  aux.aux==0
]
```

Example 5.25 (Equivalent systems with Mathematica). Example 5.7 can be solved using the command EquivalentQ defined in Program 3.1, which confirms that the two systems are equivalent:

```
In[]:= EquivalentQ[{{1,-1,2},{3,2,-4}},{3,6},
                   {{2,-7,14},{2,3,-6}},{9,3}]
Out[]= True
```

■

The following Mathematica program uses the Orthogonal Module in Chapter 2 to solve a system of linear equations in some selected variables.

Program 5.4 (Solving a system in some selected variables)

```
(* Solving a linear system of equations in selected variables *)
SolveSystemSelectedVar[Coefficients_,IndependTerm_,var_]:=
  Module[{A,B,C,H,n=Length[Coefficients[[1]]],var1=var},
    A=Append[Transpose[Coefficients],-IndependTerm];
    AppendTo[var1,n+1];
    B=A[[var1]];
    C=-A[[Complement[Range[n],var1]]];
    H=Orthogonal[Orthogonal[C].Transpose[B]];
    Transpose[Drop[Transpose[H],-1]]
]
```

Example 5.26 (Solving a linear system in selected variables with Mathematica). The solution of Example 5.8 can be obtained with the command `SolveSystemSelectedVar` defined in Program 5.4:

```
In[]:= SolveSystemSelectedVar[{{-1,1,1,2,-3,1,-1},
        {1,-1,0,-1,1,0,1},{0,0,1,0,-1,1,2},
        {0,0,-1,1,0,-1,-3}},{-2,-1,-1,-1},{1,2,3}]
Out[]= {{1, 1, 0}, {0, 0, 1}, {-3, 0, 0}}
```

■

Exercises

5.1 Find all polynomials of degree less than or equal to 5 passing through the points

$$P_1 = (5,1),\ P_2 = (6,-1),\ \text{and } P_3 = (2,3).$$

Select from the set above the unique polynomial passing through the points

$$P_4 = (2,1),\ P_5 = (3,-4),\ \text{and } P_6 = (-1,1).$$

If we remove P_4, update the above polynomial set.

5.2 Table 5.31 shows the numbers of new students accepted in four different universities and the numbers of students that will study mathematics (M), philosophy (P), economics (E), literature (L), and architecture (A).

(a) Suppose that the administrator of University 1 is interested in knowing all possible distributions of students in his University, (solving in the appropriately selected variables). Determine these distributions.

(b) Calculate all possible distributions of students in the different universities.

Table 5.31

	M	P	E	L	A	Total
University 1	x_1	x_2	x_3	x_4	x_5	520
University 2	x_6	x_7	x_8	x_9	x_{10}	660
University 3	x_{11}	x_{12}	x_{13}	x_{14}	x_{15}	700
University 4	x_{16}	x_{17}	x_{18}	x_{19}	x_{20}	590
Total	240	510	1100	350	270	

Table 5.32

	M	P	E	L	A	Total
University 1	50	84	x_1	210	94	x_{11}
University 2	72	67	x_2	190	89	x_{12}
University 3	x_6	x_7	x_3	x_8	x_9	x_{10}
University 4	32	60	x_4	175	65	x_{13}
University 5	44	70	x_5	145	70	x_{14}
Total	x_{15}	x_{16}	x_{17}	x_{18}	x_{19}	x_{20}

5.3 Consider now the data in Table 5.32 and answer the same questions posed in the previous exercise.

5.4 Discuss the compatibility of the following system of equations as a function of the values of the parameters a and b:

$$\begin{array}{rrrrcl} x_1 & +x_2 & & +x_4 & = & 0 \\ & -3x_2 & +x_3 & +2x_4 & = & b-a \\ 2x_1 & +3x_2 & & -x_4 & = & a \\ x_1 & +x_2 & -x_3 & -x_4 & = & b \end{array}$$

5.5 Using the least squares method, fit a polynomial of degree 2, that is, $f(x) = ax^2 + bx + c$, to the points

$$P_1 = (2,3),\ P_2 = (-3,-1),\ P_3 = (1,3),\ \text{and}\ P_4 = (0,2).$$

5.6 Consider Fig. 5.7 and calculate the T_1, T_2, and T_3 forces associated with the string tensions, and the acceleration of the system.

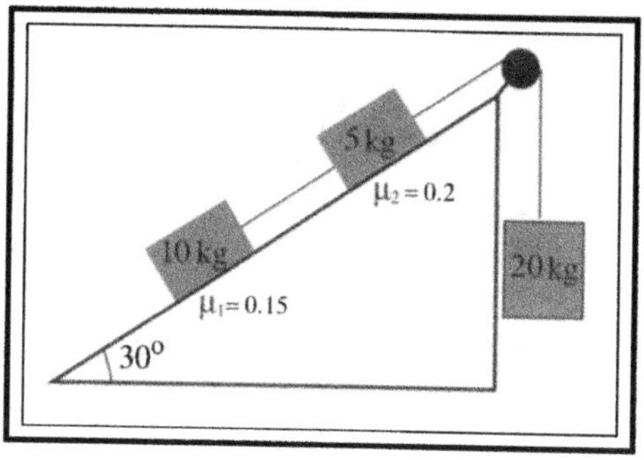

Fig. 5.7 Dynamic system.

5.7 Calculate the electric currents of the circuit in Fig. 5.8, assuming that the values associated with resistors and voltages are:

$$
\begin{aligned}
R_1 &= 2\Omega, & V_1 &= 10 \text{ V} \\
R_2 &= 4\Omega, & V_2 &= 5 \text{ V} \\
R_3 &= 2\Omega, & V_3 &= 8 \text{ V} \\
R_4 &= 1\Omega, & V_4 &= 15 \text{ V} \\
R_5 &= 4\Omega, & V_5 &= 9 \text{ V} \\
R_6 &= 3\Omega, & & \\
R_7 &= 3\Omega. & &
\end{aligned}
$$

5.8 Find which of the three following systems are equivalent.

$$
\left\{ \begin{array}{rrrcr} 4x_1 & +x_2 & -2x_3 & = & 9 \\ 2x_1 & +3x_2 & 6x_3 & = & 3 \end{array} \right.
$$

$$
\left\{ \begin{array}{rrrcr} 2x_1 & -x_2 & & = & -1 \\ x_1 & -2x_2 & -x_3 & = & 3 \end{array} \right.
$$

$$
\left\{ \begin{array}{rrrcr} \frac{1}{2}x_1 & -\frac{1}{2}x_2 & +x_3 & = & \frac{3}{2} \\ & -x_2 & -10x_3 & = & 0 \end{array} \right.
$$

5.9 Obtain the solution of the following system of equations:

$$
\begin{pmatrix} 1 & -2 & -1 & -2 & 3 \\ 0 & -1 & 0 & -3 & 1 \\ -1 & 1 & 1 & -1 & -2 \\ 0 & -2 & 0 & -6 & 2 \\ -1 & -1 & 1 & -7 & 0 \\ -2 & -2 & 1 & 1 & 0 \\ -2 & -5 & -2 & -1 & 1 \end{pmatrix} \begin{pmatrix} x_1 \\ x_2 \\ x_3 \\ x_4 \\ x_5 \end{pmatrix} = \begin{pmatrix} 1 \\ 2 \\ 1 \\ 4 \\ 5 \\ -2 \\ 1 \end{pmatrix}.
$$

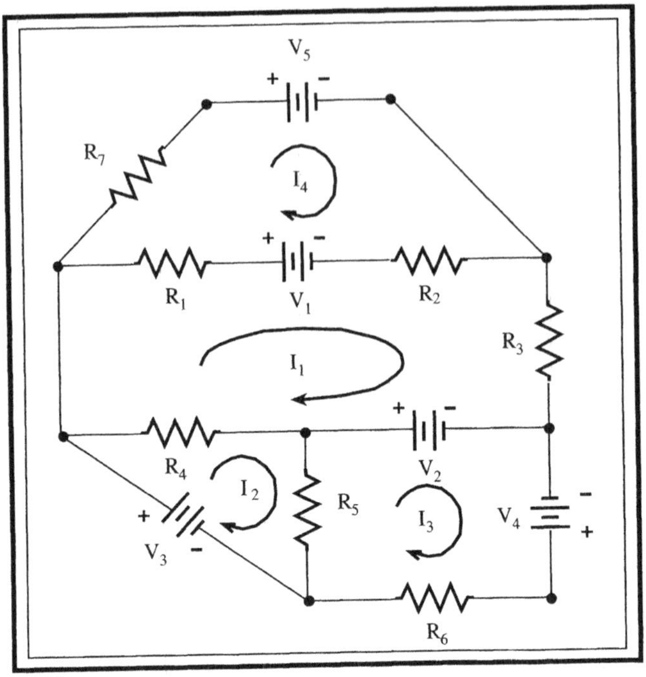

Fig. 5.8 An electric circuit.

Using the solution of the previous system, solve the system in the following cases:

(a) After adding the equation $2x_1 - 3x_2 + 2x_5 = 2$.

(b) After removing equation 3.

(c) After removing equation 6.

(d) After sequentially removing equations $1, 3, 7$, and 6.

Appendix: Proof of Lemma 5.2

The proof is obvious in cases 1 and 3 and for the ϕ and $\mathbf{w}_j$ in case 2, but it is a lot more involved for the ψ functions in case 2. In case 2 we have $\exists \ell \in \{1, \ldots, n\} | \phi(\ell) = k \wedge \psi(\ell) \neq \emptyset$. We initially consider three cases depending on the values of j and r:

1. If $j = \ell$ we simply eliminate r from $\psi(j)$, that is,

$$\psi^*(j) = \psi(j) - \{r\}. \tag{A.1}$$

2. If $j \neq \ell$ and $r \notin \psi(j)$, removing the equation does not affect the $\mathbf{w}_j$ column and then
$$\psi^*(j) = \psi(j). \tag{A.2}$$

3. If $j \neq \ell$ and $r \in \psi(j)$ things are much more complex. In the following, we determine whether or not a given $v \in \{1, \ldots, m\} - \{k\}$ belongs to $\psi^*(j)$. To that aim we remind the reader that, according to (5.87), we have
$$v \in \psi^*(j) \Leftrightarrow \left\{ \begin{array}{lcl} \mathbf{u}_v^T \bullet \mathbf{w}_j^* & = & \mathbf{u}_v^T \bullet (\mathbf{w}_j - t_r^j \mathbf{w}_\ell / t_r^\ell) = t_v^j - t_r^j t_v^\ell / t_r^\ell \neq 0, \\ \phi^*(j) & \neq & v. \end{array} \right. \tag{A.3}$$

We have the following four cases:

(a) $v \notin \psi(j) \cup \psi(\ell)$. In this case we have
$$\left. \begin{array}{lcl} v \notin \psi(\ell) & & \\ v \neq k & \Rightarrow & \phi(\ell) = k \neq v \end{array} \right\} \Rightarrow t_v^\ell = 0, \tag{A.4}$$
and using (A.3) we get
$$t_v^{*j} = t_v^j.$$
In addition, we have
$$v \notin \psi(j) \Rightarrow \left\{ \begin{array}{lcl} t_v^j - 0 & \Rightarrow & t_v^{*j} = 0 \quad \Rightarrow v \notin \psi^*(j) \\ \text{or} & & \\ \phi(j) = v & \Rightarrow & \phi^*(j) = v \quad \Rightarrow v \notin \psi^*(j) \end{array} \right. \tag{A.5}$$

(b) $v \in \psi(\ell) - \psi(j)$. In this case we obtain
$$v \in \psi(\ell) \Rightarrow t_v^\ell \neq 0 \wedge \phi(\ell) \neq v \tag{A.6}$$
and
$$v \notin \psi(j) \Rightarrow t_v^j = 0 \text{ or } \phi(j) = v. \tag{A.7}$$
If $t_v^j = 0$, since $t_v^\ell \neq 0$ and $t_r^j \neq 0$, we have $t_v^{*j} = -t_r^j t_v^\ell / t_r^\ell \neq 0$. In addition, $\phi^*(j) = \phi(j) \neq v$, because if $\phi(j) = v$, v is a basic equation and then $v \notin \psi(\ell)$, in contradiction with the above assumption $v \in \psi(\ell)$. Thus, $v \in \psi^*(j)$.

(c) $v \in \psi(j) - \psi(\ell)$. In this case we have
$$v \notin \psi(\ell) \text{ and } \phi(\ell) = k \neq v \Rightarrow t_v^\ell = 0 \Rightarrow t_v^{*j} = t_v^j; \tag{A.8}$$
and then we can write
$$v \in \psi(j) \Rightarrow \left\{ \begin{array}{lcl} t_v^j \neq 0 & \Rightarrow & t_v^{*j} = t_v^j \neq 0, \\ \phi(j) \neq v & \Rightarrow & \phi^*(j) = \left\{ \begin{array}{ll} \phi(j) \neq v & \text{if } \ j \neq \ell, \\ r \neq v & \text{if } \ j = \ell. \end{array} \right. \end{array} \right. \tag{A.9}$$

Thus, we get $v \in \psi^*(j)$.

(d) $v \in \psi(j) \cap \psi(\ell)$. In this case we have

$$v \in \psi(j) \Rightarrow t_v^j \neq 0 \wedge \phi(j) \neq v, \tag{A.10}$$

$$v \in \psi(\ell) \Rightarrow t_v^\ell = 0 \text{ or } \phi(\ell) \neq v, \tag{A.11}$$

and

$$\phi^*(j) = \begin{cases} \phi(j) \neq v, & \text{if } j \neq \ell \\ r, & \text{if } j = \ell \end{cases} \tag{A.12}$$

and

$$t_v^{*j} = t_v^j - t_r^j t_v^\ell / t_r^\ell \neq 0 \Leftrightarrow t_v^j t_r^\ell \neq t_r^j t_v^\ell. \tag{A.13}$$

In particular, $t_v^{*j} \neq 0$ implies $v \neq r$.

Thus, we get

$$\psi^*(j) \equiv A_{\ell j}, \tag{A.14}$$

where $A_{\ell j}$ is given in (5.95). ∎

Part II

Cones and Systems of Inequalities

Introduction to Part II

Linear systems of equations allow solving interesting problems of real life. However, many real problems lead not to linear equations but to linear inequalities. Unfortunately, systems of linear inequalities, in spite of their importance, are not well known. In this part of the book we analyze the problem of systems of inequalities, discussing the problems of compatibility and uniqueness, and giving some methods for obtaining the solution set, when it exists.

Inequalities are closely related to the concepts of cone, polytope, and polyhedron. Thus, we dedicate Chapters 6 and 7 to these concepts and the main theorems to be used later. The most important of these concepts is that of the polyhedral convex cone, which is the fundamental structure to be used in this part of the book. We also introduce several types of linear combinations which appear in the theoretical developments. Among them are those with non-negative coefficients and the linear convex combinations.

We also show that a linear space of finite dimension is nothing but a particular case of a cone. Thus, the cone concept is broader than the linear space concept and, more importantly, allows us to replace the standard methods of classical linear algebra with the specific methods of cones, which are more efficient.

The concept of a polyhedron will arise in a natural form as a subset of a polyhedral convex cone and the concept of the binding set of a cone will allow us to define interior, exterior, and boundary of a cone.

Finally, we shall introduce the concept of dual cone, leading to a very important theorem by Farkas, which is stated in a new form that allows one to consider different problems from a new and simpler point of view.

In the introduction to Part I, we discussed the impact of linear systems of equations on economics and mentioned problems dealing with systems of linear equations of large size. Leontief (twentieth century), with his spectral analysis, which is known as "Input–output" analysis, concluded with systems of equations of a similar format, with the peculiarity that now the economic variables must be positive. This gives rise to the concept of systems of equations with restricted variables and is the starting point for mathematical programming, an area of continuous research and development.

In Chapter 8 we deal with systems of equations and inequalities. We start by analyzing the problems of compatibility and uniqueness of solutions. These are two key conditions needed to have physically well–defined problems, and they can be exploited to perform a deep analysis of the physical realities existing behind mathematical models. It is shown that the polar or dual cone allows one to solve these two problems in an easy way.

We also provide dual methods to solve systems of equations and inequalities, and once again the polar cone gives the answer.

Our analysis of linear inequalities in this part of the book also includes the case of strictly restricted variables, a problem that has been studied by outstanding mathematicians, such as Khun, Fourier, and Motzkin, though in a nonoptimized form. Here the cone concept clearly elucidates the depth of these problems.

6

Polyhedral Convex Cones

6.1 INTRODUCTION

Cones, polytopes, and polyhedra are essential concepts for understanding the problems associated with systems of inequalities. In the following two chapters we give the most important concepts and tools required for working with inequalities. This prepares the reader to understand the powerful methods to be developed in Chapter 8.

This chapter is devoted to cones. In Section 6.2 we deal with convex sets, giving a definition and some properties. In Section 6.3 we define several types of linear combinations that are needed in the rest of the book. Section 6.4 introduces the concept of polyhedral convex cone and shows that a linear space is a particular case of a cone. This allows us to replace the classical methods of linear algebra with the powerful methods of cones. We also define the general form of a cone and introduce the polar cone and other related concepts, which are the bases for the methods to be developed later. However, the star in this chapter is the powerful Γ-algorithm, which allows obtaining the polar or dual cone and is the most important tool to solve systems of inequalities.

6.2 CONVEX SETS

Convex sets are the natural and simplest type of sets arising from systems of linear inequalities.

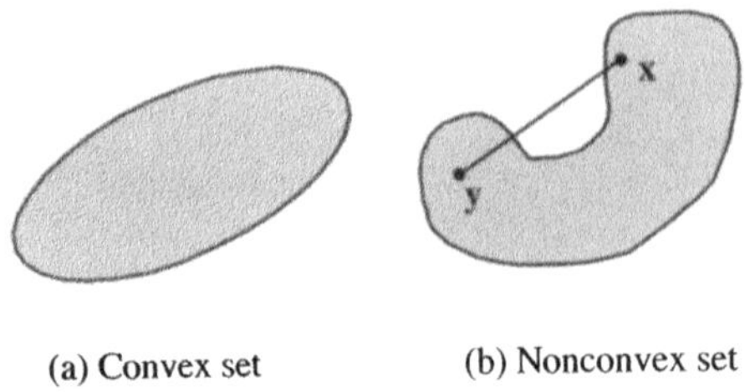

Fig. 6.1 Illustration of convex and nonconvex sets in E^2.

Definition 6.1 (Convex set). *A nonempty set S in E^n is said to be convex iff*

$$\lambda \mathbf{x} + (1 - \lambda)\mathbf{y} \in S \tag{6.1}$$

for every $\lambda \in [0, 1]$ and $\mathbf{x}, \mathbf{y} \in S$.

Remark 6.1 *A set S is convex if the line segment joining every two points of S lies entirely in S. Fig. 6.1 shows a convex set and a nonconvex set in E^2.*

Remark 6.2 *Hyperplanes and halfspaces are particular cases of convex sets.*

The following properties are immediate consequences of the definition of convexity:

- If S_1 and S_2 are convex sets, then the set

$$S_1 \cap S_2 \equiv \{\mathbf{x} \mid \mathbf{x} \in S_1 \text{ and } \mathbf{x} \in S_2\}$$

  also is a convex set.

- If S_1 and S_2 are convex sets, then the set

$$S_1 + S_2 \equiv \{\mathbf{y} = \mathbf{x}_1 + \mathbf{x}_2 \mid \mathbf{x}_1 \in S_1 \text{ and } \mathbf{x}_2 \in S_2\}$$

  also is a convex set.

6.3 TYPES OF LINEAR COMBINATIONS

Let $\{\mathbf{a}_1, \ldots, \mathbf{a}_m\}$ be an ordered set of m vectors in E^n. We denote by $\mathbf{A}$ the matrix whose columns are the components of the vectors in the set above.

In Chapter 1, Definition 1.3, we defined unrestricted linear combinations and the set $\mathbf{A}_\rho$ as the set of all linear combinations of the vectors $\mathbf{a}_1, \ldots, \mathbf{a}_m$. In this section we define other types of linear combinations and their associated sets, to be used in later chapters.

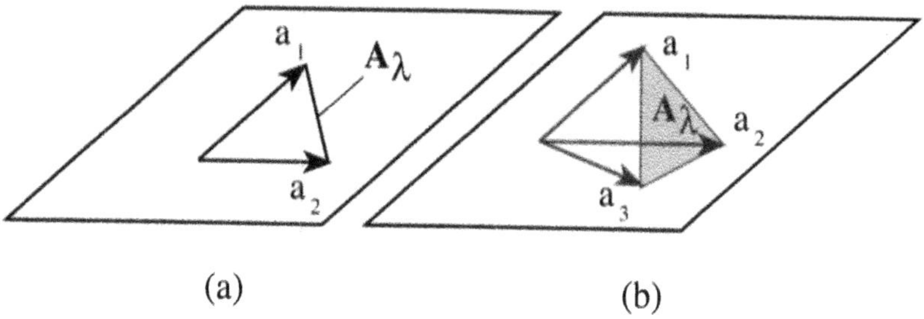

Fig. 6.2 Examples of sets $\mathbf{A}_\lambda$ in E^2.

Definition 6.2 (Linear convex combination). *We say that a vector* $\mathbf{x}$ *is a linear convex combination of vectors of* $\mathbf{A} = (\mathbf{a}_1, \ldots, \mathbf{a}_m)$ *iff*

$$\mathbf{x} = \lambda_1 \mathbf{a}_1 + \cdots + \lambda_m \mathbf{a}_m, \tag{6.2}$$

where $\mathbb{R}^+$ *is the set of nonnegative real numbers,* $\lambda_i \in \mathbb{R}^+, \forall i$ *and* $\sum_{i=1}^{m} \lambda_i = 1$.

We denote the set of all linear convex combinations of vectors of $\mathbf{A}$ by $\mathbf{A}_\lambda$.

Example 6.1 (The set $\mathbf{A}_\lambda$). Fig. 6.2 shows the sets $\mathbf{A}_\lambda$ associated with two and three vectors in E^2. In case (a), the set $\mathbf{A}_\lambda$ is a line segment and in case (b) it is a triangle.

■

Definition 6.3 (Nonnegative linear combination). *We say that a vector* $\mathbf{x}$ *is a nonnegative linear combination of vectors of* $\mathbf{A}$ *iff*

$$\mathbf{x} = \pi_1 \mathbf{a}_1 + \cdots + \pi_m \mathbf{a}_m, \tag{6.3}$$

where $\pi_i \in \mathbb{R}^+, \forall i$.

We denote the set of all nonnegative linear combinations of the vectors in $\mathbf{A}$ by $\mathbf{A}_\pi$.

Example 6.2 (The set $\mathbf{A}_\pi$). Fig. 6.3 shows the sets $\mathbf{A}_\pi$ (shaded regions) associated with two and three vectors in E^2. Note that these sets are different since $\mathbf{a}_3$ is not a nonnegative linear combination of vectors $\mathbf{a}_1$ and $\mathbf{a}_2$. ■

Definition 6.4 (All positive linear combination). *We say that a vector* $\mathbf{x}$ *is an all positive linear combination of vectors of* $\mathbf{A}$ *iff*

$$\mathbf{x} = \sigma_1 \mathbf{a}_1 + \cdots + \sigma_m \mathbf{a}_m, \tag{6.4}$$

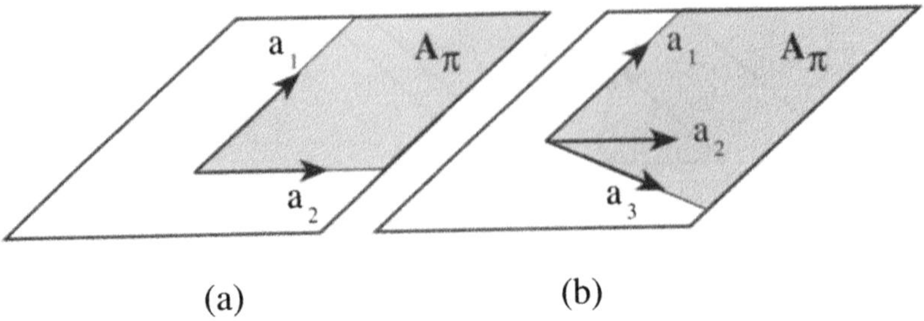

Fig. 6.3 Examples of sets $\mathbf{A}_\pi$ in E^2.

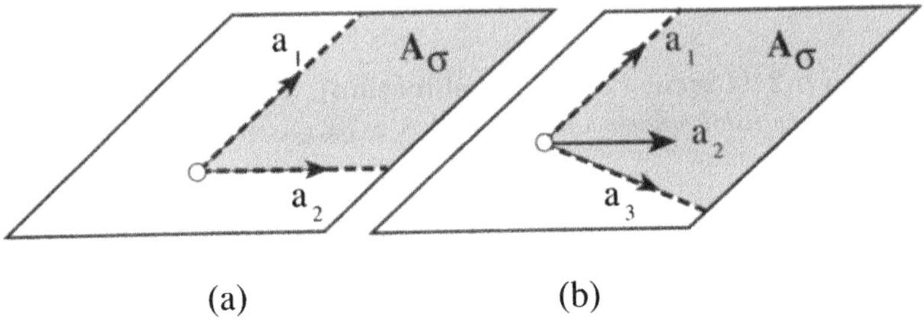

Fig. 6.4 Examples of sets $\mathbf{A}_\sigma$ in E^2.

where $\sigma_i \in \mathbb{R}^{++}, \forall i$, and $\mathbb{R}^{++}$ is the set of positive real numbers.

We denote by $\mathbf{A}_\sigma$ the set of all positive linear combinations of vectors of $\mathbf{A}$.

Example 6.3 (The set $\mathbf{A}_\sigma$). Fig. 6.4 shows the set $\mathbf{A}_\sigma$ associated with two and three vectors in E^2. In case (a), the two vectors do not belong to $\mathbf{A}_\sigma$ but in case (b), the vector $\mathbf{a}_2$ belongs to $\mathbf{A}_\sigma$. In both cases, the vertex of the cone does not belong to $\mathbf{A}_\sigma$. ■

Example 6.4 (Selecting the adequate linear combination in a real case). Suppose that we wish to distribute a given amount π of food among m individuals. Let $\boldsymbol{\pi} = (\pi_1, \ldots, \pi_m)^T$ be the vector of the amounts received by each individual. Let $\mathbf{e}_j = (0, 0, \ldots, 1, \ldots, 0)^T$, where the 1 occupies the jth place. Then we have

$$\pi = \pi_1 + \cdots + \pi_m$$

and

$$\boldsymbol{\pi} = \pi_1 \mathbf{e}_1 + \cdots + \pi_m \mathbf{e}_m,$$

which is a nonnegative linear combination of $\mathbf{e}_1, \ldots, \mathbf{e}_m$.

If the amount to be distributed is unlimited, any possible solution is a nonnegative linear combination of the **e** vectors. If we establish that all individuals must benefit from the distribution, then the linear combination must be positive.

If the amount to be distributed is limited, it can be assumed, without loss of generality, that it is unity, and then we have a linear convex combination. ■

6.4 POLYHEDRAL CONVEX CONES

Definition 6.5 (Polyhedral convex cone). *Let* $\mathbf{A} = (\mathbf{a}_1, \ldots, \mathbf{a}_m)$. *The set*

$$\mathbf{A}_\pi \equiv \{\mathbf{x} \in E^n \mid \mathbf{x} = \pi_1\mathbf{a}_1 + \ldots + \pi_m\mathbf{a}_m \quad \text{with} \quad \pi_i \geq 0; i = 1, \ldots, m\}$$

of all nonnegative linear combinations of vectors of $\mathbf{A}$ *is known as a polyhedral convex cone or simply a cone. The vectors* $\mathbf{a}_1, \ldots, \mathbf{a}_m$ *are the cone generators.*

It will be shown in Chapter 8 that this definition is equivalent to the following one.

Definition 6.6 (Alternative definition of a cone). *A cone is the intersection of a finite set of halfspaces having a common point* $\mathbf{x}_0$ *(the cone vertex), that is,*

$$\{\mathbf{x}_0 + \mathbf{x} \in E^n \mid \mathbf{H}\mathbf{x} \leq \mathbf{0}\}. \tag{6.5}$$

Example 6.5 (Some cones in E^2). Fig. 6.5 shows three cones in E^2. Cone (a) is generated by two linearly independent vectors; cone (b) has three linear dependent generators and coincides with the cone generated by $\mathbf{a}_1$ and $\mathbf{a}_3$, which are independent. Finally, cone (c) has three linearly dependent generators but its set of generators can't be reduced (removing any one of them causes the new generated cone to be different). ■

Example 6.6 (Some cones in E^3). Fig. 6.6 shows three cones in E^3. Cone (a) has three linearly independent generators, cone (b) has three linearly dependent generators, and, finally, cone (c) has three linear dependent generators with a nonreducible generator. ■

Remark 6.3 *A linear space* $\mathbf{A}_\rho$ *is a particular case of a cone:*

$$\mathbf{A}_\rho \equiv (\mathbf{A} : -\mathbf{A})_\pi, \tag{6.6}$$

where $-\mathbf{A}$ *is the negative of* $\mathbf{A}$. *In other words, a linear space is the cone generated by its generators and their opposite vectors.*

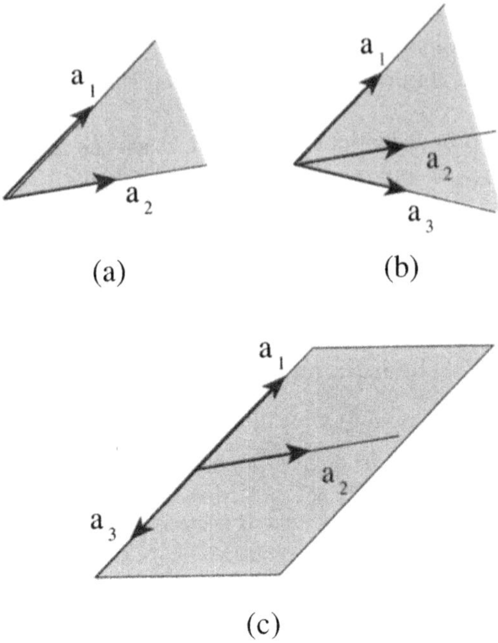

Fig. 6.5 Examples of cones in E^2.

Expression (6.6) shows that the cone concept is broader than the space concept and, more importantly, allows us to replace the usual methods of classical linear algebra with the specific methods of cones, which are more efficient.

Theorem 6.1 *Given a linear space* $\mathbf{A}_\rho$ *and a vector* $\mathbf{x} \in \mathbf{A}_\sigma$, *then* $\mathbf{A}_\rho \equiv (\mathbf{A} : -\mathbf{x})_\pi$.

Proof: Since $(\mathbf{A} : -\mathbf{x})_\pi \subseteq \mathbf{A}_\rho$, we only need to prove that $\mathbf{A}_\rho \subseteq (\mathbf{A} : -\mathbf{x})_\pi$. If $\mathbf{x} \in \mathbf{A}_\sigma$, $\mathbf{x}$ can be written as

$$\mathbf{x} = \sum_{i=1}^{m} \sigma_i \mathbf{a}_i; \sigma_i > 0; i = 1, \ldots, m. \tag{6.7}$$

Similarly, if $\mathbf{y} \in \mathbf{A}_\rho$, for an arbitrary real π, we have

$$\mathbf{y} = \sum_{i=1}^{m} \rho_i \mathbf{a}_i = \sum_{i=1}^{m} \rho_i \mathbf{a}_i + \pi \mathbf{x} - \pi \mathbf{x} = \sum_{i=1}^{m} (\rho_i + \pi \sigma_i) \mathbf{a}_i + \pi(-\mathbf{x}). \tag{6.8}$$

Choosing

$$\pi = \max_{i=1,\ldots,m} \left| \frac{\rho_i}{\sigma_i} \right| \geq 0, \tag{6.9}$$

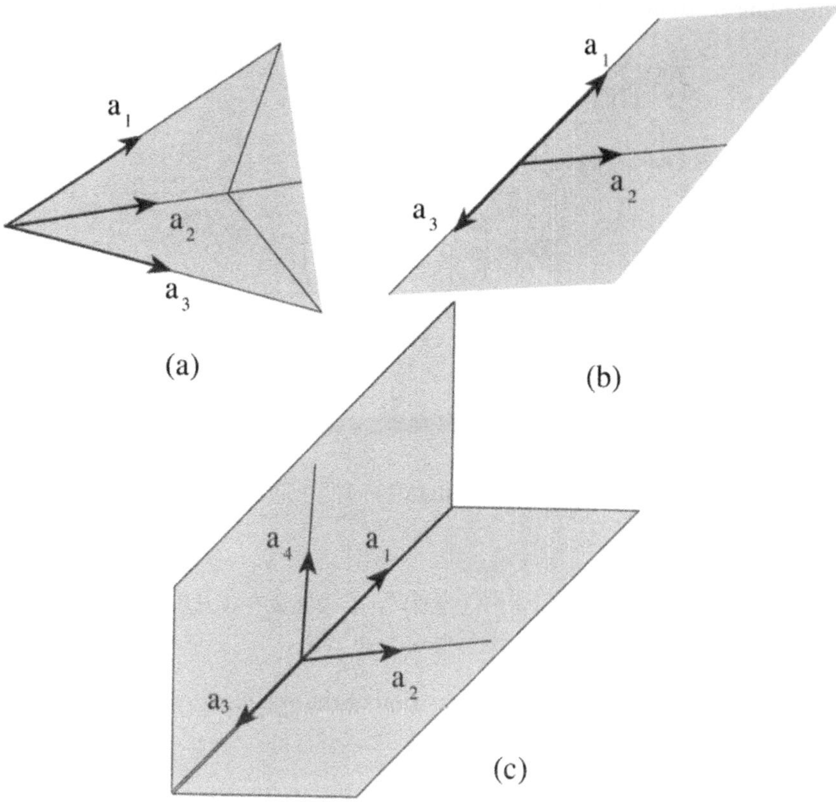

Fig. 6.6 Three examples of cones in E^3.

we have $\rho_i + \pi\sigma_i \geq 0; i = 1, \ldots, m$, and then $\mathbf{y} \in (\mathbf{A} : -\mathbf{x})_\pi$. ∎

Remark 6.4 *The important practical implication of Theorem 6.1 is that any linear space generated by m vectors can be considered as a cone generated by m+1 vectors. This improves the weak result in Remark 6.3, Expression (6.6), which required 2m generators.*

If one is interested in writing the linear space as a cone, any selection of $\mathbf{x} \in \mathbf{A}_\sigma$ *is valid; however,* $\mathbf{x} = \sum_{i=1}^{m} \mathbf{a}_i$, *which leads to* $\pi = \max_{i=1,\ldots,m} |\rho_i|$, *is an advantageous selection of* $\mathbf{x}$.

Example 6.7 (Linear space as a cone). Consider the linear space $\mathbf{A}_\rho$ with

$$\mathbf{A} = \begin{pmatrix} 1 & -1 \\ 2 & 0 \\ 3 & 1 \end{pmatrix}.$$

According to Remark 6.4, we can use Theorem 6.1, with

$$\mathbf{x} = \sum_{i=1,2} \mathbf{a}_i = (1,2,3)^T + (-1,0,1)^T = (0,2,4)^T,$$

and get $\mathbf{A}_\rho \equiv (\mathbf{A} : -\mathbf{x})_\pi \equiv \mathbf{B}_\pi$, with

$$\mathbf{B} = \begin{pmatrix} 1 & -1 & 0 \\ 2 & 0 & -2 \\ 3 & 1 & -4 \end{pmatrix}.$$

Now, given a vector $\mathbf{y}$ of $\mathbf{A}_\rho$, for example,

$$\mathbf{y} = -(1,2,3)^T - 2(-1,0,1)^T = (1,-2,-5)^T,$$

we can choose

$$\pi = \max_{i=1,2} |\rho_i| = \max(|-1|, |-2|) = 2,$$

and then

$$\begin{aligned} \mathbf{y} &= (1,-2,-5)^T = (-1+2)(1,2,3)^T + (-2+2)(-1,0,1)^T + 2(0,-2,-4)^T \\ &= (1,2,3)^T + 2(0,-2,-4)^T = \mathbf{a}_1 + 2(-\mathbf{x}), \end{aligned} \tag{6.10}$$

which is a nonnegative linear combination of the vector $\mathbf{a}_1$ and the vector $-\mathbf{x}$. ■

Remark 6.5 *The sum of two cones is another cone, since*

$$\mathbf{A}_\pi + \mathbf{B}_\pi \equiv (\mathbf{A} : \mathbf{B})_\pi.$$

Consequently, the sum of a cone and a linear space is a cone too. Indeed, the sum of $\mathbf{A}_\pi$ *and* $\mathbf{B}_\rho$ *is* $(\mathbf{A} : \mathbf{B} : -\mathbf{B})_\pi$.

Proposition 6.1 *A vector* $\mathbf{x}$ *belongs to the cone* $\mathbf{A}_\pi$ *iff it coincides with the cone generated by* $\mathbf{A}$ *and* $\mathbf{x}$*; that is,*

$$\mathbf{x} \in \mathbf{A}_\pi \Leftrightarrow \mathbf{A}_\pi \equiv (\mathbf{A} : \mathbf{x})_\pi, \tag{6.11}$$

where $(\mathbf{A} : \mathbf{x})_\pi$ *is the cone generated by* $\mathbf{a}_1, \ldots, \mathbf{a}_m$ *and* $\mathbf{x}$.

Proof: [⇒] Suppose $\mathbf{x} \in \mathbf{A}_\pi$. Then there exist m nonnegative numbers $\pi_i^x \geq 0$, $i = 1, \ldots, m$, such that

$$\mathbf{x} = \pi_1^x \mathbf{a}_1 + \cdots + \pi_m^x \mathbf{a}_m. \tag{6.12}$$

Obviously, $\mathbf{A}_\pi \subseteq (\mathbf{A} : \mathbf{x})_\pi$, and conversely if a vector $\mathbf{y} \in (\mathbf{A} : \mathbf{x})_\pi$ then

$$\mathbf{y} = \pi_1 \mathbf{a}_1 + \cdots + \pi_m \mathbf{a}_m + \pi_{m+1} \mathbf{x} \text{ with } \pi_i \geq 0.$$

Using (6.12), we get

$$\mathbf{y} = (\pi_1 + \pi_1^x \pi_{m+1})\mathbf{a}_1 + \cdots + (\pi_m + \pi_m^x \pi_{m+1})\mathbf{a}_m$$

and $\pi_i + \pi_i^x \pi_{m+1} \geq 0$ for all $i = 1, \ldots, m$. Thus, $\mathbf{y} \in \mathbf{A}_\pi$.

[$\Leftarrow$] The reciprocal implication is obvious because $\mathbf{x} \in (\mathbf{A} : \mathbf{x})_\pi \equiv \mathbf{A}_\pi$. ∎

The preceding proposition can be generalized as follows.

Proposition 6.2 *A set of vectors* $\mathbf{X} = \{\mathbf{x}_1, \ldots, \mathbf{x}_k\}$ *is included in the cone* $\mathbf{A}_\pi$ *iff* $\mathbf{A}_\pi \equiv (\mathbf{A} : \mathbf{X})_\pi$*; that is,*

$$\mathbf{X} \subseteq \mathbf{A}_\pi \Leftrightarrow \mathbf{A}_\pi \equiv (\mathbf{A} : \mathbf{X})_\pi,$$

where $(\mathbf{A} : \mathbf{X})_\pi$ *is the cone generated by* $\mathbf{a}_1, \ldots, \mathbf{a}_m$ *and* $\mathbf{x}_1, \ldots, \mathbf{x}_k$.

This proposition can be proved by induction, using Proposition 6.1.

Proposition 6.3 *A cone* $\mathbf{A}_\pi$ *is included in another cone* $\mathbf{B}_\pi$ *iff the generators of the first cone belong to the second.*

Definition 6.7 (General form of a cone). *If we have a cone* $\mathbf{A}_\pi$*, we can classify its generators* $\mathbf{A} = (\mathbf{a}_1, \ldots, \mathbf{a}_m)$ *into two groups:*

1. *The generators whose opposite vectors belong to the cone, that is,*

$$\mathbf{B} \equiv \{\mathbf{a}_i \mid -\mathbf{a}_i \in \mathbf{A}_\pi\}.$$

2. *The generators whose opposite vectors do not belong to the cone, that is,*

$$\mathbf{C} \equiv \{\mathbf{a}_i \mid -\mathbf{a}_i \notin \mathbf{A}_\pi\}.$$

Thus, we can express the cone in the following form:

$$\mathbf{A}_\pi \equiv (\mathbf{B} : -\mathbf{B} : \mathbf{C})_\pi \equiv \mathbf{B}_\rho + \mathbf{C}_\pi, \quad (6.13)$$

which is known as the general form of a cone.

Example 6.8 (General form of a cone). For the cones in Fig. 6.5 we have

(a) $\mathbf{B} \equiv \emptyset$; $\mathbf{C} \equiv \{\mathbf{a}_1, \mathbf{a}_2\}$
(b) $\mathbf{B} \equiv \emptyset$; $\mathbf{C} \equiv \{\mathbf{a}_1, \mathbf{a}_2, \mathbf{a}_3\}$
(c) $\mathbf{B} \equiv \{\mathbf{a}_1, \mathbf{a}_3\}$; $\mathbf{C} \equiv \{\mathbf{a}_2\}$.

Similarly, for the cones in Fig. 6.6 we have

(a) $\mathbf{B} \equiv \emptyset$; $\mathbf{C} \equiv \{\mathbf{a}_1, \mathbf{a}_2, \mathbf{a}_3\}$
(b) $\mathbf{B} \equiv \{\mathbf{a}_1, \mathbf{a}_3\}$; $\mathbf{C} \equiv \{\mathbf{a}_2\}$
(c) $\mathbf{B} \equiv \{\mathbf{a}_1, \mathbf{a}_3\}$; $\mathbf{C} \equiv \{\mathbf{a}_2, \mathbf{a}_4\}$.

∎

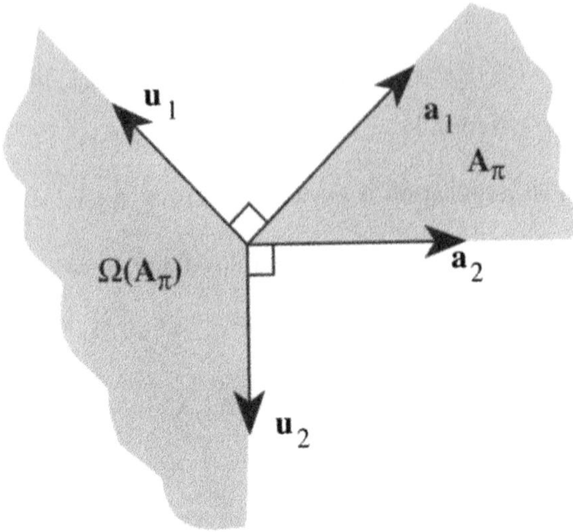

Fig. 6.7 Example of a cone $\mathbf{A}_\pi$ and its polar cone $\Omega(\mathbf{A}_\pi)$.

6.4.1 The Polar Cone

Definition 6.8 (Nonpositive dual or polar cone). *Let $\mathbf{A}_\pi$ be a cone in E^n with generators $\mathbf{a}_1, \ldots, \mathbf{a}_m$. The nonpositive dual $\Omega(\mathbf{A}_\pi)$ of $\mathbf{A}_\pi$ is defined as the set*

$$\Omega(\mathbf{A}_\pi) \equiv \left\{\mathbf{u} \in E^n \mid \mathbf{A}^T\mathbf{u} \leq \mathbf{0}\right\} \equiv \left\{\mathbf{u} \in E^n \mid \mathbf{a}_i \bullet \mathbf{u} \leq 0;\ i = 1, \ldots, m\right\}.$$

Minkowski (1911) proved that $\Omega(\mathbf{A}_\pi)$ is a cone too, which is known as the *polar cone.*

Remark 6.6 *The nonpositive dual of a linear space coincides with its orthogonal:*

$$\begin{aligned}\Omega(\mathbf{A}_\rho) &\equiv \Omega((\mathbf{A}, -\mathbf{A})_\pi) \equiv \{\mathbf{u} \in E^n | \mathbf{a}_i \bullet \mathbf{u} \leq 0 \wedge -\mathbf{a}_i \bullet \mathbf{u} \leq 0; i = 1, \ldots, m\} \\ &\equiv \{\mathbf{u} \in E^n | \mathbf{a}_i \bullet \mathbf{u} = 0; i = 1, \ldots, m\}.\end{aligned}$$

Example 6.9 (Polar cone). Fig. 6.7 shows the polar cone $\Omega(\mathbf{A}_\pi) \equiv (\mathbf{u}_1, \mathbf{u}_2)_\pi$ associated with the cone $(\mathbf{a}_1, \mathbf{a}_2)_\pi$.

■

The dual or polar cone satisfies the following properties:

- The dual of the dual of $\mathbf{A}_\pi$ is $\mathbf{A}_\pi$ itself: $\Omega(\Omega(\mathbf{A}_\pi)) \equiv \mathbf{A}_\pi$.
- The dual is unique: $\mathbf{A}_\pi \equiv \mathbf{B}_\pi \iff \Omega(\mathbf{A}_\pi) \equiv \Omega(\mathbf{B}_\pi)$.

- The dual of a subset contains the dual of the superset:

$$\mathbf{E}_\pi \subseteq \mathbf{A}_\pi \iff \Omega(\mathbf{A}_\pi) \subseteq \Omega(\mathbf{E}_\pi).$$

The following lemma is due to Farkas (1902) and is the basis for the duality theory.

Lemma 6.1 *(Farkas). If Ω $(\mathbf{A}_\pi)$ is the nonpositive dual of $\mathbf{A}_\pi$, then*

$$\mathbf{x} \in \mathbf{A}_\pi \Leftrightarrow \boldsymbol{\mu} \bullet \mathbf{x} \leq 0; \forall \boldsymbol{\mu} \in \Omega(\mathbf{A}_\pi). \tag{6.14}$$

Theorem 6.2 *The vector $\mathbf{x}$ belongs to the cone $\mathbf{A}_\pi$ iff $\mathbf{V}^T\mathbf{x} = \mathbf{0} \wedge \mathbf{W}^T\mathbf{x} \leq \mathbf{0}$; that is,*

$$\boxed{\mathbf{x} \in \mathbf{A}_\pi \Leftrightarrow \mathbf{V}^T\mathbf{x} = \mathbf{0} \wedge \mathbf{W}^T\mathbf{x} \leq \mathbf{0},} \tag{6.15}$$

where $\mathbf{V}_\rho + \mathbf{W}_\pi$ is the general form of the cone $\Omega(\mathbf{A}_\pi)$.

Proof: The dual cone $\Omega(\mathbf{A}_\pi)$ can be written as

$$\Omega(\mathbf{A}_\pi) \equiv \mathbf{U}_\pi \equiv \mathbf{V}_\rho + \mathbf{W}_\pi \equiv (\mathbf{V} : -\mathbf{V} : \mathbf{W})_\pi. \tag{6.16}$$

According to the Farkas lemma, we have

$$\begin{array}{rcl} \mathbf{x} \in \mathbf{A}_\pi & \Leftrightarrow & \mathbf{U}^T\mathbf{x} \leq \mathbf{0} \Leftrightarrow \mathbf{V}^T\mathbf{x} \leq \mathbf{0} \wedge -\mathbf{V}^T\mathbf{x} \leq \mathbf{0} \wedge \mathbf{W}^T\mathbf{x} \leq \mathbf{0} \\ & \Leftrightarrow & \mathbf{V}^T\mathbf{x} = \mathbf{0} \wedge \mathbf{W}^T\mathbf{x} \leq \mathbf{0}. \end{array} \tag{6.17}$$

■

In the next section we give the Γ-algorithm, which allows calculating, with a minimal sets of generators, the dual or polar cone $\Omega(\mathbf{A}_\pi)$ of $\mathbf{A}_\pi$ in its general form.

6.5 THE Γ-PROCESS

The computational process associated with what has been named the Γ-algorithm has been treated by Jubete (1987, 1991). This process allows us to obtain the nonpositive dual $\Omega(\mathbf{A}_\pi) \equiv \mathbf{V}_\rho + \mathbf{W}_\pi$ of $\mathbf{A}_\pi \equiv \mathbf{B}_\rho + \mathbf{C}_\pi$, that is,

$$\begin{array}{rcl} \Omega(\mathbf{A}_\pi) & \equiv & \{\boldsymbol{\mu} \in E^n | \boldsymbol{\mu} \bullet \mathbf{a}_i \leq 0; i = 1, \ldots, m\} \\ & \equiv & \{\boldsymbol{\mu} \in E^n | \boldsymbol{\mu} \bullet \mathbf{b}_i = 0, \boldsymbol{\mu} \bullet \mathbf{c}_j \leq 0, \forall i, j\}, \end{array} \tag{6.18}$$

with a minimum number of generators based on the analysis of the two different situations that can occur: Γ_I, which is a simple substitution, and Γ_{II}, which deserves a more careful analysis.

6.5.1 The Γ_I-Process

In this process we start with the most general vector in E^n:

$$\boldsymbol{\mu} = \rho_1(1,\ldots,0)^T + \cdots + \rho_n(0,\ldots,1)^T. \tag{6.19}$$

Since there is no restriction on $\boldsymbol{\mu}$, we start with all E^n, that is,

$$E^n \equiv \Omega(\mathbf{A}_\pi^0) \equiv \mathbf{U}_\pi^0 \equiv \mathbf{V}_\rho^0. \tag{6.20}$$

Next, we sequentially impose the constraints $\boldsymbol{\mu} \bullet \mathbf{a}_i \leq 0$ for $i = 1,\ldots,m$, which are used to eliminate the ρ's, until no more ρ_j exist. In order to determine $\Omega(\mathbf{A}_\pi)$, we define $\mathbf{A}^h \equiv \{\mathbf{a}_1,\ldots,\mathbf{a}_h\}$, the first h columns of $\mathbf{A}$, and we sequentially determine $\Omega(\mathbf{A}_\pi^h) \equiv \mathbf{U}_\pi^h \equiv \mathbf{V}_\rho^h + \mathbf{W}_\pi^h$, where $\mathbf{V}_\rho^h$ and $\mathbf{W}_\pi^h$ refer to the set of generators of $\mathbf{U}_\pi^h$ with unrestricted (ρ) and nonnegative (π) coefficients, respectively. In the first iteration, we start from E^n in canonical form, that is,

$$E^n \equiv \Omega(\mathbf{A}_\pi^0) \equiv \mathbf{U}_\pi^0 \equiv \mathbf{V}_\rho^0 \equiv \boldsymbol{\mu} = \rho_1(1,\ldots,0)^T + \cdots + \rho_n(0,\ldots,1)^T, \tag{6.21}$$

and we impose the condition

$$\mathbf{V}_\rho^0 \bullet \mathbf{a}^1 = -\pi_1, \tag{6.22}$$

where $\pi_1 \geq 0$.

Two different cases are considered:

- **Case 1: $\mathbf{a}_1$ is a column of the C matrix.** Here, $-\mathbf{a}_1 \notin \mathbf{A}_\pi$. Then we eliminate from (6.22) any ρ_j and obtain

$$\Omega(\mathbf{A}_\pi^1) \equiv \mathbf{U}_\pi^1 \equiv \mathbf{V}_\rho^1 + \mathbf{W}_\pi^1. \tag{6.23}$$

- **Case 2: $\mathbf{a}_1$ is a column of the B matrix.** In this case, we need to introduce two vectors $\mathbf{a}_1$ and $-\mathbf{a}_1$. However, this is equivalent to letting $\pi_1 = 0$. Thus, we let $\pi_1 = 0$, we eliminate any ρ_j, and we obtain $\Omega(\mathbf{A}_\pi^1)$ as in Case 1.

For iterations $h > 1$, we repeat the same process as above and impose the condition

$$\mathbf{U}_\pi^{h-1} \bullet \mathbf{a}_h = -\pi_h \leq 0.$$

At the end of the $(h-1)$th step we have

$$\Omega(\mathbf{A}_\pi^{h-1}) \equiv \mathbf{U}_\pi^{h-1} \equiv \mathbf{V}_\rho^{h-1} + \mathbf{W}_\pi^{h-1} \equiv \sum_{i=p}^{q} \rho_i \mathbf{v}_i + \sum_{i=1}^{p-1} \pi_i \mathbf{w}_i, \tag{6.24}$$

and then

$$\mathbf{U}_\pi^{h-1} \bullet \mathbf{a}_h = \sum_{i=p}^{q} \rho_i t_i + \sum_{i=1}^{p-1} \pi_i t_i = -\pi_p. \tag{6.25}$$

If not all $t_i = \mathbf{v}_i \bullet \mathbf{a}_h; i = p, \ldots, q$ values are null, we assume, without loss of generality, that $t_p \neq 0$ (otherwise, we can permute vectors). Thus, we can eliminate ρ_p:

$$\rho_p = \left(-\pi_p - \sum_{i=1}^{p-1} \pi_i t_i - \sum_{i=p+1}^{q} \rho_i t_i\right) / t_p. \tag{6.26}$$

Replacing (6.26) in (6.24), we obtain

$$\mathbf{U}_\pi^h \equiv \sum_{i=p+1}^{q} \rho_i \left(\mathbf{v}_i - \frac{t_i}{t_p}\mathbf{v}_p\right) + \sum_{i=1}^{p-1} \pi_i \left(\mathbf{w}_i - \frac{t_i}{t_p}\mathbf{v}_p\right) - \pi_p \frac{\mathbf{v}_p}{t_p}. \tag{6.27}$$

An important practical consequence of (6.27) is that the standard normalization and pivoting process can be used to eliminate the ρ_i, by changing the sign of the pivot column after the normalization process.

6.5.2 The Γ_{II}-Process

The Γ_{II}-process must be used when $\mathbf{V}^{h-1} \equiv \emptyset$ or $\mathbf{v}_i^{h-1} \bullet \mathbf{a}_h = 0, \forall i$ for any $\mathbf{a}_h$ with

$$\Omega(\mathbf{A}_\pi^{h-1}) \equiv \mathbf{V}_\rho^{h-1} + \mathbf{W}_\pi^{h-1}.$$

In this case, the following sets must be considered:

$$\begin{array}{lcl} \mathbf{W}^- & \equiv & \{\mathbf{w}_i^{h-1} | \mathbf{w}_i^{h-1} \bullet \mathbf{a}_h < 0\}, \\ \mathbf{W}^+ & \equiv & \{\mathbf{w}_i^{h-1} | \mathbf{w}_i^{h-1} \bullet \mathbf{a}_h > 0\}, \\ \mathbf{W}^0 & \equiv & \{\mathbf{w}_i^{h-1} | \mathbf{w}_i^{h-1} \bullet \mathbf{a}_h = 0\}, \end{array} \tag{6.28}$$

and we proceed as follows:

- If $\mathbf{W}^- \equiv \emptyset$ then

$$\Omega(\mathbf{A}_\pi^h) \equiv \mathbf{V}_\rho^{h-1} + \mathbf{W}_\pi^0. \tag{6.29}$$

- If $\mathbf{W}^+ \equiv \emptyset$ then

$$\Omega(\mathbf{A}_\pi^h) \equiv \Omega(\mathbf{A}_\pi^{h-1}). \tag{6.30}$$

- If $\mathbf{W}^- \not\equiv \emptyset \wedge \mathbf{W}^+ \not\equiv \emptyset$ we build the set

$$\mathbf{W}_c^{-+} \equiv \{\mathbf{w} = t_j \mathbf{w}_i^{h-1} - t_i \mathbf{w}_j^{h-1} | \mathbf{w}_i \in \mathbf{W}^-, \mathbf{w}_j \in \mathbf{W}^+, t_k = \mathbf{w}_k^{h-1} \bullet \mathbf{a}_h\}$$

  and then

$$\Omega(\mathbf{A}_\pi^h) \equiv \mathbf{V}_\rho^{h-1} + \mathbf{W}_\pi^0 + \mathbf{W}_\pi^- + (\mathbf{W}_c^{-+})_\pi. \tag{6.31}$$

  If, in addition, we want to eliminate the redundant vectors in $\mathbf{W}_c^{-+}$, then we eliminate those vectors $\mathbf{w} \in \mathbf{W}_c^{-+}$ such that

$$\begin{array}{lll} A^{h-1}(\mathbf{w}) \subseteq A(\mathbf{w}_k^{h-1}), & \text{for some } \mathbf{w}_k^{h-1} \in \mathbf{W}^0, & \textit{or} \\ A^{h-1}(\mathbf{w}) \subseteq A(\mathbf{w}^*), & \text{for some } \mathbf{w}^* \in \mathbf{W}_c^{-+}, & \end{array}$$

where $A(\mathbf{w}) \equiv \{\mathbf{a}_j | \mathbf{w} \bullet \mathbf{a}_j = 0\}$ is the subset of A orthogonal to $\mathbf{w}$, and $\mathbf{W}^{-+}$ is the set $\mathbf{W}_c^{-+}$ after the elimination of redundant vectors process. We shall use the notation $A(\mathbf{U})$ to refer to the set of all sets

$$A_{\mathbf{u}_j} \equiv \{i | \mathbf{a}_i \in A(\mathbf{u}_j^h)\}.$$

Then we write

$$\Omega(\mathbf{A}_\pi^h) \equiv \mathbf{V}_\rho^{h-1} + \mathbf{W}_\pi^0 + \mathbf{W}_\pi^- + \mathbf{W}_\pi^{-+}. \tag{6.32}$$

Example 6.10 (Dual). Suppose we wish to find the dual $\Omega(\mathbf{B}_\rho + \mathbf{C}_\pi)$, where

$$\mathbf{B} = \begin{pmatrix} 1 \\ 2 \\ -1 \end{pmatrix}; \ \mathbf{C} = \begin{pmatrix} 1 & 2 & 0 & 0 \\ -1 & 1 & -1 & 0 \\ 3 & -2 & 0 & 1 \end{pmatrix}. \tag{6.33}$$

To this end, the steps of the Γ-process are as follows: We start with the most general vector in E^3:

$$\mathbf{V}_\rho^0 = \rho_1(1,0,0)^T + \rho_2(0,1,0)^T + \rho_3(0,0,1)^T. \tag{6.34}$$

Next, we impose

$$\mathbf{V}_\rho^0 \bullet \mathbf{b}_1 = \rho_1 + 2\rho_2 - \rho_3 = 0, \tag{6.35}$$

which is equivalent to using the vectors $\mathbf{b}_1 = (1,2,-1)^T$ and $-\mathbf{b}_1$ and imposing the nonpositive dot products of both. Eliminating ρ_3 from (6.35) and (6.34), the dual $\Omega(\mathbf{B}_\rho)$ becomes

$$\Omega(\mathbf{B}_\rho) \equiv \mathbf{U}_\pi^1 \equiv \mathbf{V}_\rho^1 \equiv \rho_1(1,0,1)^T + \rho_2(0,1,2)^T. \tag{6.36}$$

Then we use the vector $\mathbf{c}_1 = (1,-1,3)^T$ and impose

$$\mathbf{V}_\rho^1 \bullet \mathbf{c}_1 = 4\rho_1 + 5\rho_2 = -\pi_1 \leq 0. \tag{6.37}$$

Eliminating ρ_2 from (6.37) and (6.36), we obtain

$$\Omega(\mathbf{B}_\rho + \mathbf{c}_{1\pi}) \equiv \mathbf{U}_\pi^2 \equiv \rho_1(5,-4,-3)^T + \pi_1(0,-1,-2)^T, \tag{6.38}$$

and, similarly, using the vector $\mathbf{c}_2 = (2,1,-2)^T$, we get

$$\Omega(\mathbf{B}_\rho + (\mathbf{c}_1, \mathbf{c}_2)_\pi) \equiv \mathbf{U}_\pi^3 \equiv \pi_1(-1,0,-1)^T + \pi_2(-5,4,3)^T. \tag{6.39}$$

Next, we try to introduce the vector $\mathbf{c}_3 = (0,-1,0)^T$ and obtain the dot products $\mathbf{w}_1^3 \bullet \mathbf{c}_3 = 0 \wedge \mathbf{w}_2^3 \bullet \mathbf{c}_3 = -4$; but we have no more ρ_i. Thus, we need to use the Γ_{II}-process.

From (6.28), $\mathbf{W}^+ \equiv \emptyset$ and by (6.30), $\mathbf{U}_\pi^4 \equiv \mathbf{U}_\pi^3$; that is,

$$\mathbf{U}_\pi^4 \equiv \Omega(\mathbf{B}_\rho + (\mathbf{c}_1, \mathbf{c}_2, \mathbf{c}_3)_\pi) \equiv \pi_1(-1,0,-1)^T + \pi_2(-5,4,3)^T \equiv \mathbf{W}_\pi^4. \tag{6.40}$$

Similarly, introducing the vector $\mathbf{c}_4 = (0,0,1)^T$, the following dot products are obtained $\mathbf{w}_1^4 \bullet \mathbf{c}_4 = -1 \wedge \mathbf{w}_2^4 \bullet \mathbf{c}_4 = 3$. From (6.28), we have

$$\mathbf{W}^- \equiv \{\mathbf{w}_1^4\}, \mathbf{W}^+ \equiv \{\mathbf{w}_2^4\}. \tag{6.41}$$

Then

$$\mathbf{W}_c^{-+} \equiv \mathbf{W}^{-+} \equiv \{3(-1,0,-1)^T - (-1)(-5,4,3)^T\} \equiv \{(-8,4,0)^T\}. \tag{6.42}$$

Finally, by (6.31), we get the dual $\Omega(\mathbf{B}_\rho + \mathbf{C}_\pi)$:

$$\mathbf{U}_\pi^5 \equiv \pi_1(-1,0,-1)^T + \pi_2(-2,1,0)^T \equiv \pi_1\mathbf{w}_1 + \pi_2\mathbf{w}_2. \tag{6.43}$$

■

Next, we give a simplified (some details are omitted) version of the Γ-algorithm. For those readers interested in a computer implementation, a complete version is given in Section 6.6.

Algorithm 6.1 The Γ-algorithm.

- **Input:** A set of generators $\{\mathbf{b}_1, \ldots, \mathbf{b}_{n_b}\}$ of the linear space $\mathbf{B}_\rho$ and a set of generators $\{\mathbf{c}_1, \ldots, \mathbf{c}_{n_c}\}$ of the cone $\mathbf{C}_\pi$, both defined in E^n.
- **Output:** The general form of the dual $\Omega(\mathbf{A}_\pi)$, given as two lists of vectors, corresponding to the $\mathbf{V}_\rho$ and the $\mathbf{W}_\pi$ components of its general form.

Initialization:

- Let $\mathbf{A} = (\mathbf{B} : \mathbf{C})$, $\mathbf{U} = \mathbf{I}_n$, where $\mathbf{I}_n$ is the identity matrix of dimension n, $\mathbf{V}^0 = \mathbf{U}, \mathbf{W}^0 = \emptyset$, and $m = n_b + n_c$.
- Initialize the n orthogonal sets $A(\mathbf{U})$ to empty sets and let $s = 0$.

Repeat for h=1 to m:

Step 1: Calculate the dot products. We introduce the $\mathbf{a}_h$ constraint in the process and calculate $\mathbf{t} = \mathbf{a}_h^T \bullet \mathbf{U}$.

Step 2: Update the orthogonal sets. We have $A(\mathbf{U}) \equiv \{A_{\mathbf{u}_j}; j = 1, \ldots, n\}$, adding index h to $A_{\mathbf{u}_j}$ when $t_j = 0$.

Step 3: Look for the pivot. Find a pivot column ($t_{pivot} \neq 0$) among those columns not previously used for pivot.

Step 4: Test for Γ_{I} or Γ_{II}. If $\mathbf{V}^{h-1} \equiv \emptyset$ or a pivot is not found, go to the Γ_{II}-process. Otherwise go to the Γ_I-process.

Step 5: Γ_{I}-process. Normalize the pivot column and perform the pivoting process to eliminate the corresponding ρ coefficient. If $h < n_b$, remove the pivot column. Otherwise, add the pivot column to $\mathbf{W}^{h-1}$ and remove it from $\mathbf{V}^{h-1}$ to obtain $\mathbf{W}^h$ and $\mathbf{V}^h$, respectively. Then, go to Step 1.

Step 6: Γ_{II}-process.

6.1 Determine the sets

$$I(\mathbf{W}^-) \equiv \{i \le s \mid t_i < 0\},$$

$$I(\mathbf{W}^+) \equiv \{i \le s \mid t_i > 0\},$$

$$I(\mathbf{W}^0) \equiv \{i \le s \mid t_i = 0\},$$

and the corresponding sets $\mathbf{W}^-, \mathbf{W}^+$, and $\mathbf{W}^0$ given in (6.28).

6.2 If $\mathbf{W}^- \equiv \emptyset$ then

$$\Omega\left(\mathbf{A}_\pi^h\right) \equiv \mathbf{V}_\rho^{h-1} + \mathbf{W}_\pi^0, \tag{6.44}$$

and go to Step 1.

6.3 If $I(\mathbf{W}^-) \not\equiv \emptyset \wedge I(\mathbf{W}^+) \not\equiv \emptyset$ build the sets

$$\begin{aligned} A(\mathbf{W}_c^{-+}) &\equiv \{A_{\mathbf{w}_j} \cap A_{\mathbf{w}_k} | j \in I(\mathbf{W}^-), k \in I(\mathbf{W}^+)\}, \\ \mathbf{W}_c^{-+} &\equiv \{t_k \mathbf{w}_j - t_j \mathbf{w}_k | j \in I(\mathbf{W}^-), k \in I(\mathbf{W}^+), t_k = \mathbf{w}_k \bullet \mathbf{a}_h\}. \end{aligned}$$

Step 7: Obtain the set $\mathbf{W}^{-+}$ by eliminating from $\mathbf{W}_c^{-+}$ those vectors $\mathbf{w} \in \mathbf{W}_c^{-+}$ such that

$$\begin{aligned} A^{h-1}(\mathbf{w}) &\subseteq A(\mathbf{w}_k^{h-1}), && \text{for some } \mathbf{w}_k^{h-1} \in \mathbf{W}^0 \\ A^{h-1}(\mathbf{w}) &\subseteq A(\mathbf{w}^*), && \text{for some } \mathbf{w}^* \in \mathbf{W}^{-+}, \end{aligned}$$

and then

$$\Omega(\mathbf{A}_\pi^h) \equiv \mathbf{V}_\rho^{h-1} + \mathbf{W}_\pi^0 + \mathbf{W}_\pi^- + \mathbf{W}_\pi^{-+}. \tag{6.45}$$

Step 8: Test for empty dual. If $\Omega(\mathbf{A}_\pi^h) \equiv \emptyset$ return empty set and exit the algorithm. Otherwise, if $h < m$, let $h = h + 1$ and go to Step 1; otherwise, go to Step 9.

Step 9: Return solution.

The Γ-algorithm is illustrated in the following example. To this aim, we use tables corresponding to all the algorithm iterations. The first column of the table in the h-th iteration is for the vector $\mathbf{a}_h$. In the main part of the tables we show the matrix $\mathbf{U}$, where we distinguish between the vectors in $\mathbf{V}^h$ and those in $\mathbf{W}^h$. The last row is reserved for the dot products of $\mathbf{a}_h$ and the column vectors of $\mathbf{U}$. Finally, boldfaced numbers are used for the pivot column.

Table 6.1 Added constraint, **U** matrix, and associated dot products.

Iteration 1				
$\mathbf{a}_1$	$\mathbf{v}_1^1$	$\mathbf{v}_2^1$	$\mathbf{v}_3^1$	$\mathbf{v}_4^1$
–1	**1**	0	0	0
0	**0**	1	0	0
0	**0**	0	1	0
0	**0**	0	0	1
$\mathbf{t}^1$	**–1**	0	0	0

Example 6.11 (Dual using the Γ-algorithm). Consider the cone $\mathbf{A}_\pi$, where

$$\mathbf{A} = \begin{pmatrix} -1 & 0 & 0 & 0 & -1 & 0 & 2 \\ 0 & -1 & 0 & 0 & 2 & -1 & 0 \\ 0 & 0 & -1 & 0 & -1 & 1 & -1 \\ 0 & 0 & 0 & -1 & 0 & -1 & -1 \end{pmatrix}. \tag{6.46}$$

To obtain the dual $\Omega(\mathbf{A}_\pi)$ we use the Γ-algorithm and, after the obvious initializations, we proceed as follows.

Step 1: For $h = 1$, constraint 1 is introduced, and the dot products are calculated, as shown in Table 6.1.

Step 2: The orthogonal sets $A(\mathbf{U})$ are updated:

$$A_{\mathbf{v}_1} \equiv \{\}, A_{\mathbf{v}_2} \equiv \{1\}, A_{\mathbf{v}_3} \equiv \{1\}, A_{\mathbf{v}_4} \equiv \{1\}.$$

Step 3: The pivot is column vector 1.

Step 4: Since $s < 4$ and *pivot* > 0, we go to the Γ_I-process.

Step 5: The pivot column is normalized and the pivoting process performed as shown in Table 6.2. Index h is increased to 2 and we return to Step 1.

Step 1: Constraint 2 is introduced, and the dot products are calculated, as shown in Table 6.2.

Step 2: The orthogonal sets $A(\mathbf{U})$ are updated:

$$A_{\mathbf{w}_1} \equiv \{2\}, A_{\mathbf{v}_1} \equiv \{1\}, A_{\mathbf{v}_2} \equiv \{1,2\}, A_{\mathbf{v}_3} \equiv \{1,2\}.$$

Step 3: The pivot is column vector 2.

Step 4: Since $s < 4$ and *pivot* > 0, we go to the Γ_I-process.

Step 5: The pivot column is normalized and the pivoting process performed as shown in Table 6.3. Index h is increased to 3 and we return to Step 1.

Step 1: Constraint 3 is introduced, and the dot products are calculated, as shown in Table 6.3.

Step 2: The orthogonal sets $A(\mathbf{U})$ are updated:

$$A_{\mathbf{w}_1} \equiv \{2,3\}, A_{\mathbf{w}_2} \equiv \{1,3\}, A_{\mathbf{v}_1} \equiv \{1,2\}, A_{\mathbf{v}_2} \equiv \{1,2,3\}.$$

Table 6.2 Added constraint, **U** matrix, and associated dot products.

	Iteration 2			
$\mathbf{a}_2$	$\mathbf{w}_1^2$	$\mathbf{v}_1^2$	$\mathbf{v}_2^2$	$\mathbf{v}_3^2$
0	1	**0**	0	0
–1	0	**1**	0	0
0	0	**0**	1	0
0	0	**0**	0	1
$\mathbf{t}^2$	0	**–1**	0	0

Table 6.3 Added constraint, **U** matrix, and associated dot products.

	Iteration 3			
$\mathbf{a}_3$	$\mathbf{w}_1^3$	$\mathbf{w}_2^3$	$\mathbf{v}_1^3$	$\mathbf{v}_2^3$
0	1	0	**0**	0
0	0	1	**0**	0
–1	0	0	**1**	0
0	0	0	**0**	1
$\mathbf{t}^3$	0	0	**–1**	0

Step 3: The pivot is column vector 3.

Step 4: Since $s < 4$ and *pivot* > 0, we go to the Γ_I-process.

Step 5: The pivot column is normalized and the pivoting process performed as shown in Table 6.4. Index h is increased to 4 and we return to Step 1.

Step 1: Constraint 4 is introduced, and the dot products are calculated, as shown in Table 6.4.

Step 2: The orthogonal sets $A(\mathbf{U})$ are updated:

$$A_{\mathbf{w}_1} \equiv \{2,3,4\}, A_{\mathbf{w}_2} \equiv \{1,3,4\}, A_{\mathbf{w}_3} \equiv \{1,2,4\}, A_{\mathbf{v}_1} \equiv \{1,2,3\}.$$

Step 3: The pivot is column vector 4.

Step 4: Since $s < 4$ and *pivot* > 0, we go to the Γ_I-process.

Step 5: The pivot column is normalized and the pivoting process performed as shown in Table 6.5. Index h is increased to 5 and we return to Step 1.

Step 1: Constraint 5 is introduced, and the dot products are calculated, as shown in Table 6.5.

Step 2: The orthogonal sets $A(\mathbf{U})$ are updated:

$$A_{\mathbf{w}_1} \equiv \{2,3,4\}, A_{\mathbf{w}_2} \equiv \{1,3,4\}, A_{\mathbf{w}_3} \equiv \{1,2,4\}, A_{\mathbf{w}_4} \equiv \{1,2,3,5\}.$$

Step 3: No pivot column can be obtained.

Table 6.4 Added constraint, **U** matrix, and associated dot products.

Iteration 4				
$\mathbf{a}_4$	$\mathbf{w}_1^4$	$\mathbf{w}_2^4$	$\mathbf{w}_3^4$	$\mathbf{v}_1^4$
0	1	0	0	**0**
0	0	1	0	**0**
0	0	0	1	**0**
–1	0	0	0	**1**
$\mathbf{t}^4$	0	0	0	**–1**

Table 6.5 Added constraint, **U** matrix, and associated dot products.

Iteration 5				
$\mathbf{a}_5$	$\mathbf{w}_1^5$	$\mathbf{w}_2^5$	$\mathbf{w}_3^5$	$\mathbf{w}_4^5$
–1	1	0	0	0
2	0	1	0	0
–1	0	0	1	0
0	0	0	0	1
$\mathbf{t}^5$	–1	2	–1	0

Step 4: Go to the Γ_{II}-process.
Step 6.1: $I(\mathbf{W}^-) \equiv \{1,3\}$; $I(\mathbf{W}^+) \equiv \{2\}$; $I(\mathbf{W}^0) \equiv \{4\}$.
Step 6.3: The initial list $A(\mathbf{W}_c^{-+})$ of vectors is

$$\{\{3,4\},\{1,4\}\}.$$

The corresponding list of $\mathbf{W}_c^{-+}$ vectors is

$$\{(2,1,0,0)^T,(0,1,2,0)^T\}.$$

Step 7: The elimination of redundant vectors in $\mathbf{W}_c^{-+}$ is performed, as shown in Table 6.6. Next, a new **U** matrix including $\mathbf{W}^0, \mathbf{W}^-, \mathbf{W}^{-+}$ is built.
Step 8: Since the dual is not empty, we go to Step 1.
Step 1: Constraint 6 is introduced, and the dot products are calculated, as shown in Table 6.7.

Table 6.6 Removing redundant vectors from the $\mathbf{W}_c^{-+}$ list.

i	$A_{\mathbf{w}_i}$	$A(\mathbf{W}_c^{-+})$	Removed by
1	{3,4}	{{3,4},{1,4}}	None
2	{1,4}	{{3,4},{1,4}}	None

Table 6.7 Added constraint, **U** matrix, and associated dot products.

Iteration 6					
$\mathbf{a}_6$	$\mathbf{w}_1^6$	$\mathbf{w}_2^6$	$\mathbf{w}_3^6$	$\mathbf{w}_4^6$	$\mathbf{w}_5^6$
0	0	1	0	2	0
–1	0	0	0	1	1
1	0	0	1	0	2
–1	1	0	0	0	0
$\mathbf{t}^6$	–1	0	1	–1	1

Table 6.8 Removing redundant vectors from the $\mathbf{W}_c^{-+}$ list.

i	$A_{\mathbf{w}_i}$	$A(\mathbf{W}_c^{-+})$	Removed by
1	{1,2}	{{1,2},{1,5},{4},{4,5}}	None
2	{1,5}	{{1,2},{1,5},{4},{4,5}}	None
3	{4}	{{1,2},{1,5},{4,5}}	$A(\mathbf{W}^0)$
4	{4,5}	{{1,2},{1,5},{4,5}}	None

Step 2: The orthogonal sets $A(\mathbf{U})$ are updated:

$$A_{\mathbf{w}_1} \equiv \{1,2,3,5\}, A_{\mathbf{w}_2} \equiv \{2,3,4,6\}, A_{\mathbf{w}_3} \equiv \{1,2,4\},$$

$$A_{\mathbf{w}_4} \equiv \{3,4,5\}, A_{\mathbf{w}_5} \equiv \{1,4,5\}.$$

Step 3: No pivot column can be obtained.
Step 4: Go to the Γ_{II}-process.
Step 6.1: $I(\mathbf{W}^-) \equiv \{1,4\}$; $I(\mathbf{W}^+) \equiv \{3,5\}$; $I(\mathbf{W}^0) \equiv \{2\}$.
Step 6.3: The initial list $A(\mathbf{W}_c^{-+})$ of vectors is

$$\{\{1,2\},\{1,5\},\{4\},\{4,5\}\}.$$

The corresponding list of $\mathbf{W}_c^{-+}$ vectors is

$$\{(0,0,1,1)^T,(0,1,2,1)^T,(2,1,1,0)^T,(2,2,2,0)^T\}.$$

Step 7: The elimination of redundant vectors in $\mathbf{W}_c^{-+}$ is performed, as shown in Table 6.8. The set $\{4\}$ is excluded from the $A(\mathbf{W}_c^{-+})$ list because it is contained in the set $A(\mathbf{W}^0) \equiv \{2,3,4,6\}$.

Step 7.5: A new **U** matrix including $\mathbf{W}^0, \mathbf{W}^-, \mathbf{W}^{-+}$ is built and we make $s = 6$.

Step 8: Since the dual is not empty, we go to Step 1.

Step 1: Constraint 7 is introduced, and the dot products are calculated, as shown in Table 6.9.

Table 6.9 Added constraint, **U** matrix, and associated dot products.

Iteration 7						
$\mathbf{a}_7$	$\mathbf{w}_1^7$	$\mathbf{w}_2^7$	$\mathbf{w}_3^7$	$\mathbf{w}_4^7$	$\mathbf{w}_5^7$	$\mathbf{w}_6^7$
2	1	0	2	0	0	2
0	0	0	1	0	1	2
-1	0	0	0	1	2	2
-1	0	1	0	1	1	0
$\mathbf{t}^7$	2	-1	4	-2	-3	2

Step 2: The orthogonal sets $A(\mathbf{U})$ are updated:

$$A_{\mathbf{w}_1} \equiv \{2,3,4,6\}, A_{\mathbf{w}_2} \equiv \{1,2,3,5\}, A_{\mathbf{w}_3} \equiv \{3,4,5\},$$

$$A_{\mathbf{w}_4} \equiv \{1,2,6\}, A_{\mathbf{w}_5} \equiv \{1,5,6\}, A_{\mathbf{w}_6} \equiv \{4,5,6\}.$$

Step 3: No pivot column can be obtained.
Step 4: Go to the Γ_{II}-process.
Step 6: $I(\mathbf{W}^-) = \{2,4,5\}$; $I(\mathbf{W}^+) = \{1,3,6\}$; $I(\mathbf{W}^0) = \{\}$.
Step 6.3: The initial list $A(\mathbf{W}_c^{-+})$ of vectors is

$$\{\{2,3\},\{3,5\},\{5\},\{2,6\},\{\},\{6\},\{6\},\{5\},\{5,6\}\}.$$

The corresponding list of $\mathbf{W}_c^{-+}$ vectors is

$$\{(1,0,0,2)^T,(2,1,0,4)^T,(2,2,2,2)^T,(2,0,2,2)^T,(4,2,4,4)^T,$$
$$(4,4,6,2)^T,(3,2,4,2)^T,(6,7,8,4)^T,(6,8,10,2)^T\}.$$

Step 7: The elimination of redundant vectors in $\mathbf{W}_c^{-+}$ is performed, as shown in Table 6.10. The set $\{5\}$ is excluded from the $A_{\mathbf{w}_i}$ list because it is contained in the set $A_{\mathbf{w}_2} \equiv \{3,5\}$. The set $\{\}$ is excluded from the $A(\mathbf{W}_c^{-+})$ list because it is contained in the set $A_{\mathbf{w}_1} \equiv \{2,3\}$. The set $\{6\}$ is excluded from the $A(\mathbf{W}_c^{-+})$ list because it is contained in the set $A_{\mathbf{w}_4} \equiv \{2,6\}$. The set $\{6\}$ is excluded from the $A(\mathbf{W}_c^{-+})$ list because it is contained in the set $A_{\mathbf{w}_4} \equiv \{2,6\}$. The set $\{5\}$ is excluded from the $A(\mathbf{W}_c^{-+})$ list because it is contained in the set $A_{\mathbf{w}_2} \equiv \{3,5\}$. A new **U** matrix including $\mathbf{W}^0, \mathbf{W}^-, \mathbf{W}^{-+}$ is built.

Step 8: Since the dual is not empty, we go to Step 1, but since there are no more constraints we go to Step 9.

Step 9: We return the solution $\Omega(\mathbf{A}_\pi) \equiv \mathbf{V}_\rho + \mathbf{W}_\pi$, where **V** is the empty set, and a set of generators of **W** is (see Table 6.11)

$$\{(0,0,0,1)^T,(0,0,1,1)^T,(0,1,2,1)^T,(1,0,0,2)^T,$$

$$(2,1,0,4)^T,(1,0,1,1)^T,(3,4,5,1)^T\}.$$

Table 6.10 Removing redundant vectors from the $\mathbf{W}_c^{-+}$ list.

i	$A_{\mathbf{w}_i}$	$A(\mathbf{W}_c^{-+})$	Removed by
1	{2,3}	{{2,3},{3,5},{5},{2,6},{},{6},{6},{5},{5,6}}	None
2	{3,5}	{{2,3},{3,5},{5},{2,6},{},{6},{6},{5},{5,6}}	None
3	{5}	{{2,3},{3,5},{2,6},{},{6},{6},{5},{5,6}}	$A_{\mathbf{w}_2}$
4	{2,6}	{{2,3},{3,5},{2,6},{},{6},{6},{5},{5,6}}	None
5	{}	{{2,3},{3,5},{2,6},{6},{6},{5},{5,6}}	$A_{\mathbf{w}_1}$
6	{6}	{{2,3},{3,5},{2,6},{6},{5},{5,6}}	$A_{\mathbf{w}_4}$
7	{6}	{{2,3},{3,5},{2,6},{5},{5,6}}	$A_{\mathbf{w}_4}$
8	{5}	{{2,3},{3,5},{2,6},{5,6}}	$A_{\mathbf{w}_2}$
9	{5,6}	{{2,3},{3,5},{2,6},{5,6}}	None

Table 6.11 Final table with the dual generators.

Dual						
$\mathbf{w}_1$	$\mathbf{w}_2$	$\mathbf{w}_3$	$\mathbf{w}_4$	$\mathbf{w}_5$	$\mathbf{w}_6$	$\mathbf{w}_7$
0	0	0	1	2	2	6
0	0	1	0	1	0	8
0	1	2	0	0	2	10
1	1	1	2	4	2	2

■

Example 6.12 (Vectors in a cone). Consider the cone $\mathbf{A}_\pi$ in Fig. 6.8, where

$$\mathbf{A} = \begin{pmatrix} 1 & 1 & 0 \\ 1 & 0 & 1 \\ 0 & 1 & 1 \end{pmatrix}, \tag{6.47}$$

and suppose that we wish to know whether or not the vectors $(1,1,1)^T$ and $(1,0,0)^T$ belong to $\mathbf{A}_\pi$. To this aim, the dual $\Omega(\mathbf{A}_\pi)$ is obtained using the Γ-algorithm, as illustrated in Table 6.12, from which we have

$$\mathbf{V} \equiv \emptyset; \mathbf{W} = \begin{pmatrix} -1/2 & -1/2 & 1/2 \\ -1/2 & 1/2 & -1/2 \\ 1/2 & -1/2 & -1/2 \end{pmatrix}. \tag{6.48}$$

We immediately check, using Equation (6.15), whether or not a vector belongs to $\mathbf{A}_\pi$. For the vector $(1,1,1)^T$, we have

$$\mathbf{W}^T \begin{pmatrix} 1 \\ 1 \\ 1 \end{pmatrix} = \begin{pmatrix} -1/2 & -1/2 & 1/2 \\ -1/2 & 1/2 & -1/2 \\ 1/2 & -1/2 & -1/2 \end{pmatrix} \begin{pmatrix} 1 \\ 1 \\ 1 \end{pmatrix} = \begin{pmatrix} -1/2 \\ -1/2 \\ -1/2 \end{pmatrix} \leq \mathbf{0}, \tag{6.49}$$

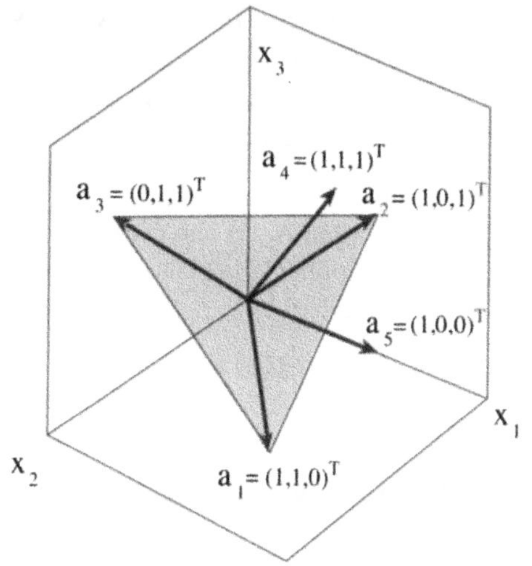

Fig. 6.8 Determining if two vectors belong to a cone.

Table 6.12 Illustration of the different iterations used with the Γ-algorithm to obtain the dual $\Omega(\mathbf{A}_\pi)$ in Example 6.12.

Iteration 1				Iteration 2			
$\mathbf{a}_1$	$\mathbf{v}_1$	$\mathbf{v}_2$	$\mathbf{v}_3$	$\mathbf{a}_2$	$\mathbf{w}_1$	$\mathbf{v}_1$	$\mathbf{v}_2$
1	**1**	0	0	1	−1	**−1**	0
1	**0**	1	0	0	0	**1**	0
0	**0**	0	1	1	0	**0**	1
$\mathbf{t}^1$	**1**	1	0	$\mathbf{t}^2$	−1	**−1**	1

Iteration 3				Dual		
$\mathbf{a}_3$	$\mathbf{w}_1$	$\mathbf{w}_2$	$\mathbf{v}_1$	$\mathbf{w}_1$	$\mathbf{w}_2$	$\mathbf{w}_3$
0	0	−1	**−1**	−1/2	−1/2	1/2
1	−1	1	**1**	−1/2	1/2	−1/2
1	0	0	**1**	1/2	−1/2	−1/2
$\mathbf{t}^3$	−1	1	**2**			

which shows that it belongs to the cone $\mathbf{A}_\pi$. Similarly, for the vector $(1,0,0)^T$, we have

$$\mathbf{W}^T \begin{pmatrix} 1 \\ 0 \\ 0 \end{pmatrix} = \begin{pmatrix} -1/2 & -1/2 & 1/2 \\ -1/2 & 1/2 & -1/2 \\ 1/2 & -1/2 & -1/2 \end{pmatrix} \begin{pmatrix} 1 \\ 0 \\ 0 \end{pmatrix} = \begin{pmatrix} -1/2 \\ -1/2 \\ 1/2 \end{pmatrix} \not\leq \mathbf{0}, \quad (6.50)$$

which shows that it does not belong to the cone $\mathbf{A}_\pi$. ∎

Corollary 6.1 *The vector* $\mathbf{x}$ *belongs to the space* $\mathbf{A}_\rho$ *iff* $\mathbf{V}^T\mathbf{x} = \mathbf{0}$*; that is,*

$$\boxed{\mathbf{x} \in \mathbf{A}_\rho \Leftrightarrow \mathbf{V}^T\mathbf{x} = \mathbf{0},} \tag{6.51}$$

where $\Omega(\mathbf{A}_\rho) \equiv \mathbf{V}_\rho$.

Proof: Taking into account that $\mathbf{A}_\rho$ is the cone $(\mathbf{A} : -\mathbf{A})_\pi$, and that the dual of $\mathbf{A}_\rho$ has only the $\mathbf{V}$ component, that is, its general form is $\Omega(\mathbf{A}_\rho) \equiv \Omega((\mathbf{A} : -\mathbf{A})_\pi) \equiv \mathbf{V}_\rho$, we can write (6.51). ■

Corollary 6.2 *The null vector* $\mathbf{0}$ *does not belong to* $\mathbf{A}_\xi$ *iff* ${\mathbf{V}^{(i)}}^T\mathbf{a}_{i+1} \neq \mathbf{0}, \forall i \leq m$*; that is,*

$$\boxed{\mathbf{0} \notin \mathbf{A}_\xi \Leftrightarrow {\mathbf{V}^{(i)}}^T\mathbf{a}_{i+1} \neq \mathbf{0}, \forall i,} \tag{6.52}$$

where $\mathbf{A}^{(0)} = \mathbf{0}$, $\mathbf{A}^{(i)} = (\mathbf{a}_1, \ldots, \mathbf{a}_i)$ *are the first* i *columns of* $\mathbf{A}$ *and* $\mathbf{V}^{(i)} \equiv \Omega(\mathbf{A}_\rho^{(i)})$.

Proof: It is well known that

$$\mathbf{0} \notin \mathbf{A}_\xi \Leftrightarrow \mathbf{a}_{i+1} \notin \mathbf{A}_\rho^{(i)}, \forall i. \tag{6.53}$$

Applying Corollary 6.1 to the vectors $\mathbf{a}_{i+1}$ and the spaces $\mathbf{A}_\rho^{(i)}$, for all i, we get the desired result. ■

Remark 6.7 *Taking into account that* $\mathbf{0} \notin \mathbf{A}_\xi$ *holds iff the set* $\{\mathbf{a}_1, \ldots, \mathbf{a}_m\}$ *of vectors are linearly independent, condition (6.52) gives a practical way for checking this condition.*

Example 6.13 (Linear independence and cones). Consider the set of vectors $\{(1,1,0)^T, (1,0,1)^T, (0,1,1)^T\}$ of Example 6.12. From Table 6.12, we obtain

$$ {\mathbf{V}^{(0)}}^T = \begin{pmatrix} 1 & 0 & 0 \\ 0 & 1 & 0 \\ 0 & 0 & 1 \end{pmatrix}, {\mathbf{V}^{(1)}}^T = \begin{pmatrix} -1 & 1 & 0 \\ 0 & 0 & 1 \end{pmatrix}, {\mathbf{V}^{(2)}}^T = \begin{pmatrix} -1 & 1 & 1 \end{pmatrix}, \tag{6.54}$$

where the resulting right–hand side values are those in the last row of Table 6.12, excluding columns before the pivots. Then we have

$$\begin{array}{rcccl} {\mathbf{V}^{(0)}}^T\mathbf{a}_1 & = & \begin{pmatrix} 1 & 0 & 0 \\ 0 & 1 & 0 \\ 0 & 0 & 1 \end{pmatrix}\begin{pmatrix} 1 \\ 1 \\ 0 \end{pmatrix} & = & \begin{pmatrix} 1 & 1 & 0 \end{pmatrix} \neq \mathbf{0}, \\ {\mathbf{V}^{(1)}}^T\mathbf{a}_2 & = & \begin{pmatrix} -1 & 1 & 0 \\ 0 & 0 & 1 \end{pmatrix}\begin{pmatrix} 1 \\ 0 \\ 1 \end{pmatrix} & = & \begin{pmatrix} -1 & 1 \end{pmatrix} \neq \mathbf{0}, \\ {\mathbf{V}^{(2)}}^T\mathbf{a}_3 & = & \begin{pmatrix} -1 & 1 & 1 \end{pmatrix}\begin{pmatrix} 0 \\ 1 \\ 1 \end{pmatrix} & = & (2) \neq \mathbf{0}. \end{array} \tag{6.55}$$

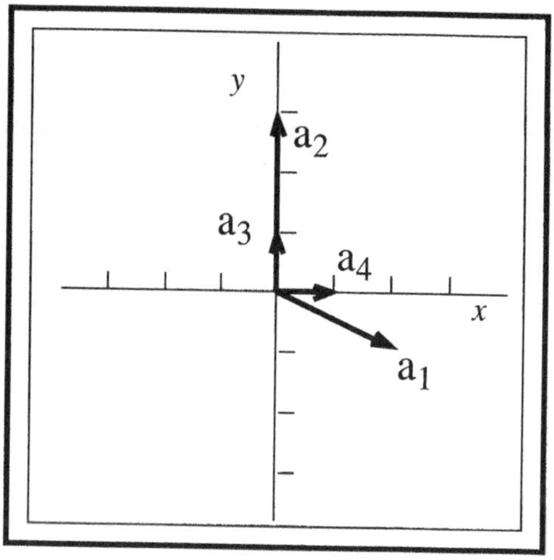

Fig. 6.9 Vectors in Example 6.14.

Thus, the Γ-algorithm process also determines whether or not a set of vectors is linearly independent.

If we now replace the last vector above $(0,1,1)^T$ by $(2,1,1)^T$, the last row in (6.55) becomes

$$\mathbf{V}^{(2)^T}\tilde{\mathbf{a}}_3 = \begin{pmatrix} -1 & 1 & 1 \end{pmatrix}\begin{pmatrix} 2 \\ 1 \\ 1 \end{pmatrix} = \mathbf{0}, \tag{6.56}$$

which shows that the vectors $\{(1,1,0)^T, (1,0,1)^T, (2,1,1)^T\}$ are linearly dependent.

■

Definition 6.9 (Minimum set of generators. Basis). *Given a cone* $\mathbf{A}_\pi$, *a set of vectors* $\{\mathbf{b}_1, \ldots, \mathbf{b}_m\}$ *is said to be a* minimal set of generators *or a* basis *of* $\mathbf{A}_\pi$ *iff*

$$\mathbf{B}_\pi \equiv \mathbf{A}_\pi \wedge \mathbf{b}_i \notin (\mathbf{B}^{\sim \mathbf{b}_i})_\pi, \forall i, \tag{6.57}$$

where $\mathbf{B}^{\sim \mathbf{b}_i}$ *is the matrix* $\mathbf{B}$ *after removing its column* $\mathbf{b}_i$.

Remark 6.8 *A minimum set of generators of a cone* $\mathbf{A}_\pi$ *can be obtained by using the identity* $\mathbf{A}_\pi \equiv \Omega(\Omega(\mathbf{A}_\pi))$ *and the* Γ*-algorithm.*

Example 6.14 (Finding a minimum set of generators for a cone). Consider the cone $\mathbf{A}_\pi$ (see Fig. 6.9), where

$$\mathbf{A} = \begin{pmatrix} 2 & 0 & 0 & 1 \\ -1 & 3 & 1 & 0 \end{pmatrix}.$$

Table 6.13 Iterations for obtaining $\Omega(\mathbf{A}_\pi)$ in Example 6.14.

Iteration 1			Iteration 2			Iteration 3		
$\mathbf{a}_1$	$\mathbf{v}_1^1$	$\mathbf{v}_2^1$	$\mathbf{a}_2$	$\mathbf{w}_1^2$	$\mathbf{v}_1^2$	$\mathbf{a}_3$	$\mathbf{w}_1^3$	$\mathbf{w}_2^3$
2	**1**	0	0	–1/2	**1/2**	0	–1/2	–1/6
–1	**0**	1	3	0	**1**	1	0	–1/3
$\mathbf{t}^1$	**2**	–1	$\mathbf{t}^2$	0	**3**	$\mathbf{t}^3$	0	–1/3

Iteration 4			Dual	
$\mathbf{a}_4$	$\mathbf{w}_1^4$	$\mathbf{w}_2^4$	$\mathbf{w}_1$	$\mathbf{w}_2$
1	–1/2	–1/6	–1/2	–1/6
0	0	–1/3	0	–1/3
$\mathbf{t}^4$	–1/2	–1/6		

Table 6.14 Iterations for obtaining $\Omega(\Omega(\mathbf{A}_\pi))$ in Example 6.14.

Iteration 1			Iteration 2			Dual	
$\mathbf{a}_1$	$\mathbf{v}_1^1$	$\mathbf{v}_2^1$	$\mathbf{a}_2$	$\mathbf{w}_1^2$	$\mathbf{v}_1^2$	$\mathbf{w}_1$	$\mathbf{w}_2$
–1/2	**1**	0	–1/6	2	**0**	2	0
0	**0**	1	–1/3	0	**1**	–1	3
$\mathbf{t}^1$	**–1/2**	0	$\mathbf{t}^2$	–1/3	**–1/3**		

Using the Γ-algorithm, we obtain the results in Table 6.13, from which we get

$$\Omega(\mathbf{A}) \equiv \mathbf{V}_\rho + \mathbf{W}_\pi, \tag{6.58}$$

where

$$\mathbf{V} \equiv \emptyset; \mathbf{W} \equiv \begin{pmatrix} -1/2 & -1/6 \\ 0 & -1/3 \end{pmatrix}. \tag{6.59}$$

Using again the Γ-algorithm, with the columns of $\mathbf{W}$, we obtain the results in Table 6.14:

$$\Omega(\Omega(\mathbf{A})) \equiv \mathbf{V}_\rho^* + \mathbf{W}_\pi^*, \tag{6.60}$$

where

$$\mathbf{V}^* = \emptyset; \mathbf{W}^* = \begin{pmatrix} 2 & 0 \\ -1 & 3 \end{pmatrix}, \tag{6.61}$$

which shows that a minimum set of generators for the cone $\mathbf{A}_\pi$ is

$$\{\mathbf{a}_1, \mathbf{a}_2\} \equiv \{(2,-1)^T, (0,3)^T\}.$$

■

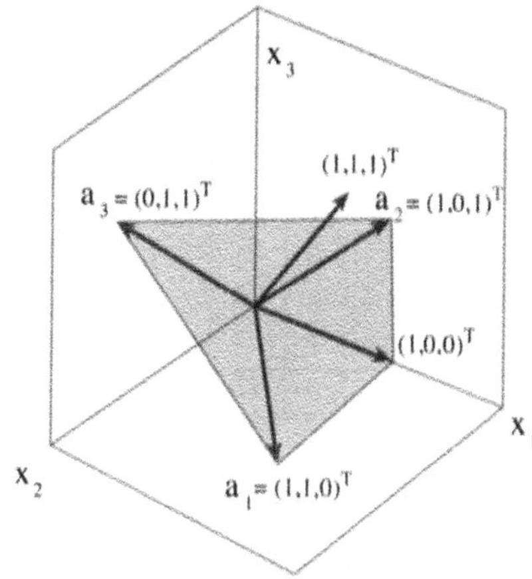

Fig. 6.10 A cone and a minimal set of generators.

Example 6.15 (Minimal set of generators). Consider the cone $\mathbf{A}_\pi$ (see Fig. 6.10), where

$$\mathbf{A} = \begin{pmatrix} 1 & 1 & 0 & 1 & 1 \\ 1 & 0 & 1 & 1 & 0 \\ 0 & 1 & 1 & 1 & 0 \end{pmatrix}.$$

Using the Γ-algorithm, we obtain the results in Table 6.15, from which we get

$$\Omega(\mathbf{A}) \equiv \mathbf{V}_\rho + \mathbf{W}_\pi, \tag{6.62}$$

where

$$\mathbf{V} = \emptyset; \mathbf{W} = \frac{1}{2}\begin{pmatrix} -1 & -1 & 0 & 0 \\ -1 & 1 & -1 & 0 \\ 1 & -1 & 0 & -1 \end{pmatrix}. \tag{6.63}$$

Using again the Γ-algorithm, with the columns of $\mathbf{W}$, we obtain the results in Table 6.16:

$$\Omega(\Omega(\mathbf{A})) \equiv \mathbf{V}^*_\rho + \mathbf{W}^*_\pi, \tag{6.64}$$

where

$$\mathbf{V}^* = \emptyset; \mathbf{W}^* = \begin{pmatrix} 1/2 & 0 & 1/2 & 1/2 \\ 0 & 1 & 0 & 1/2 \\ 1/2 & 1 & 0 & 0 \end{pmatrix}, \tag{6.65}$$

which shows that a minimum set of generators for the cone $\mathbf{A}_\pi$ is

$$\{(1/2, 0, 1/2)^T, (0, 1, 1)^T, (1/2, 0, 0)^T, (1/2, 1/2, 0)^T\},$$

or equivalently (multiplying the first, third, and fourth vectors by 2) (see Fig. 6.10):

$$\{(1, 0, 1)^T, (0, 1, 1)^T, (1, 0, 0)^T, (1, 1, 0)^T\}.$$

■

Table 6.15 Illustration of the different iterations used with the Γ-algorithm to obtain the dual $\Omega(\mathbf{A}_\pi)$ in Example 6.15.

Iteration 1				Iteration 2			
$\mathbf{a}_1$	$\mathbf{v}_1$	$\mathbf{v}_2$	$\mathbf{v}_3$	$\mathbf{a}_2$	$\mathbf{w}_1$	$\mathbf{v}_1$	$\mathbf{v}_2$
1	**1**	0	0	1	–1	**–1**	0
1	**0**	1	0	0	0	**1**	0
0	**0**	0	1	1	0	**0**	1
$\mathbf{t}^1$	**1**	1	0	$\mathbf{t}^2$	–1	**–1**	1

Iteration 3				Iteration 4			
$\mathbf{a}_3$	$\mathbf{w}_1$	$\mathbf{w}_2$	$\mathbf{v}_1$	$\mathbf{a}_4$	$\mathbf{w}_1$	$\mathbf{w}_2$	$\mathbf{w}_3$
0	0	–1	**–1**	1	–1/2	–1/2	1/2
1	–1	1	**1**	1	–1/2	1/2	–1/2
1	0	0	**1**	1	1/2	–1/2	–1/2
$\mathbf{t}^3$	–1	1	**2**	$\mathbf{t}^4$			

Iteration 5				Dual			
$\mathbf{a}_5$	$\mathbf{w}_1$	$\mathbf{w}_2$	$\mathbf{w}_3$	$\mathbf{w}_1$	$\mathbf{w}_1$	$\mathbf{w}_3$	$\mathbf{w}_4$
1	–1/2	–1/2	1/2	–1/2	–1/2	0	0
0	–1/2	1/2	–1/2	–1/2	1/2	–1/2	0
0	1/2	–1/2	–1/2	1/2	–1/2	0	–1/2
$\mathbf{t}^5$	–1/2	–1/2	1/2				

Table 6.16 Illustration of the different iterations used with the Γ-algorithm to obtain the bidual $\Omega(\Omega(\mathbf{A}_\pi))$ in Example 6.15.

Iteration 1				Iteration 2				Iteration 3			
$\mathbf{a}_1$	$\mathbf{v}_1$	$\mathbf{v}_2$	$\mathbf{v}_3$	$\mathbf{a}_2$	$\mathbf{w}_1$	$\mathbf{v}_1$	$\mathbf{v}_2$	$\mathbf{a}_3$	$\mathbf{w}_1$	$\mathbf{w}_2$	$\mathbf{v}_1$
–1	**1**	0	0	–1	1	**–1**	1	0	1/2	1/2	**0**
–1	**0**	1	0	1	0	**1**	0	–1	1/2	–1/2	**1**
1	**0**	0	1	–1	0	**0**	1	0	0	0	**1**
$\mathbf{t}^1$	**–1**	–1	1	$\mathbf{t}^2$	–1	**2**	–2	$\mathbf{t}^3$	–1/2	1/2	**–1**

Iteration 4				Bidual			
$\mathbf{a}_5$	$\mathbf{w}_1$	$\mathbf{w}_2$	$\mathbf{w}_3$	$\mathbf{w}_1$	$\mathbf{w}_1$	$\mathbf{w}_3$	$\mathbf{w}_4$
0	1/2	1/2	0	1/2	0	1/2	1/2
0	0	0	1	0	1	0	1/2
–1	–1/2	1/2	1	1/2	1	0	0
$\mathbf{t}^4$	1/2	–1/2	–1				

6.5.3 Extended, Reduced, and Notable Cones

Definition 6.10 (Extended cone). *The cones* $(\mathbf{A} : \mathbf{x})_\pi$ *and* $(\mathbf{A} : \mathbf{X})_\pi$ *defined in Propositions 6.1 and 6.2 are known as extended cones of* $\mathbf{A}_\pi$.

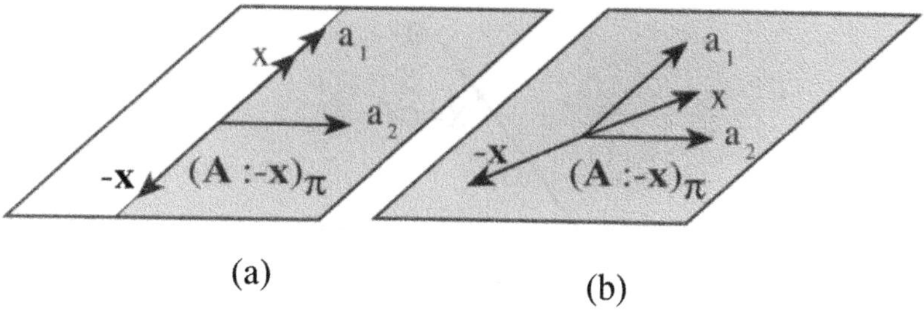

Fig. 6.11 Two examples of the notable cones of a vector.

Example 6.16 (Extended cone in E^2). Cone (b) of Fig. 6.5 is the extended cone of (a) by the vector $\mathbf{a}_3$. ■

Definition 6.11 (Reduced cone). *Given the cone $\mathbf{A}_\pi$ and one of its generators $\mathbf{a}_h$, we shall call the cone of $\mathbf{A}_\pi$ reduced by $\mathbf{a}_h$ the reduced cone $(\mathbf{A}^{\sim \mathbf{a}_h})_\pi$, where $\mathbf{A}^{\sim \mathbf{a}_h}$ is the matrix $\mathbf{A}$ without column $\mathbf{a}_h$.*

Example 6.17 (Reduced cone in E^3). The cone (b) in Fig. 6.6 is the reduced cone of (c) in the same figure by the vector $\mathbf{a}_4$. ■

Definition 6.12 (Notable cone). *Given the cone $\mathbf{A}_\pi$ and one vector $\mathbf{x}$, the cone of $\mathbf{A}_\pi$ extended by the vector $-\mathbf{x}$, that is, the cone $(\mathbf{A} : -\mathbf{x})_\pi$, is known as the* notable cone *of $\mathbf{x}$ with respect to the cone $\mathbf{A}_\pi$. Similarly, the notable cone of a set of vectors $\mathbf{X}$ with respect to the cone $\mathbf{A}_\pi$ can be defined by means of $(\mathbf{A} : -\mathbf{X})_\pi$.*

Example 6.18 (Notable cone). Fig. 6.11 shows two examples of notable cones. In the first, the cone reduces to a halfplane because $\mathbf{x}$ is proportional to one of the generators. In the second, the notable cone is the whole E^2. ■

Remark 6.9 *The linear space generated by $\mathbf{A}$ coincides with its notable cone with respect to $\mathbf{A}_\pi$:*

$$\mathbf{A}_\rho \equiv (\mathbf{A} : -\mathbf{A})_\pi. \tag{6.66}$$

Theorem 6.3 *Let $\mathbf{A}_\pi$ be a cone and $\mathbf{x}$ be an arbitrary vector. The vector $-\mathbf{x}$ belongs to the notable cone of $\mathbf{a}_i$ with respect to $\mathbf{A}_\pi$ iff it contains the notable cone of $\mathbf{x}$ with respect to $\mathbf{A}_\pi$; that is,*

$$-\mathbf{x} \in (\mathbf{A} : -\mathbf{a}_i)_\pi \Leftrightarrow (\mathbf{A} : -\mathbf{x})_\pi \subseteq (\mathbf{A} : -\mathbf{a}_i)_\pi. \tag{6.67}$$

The proof of this theorem can be seen in Jubete (1991).

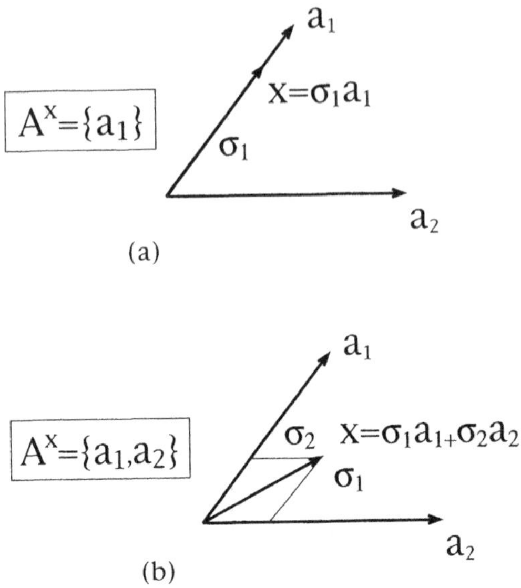

Fig. 6.12 Binding sets of the vector $\mathbf{x}$ in the cones generated by $\mathbf{a}_1$ and $\mathbf{a}_2$ in cases (a) and (b).

6.5.4 Binding Set of a Vector

Definition 6.13 (Binding set of a vector). *Let* $\mathbf{A}_\pi$ *be a cone and* $\mathbf{x}$ *be a vector. Then the binding set of* $\mathbf{x}$ *is the set of the cone generators that define* $\mathbf{x}$*, that is,*

$$\mathbf{A}^{\mathbf{x}} \equiv \{\mathbf{a}_i \in \mathbf{A} \mid \mathbf{x} \in \mathbf{A}_\pi + \sigma \mathbf{a}_i\}, \tag{6.68}$$

where $\sigma > 0$.

Note that this is equivalent to stating that $\mathbf{x}$ can be written as a linear combination of the generators with a positive component with respect to $\mathbf{a}_i$.

Remark 6.10 *Obviously, we have*

- *the binding set of a vector that doesn't belong to the cone is the empty set, and*
- *every cone's generator belongs to its binding set, that is,* $\mathbf{a}_i \in \mathbf{A}^{\mathbf{a}_i}$.

Example 6.19 (Binding set in E^2). Fig. 6.12 shows the binding sets of two vectors in a cone generated by the two vectors $\mathbf{a}_1$ and $\mathbf{a}_2$. ■

Proposition 6.4 *Every non-null vector of a cone is a strictly positive linear combination of its binding set's elements, that is,*

$$\mathbf{0} \neq \mathbf{x} \in \mathbf{C}_\pi \Rightarrow \mathbf{x} \in \mathbf{C}_\sigma^{\mathbf{x}}. \tag{6.69}$$

Proof: Let $\mathbf{x} \in \mathbf{A}_\pi$. Then

$$\mathbf{x} = \pi_1 \mathbf{a}_1 + \cdots + \pi_m \mathbf{a}_m,$$

with $\pi_i \geq 0; i = 1, 2, \ldots m$. If $\mathbf{x} \neq \mathbf{0}$ it can be assumed, without loss of generality, that there exists $k \geq 1$ such that the first k π_i coefficients are strictly positive and the last $m - k$ are null. Then

$$\mathbf{x} = \pi_1 \mathbf{a}_1 + \cdots + \pi_k \mathbf{a}_k \quad \text{with} \quad \pi_i > 0; i = 1, \ldots, k,$$

and $\mathbf{a}_1, \ldots, \mathbf{a}_k \in \mathbf{A}^{\mathbf{x}}$.

■

Proposition 6.5 *If $\mathbf{A}^{\mathbf{a}_i}$ is the binding set of $\mathbf{a}_i$, we have*

$$\mathbf{a}_i \in \mathbf{A}^{\mathbf{x}} \Leftrightarrow \mathbf{A}^{\mathbf{a}_i} \subseteq \mathbf{A}^{\mathbf{x}}.$$

Proof: The vectors $\mathbf{x}$ and $\mathbf{a}_i$ can be expressed as strictly positive linear combinations of the vectors in their binding sets:

$$\begin{aligned} \mathbf{x} &= \sum_{\mathbf{a}_k \in \mathbf{A}^{\mathbf{x}}} \mathbf{a}_k \sigma_k, \\ \mathbf{a}_i &= \sum_{\mathbf{a}_j \in \mathbf{A}^{\mathbf{a}_i}} \mathbf{a}_j \bar{\sigma_j}. \end{aligned}$$

Suppose that $\mathbf{a}_i \subset \mathbf{A}^{\mathbf{x}}$. Then

$$\mathbf{x} = \sum_{\substack{\mathbf{a}_k \in \mathbf{A}^{\mathbf{x}} \\ \mathbf{a}_k \neq \mathbf{a}_i}} \mathbf{a}_k \sigma_k + \left(\sum_{\mathbf{a}_j \in \mathbf{A}^{\mathbf{a}_i}} \mathbf{a}_j \bar{\sigma_j} \right) \sigma_i. \tag{6.70}$$

In (6.70) every vector $\mathbf{a}_j \in \mathbf{A}^{\mathbf{a}_i}$ has a strictly positive coefficient, and thus $\mathbf{a}_j \in \mathbf{A}^{\mathbf{x}}$.

The reverse implication is obvious because $\mathbf{a}_i \in \mathbf{A}^{\mathbf{a}_i} \subseteq \mathbf{A}^{\mathbf{x}}$.

■

Proposition 6.6 *Let $\mathbf{x} \in \mathbf{A}_\pi$ and $\mathbf{a}_i$ be a cone's generator. Then*

$$\mathbf{a}_i \in \mathbf{A}^{\mathbf{x}} \Leftrightarrow \mathbf{x} \in \mathbf{A}_\pi \wedge (\mathbf{A} : -\mathbf{a}_i)_\pi \subseteq (\mathbf{A} : -\mathbf{x})_\pi, \tag{6.71}$$

or its dual equivalent

$$\boxed{\mathbf{a}_i \in \mathbf{A}^{x} \Leftrightarrow \mathbf{U}^T \mathbf{x} \leq \mathbf{0} \wedge \mathbf{W}(\mathbf{x}) \subseteq \mathbf{W}(\mathbf{a}_i),} \tag{6.72}$$

where $\Omega(\mathbf{A}_\pi) \equiv \mathbf{V}_\rho + \mathbf{W}_\pi$, *and* $\mathbf{W}(\mathbf{x})$ *and* $\mathbf{W}(\mathbf{a}_i)$ *are the subsets of* $\mathbf{W}$ *orthogonal to* $\mathbf{x}$ *and* $\mathbf{a}_i$, *respectively.*

Proof: Using Proposition 6.3, we have an alternative formulation:

$$\mathbf{a}_i \in \mathbf{A}^{\mathbf{x}} \iff -\mathbf{a}_i \in (\mathbf{A} : -\mathbf{x})_\pi.$$

In effect, if $\mathbf{a}_i \in \mathbf{A}^{\mathbf{x}}$,

$$\mathbf{x} = \sum_{j \neq i} \pi_j \mathbf{a}_j + \sigma_i \mathbf{a}_i.$$

Since $\sigma_i > 0$,

$$-\mathbf{a}_i = \sum_{j \neq i} \frac{\pi_j}{\sigma_i} \mathbf{a}_j - \frac{1}{\sigma_i} \mathbf{x},$$

and then $-\mathbf{a}_i \in (\mathbf{A} : -\mathbf{x})_\pi$, because $\pi_j/\sigma_i \geq 0, \forall j \neq i$ and $1/\sigma_i > 0$.

Conversely, if $-\mathbf{a}_i \in (\mathbf{A} : -\mathbf{x})_\pi$, since $\mathbf{A}_\pi \equiv (\mathbf{A} : \mathbf{x})_\pi$, we conclude $-\mathbf{a}_i \in (\mathbf{A} : \mathbf{x} : -\mathbf{x})_\pi$ and then

$$-\mathbf{a}_i = \sum_j \pi_j \mathbf{a}_j + \pi_x \mathbf{x} - \pi_{-x} \mathbf{x} = \sum_j \mathbf{a}_j \pi_j + \mathbf{x}\rho,$$

where $\rho = \pi_x - \pi_{-x}$. Now, since every real number can be expressed as the difference of two strictly positive numbers σ_1 and σ_2, we have $\rho = \sigma_1 - \sigma_2$ and thus

$$\mathbf{x} = \sum_j \frac{\pi_j}{\sigma_2} \mathbf{a}_j + \frac{\sigma_1}{\sigma_2} \mathbf{x} + \frac{1}{\sigma_2} \mathbf{a}_i.$$

Finally, since $\mathbf{x} \in \mathbf{A}_\pi$, we have $\mathbf{x} = \sum_j \pi_j^* \mathbf{a}_j$ and thus

$$\mathbf{x} = \sum_j (\pi_j + \frac{\sigma_1}{\sigma_2} \pi_j^*) \mathbf{a}_j + \frac{1}{\sigma_2} \mathbf{a}_i,$$

which proves that $\mathbf{a}_i \in \mathbf{A}^{\mathbf{x}}$.

In relation to (6.72), since (6.71) holds, we know that

$$\Omega\left((\mathbf{A} : -\mathbf{x})_\pi\right) \subseteq \Omega\left((\mathbf{A} : -\mathbf{a}_i)_\pi\right),$$

and since $\mathbf{x} \in \mathbf{A}_\pi$, from (6.16), we get

$$\Omega\left((\mathbf{A} : -\mathbf{x})_\pi\right) \equiv \{\boldsymbol{\mu} \in \mathbf{U}_\pi | \boldsymbol{\mu} \bullet \mathbf{x} = 0\} \equiv \mathbf{V}_\rho + (\mathbf{W}(\mathbf{x}))_\pi.$$

Similarly,

$$\Omega\left((\mathbf{A} : -\mathbf{a}_i)_\pi\right) \equiv \mathbf{V}_\rho + (\mathbf{W}(\mathbf{a}_i))_\pi,$$

and thus $(\mathbf{W}(\mathbf{x}))_\pi \subseteq (\mathbf{W}(\mathbf{a}_i))_\pi$ or equivalently $\mathbf{W}(\mathbf{x}) \subseteq (\mathbf{W}(\mathbf{a})_i)$.

In addition, note that if $\mathbf{W}(\mathbf{x}) \equiv \emptyset$, by definition, $\emptyset$ belongs to any set and, in particular, $\emptyset \subseteq \emptyset$; thus,

$$\mathbf{x} \in \mathbf{A}_\pi \wedge \mathbf{W}(\mathbf{x}) \equiv \emptyset \Rightarrow \mathbf{a}_i \in \mathbf{A}^{\mathbf{x}}, \forall i. \tag{6.73}$$

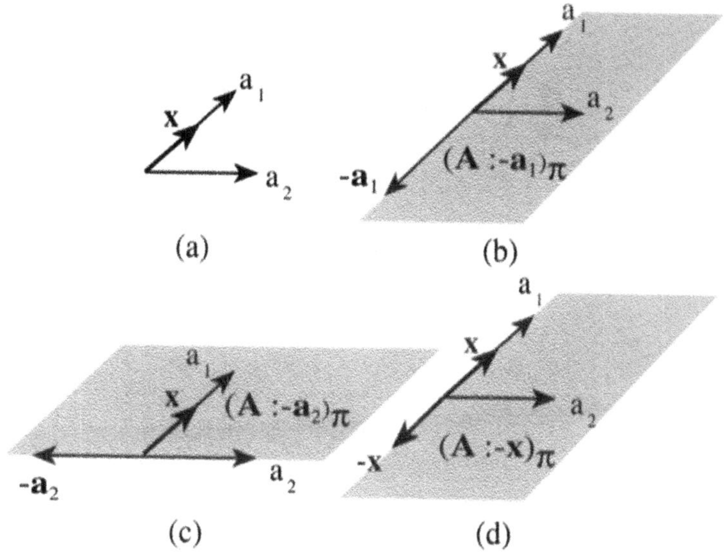

Fig. 6.13 Illustration of the process for obtaining the set $\mathbf{A}^{\mathsf{x}}$.

∎

Example 6.20 (Binding set). Fig. 6.13 shows the required steps for obtaining the set $\mathbf{A}^{\mathsf{x}}$ when the vector $\mathbf{x}$ is proportional to $\mathbf{a}_1$. To this aim, the cones $(\mathbf{A} : -\mathbf{a}_1)_\pi$, $(\mathbf{A} : -\mathbf{a}_2)_\pi$, and $(\mathbf{A} : -\mathbf{x})_\pi$ are shown in Fig. 6.13 (b), (c), and (d), respectively. Since the cone in (b) is contained (coincides) in the cone in (d), the generator $\mathbf{a}_1$ belongs to $\mathbf{A}^{\mathsf{x}}$. In contrast, since the cone in (c) is not included in the cone in (d), the generator $\mathbf{a}_2$ does not belong to $\mathbf{A}^{\mathsf{x}}$.

∎

Example 6.21 (Binding set). Similarly, Fig. 6.14 shows how to obtain the set $\mathbf{A}^{\mathsf{x}}$ in a different case. The cones $(\mathbf{A} : -\mathbf{a}_1)_\pi$, $(\mathbf{A} : -\mathbf{a}_2)_\pi$, and $(\mathbf{A} : -\mathbf{x})_\pi$ are shown in Fig. 6.14(b), (c), and (d), respectively. Since the cone in (d) coincides with the whole E^2, both $\mathbf{a}_1$ and $\mathbf{a}_2$ belong to $\mathbf{A}^{\mathsf{x}}$.

∎

Remark 6.11 *Note that Expression (6.72) gives a practical method for obtaining the binding set of a vector.*

Example 6.22 (Obtaining the binding set). Here we give an example of how to obtain the binding set of a vector using Proposition 6.6. Consider the cone in Example 6.15, where it was shown that

$$\mathbf{V} = \emptyset; \mathbf{W} = \begin{pmatrix} -1/2 & -1/2 & 0 & 0 \\ -1/2 & 1/2 & -1/2 & 0 \\ 1/2 & -1/2 & 0 & -1/2 \end{pmatrix}. \tag{6.74}$$

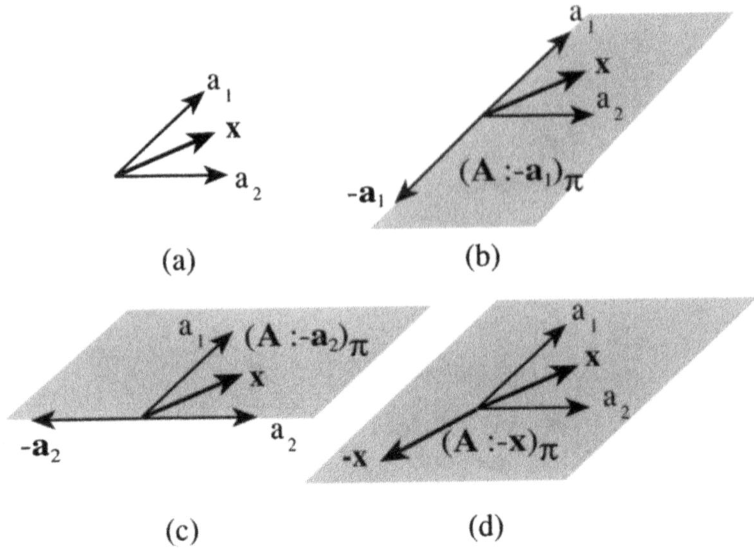

Fig. 6.14 Illustration of the process for obtaining the set $\mathbf{A}^{\mathbf{x}}$.

To obtain the binding set of the vector $\mathbf{x}_1 = (1,1,1)^T$, we calculate

$$\mathbf{U}^T\mathbf{x}_1 = \begin{pmatrix} -1/2 & -1/2 & 1/2 \\ -1/2 & 1/2 & -1/2 \\ 0 & -1/2 & 0 \\ 0 & 0 & -1/2 \end{pmatrix} \begin{pmatrix} 1 \\ 1 \\ 1 \end{pmatrix} = \begin{pmatrix} -1/2 \\ -1/2 \\ -1/2 \\ -1/2 \end{pmatrix} \le \mathbf{0}, \tag{6.75}$$

and

$$\begin{array}{l} \mathbf{W}(\mathbf{x}_1) \equiv \emptyset; \mathbf{W}(\mathbf{a}_1) \equiv \{\mathbf{w}_2, \mathbf{w}_4\}; \mathbf{W}(\mathbf{a}_2) \equiv \{\mathbf{w}_1, \mathbf{w}_3\}; \\ \mathbf{W}(\mathbf{a}_3) \equiv \{\mathbf{w}_1, \mathbf{w}_2\}; \mathbf{W}(\mathbf{a}_4) \equiv \emptyset; \mathbf{W}(\mathbf{a}_5) \equiv \{\mathbf{w}_3, \mathbf{w}_4\}. \end{array} \tag{6.76}$$

Now using Proposition 6.6, we can conclude that the binding set of $(1,1,1)^T$ is

$\mathbf{A}^{\mathbf{x}_1} \equiv \{\mathbf{a}_1, \mathbf{a}_2, \mathbf{a}_3, \mathbf{a}_4, \mathbf{a}_5\} \equiv \{(1,1,0)^T, (1,0,1)^T, (0,1,1)^T, (1,1,1)^T, (1,0,0)^T\}$.

Similarly, the binding set of $\mathbf{x}_2 = (1,1,2)^T$, taking into account that

$$\mathbf{U}^T\mathbf{x}_2 = \begin{pmatrix} -1/2 & -1/2 & 1/2 \\ -1/2 & 1/2 & -1/2 \\ 0 & -1/2 & 0 \\ 0 & 0 & -1/2 \end{pmatrix} \begin{pmatrix} 1 \\ 1 \\ 2 \end{pmatrix} = \begin{pmatrix} 0 \\ -1 \\ -1/2 \\ -1 \end{pmatrix} \le \mathbf{0} \tag{6.77}$$

and

$$\mathbf{W}(\mathbf{x}_2) \equiv \{\mathbf{w}_1\}, \tag{6.78}$$

is

$$\mathbf{A}^{\mathbf{x}_2} \equiv \{\mathbf{a}_2, \mathbf{a}_3\} \equiv \{(1,0,1)^T, (0,1,1)^T\}.$$

■

6.5.5 Interior, Exterior, and Boundary of a Cone

Definition 6.14 (Interior of a cone). *The vector* $\mathbf{x}$ *is said to be interior to a cone* $\mathbf{A}_\pi$ *iff its binding set coincides with the set of the cone generators, that is,*

$$\mathbf{x} \in \mathbf{A}_\pi^{in} \Leftrightarrow \mathbf{A}^x \equiv \mathbf{A}. \tag{6.79}$$

The set of interior points in a cone is called its interior set *and is denoted as* $\mathbf{A}_\pi^{in}$.

Example 6.23 (Interior of a cone). Consider the cone in Example 6.22. The vector $(1,1,1)^T$ belongs to the interior of the cone $\mathbf{A}_\pi$ because, as shown in that example,

$$\mathbf{A}^{\mathbf{x}_1} \equiv \mathbf{A}^{(1,1,1)^T} \equiv \{(1,1,0)^T, (1,0,1)^T, (0,1,1)^T, (1,1,1)^T, (1,0,0)^T\} = \mathbf{A}.$$

Similarly, the vector $(1,1,2)^T$ does not belong to the interior of the cone $\mathbf{A}_\pi$ because

$$\mathbf{A}^{\mathbf{x}_2} \equiv \mathbf{A}^{(1,1,2)^T} \equiv \{\mathbf{a}_2, \mathbf{a}_3\} \equiv \{(1,0,1)^T, (0,1,1)^T\} \neq \mathbf{A}.$$

■

Proposition 6.7 *The vector* $\mathbf{x}$ *belongs to the interior of the cone* $\mathbf{A}_\pi$ *iff it belongs to* $\mathbf{A}_\pi$ *and* $\mathbf{A} \equiv \mathbf{A}^{\mathbf{x}}$.

Proposition 6.8 *A generator* $\mathbf{a}_i$ *belongs to the cone's interior* $\mathbf{A}_\pi$ *iff* $\mathbf{A}^{a_i} \equiv \mathbf{A}$.

Definition 6.15 (Boundary of a cone). *The vector* $\mathbf{x}$ *is said to belong to the boundary of the cone* $\mathbf{A}_\pi$ *iff its binding set does not coincide with the set of its generators. The set of boundary vectors in a cone is called its* boundary set *and is denoted as* $\mathbf{A}_\pi^{bd}$

Example 6.24 (Boundary of a cone). Consider the cone in Example 6.22. The vector $(1,1,1)^T$ does not belong to the boundary of the cone $\mathbf{A}_\pi$ because

$$\mathbf{A}^{\mathbf{x}_1} \equiv \mathbf{A}^{(1,1,1)^T} \equiv \{(1,1,0)^T, (1,0,1)^T, (0,1,1)^T, (1,1,1)^T, (1,0,0)^T\} = \mathbf{A}.$$

Similarly, the vector $(1,1,2)^T$ belongs to the boundary of the cone $\mathbf{A}_\pi$, because

$$\mathbf{A}^{\mathbf{x}_2} \equiv \mathbf{A}^{(1,1,2)^T} \equiv \{\mathbf{a}_2, \mathbf{a}_3\} \equiv \{(1,0,1)^T, (0,1,1)^T\} \neq \mathbf{A}.$$

■

Proposition 6.9 *Every non-null vector* $\mathbf{x}$ *of the cone* $\mathbf{A}_\pi$ *belongs to* $\mathbf{A}_\sigma^{\mathbf{x}}$.

Proposition 6.10 *Every vector* $\mathbf{x}$ *in the boundary of the cone* $\mathbf{A}_\pi$ *belongs to* $\mathbf{A}_\sigma^{\mathbf{x}}$.

Definition 6.16 (Exterior of a cone). *The vector* $\mathbf{x}$ *is said to be exterior to the cone* $\mathbf{A}_\pi$ *if it does not belong to it. The set of exterior vectors to a cone is called its* exterior set *and is denoted as* $\mathbf{A}_\pi^{ex}$.

6.6 THE COMPLETE Γ-ALGORITHM

In this section we give the complete Γ-algorithm for those readers interested in a computer implementation.

Algorithm 6.2 The Γ-algorithm.

- **Input:** A set of generators $\{\mathbf{b}_1, \ldots, \mathbf{b}_{n_b}\}$ of the linear space $\mathbf{B}_\rho$ and a set of generators $\{\mathbf{c}_1, \ldots, \mathbf{c}_{n_c}\}$ of the cone $\mathbf{C}_\pi$, both defined in E^n.
- **Output:** The general form of the dual $\Omega(\mathbf{A}_\pi)$, given as two lists of vectors, corresponding to the $\mathbf{V}_\rho$ and $\mathbf{W}_\pi$ components of its general form.

Initialization:

- Let $\mathbf{A} = (\mathbf{B} : \mathbf{C})$, $\mathbf{U} = \mathbf{I}_n$, where $\mathbf{I}_n$ is the identity matrix of dimension n, and $m = n_b + n_c$.
- Let $A(\mathbf{U})$ equal a list of n empty lists, and $s = 0$.

Repeat for h=1 to m:

1. **Introduce the $\mathbf{a}_h^T$ vector and calculate the dot products:**

   Let $\mathbf{t} = (t_1, \ldots, t_n) = \mathbf{a}_h^T \bullet \mathbf{U}$.

2. **Update the orthogonal sets** $A(\mathbf{U}) = \{A_{\mathbf{u}_j}; j = 1, \ldots, n\}$:

   Repeat for $j = 1$ to n: If $t_j = 0$ then add index h to $A_{\mathbf{u}_j}$.

3. **Look for the pivot:**

   3.1 Obtain $pivot = \min\{j | t_j \neq 0, s + 1 \leq j \leq n\}$ or $pivot = 0$ if this set is empty.

   3.2 If $\min(pivot, s + 1) = n$ and $h \leq n_b$, then output "$\mathbf{V}_\rho = \emptyset$ and $\mathbf{W}_\pi = \emptyset$" and exit the algorithm.

4. **Test for $\Gamma_{\mathbf{I}}$ or $\Gamma_{\mathbf{II}}$:** If $s < n$ and $pivot > 0$ then go to the Γ_I-process. Otherwise, go to the $\Gamma_{\mathbf{II}}$-process.

$\Gamma_{\mathbf{I}}$-process:

5.1 $s = s + 1$.

5.2 If $s \neq pivot$, then permute the s and $pivot$ columns of $\mathbf{U}$, the corresponding elements of $\mathbf{t}$ and $A(\mathbf{U})$, and let $pivot = s$.

5.3 Normalize the $pivot$ column: $\mathbf{u}_{pivot} = -\mathbf{u}_{pivot}/t_{pivot}$.

5.4 Pivoting process: $\mathbf{u}_j = \mathbf{u}_j + t_j \mathbf{u}_{pivot}; j = 1, \ldots, n; j \neq pivot$.

5.5 If $h \leq n_b$ then remove the first column of $\mathbf{U}$ and the first list of $A(\mathbf{U})$ and let $s = s - 1$ and $n = n - 1$.

5.6 Let $h = h + 1$ and go to Step 1.

Γ_{II}-process:

6.1 Determine the sets $I(\mathbf{W}^-), I(\mathbf{W}^+)$ and $I(\mathbf{W}^0)$:

$$I(\mathbf{W}^-) \equiv \{i \leq s \mid t_i < 0\},$$
$$I(\mathbf{W}^+) \equiv \{i \leq s \mid t_i > 0\},$$
$$I(\mathbf{W}^0) \equiv \{i \leq s \mid t_i = 0\}.$$

6.2 If $I(\mathbf{W}^-) \equiv \emptyset$ then:

Build a new matrix $\mathbf{U}$ including the columns of the old $\mathbf{U}$ corresponding to the set of indices $I(\mathbf{W}^0) \cup \{s+1, \ldots, n\}$.

Build a new list $A(\mathbf{U})$ including the lists of the old $A(\mathbf{U})$ corresponding to the set of indices $I(\mathbf{W}^0) \cup \{s+1, \ldots, n\}$.

Let $s = |\mathbf{W}^0|$ and $n =$ number of columns of $\mathbf{U}$.

Let $h = h + 1$ and go to Step 1.

6.3 If $I(\mathbf{W}^+) \not\equiv \emptyset$ then:

Build the list $A(\mathbf{W}_c^{-+}) \equiv \{A_{\mathbf{w}_j} \cap A_{\mathbf{w}_k} | j \in I(\mathbf{W}^-), k \in I(\mathbf{W}^+)\}$.

Build the list $\mathbf{W}_c^{-+} \equiv \{t_k \mathbf{w}_j - t_j \mathbf{w}_k | j \in I(\mathbf{W}^-), k \in I(\mathbf{W}^+)\}$.

Build the sets $\mathbf{A}(\mathbf{W}_c^{-+})$ and $\mathbf{W}_c^{-+}$:

7.1 Initialize the sets $\mathbf{A}(\mathbf{W}_c^{-+}) \equiv \emptyset$ and $\mathbf{W}_c^{-+} \equiv \emptyset$, and make $j = 1$.

7.2 While $j \leq |\mathbf{A}_c(\mathbf{W}^{-+})|$ do:

Excluded $= false$ and $k = 1$.

While $k \leq |\mathbf{W}^0|$ and not *Excluded* do:

If $A_{\mathbf{w}_j} \subseteq A_{\mathbf{w}^0}$ then *Excluded* $=$ *true*.
$k = k + 1$.

If not *Excluded* then:

$k = 1$.

While $k \leq |\mathbf{A}(\mathbf{W}_c^{-+})|$ and not *Excluded* do:
- If $k \neq j$ and $A_{\mathbf{w}_j} \subseteq A_{\mathbf{w}_k}$ then make *Excluded* $=$ *true*.
- Let $k = k + 1$.

If *Excluded* make $A_{\mathbf{w}_j} \equiv \emptyset$.

Let $j = j + 1$.

7.3 Build the new $\mathbf{A}(\mathbf{W}_c^{-+})$ and $\mathbf{W}_c^{-+}$ by including in $\mathbf{A}(\mathbf{W}_c^{-+})$ the list of all $A_{\mathbf{w}_j} \neq \emptyset$ and in $\mathbf{W}_c^{-+}$ the corresponding vectors of $\mathbf{w}_j$.

7.4 Add to all lists in $\mathbf{A}(\mathbf{W}_c^{-+})$ index h.

7.5 Build a new matrix $\mathbf{U}$ including the columns of the old $\mathbf{U}$ corresponding to the set of indices $I(\mathbf{W}^0) \cup I(\mathbf{W}^-)$, the columns associated with the vectors $\mathbf{W}^{-+}$, and the columns of the old $\mathbf{U}$ corresponding to the set of indices $\{s+1, \ldots, n\}$.

7.6 Build a new list $A(\mathbf{U})$ including the lists of the old $A(\mathbf{U})$ corresponding to the set of indices $I(\mathbf{W}^0) \cup I(\mathbf{W}^-)$, the lists associated with the vectors $\mathbf{A}(\mathbf{W}_c^{-+})$, and the lists of the old $A_{\mathbf{w}_i}$ corresponding to the set of indices $\{s+1, \ldots, n\}$.

7.7 Let $s = |\mathbf{W}^0| + |\mathbf{W}^-| + |\mathbf{W}^{-+}|$ and $n =$ number of columns of $\mathbf{U}$.

8 **Check end**. If $n = 0$ then output $\emptyset$ and exit the algorithm. Otherwise, let $h = h + 1$ and go to Step 1.
9 **Return output**. Return the last $n - s$ columns of $\mathbf{U}$ as $\mathbf{V}$ and the first s columns of $\mathbf{U}$ as $\mathbf{W}$.

6.7 MATHEMATICA PROGRAM

The following Mathematica program implements the Γ-algorithm described in this chapter.

Program 6.1 (Obtaining the dual cone using the Γ-algorithm)

```
DualCone[B1_List,C1_List]:= Module[{gammaII,aux,SepIndex=0,
  i,BB,CC,j,jj,k,kk,kkk,ii,n,nb=Length[B1],
 nc=Length[C1],Eq,V,W,T,TT,wi,wj,wk,wij,ti,tj,AWij,pivot,
  rho,pi={},lastComponent,NonnullIndex,
  NonnullComponentVector,aux1,aux2},BB=B1; CC=C1;
  For[kk=1,kk<=nb,kk++,BB[[kk]]=AppendTo[BB[[kk]],0]];
  For[kkk=1,kkk<=nc,kkk++,
    CC[[kkk]]=AppendTo[CC[[kkk]],0]
  ];
  Eq=Join[BB,CC];
  n=Length[Eq[[1]]];
  T=IdentityMatrix[n];
  AW=Table[{},{i,1,n}];
 Do[gammaII=False;

 (* Dot products  *)
  TT=T.Eq[[i]];
```

```
(* Determine sets AW *)
 Do[
  If[TT[[j]]==0,AppendTo[AW[[j]],i]]
,{j,1,Length[TT]}];

(* Look for pivot *)
  pivot=0;
 Do[If[TT[[j]]!=0,pivot=j;Break[]]
   ,{j,SepIndex+1,n}];
 If[(Min[pivot,SepIndex+1]==n && i<=nb),T={};Break[]];

(* Gamma I *)
If[(SepIndex<n && pivot>0),
  SepIndex++;
 If[SepIndex!=pivot,
   (* Permute vectors *)
    aux=T[[SepIndex]];T[[SepIndex]]=T[[pivot]];
        T[[pivot]]=aux;
       aux=TT[[SepIndex]];TT[[SepIndex]]=TT[[pivot]];
        TT[[pivot]]=aux;
       aux=AW[[SepIndex]];AW[[SepIndex]]=AW[[pivot]];
        AW[[pivot]]=aux;
      ];

      (* Normalize the pivot vector *)
      T[[SepIndex]]=Map[(#/TT[[SepIndex]])&,
          -T[[SepIndex]]];

      (* Pivoting *)
      Do[If[(j!=SepIndex && TT[[j]]!=0),
       T[[j]]+=TT[[j]]*T[[SepIndex]];AppendTo[AW[[j]],i]
      ],{j,1,n}]
     If[i<=nb,
     T=Drop[T,1];AW=Drop[AW,1];SepIndex--;n--],
  gammaII=True;
];

(* Gamma II *)
If[gammaII,

   (* Determine sets W-,W+ and W0 *)
wi={};wj={};wk={};ti={};tj={};
Do[
   If[TT[[j]]>0,
```

```
  AppendTo[wj,j];AppendTo[tj,TT[[j]]],
If[TT[[j]]<0,
    AppendTo[wi,j];AppendTo[ti,TT[[j]]],
    AppendTo[wk,j]]
],{j,1,SepIndex}];
If[Length[wi]==0,
      T=Join[T[[wk]],T[[Range[SepIndex+1,n]]]];
      AW=Join[AW[[wk]],AW[[Range[SepIndex+1,n]]]];
   SepIndex=Length[wk];n=Length[T],
   If[Length[wj]!=0,
    wij=Flatten[Table[tj[[jj]]*T[[wi[[ii]]]]
                -ti[[ii]]*T[[wj[[jj]]]],
          {ii,1,Length[wi]},{jj,1,Length[wj]}],1];
   AWij=Flatten[Table[Intersection[AW[[wi[[ii]]]],
        AW[[wj[[jj]]]]],{ii,1,Length[wi]},
        {jj,1,Length[wj]}],1];

   (* Elimination Process *)
  ii=1;
  While[ii<=Length[AWij],
     Excluded=False;
  k=1;
 While[(k<=Length[wk] && !Excluded),
   If[TrueQ[Complement[AWij[[ii]],AW[[wk[[k]]]]]=={}],
     Excluded=True;
     ];
   k++
 ];
 If[!Excluded,
   k=1;
   While[(k<=Length[AWij] && !Excluded),
   If[(TrueQ[Complement[AWij[[ii]],AWij[[k]]]=={}]
                   && k!=ii),
      Excluded=True;
  ];
 k++
 ]
 ];
 If[Excluded,AWij[[ii]]={-1}];
 ii++
 ];
  aux1={};aux2={};
   Do[If[AWij[[ii]]!={-1},AppendTo[aux1,AWij[[ii]]];
        AppendTo[aux2,wij[[ii]]]]
     ,{ii,1,Length[AWij]}];
```

```
    AWij=aux1;wij=aux2;
    Do[AppendTo[AWij[[ii]],i],{ii,1,Length[AWij]}];
    T=Join[T[[wk]],T[[wi]],wij,T[[Range[SepIndex+1,n]]]];
     AW=Join[AW[[wk]],AW[[wi]],AWij,AW[[Range[SepIndex+1,n]]]];
    SepIndex=Length[wk]+Length[wi]+Length[wij];n=Length[T];
    ];
  ];
    If[T=={},Break[]];
];
  ,{i,1,nb+nc}];

   (* Separate components of T *)
  If[T!={},
    lastComponent=Length[T[[1]]];

    (* Normalize vectors (last component = 1 if non null *)
     Do[
       If[T[[i,lastComponent]]!=0,
         T[[i]]/=T[[i,lastComponent]];
       ]
     ,{i,1,n}];
     rho=T[[Range[SepIndex+1,n]]];
        If[rho!={},

        (* Look for some rho with non-null last component *)
             NonnullIndex=0;
              Do[If[rho[[i,lastComponent]]!=0,
                   NonnullIndex=i;Break[]]
              ,{i,1,Length[rho]}];

        (* Substract the non-null component rho from the
           rest rho and pi *)
          If[NonnullIndex!=0,
            NonnullComponentVector=rho[[NonnullIndex]];
            rho=Drop[rho,{NonnullIndex}];
            Do[If[rho[[i,lastComponent]]!=0,
               rho[[i]]-=NonnullComponentVector],
                       {i,1,Length[rho]}];
            AppendTo[pi,Drop[NonnullComponentVector,-1]];
            Do[If[T[[i,lastComponent]]==0,
                 AppendTo[pi,Drop[T[[i]],-1]],
                 AppendTo[pi,Drop[T[[i]]
                     -NonnullComponentVector,-1]]
                      ]
                 ,{i,1,SepIndex}],
```

```
            Do[If[T[[i,lastComponent]]==0,
                AppendTo[pi,Drop[T[[i]],-1]]
                   ],{i,1,SepIndex}];
            ],
         ];
        If[rho!={},rho=Transpose[Drop[Transpose[rho],-1]]];
        If[Length[pi]!=0 & pi[[1]]==Table[0,
         {i,1,Length[pi[[1]]]}],pi=Drop[pi,1]];
        Print["\n","DUAL:","\n"];
        Print["V = ",rho];
        Print["W = ",pi];

       (* Normalized Dual *)
         Print["\n","NORMALIZED DUAL:","\n"];
    Do[
        If[(Length[rho]!=0 && rho[[i,lastComponent-1]]!=0),
          rho[[i]]/=rho[[i,lastComponent-1]]
           ],{i,1,Length[rho]}];
         Do[
          If[(Length[pi]!=0 && pi[[i,lastComponent-1]]!=0) ,
          pi[[i]]/=pi[[i,lastComponent-1]]
           ]
          ,{i,1,Length[pi]}];
         Print["V = ",rho];
        Print["W = ",pi],
     Print["No Solution"];
 ];
]
```

Example 6.25 (Dual cone using Mathematica). Here we illustrate the use of the above program. Two lists of vectors are given: the first one corresponds to $\mathbf{B}_\rho$ generators and the second to the $\mathbf{C}_\pi$ generators, where $\mathbf{A}_\pi \equiv \mathbf{B}_\rho + \mathbf{C}_\pi$ is the general form of the cone.

```
In[]:=
DualCone[{},{{2,2,-1,4},{1,-1,0,-3},{2,-1,1,-1},{0,-1,1,1},
{0,1,-1,2},{0,-1,1,-1}}]


DUAL:
V = {}
          3      1        3       3       3               3    3  3
W = {{0, -, 1, -(-)}, {-(--), -(--), -(--), 0}, {-(--), -, -, 0},
          2      2        10      10      10              10   5  5

      1     4  1     3
   {-(--), -, -, -(--)}}
      10    5  5     10
```

```
NORMALIZED DUAL:
V = {}
                           3       3       3          3    3  3
W = {{0, -3, -2, 1}, {-(--), -(--), -(--), 0}, {-(--), -, -, 0},
                          10      10      10         10   5  5
    1   8    2
  {-, -(-), -(-), 1}}
    3   3    3
```

The program returns two lists of vectors corresponding to $\mathbf{V}_\rho$ (in this example it is empty) and $\mathbf{W}_\pi$ of the dual cone, that is, $\Omega(\mathbf{A}_\pi) = \mathbf{V}_\rho + \mathbf{W}_\pi$. ∎

Exercises

6.1 Show that the null vector is interior to a cone $\mathbf{A}_\pi$ iff all vectors in the cone are interior too.

6.2 Given a cone $\mathbf{A}_\pi$ show that $-\mathbf{v}$ cannot belong to the boundary of $\mathbf{A}_\pi$ if $\mathbf{v} \in \mathbf{A}_\pi^{in}$.

6.3 Given a cone $\mathbf{A}_\pi$ show that $-\mathbf{v}$ cannot belong to $\mathbf{A}_\pi^{in}$ if $\mathbf{v}$ belongs to the boundary of $\mathbf{A}_\pi$.

6.4 Show that the only common point of a cone and its dual cone is the vertex.

6.5 Consider the cone $\mathbf{A}_\pi$, with

$$\mathbf{A} = \begin{pmatrix} 0 & 2 & 0 \\ 1 & 1 & 1 \\ 2 & 1 & 0 \\ 1 & 1 & -1 \end{pmatrix} \tag{6.80}$$

Determine whether or not $(1,1,1)^T$ and $(1,0,0)^T$ belong to the cone $\mathbf{A}_\pi$.

6.6 Obtain the binding set of the vector $(1,3,-1)^T$ in the cone generated by $\mathbf{A}$ in (6.80).

6.7 Let $\mathbf{A} = \{(1,1,0)^T, (-1,1,0)^T, (0,0,1)^T, (2,-1,1)^T\}$ be a set of vectors in E^3, and consider

$$\begin{aligned} \mathbf{P}_1 &= (5,-1,0)^T, \\ \mathbf{P}_2 &= (4,1,-2)^T, \\ \mathbf{P}_3 &= (1,-1,0)^T, \\ \mathbf{P}_4 &= (1,0,1)^T. \end{aligned}$$

Obtain all possible linear combinations of $\mathbf{P}_i$, $i = 1, \ldots, 4$ in $\mathbf{A}$ indicating the type of linear combination.

6.8 Calculate the general form of the cones in E^2 generated by

$$\mathbf{A} = \begin{pmatrix} 3/2 & 3 & 1 \\ -2 & 1 & 4 \end{pmatrix},$$

$$\mathbf{B} = \begin{pmatrix} 1 & -1 & 2 \\ 1 & -1 & 0 \end{pmatrix},$$

$$\mathbf{C} = \begin{pmatrix} -2 & -1 \\ -3 & 1 \end{pmatrix}.$$

6.9 Using the Γ-algorithm, determine whether or not the set of vectors $\mathbf{B} = \{(-2,2,1)^T, (1,0,1)^T, (-1,2,2)^T\}$ in E^3 is linearly independent.

6.10 Consider the cone $\mathbf{A}_\pi$, where

$$\mathbf{A} = \begin{pmatrix} 1 & 0 & 0 & 1 & 1 \\ 0 & 1 & 0 & 1 & 0 \\ 1 & 1 & 1 & 0 & 1 \end{pmatrix}.$$

Use the Γ-algorithm, to obtain the polar cone $\Omega(\mathbf{A}_\pi)$.

6.11 Consider the cone in Exercise 6.10 and obtain the binding set of the vector $\mathbf{x} = (1,-1,2)^T$.

6.12 Let $\mathbf{B}_\pi$ be a cone, where

$$\mathbf{B} = \begin{pmatrix} 2 & 1 & 0 & 0 & -1 \\ 0 & 2 & -1 & 1 & 2 \\ 1 & -1 & 1 & 1 & 1 \end{pmatrix}.$$

Determine whether or not the vectors $\mathbf{x}_1 = (0,-1,1)^T$ and $\mathbf{x}_2 = (1,-1,-1)^T$ belong to the interior of the cone $\mathbf{B}_\pi$.

7

Polytopes and Polyhedra

7.1 INTRODUCTION

This chapter deals with polytopes and polyhedra. Section 7.2 deals with polytopes and shows how the cone associated with a polytope and the intersection of several polytopes can be obtained, using the Γ-algorithm. Section 7.3 introduces polyhedra and the concepts of vertex and edge of a polyhedron. It also gives a method for obtaining vertices and edges based on polar cones.

7.2 POLYTOPES

The polygon has played an important role in the history of geometry and occupied mathematicians for many years. Its generalization to higher dimensions leads to the concept of the polytope, which is given in this chapter.

Definition 7.1 (Polytope generated by a set of vectors). *Let*

$$\mathbf{P} \equiv \{\mathbf{p}_1, \ldots, \mathbf{p}_m\}$$

be a set of m vectors of E^n. The set

$$\mathbf{P}_\lambda \equiv \left\{ \mathbf{x} \in E^n \mid \mathbf{x} = \lambda_1 \mathbf{p}_1 + \cdots + \lambda_m \mathbf{p}_m \quad with \quad \lambda_i \geq 0\,;\ \sum_{i=1}^{m} \lambda_i = 1 \right\}$$

of convex linear combinations of the vectors $\mathbf{p}_1, \ldots, \mathbf{p}_m$ is known as the polytope or the convex polyhedron generated by these vectors.

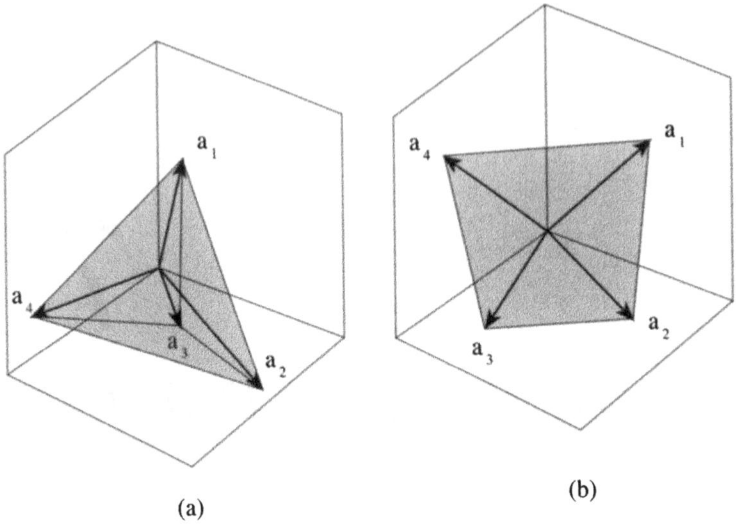

Fig. 7.1 Two examples of convex polytopes.

Example 7.1 (Some polytopes). Fig. 7.1 shows two examples of convex polytopes. Case (a) is a tetrahedron, because the points associated with the generators $\mathbf{a}_1, \mathbf{a}_2, \mathbf{a}_3$ and $\mathbf{a}_4$ do not belong to the same plane. However, in case (b) these points are coplanar and the polytope degenerates to a simple quadrilateral.

■

Remark 7.1 *A polytope can also be defined by a set of inequalities, as shown in the following example.*

Example 7.2 (A polytope defined by a set of inequalities). The polytope generated by the set of vectors

$$\{(0,0,0)^T, (1,0,0)^T, (0,1,0)^T, (0,0,1)^T\}$$

coincides with the polytope defined by the set of inequalities

$$\begin{array}{rrrcl} & & x_1 & \geq & 0 \\ & & x_2 & \geq & 0 \\ & & x_3 & \geq & 0 \\ x_1 & +x_2 & +x_3 & \leq & 1 \end{array} \qquad (7.1)$$

as illustrated in Fig. 7.2. ■

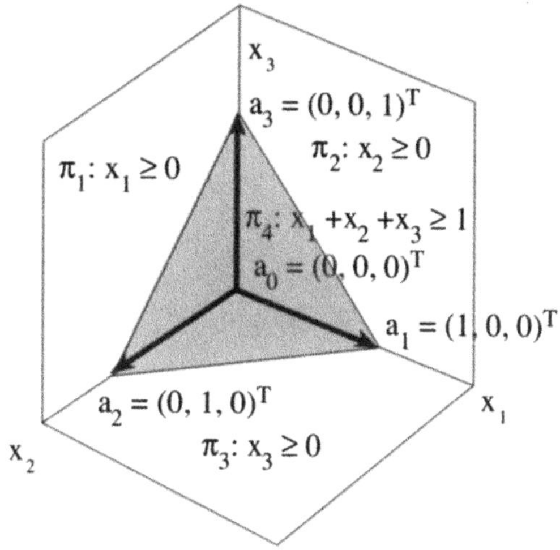

Fig. 7.2 Illustration of two different definitions of a polytope: by its generators, and by a set of inequalities.

7.2.1 Cone Associated with a Polytope

Theorem 7.1 *(Cone associated with a polytope). Given the polytope* $\mathbf{A}_\lambda$, *there exists a one-to-one correspondence between its elements and those of the cone* $\begin{pmatrix}\mathbf{A}\\ \mathbf{1}\end{pmatrix}_\pi$, *where* $\begin{pmatrix}\mathbf{A}\\ \mathbf{1}\end{pmatrix}$ *is the matrix extended with a row of ones. This can be expressed by means of*

$$\mathbf{x} \in \mathbf{A}_\lambda \Leftrightarrow \begin{pmatrix}\mathbf{x}\\ 1\end{pmatrix} \subset \begin{pmatrix}\mathbf{A}\\ \mathbf{1}\end{pmatrix}_\pi . \tag{7.2}$$

Example 7.3 (Cone associated with a convex polytope). Fig. 7.3 shows how the cone associated with a convex polytope can be generated. The idea consists of increasing the dimension of E^n by one (vertical in Fig. 7.3). To this end we obtain the new origin O_3, by subtracting the unit vector $\mathbf{e}_{n+1}$ from the old origin O_2, such that the last component of any of the old vectors is one. Next, we consider the cone generated by the vectors $\mathbf{a}_1+\overline{O_3O_1}, \ldots, \mathbf{a}_5+\overline{O_3O_1}$, which is the desired cone. ∎

7.2.2 Intersection of Several Polytopes

In this section we use the Γ–algorithm to obtain the intersection of two or more polytopes. Assume that we have m polytopes $\mathbf{P}^1_\lambda, \ldots, \mathbf{P}^m_\lambda$ in E^n, which are not assumed to be given by minimal sets of generators. Let $\mathbf{Q}^1_\pi, \ldots, \mathbf{Q}^m_\pi$

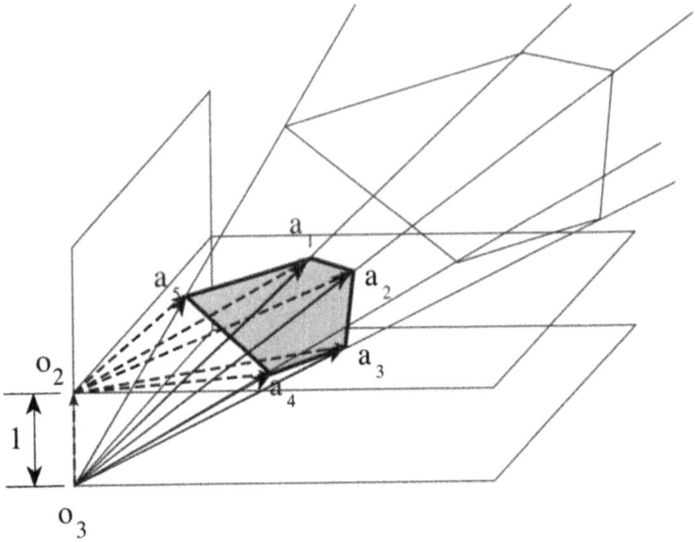

Fig. 7.3 Illustration of how the cone associated with a polytope can be generated.

be the corresponding associated cones, that is,

$$\mathbf{Q}^1 = \begin{pmatrix} \mathbf{P}^1 \\ \mathbf{1} \end{pmatrix}, \ldots, \mathbf{Q}^m = \begin{pmatrix} \mathbf{P}^m \\ \mathbf{1} \end{pmatrix}. \tag{7.3}$$

To obtain the intersection of the initial polytopes, we obtain first the intersection of their associated cones, as follows. According to the lemma developed by Farkas, Lemma 6.1, we have

$$\begin{pmatrix} \mathbf{x} \\ 1 \end{pmatrix} \in \mathbf{Q}^i_\pi \quad \Leftrightarrow \quad \mathbf{V}^{iT} \begin{pmatrix} \mathbf{x} \\ 1 \end{pmatrix} = \mathbf{0} \wedge \mathbf{W}^{iT} \begin{pmatrix} \mathbf{x} \\ 1 \end{pmatrix} \leq \mathbf{0}; \; i = 1, \ldots, m, \tag{7.4}$$

where $\begin{pmatrix} \mathbf{x} \\ 1 \end{pmatrix} \in E^{n+1}$, and $\Omega(\mathbf{Q}^i_\pi) \equiv \mathbf{V}^i_\rho + \mathbf{W}^i_\pi$ is the dual of $\mathbf{Q}^i_\pi$; $i = 1, \ldots, m$, in standard form, we have

$$\begin{aligned} \mathbf{x} \in \bigcap_{i=1}^{m} \mathbf{P}^i_\lambda \Leftrightarrow \begin{pmatrix} \mathbf{x} \\ 1 \end{pmatrix} \in \mathbf{Q}^i_\pi; \forall i \Leftrightarrow \\ \mathbf{V}^{iT} \begin{pmatrix} \mathbf{x} \\ 1 \end{pmatrix} = \mathbf{0} \wedge \mathbf{W}^{iT} \begin{pmatrix} \mathbf{x} \\ 1 \end{pmatrix} \leq \mathbf{0}; i = 1, \ldots, m, \end{aligned} \tag{7.5}$$

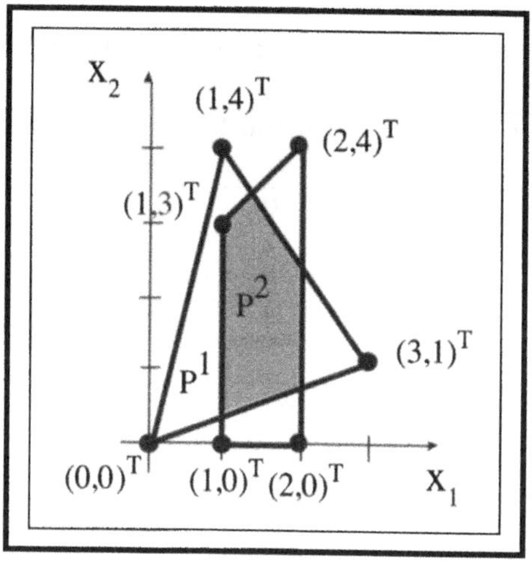

Fig. 7.4 Intersection of two polytopes.

which implies

$$\mathbf{x} \in \bigcap_{i=1}^{m} \mathbf{P}_\lambda^i \Leftrightarrow \begin{cases} \begin{pmatrix} \mathbf{x} \\ x_{n+1} \end{pmatrix} \in \Omega\left[(\mathbf{V}^1, \ldots, \mathbf{V}^m)_\rho + (\mathbf{W}^1, \ldots, \mathbf{W}^m, -\mathbf{e}_{n+1})_\pi\right], \\ x_{n+1} = 1. \end{cases} \tag{7.6}$$

Thus, to obtain the intersection of the m polytopes, we use the following algorithm.

Algorithm 7.1 Intersection of polytopes.

- **Input:** m polytopes $\mathbf{P}_\lambda^1, \ldots, \mathbf{P}_\lambda^m$.
- **Output:** The intersection polytope.

Step 1: We obtain the dual cones $\Omega(\mathbf{Q}_\pi^i) \equiv \mathbf{V}_\rho^i + \mathbf{W}_\pi^i$ of $\mathbf{Q}_\pi^i$, for $i = 1, \ldots, m$.

Step 2: We obtain the dual cone $\mathbf{H}_\pi$ of

$$(\mathbf{V}^1, \ldots, \mathbf{V}^m)_\rho + (\mathbf{W}^1, \ldots, \mathbf{W}^m, -\mathbf{e}_{n+1})_\pi.$$

Step 3: We return the polytope such that its associated cone is $\mathbf{H}_\pi$.

Example 7.4 (Intersection of two polytopes). Consider the polytopes $\mathbf{P}_\lambda^1$ and $\mathbf{P}_\lambda^2$ in Fig. 7.4, where

$$\mathbf{P}^1 = \begin{pmatrix} 0 & 1 & 3 & 1 \\ 0 & 4 & 1 & 1 \end{pmatrix}; \; \mathbf{P}^2 = \begin{pmatrix} 1 & 2 & 1 & 2 \\ 0 & 0 & 3 & 4 \end{pmatrix}. \tag{7.7}$$

Table 7.1 Iterations for obtaining the dual of $\mathbf{Q}_\pi^1$ in Example 7.4.

	Iteration 1				Iteration 2				Iteration 3		
$\mathbf{a}_1$	$\mathbf{v}_1^1$	$\mathbf{v}_2^1$	$\mathbf{v}_3^1$	$\mathbf{a}_2$	$\mathbf{w}_1^2$	$\mathbf{v}_1^2$	$\mathbf{v}_2^2$	$\mathbf{a}_3$	$\mathbf{w}_1^3$	$\mathbf{w}_2^3$	$\mathbf{v}_1^3$
0	1	0	**0**	1	0	**0**	1	3	0	0	**1**
0	0	1	**0**	4	0	**1**	0	1	1/4	–1/4	**–1/4**
1	0	0	**1**	1	–1	**0**	0	1	–1	0	**0**
$\mathbf{t}^1$	0	0	**1**	$\mathbf{t}^2$	–1	**4**	1	$\mathbf{t}^3$	–3/4	–1/4	**11/4**

	Iteration 4			Dual		
$\mathbf{a}_4$	$\mathbf{w}_1^4$	$\mathbf{w}_2^4$	$\mathbf{w}_3^4$	$\mathbf{w}_1$	$\mathbf{w}_2$	$\mathbf{w}_3$
1	3/11	1/11	–4/11	3/11	1/11	–4/11
1	2/11	–3/11	1/11	2/11	–3/11	1/11
1	–1	0	0	–1	0	0
$\mathbf{t}^4$	–6/11	–2/11	–3/11			

The associated cones become $\mathbf{Q}_\pi^1$ and $\mathbf{Q}_\pi^2$, where

$$\mathbf{Q}^1 = \begin{pmatrix} 0 & 1 & 3 & 1 \\ 0 & 4 & 1 & 1 \\ 1 & 1 & 1 & 1 \end{pmatrix}; \; \mathbf{Q}^2 = \begin{pmatrix} 1 & 2 & 1 & 2 \\ 0 & 0 & 3 & 4 \\ 1 & 1 & 1 & 1 \end{pmatrix}. \tag{7.8}$$

Using Algorithm 7.1, we proceed as follows.
Step 1: We obtain the dual cone $\Omega(\mathbf{Q}_\pi^1) \equiv \mathbf{V}_\rho^1 + \mathbf{W}_\pi^1$ of $\mathbf{Q}_\pi^1$, as indicated in Table 7.1. Thus, we get

$$\mathbf{V}^1 = \emptyset; \; \mathbf{W}^1 = \begin{pmatrix} 3/11 & 1/11 & -4/11 \\ 2/11 & -3/11 & 1/11 \\ -1 & 0 & 0 \end{pmatrix}.$$

We obtain the dual cone $\Omega(\mathbf{Q}_\pi^2) \equiv \mathbf{V}_\rho^2 + \mathbf{W}_\pi^2$ of $\mathbf{Q}_\pi^2$, as indicated in Table 7.2. Thus, we get

$$\mathbf{V}^2 = \emptyset; \; \mathbf{W}^2 = \begin{pmatrix} -1 & 0 & -1/3 & 4/3 \\ 0 & -1/3 & 1/3 & 0 \\ 1 & 0 & -2/3 & -8/3 \end{pmatrix}.$$

Step 2: We obtain the dual cone $\mathbf{H}_\pi \equiv \Omega(\mathbf{A}_\pi)$, where

$$\mathbf{A} = \begin{pmatrix} 3/11 & 1/11 & -4/11 & -1 & 0 & -1/3 & 4/3 & 0 \\ 2/11 & -3/11 & 1/11 & 0 & -1/3 & 1/3 & 0 & 0 \\ -1 & 0 & 0 & 1 & 0 & -2/3 & -8/3 & -1 \end{pmatrix},$$

following the process indicated in Table 7.3.
Step 3: Finally, we normalize the dual as shown in Table 7.4 and we obtain the polytope such that its associated cone is $\mathbf{H}_\pi$, which becomes

$$\mathbf{H}_\lambda \equiv \begin{pmatrix} 1 & 7/5 & 1 & 2 & 2 \\ 1/3 & 17/5 & 3 & 2/3 & 5/2 \end{pmatrix}_\lambda$$

Table 7.2 Iterations for obtaining the dual of $\mathbf{Q}_\pi^2$ in Example 7.4.

Iteration 1				Iteration 2				Iteration 3			
$\mathbf{a}_1$	$\mathbf{v}_1^1$	$\mathbf{v}_2^1$	$\mathbf{v}_3^1$	$\mathbf{a}_2$	$\mathbf{w}_1^2$	$\mathbf{v}_1^2$	$\mathbf{v}_2^2$	$\mathbf{a}_3$	$\mathbf{w}_1^3$	$\mathbf{w}_2^3$	$\mathbf{v}_1^3$
1	1	0	0	2	–1	0	–1	1	1	–1	0
0	0	1	0	0	0	1	0	3	0	0	1
1	0	0	1	1	0	0	1	1	–2	1	0
$\mathbf{t}^1$	1	0	1	$\mathbf{t}^2$	–2	0	–1	$\mathbf{t}^3$	–1	0	3

Iteration 4				Dual			
$\mathbf{a}_4$	$\mathbf{w}_1^4$	$\mathbf{w}_2^4$	$\mathbf{w}_3^4$	$\mathbf{w}_1$	$\mathbf{w}_2$	$\mathbf{w}_3$	$\mathbf{w}_4$
2	1	–1	0	–1	0	–1/3	4/3
4	1/3	0	–1/3	0	–1/3	1/3	0
1	–2	1	0	1	0	–2/3	–8/3
$\mathbf{t}^4$	4/3	–1	–4/3				

and which is the intersection of the two polytopes. ■

Example 7.5 (The two-dimensional moving polytopes). Consider three point triplets $\mathbf{P}^1, \mathbf{P}^2$, and $\mathbf{P}^3$, such that they simultaneously start moving with uniform speeds. The first triplet starts from the point $(0,0)$, the second from $(5,0)$, and the third from $(0,5)$. After one hour their positions are given by

$$\begin{aligned}
\mathbf{P}^1 &\equiv \{(1,0)^T, (0,1)^T, (1,1)^T\},\\
\mathbf{P}^2 &\equiv \{(4,0)^T, (3,2)^T, (3,-1)^T\},\\
\mathbf{P}^3 &\equiv \{(0,4)^T, (2,4)^T, (-1,3)^T\}.
\end{aligned}$$

After t hours, the positions of $\mathbf{P}^1, \mathbf{P}^2$, and $\mathbf{P}^3$ are

$$\begin{aligned}
\mathbf{P}^1(t) &\equiv \{(t,0)^T, (0,t)^T, (t,t)^T\},\\
\mathbf{P}^2(t) &\equiv \{(5-t,0)^T, (5-2t,2t)^T, (5-2t,-t)^T\},\\
\mathbf{P}^3(t) &\equiv \{(0,5-t)^T, (2t,5-t)^T, (-t,5-2t)^T\}.
\end{aligned}$$

Assume that we are interested in determining the instants at which there is a single common point to the three polytopes, $\mathbf{P}_\lambda^i; i = 1,2,3$, generated by each of the triplets. To this aim we first determine the intersection $\mathbf{A}$ of the polytopes at instant t, using Algorithm 7.1.

Step 1: First, we obtain the dual cones, $\Omega(\mathbf{Q}_\pi^i(t)); i = 1,2,3$, of the cones

Table 7.3 Iterations for obtaining the dual of $\mathbf{H}_\pi$ in Example 7.4.

Iteration 1				Iteration 2			
$\mathbf{a}_1$	$\mathbf{v}_1^1$	$\mathbf{v}_2^1$	$\mathbf{v}_3^1$	$\mathbf{a}_2$	$\mathbf{w}_1^2$	$\mathbf{v}_1^2$	$\mathbf{v}_2^2$
3/11	**1**	0	0	1/11	–11/3	**–2/3**	11/3
2/11	**0**	1	0	–3/11	0	**1**	0
–1	**0**	0	1	0	0	**0**	1
$\mathbf{t}^1$	**3/11**	2/11	–1	$\mathbf{t}^2$	–1/3	**–1/3**	1/3

Iteration 3				Iteration 4			
$\mathbf{a}_3$	$\mathbf{w}_1^3$	$\mathbf{w}_2^3$	$\mathbf{v}_1^3$	$\mathbf{a}_4$	$\mathbf{w}_1^4$	$\mathbf{w}_2^4$	$\mathbf{w}_3^4$
–4/11	–3	–2	**3**	–1	0	1	3
1/11	–1	3	**1**	0	0	4	1
0	0	0	**1**	1	1	1	1
$\mathbf{t}^3$	1	1	**–1**	$\mathbf{t}^4$	1	0	–2

Iteration 5				Iteration 6			
$\mathbf{a}_5$	$\mathbf{w}_1^5$	$\mathbf{w}_2^5$	$\mathbf{w}_3^5$	$\mathbf{a}_6$	$\mathbf{w}_1^6$	$\mathbf{w}_2^6$	$\mathbf{w}_3^6$
0	1	3	3	–1/3	1	3	3
–1/3	4	1	1	1/3	4	1	1
0	1	1	3	–2/3	1	1	3
$\mathbf{t}^5$	–4/3	–1/3	–1/3	$\mathbf{t}^6$	1/3	–4/3	–8/3

Iteration 7					Dual					
$\mathbf{a}_7$	$\mathbf{w}_1^7$	$\mathbf{w}_2^7$	$\mathbf{w}_3^7$	$\mathbf{w}_4^7$	$\mathbf{a}_8$	$\mathbf{w}_1^8$	$\mathbf{w}_2^8$	$\mathbf{w}_3^8$	$\mathbf{w}_4^8$	$\mathbf{w}_5^8$
4/3	3	3	7/3	11/3	0	3	7/3	11/3	16	64/9
0	1	1	17/3	11	0	1	17/3	11	16/3	80/9
–8/3	1	3	5/3	11/3	–1	3	5/3	11/3	8	32/9
$\mathbf{t}^7$	4/3	–4	–4/3	–44/9	$\mathbf{t}^8$	–3	–5/3	–11/3	–8	–32/9

Table 7.4 Normalizing the dual of $\mathbf{H}_\pi$ in Example 7.4.

Dual					Normalized dual				
$\mathbf{w}_1$	$\mathbf{w}_2$	$\mathbf{w}_3$	$\mathbf{w}_4$	$\mathbf{w}_5$	$\mathbf{w}_1$	$\mathbf{w}_2$	$\mathbf{w}_3$	$\mathbf{w}_4$	$\mathbf{w}_5$
3	7/3	11/3	16	64/9	1	7/5	1	2	2
1	17/3	11	16/3	80/9	1/3	17/5	3	2/3	5/2
3	5/3	11/3	8	32/9	1	1	1	1	1

$$\mathbf{Q}_\pi^1(t) = \begin{pmatrix} t & 0 & t \\ 0 & t & t \\ 1 & 1 & 1 \end{pmatrix}_\pi ,$$

$$\mathbf{Q}_\pi^2(t) = \begin{pmatrix} 5-t & 5-2t & 5-2t \\ 0 & 2t & -t \\ 1 & 1 & 1 \end{pmatrix}_\pi ,$$

$$\mathbf{Q}_\pi^3(t) = \begin{pmatrix} 0 & 2t & -t \\ 5-t & 5-t & 5-2t \\ 1 & 1 & 1 \end{pmatrix}_\pi ,$$

Table 7.5 Steps for obtaining $\Omega(\mathbf{Q}_\pi^1(t))$.

Iteration 1				Iteration 2			
$\mathbf{a}_1$	$\mathbf{v}_1^1$	$\mathbf{v}_2^1$	$\mathbf{v}_3^1$	$\mathbf{a}_2$	$\mathbf{w}_1^2$	$\mathbf{v}_1^2$	$\mathbf{v}_2^2$
t	**1**	0	0	0	-1/t	**0**	-1/t
0	**0**	1	0	t	0	**1**	0
1	**0**	0	1	1	0	**0**	1
$\mathbf{t}^1$	**t**	0	1	$\mathbf{t}^2$	0	**t**	1

Iteration 3				Dual		
$\mathbf{a}_3$	$\mathbf{w}_1^3$	$\mathbf{w}_2^3$	$\mathbf{v}_1^3$	$\mathbf{w}_1$	$\mathbf{w}_2$	$\mathbf{w}_3$
t	-1/t	0	**-1/t**	0	1/t	-1/t
t	0	-1/t	**-1/t**	1/t	0	-1/t
1	0	0	**1**	-1	-1	1
$\mathbf{t}^3$	-1	-1	-1			

associated with the polytopes $\mathbf{P}_\lambda^i; i = 1, 2, 3$:

$$\Omega(\mathbf{Q}_\pi^1(t)) \equiv \mathbf{W}_\pi^1 \equiv \begin{pmatrix} 0 & 1 & -1 \\ 1 & 0 & -1 \\ -t & -t & t \end{pmatrix}_\pi,$$

$$\Omega(\mathbf{Q}_\pi^2(t)) \equiv \mathbf{W}_\pi^2 \equiv \begin{pmatrix} -1 & 1 & 2 \\ 0 & -1 & 1 \\ 5-2t & t-5 & 2t-10 \end{pmatrix}_\pi,$$

$$\Omega(\mathbf{Q}_\pi^3(t)) \equiv \mathbf{W}_\pi^3 \equiv \begin{pmatrix} 1 & -1 & 0 \\ -3 & 1 & 1 \\ -5t+15 & t-5 & t-5 \end{pmatrix}_\pi.$$

Table 7.5 shows how to obtain $\Omega(\mathbf{Q}_\pi^1(t))$. The other two are obtained in a similar way. Note that, without changing the duals, we have multiplied the vector components by the non-null constant $t > 0$.

Steps 2 and 3: We could obtain the dual cone $\mathbf{H}_\pi = \Omega(\mathbf{C}_\pi)$, where $\mathbf{C}$ is

$$\begin{pmatrix} 0 & 1 & -1 & -1 & 1 & 2 & 1 & -1 & 0 & 0 \\ 1 & 0 & -1 & 0 & -1 & 1 & -3 & 1 & 1 & 0 \\ -t & -t & t & 5-2t & t-5 & 2t-10 & -5t+15 & t-5 & t-5 & -1 \end{pmatrix}.$$

Alternatively, a vector $(x_1, x_2, x_3)^T$ belongs to $\mathbf{H}_\pi \equiv \Omega(\mathbf{C}_\pi)$ with $x_3 = 1$ iff it satisfies the system of inequalities $\mathbf{W}^{i^T}\mathbf{x} \le 0; i = 1, 2, 3$: that is,

$$\begin{array}{rrrrcl} & +x_2 & -t & & \le & 0 \\ +x_1 & & -t & & \le & 0 \\ -x_1 & -x_2 & +t & & \le & 0 \\ -x_1 & & -2t & +5x_3 & \le & 0 \\ +x_1 & -x_2 & +t & -5x_3 & \le & 0 \\ +2x_1 & +x_2 & +2t & -10x_3 & \le & 0 \\ +x_1 & -3x_2 & -5t & 15x_3 & \le & 0 \\ -x_1 & +x_2 & +t & -5x_3 & \le & 0 \\ & +x_2 & +t & -5x_3 & \le & 0 \\ & & & -x_3 & \le & 0 \end{array} \tag{7.9}$$

with $x_3 = 1$.

However, (7.9) shows that $(x_1, x_2, t, x_3)^T$ belongs to the dual $\Omega(\mathbf{B}_\pi)$ of $\mathbf{B}_\pi$, where

$$\mathbf{B} = \begin{pmatrix} 0 & 1 & -1 & -1 & 1 & 2 & 1 & -1 & 0 & 0 \\ 1 & 0 & -1 & 0 & -1 & 1 & -3 & 1 & 1 & 0 \\ -1 & -1 & 1 & -2 & 1 & 2 & -5 & 1 & 1 & 0 \\ 0 & 0 & 0 & 5 & -5 & -10 & 15 & -5 & -5 & -1 \end{pmatrix}.$$

From Table 7.6, we get

$$\Omega(\mathbf{B}_\pi) = \begin{pmatrix} 0 & 5/8 & 1 & 7/4 & 35/19 & 1 & 5/4 \\ 5/2 & 25/16 & 2 & 1 & 40/19 & 2 & 5/2 \\ 5/2 & 35/16 & 2 & 11/4 & 40/19 & 3 & 5/2 \\ 1 & 1 & 1 & 1 & 1 & 1 & 1 \end{pmatrix}_\pi.$$

Forcing $x_3 = 1$, we get

$$(x_1, x_2, t) \in \begin{pmatrix} 0 & 5/8 & 1 & 7/4 & 35/19 & 1 & 5/4 \\ 5/2 & 25/16 & 2 & 1 & 40/19 & 2 & 5/2 \\ 5/2 & 35/16 & 2 & 11/4 & 40/19 & 3 & 5/2 \end{pmatrix}_\lambda,$$

which implies

$$t = (5/2)\lambda_1 + (35/16)\lambda_2 + 2\lambda_3 + (11/4)\lambda_4 + (40/19)\lambda_5 + 3\lambda_6 + (5/2)\lambda_7.$$

The first common point is obtained for the minimum value of t, that is,

$$\lambda_3 = 1,\ \lambda_1 = \lambda_2 = \lambda_4 = \lambda_5 = \lambda_6 = \lambda_7 = 0 \Rightarrow \begin{cases} t = 2 \\ \mathbf{x} = (1, 2)^T. \end{cases}$$

Similarly, the last common point is obtained for the maximum value of t, that is,

$$\lambda_6 = 1,\ \lambda_1 = \lambda_2 = \lambda_3 = \lambda_4 = \lambda_5 = \lambda_7 = 0 \Rightarrow \begin{cases} t = 3 \\ \mathbf{x} = (1, 2)^T, \end{cases}$$

Table 7.6 Steps for obtaining $\Omega(\mathbf{A}_\pi)$.

Iteration 1					Iteration 2				
$\mathbf{a}_1$	$\mathbf{v}_1^1$	$\mathbf{v}_2^1$	$\mathbf{v}_3^1$	$\mathbf{v}_4^1$	$\mathbf{a}_2$	$\mathbf{w}_1^2$	$\mathbf{v}_1^2$	$\mathbf{v}_2^2$	$\mathbf{v}_3^2$
0	1	**0**	0	0	1	0	**1**	0	0
1	0	**1**	0	0	0	–1	**0**	1	0
–1	0	**0**	1	0	–1	0	**0**	1	0
0	0	**0**	0	1	0	0	**0**	0	1
$\mathbf{t}^1$	0	**1**	–1	0	$\mathbf{t}^2$	0	**1**	–1	0

Iteration 3					Iteration 4				
$\mathbf{a}_3$	$\mathbf{w}_1^3$	$\mathbf{w}_2^3$	$\mathbf{v}_1^3$	$\mathbf{v}_2^3$	$\mathbf{a}_4$	$\mathbf{w}_1^4$	$\mathbf{w}_2^4$	$\mathbf{w}_3^4$	$\mathbf{v}_1^4$
–1	0	–1	**1**	0	–1	1	0	1	**0**
–1	–1	0	**1**	0	0	0	1	1	**0**
1	0	0	**1**	0	–2	1	1	1	**0**
0	0	0	**0**	1	5	0	0	0	**1**
$\mathbf{t}^3$	1	1	**–1**	0	$\mathbf{t}^4$	–3	–2	–3	**5**

Iteration 5				
$\mathbf{a}_5$	$\mathbf{w}_1^5$	$\mathbf{w}_2^5$	$\mathbf{w}_3^5$	$\mathbf{w}_4^5$
1	1	0	1	0
–1	0	1	1	0
1	1	1	1	0
–5	3/5	2/5	3/5	–1/5
$\mathbf{t}^5$	–1	–2	–2	1

Iteration 6						
$\mathbf{a}_6$	$\mathbf{w}_1^6$	$\mathbf{w}_2^6$	$\mathbf{w}_3^6$	$\mathbf{w}_4^6$	$\mathbf{w}_5^6$	$\mathbf{w}_6^6$
2	1	0	1	1	0	1
1	0	1	1	0	1	1
2	1	1	1	1	1	1
–10	3/5	2/5	3/5	2/5	0	1/5
$\mathbf{t}^6$	–2	–1	–1	0	3	3

Iteration 7						
$\mathbf{a}_7$	$\mathbf{w}_1^7$	$\mathbf{w}_2^7$	$\mathbf{w}_3^7$	$\mathbf{w}_4^7$	$\mathbf{w}_5^7$	$\mathbf{w}_6^7$
1	1	1	0	1	0	4
–3	0	0	1	1	4	4
–5	1	1	1	1	4	4
15	2/5	3/5	2/5	3/5	6/5	2
$\mathbf{t}^7$	2	5	–2	2	–14	2

Table 7.6 (Continued).

Iteration 8

a_8	w_1^8	w_2^8	w_3^8	w_4^8	w_5^8	w_6^8
–1	0	0	2	2	14	56
1	1	4	5	4	8	64
1	1	4	7	4	22	64
–5	2/5	6/5	16/5	2	8	152/5
t^8	0	2	–6	–4	–24	–80

Iteration 9

a_9	w_1^9	w_2^9	w_3^9	w_4^9	w_5^9	w_6^9	w_7^9
0	0	2	2	14	56	28	112
1	1	5	4	8	64	112	448
1	1	7	4	22	64	140	448
–5	2/5	16/5	2	8	152/5	224/5	784/5
t^9	0	–4	–2	–10	–24	28	112

Iteration 10

a_{10}	w_1^{10}	w_2^{10}	w_3^{10}	w_4^{10}	w_5^{10}	w_6^{10}	w_7^{10}
0	0	2	2	14	56	672	8960
0	1	5	4	8	64	1344	17920
0	1	7	4	22	64	2016	17920
–1	2/5	16/5	2	8	152/5	672	7168
t^{10}	–2/5	–16/5	–2	–8	–152/5	–672	–7168

Dual

w_1	w_2	w_3	w_4	w_5	w_6	w_7
0	2	2	14	56	672	8960
1	5	4	8	64	1344	17920
1	7	4	22	64	2016	17920
2/5	16/5	2	8	152/5	672	7168

Normalized dual

w_1	w_2	w_3	w_4	w_5	w_6	w_7
0	5/8	1	7/4	35/19	1	5/4
5/2	25/16	2	1	40/19	2	5/2
5/2	35/16	2	11/4	40/19	3	5/2
1	1	1	1	1	1	1

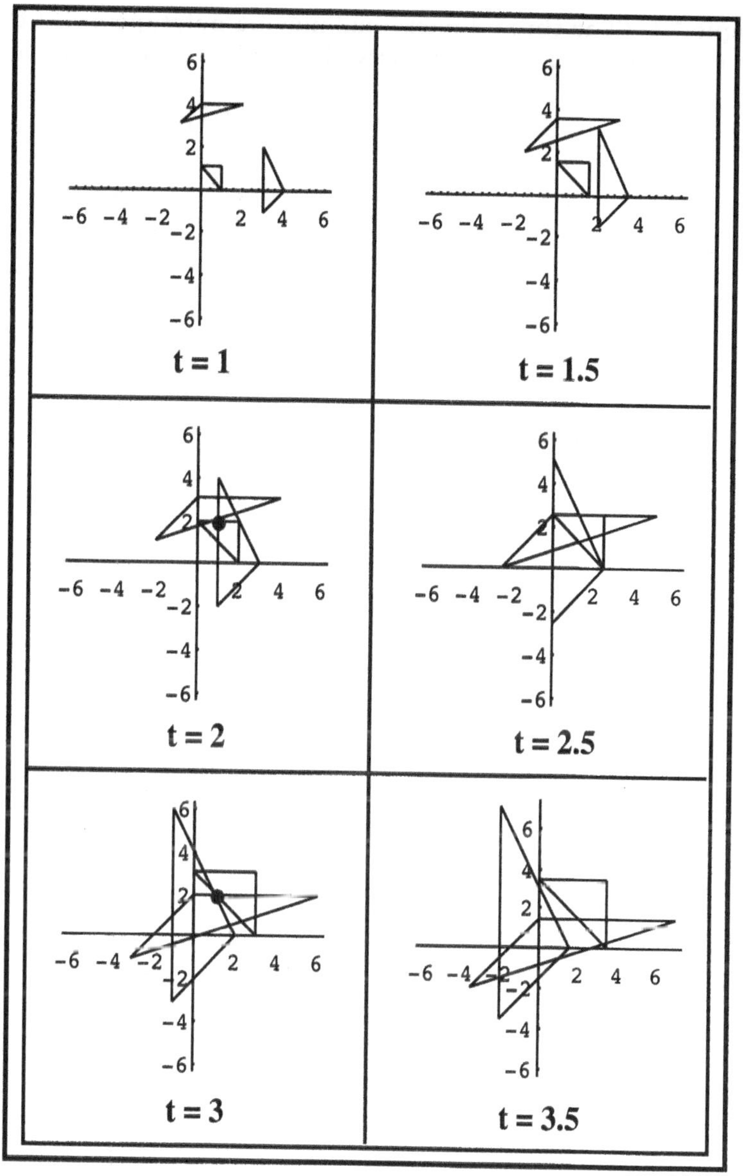

Fig. 7.5 Polytopes generated by the three moving triplets at six different times.

which is exactly the same as the first.

Fig. 7.5 shows the polyhedra at six different instants, including the two instants above. ∎

Example 7.6 (The three-dimensional moving polytopes). Let $\mathbf{P}^1$ and $\mathbf{P}^2$ be two triplets in E^3. After t hours, the positions of triplets $\mathbf{P}^1$ and $\mathbf{P}^2$ are

$$\begin{array}{rcl}\mathbf{P}^1(t) & = & \{(3t,0,t)^T,(2t,-t,0)^T,(t,t,-t)^T\},\\ \mathbf{P}^2(t) & = & \{(5-2t,t,t)^T,(5-t,0,-t)^T,(5-t,-t,0)^T\}.\end{array}$$

Suppose that we are interested in determining the instants at which there is a single common point of the two polytopes, $\mathbf{P}^1_\lambda$ and $\mathbf{P}^2_\lambda$, generated by each of the triplets. To this aim we first determine the intersection of the polytopes at instant t, using Algorithm 7.1.

Step 1: First, we obtain the dual cones, $\Omega(\mathbf{Q}^i_\pi(t)); i=1,2$, of the cones

$$\mathbf{Q}^1_\pi(t) = \begin{pmatrix} 3t & 2t & t \\ 0 & -t & t \\ t & 0 & -t \\ 1 & 1 & 1 \end{pmatrix}_\pi,$$

$$\mathbf{Q}^2_\pi(t) = \begin{pmatrix} 5-2t & 5-t & 5-t \\ t & 0 & -t \\ t & -t & 0 \\ 1 & 1 & 1 \end{pmatrix}_\pi,$$

associated with the polytopes $\mathbf{P}^1_\lambda$ and $\mathbf{P}^2_\lambda$:

$$\Omega(\mathbf{Q}^1_\pi(t)) \equiv \mathbf{V}^1_\rho + \mathbf{W}^1_\pi \equiv \begin{pmatrix} 1 \\ 0 \\ -1 \\ -2t \end{pmatrix}_\rho + \begin{pmatrix} 0 & 0 & 0 \\ -1 & 2 & -1 \\ -2 & 1 & 1 \\ -t & -t & -t \end{pmatrix}_\pi \equiv \mathbf{U}^1_\pi,$$

$$\Omega(\mathbf{Q}^2_\pi(t)) \equiv \mathbf{V}^2_\rho + \mathbf{W}^2_\pi \equiv \begin{pmatrix} 3 \\ 1 \\ 1 \\ 4t-15 \end{pmatrix}_\rho + \begin{pmatrix} 0 & 0 & 0 \\ -1 & -1 & 2 \\ -1 & 2 & -1 \\ -t & -t & -t \end{pmatrix}_\pi \equiv \mathbf{U}^2_\pi.$$

Steps 2 and 3: We obtain the dual cone $\mathbf{H}_\pi \equiv \Omega(\mathbf{B}_\rho + \mathbf{C}_\pi)$, where

$$\mathbf{B} = \begin{pmatrix} 1 & 3 \\ 0 & 1 \\ -1 & 1 \\ -2t & 4t-15 \end{pmatrix}$$

and

$$\mathbf{C} = \begin{pmatrix} 0 & 0 & 0 & 0 & 0 & 0 \\ -1 & 2 & -1 & -1 & -1 & 2 \\ -2 & 1 & 1 & -1 & 2 & -1 \\ -t & -t & -t & -t & -t & -t \end{pmatrix}.$$

But a vector $(x_1, x_2, x_3, x_4)^T$ belongs to $\mathbf{H}_\pi \equiv \Omega(\mathbf{B}_\rho + \mathbf{C}_\pi)$ iff it satisfies the system of inequalities $\mathbf{U}^{i^T}\mathbf{x} \leq \mathbf{0}; i = 1, 2$, that is, the system

$$\begin{array}{llllll} x_1 & & -x_3 & -2t & & = 0 \\ 3x_1 & +x_2 & +x_3 & +4t & +15x_4 & = 0 \\ & -x_2 & -2x_3 & -t & & \leq 0 \\ & 2x_2 & +x_3 & -t & & \leq 0 \\ & -x_2 & +x_3 & -t & & \leq 0 \\ & -x_2 & -x_3 & -t & & \leq 0 \\ & -x_2 & +2x_3 & -t & & \leq 0 \\ & 2x_2 & -x_3 & -t & & \leq 0 \end{array} \tag{7.10}$$

with $x_4 = 1$.

However, (7.10) shows that $(x_1, x_2, x_3, t, x_4)^T$ belongs to the dual space $\Omega(\mathbf{U}_\rho + \mathbf{V}_\pi)$, where

$$\mathbf{U} = \begin{pmatrix} 1 & 3 \\ 0 & 1 \\ -1 & 1 \\ -2 & 4 \\ 0 & -15 \end{pmatrix}$$

and

$$\mathbf{V} = \begin{pmatrix} 0 & 0 & 0 & 0 & 0 & 0 & 0 & 0 \\ -1 & 2 & -1 & -1 & -1 & 2 & 0 & 0 \\ -2 & 1 & 1 & -1 & 2 & -1 & 0 & 0 \\ -1 & -1 & -1 & -1 & -1 & -1 & 0 & -1 \\ 0 & 0 & 0 & 0 & 0 & 0 & -1 & 0 \end{pmatrix}.$$

From Table 7.7, we get

$$\Omega(\mathbf{U}_\rho + \mathbf{V}_\pi) \quad = \quad \begin{pmatrix} 10/3 & 65/21 & 35/13 & 20/7 \\ -5/3 & 5/21 & 5/13 & 5/7 \\ 0 & 5/7 & -15/13 & 0 \\ 5/3 & 25/21 & 25/13 & 10/7 \\ 1 & 1 & 1 & 1 \end{pmatrix}_\pi,$$

Forcing $x_4 = 1$, we get

$$(x_1, x_2, x_3, t) \in \begin{pmatrix} 10/3 & 65/21 & 35/13 & 20/7 \\ -5/3 & 5/21 & 5/13 & 5/7 \\ 0 & 5/7 & -15/13 & 0 \\ 5/3 & 25/21 & 25/13 & 10/7 \end{pmatrix}_\lambda,$$

which implies

$$t = (5/3)\lambda_1 + (25/21)\lambda_2 + (25/13)\lambda_3 + (10/7)\lambda_4.$$

The first common point is obtained for the minimum value of t, that is,

$$\lambda_2 = 1, \quad \lambda_1 = \lambda_3 = \lambda_4 = 0 \Rightarrow \left\{ \begin{array}{l} t = 25/21, \\ \mathbf{x} = (65/21, 5/21, 5/21)^T. \end{array} \right.$$

Table 7.7 Steps for obtaining $\Omega(\mathbf{U}_\rho + \mathbf{V}_\pi)$.

Iteration 1					
$\mathbf{a}_1$	$\mathbf{v}_1^1$	$\mathbf{v}_2^1$	$\mathbf{v}_3^1$	$\mathbf{v}_4^1$	$\mathbf{v}_5^1$
1	**1**	0	0	0	0
0	**0**	1	0	0	0
-1	**0**	0	1	0	0
-2	**0**	0	0	1	0
0	**0**	0	0	0	1
$\mathbf{t}^1$	**1**	0	-1	-2	0

Iteration 2				
$\mathbf{a}_2$	$\mathbf{v}_1^2$	$\mathbf{v}_2^2$	$\mathbf{v}_3^2$	$\mathbf{v}_4^2$
3	**0**	1	2	0
1	**1**	0	0	0
1	**0**	1	0	0
4	**0**	0	1	0
-15	**0**	0	0	1
$\mathbf{t}^2$	**1**	4	10	-15

Iteration 3			
$\mathbf{a}_3$	$\mathbf{v}_1^3$	$\mathbf{v}_2^3$	$\mathbf{v}_3^3$
0	**1**	2	0
-1	**-4**	-10	15
-2	**1**	0	0
-1	**0**	1	0
0	**0**	0	1
$\mathbf{t}^3$	**2**	9	-15

Iteration 4			
$\mathbf{a}_4$	$\mathbf{w}_1^4$	$\mathbf{v}_1^4$	$\mathbf{v}_2^4$
0	-1/2	**-5/2**	15/2
2	2	**8**	-15
1	-1/2	**-9/2**	15/2
-1	0	**1**	0
0	0	**0**	1
$\mathbf{t}^4$	7/2	**21/2**	-45/2

Iteration 5			
$\mathbf{a}_5$	$\mathbf{w}_1^5$	$\mathbf{w}_2^5$	$\mathbf{v}_1^5$
0	1/3	5/21	**15/7**
-1	-2/3	-16/21	**15/7**
1	1	3/7	**-15/7**
-1	-1/3	-2/21	**15/7**
0	0	0	**1**
$\mathbf{t}^5$	2	9/7	**-45/7**

Iteration 6			
$\mathbf{a}_6$	$\mathbf{w}_1^6$	$\mathbf{w}_2^6$	$\mathbf{w}_3^6$
0	1	2/3	1/3
-1	0	-1/3	1/3
-1	1/3	0	-1/3
-1	1/3	1/3	1/3
0	14/45	1/5	7/45
$\mathbf{t}^6$	-2/3	0	-1/3

Iteration 7			
$\mathbf{a}_7$	$\mathbf{w}_1^7$	$\mathbf{w}_2^7$	$\mathbf{w}_3^7$
0	1	2/3	1/3
-1	0	-1/3	1/3
2	1/3	0	-1/3
-1	1/3	1/3	1/3
0	14/45	1/5	7/45
$\mathbf{t}^7$	1/3	0	-4/3

Iteration 8			
$\mathbf{a}_8$	$\mathbf{w}_1^8$	$\mathbf{w}_2^8$	$\mathbf{w}_3^8$
0	2/3	1/3	13/9
2	-1/3	1/3	1/9
-1	0	-1/3	1/3
-1	1/3	1/3	5/9
0	1/5	7/45	7/15
$\mathbf{t}^8$	-1	2/3	-2/3

Iteration 9				
$\mathbf{a}_9$	$\mathbf{w}_1^9$	$\mathbf{w}_2^9$	$\mathbf{w}_3^9$	$\mathbf{w}_4^9$
0	2/3	13/9	7/9	32/27
0	-1/3	1/9	1/9	8/27
0	0	1/3	-1/3	0
0	1/3	5/9	5/9	16/27
-1	1/5	7/15	13/45	56/135
$\mathbf{t}^9$	-1/5	-7/15	-13/45	-56/135

Table 7.7 (Continued).

Dual				Normalized dual			
$\mathbf{w}_1$	$\mathbf{w}_2$	$\mathbf{w}_3$	$\mathbf{w}_4$	$\mathbf{w}_1$	$\mathbf{w}_2$	$\mathbf{w}_3$	$\mathbf{w}_4$
2/3	13/9	7/9	32/27	10/3	65/21	35/13	20/7
-1/3	1/9	1/9	8/27	-5/3	5/21	5/13	5/7
0	1/3	-1/3	0	0	5/7	-15/13	0
1/3	5/9	5/9	16/27	5/3	25/21	25/13	10/7
1/5	7/15	13/45	56/135	1	1	1	1

Similarly, the last common point is obtained for the maximum value of t, that is,

$$\lambda_3 = 1, \;\; \lambda_1 = \lambda_2 = \lambda_4 = 0 \Rightarrow \begin{cases} t = 25/13, \\ \mathbf{x} = (35/13, 5/13, -15/13)^T. \end{cases}$$

Fig. 7.6 shows the polyhedra at six different instants, including the two instants above. ■

7.3 POLYHEDRA

Definition 7.2 (Polyhedron). *The sum of a linear space, a cone, and a polytope is called a polyhedron.*

In Chapter 8 it will be shown that this definition is equivalent to the following.

Definition 7.3 (Alternative definition of a polyhedron). *A polyhedron is the intersection of a finite set of halfspaces, that is,*

$$\{\mathbf{x} \in E^n | \mathbf{H}\mathbf{x} \le \mathbf{b}\}. \tag{7.11}$$

Remark 7.2 *A polyhedron is a convex set.*

7.3.1 Vertices and Edges

Definition 7.4 (Vertex). *A vector* $\mathbf{x}$ *of a convex set* S *is said to be a vertex iff it does not admit a representation as a linear convex combination of two distinct vectors of* S.

Example 7.7 (Vertex). Fig. 7.7 shows with black dots the vertices of a polyhedron. Note that they are the only vectors that cannot be written as convex linear combinations of two distinct vectors in the polyhedron. The

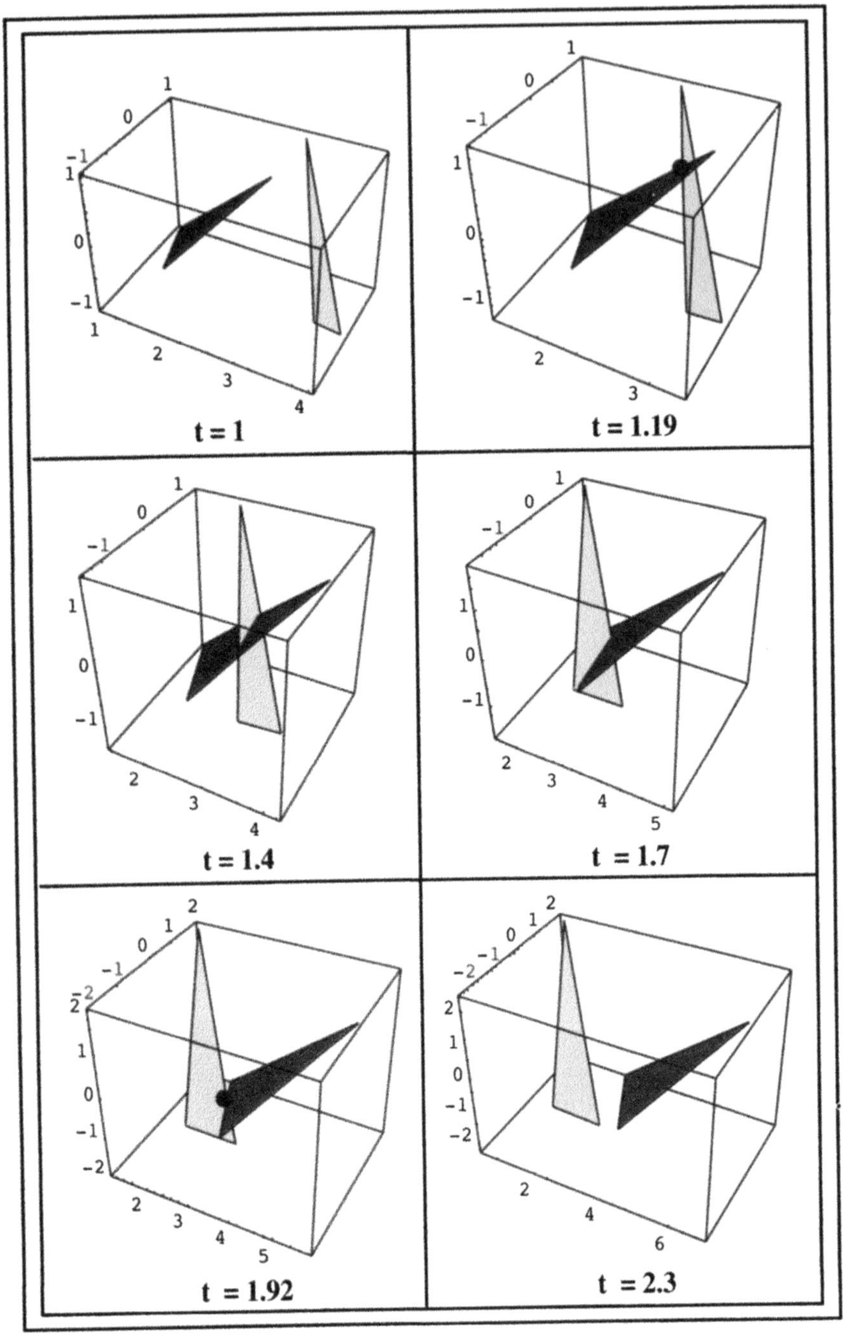

Fig. 7.6 Polyhedra generated by the two moving triplets at six different times.

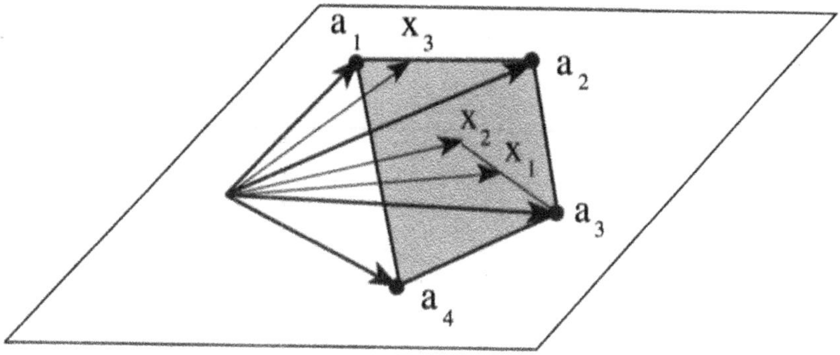

Fig. 7.7 Vertices of a polyhedron.

figure illustrates how the interior vector $\mathbf{x}_1$ and the boundary vector $\mathbf{x}_3$ can in fact be written as a linear convex combination of $\mathbf{x}_2$ *and* $\mathbf{a}_3$ and $\mathbf{a}_1$ and $\mathbf{a}_2$, respectively.

■

Remark 7.3 *The problem of finding the vertices of a polyhedron can be reduced to the problem of finding a set of minimal generators. To this aim, we add an extra unit component to all vectors and obtain this set of minimal generators using the Γ-algorithm.*

Example 7.8 (Vertices). Consider the polytope generated by the set

$$\{(0,0)^T, (1,2)^T, (2,1)^T, (2,-1)^T, (2,-2)^T, (4,2)^T\},$$

as shown in Fig. 7.8. To obtain its vertices, we consider the cone generated by the set

$$\{(0,0,1)^T, (1,2,1)^T, (2,1,1)^T, (2,-1,1)^T, (2,-2,1)^T, (4,2,1)^T\}$$

and obtain the dual of its dual, using the Γ-algorithm, as shown in Tables 7.8 and 7.9.

The dual $\Omega(\mathbf{A}_\pi)$ of $\mathbf{A}_\pi$, according to Table 7.8, is $\mathbf{W}_\pi$, with

$$\mathbf{W} = \begin{pmatrix} -2/3 & -10/9 & 0 & 80/27 \\ 1/3 & -10/9 & 1 & -40/27 \\ 0 & 0 & -2 & -80/9 \end{pmatrix}, \tag{7.12}$$

which, changing scales, is equivalent to

$$\mathbf{W} = \begin{pmatrix} -2 & -1 & 0 & 2 \\ 1 & -1 & 1 & -1 \\ 0 & 0 & -2 & -6 \end{pmatrix}. \tag{7.13}$$

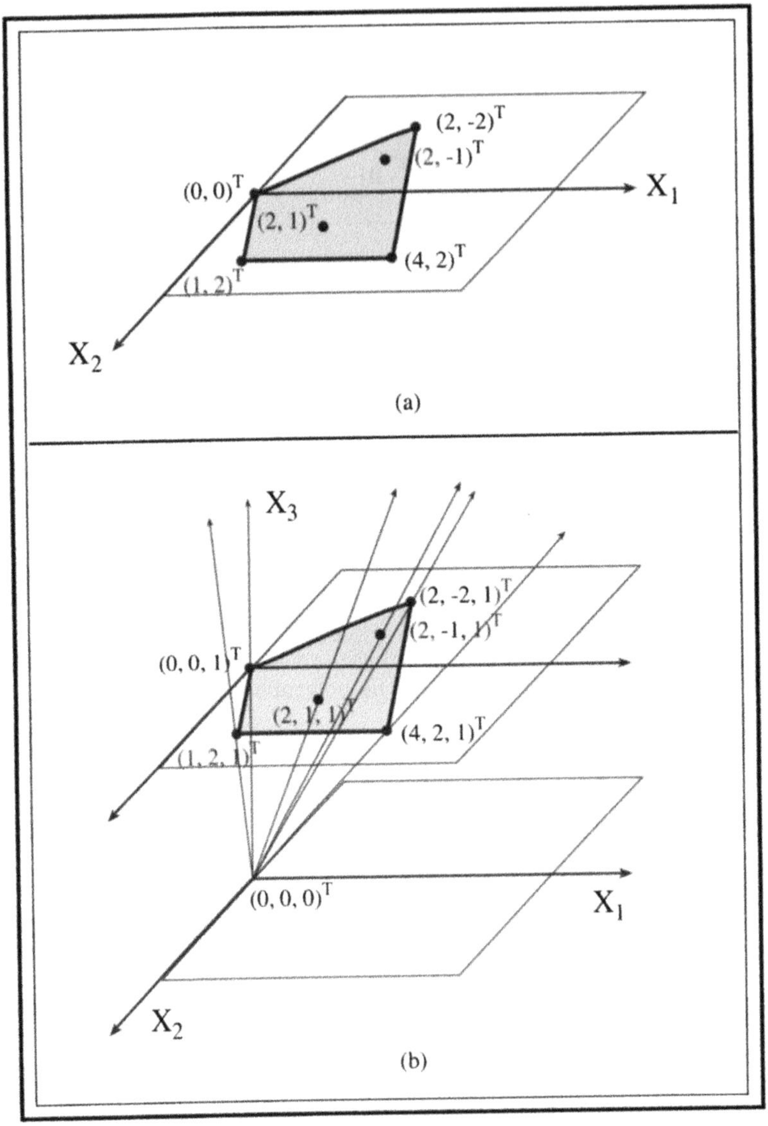

Fig. 7.8 Vertices of a polytope and the associated cone.

Table 7.8 Illustration of the different iterations of the Γ-algorithm to obtain the dual $\Omega(\mathbf{A}_\pi)$ in Example 7.8.

Iteration 1				Iteration 2				Iteration 3			
$\mathbf{a}_1$	$\mathbf{v}_1^1$	$\mathbf{v}_2^1$	$\mathbf{v}_3^1$	$\mathbf{a}_2$	$\mathbf{w}_1^2$	$\mathbf{v}_1^2$	$\mathbf{v}_2^2$	$\mathbf{a}_3$	$\mathbf{w}_1^3$	$\mathbf{w}_2^3$	$\mathbf{v}_1^3$
0	1	0	0	1	0	0	1	2	0	0	1
0	0	1	0	2	0	1	0	1	1/2	−1/2	−1/2
1	0	0	1	1	−1	0	0	1	−1	0	0
$\mathbf{t}^1$	0	0	1	$\mathbf{t}^2$	−1	2	1	$\mathbf{t}^3$	−1/2	−1/2	3/2

Iteration 4				Iteration 5				
$\mathbf{a}_4$	$\mathbf{w}_1^3$	$\mathbf{w}_2^3$	$\mathbf{w}_3^3$	$\mathbf{a}_5$	$\mathbf{w}_1^4$	$\mathbf{w}_2^4$	$\mathbf{w}_3^4$	$\mathbf{w}_4^4$
2	1/3	1/3	−2/3	2	1/3	−2/3	2/3	−1/3
−1	1/3	−2/3	1/3	−2	1/3	1/3	0	−2/3
1	−1	0	0	1	−1	0	−4/3	0
$\mathbf{t}^4$	−2/3	4/3	−5/3	$\mathbf{t}^5$	−1	−2	0	2/3

Iteration 6					Dual			
$\mathbf{a}_5$	$\mathbf{w}_1^5$	$\mathbf{w}_2^5$	$\mathbf{w}_3^5$	$\mathbf{w}_4^5$	$\mathbf{w}_1^6$	$\mathbf{w}_2^6$	$\mathbf{w}_3^6$	$\mathbf{w}_4^6$
4	2/3	1/3	−2/3	−10/9	−2/3	−10/9	0	80/27
2	0	1/3	1/3	−10/9	1/3	−10/9	1	−40/27
1	−4/3	−1	0	0	0	0	−2	−80/9
$\mathbf{t}^6$	4/3	1	−2	−20/3				

Table 7.9 Illustration of the different iterations used with the Γ-algorithm to obtain the bidual $\Omega(\Omega(\mathbf{A}_\pi))$ in Example 7.8.

Iteration 1				Iteration 2				Iteration 3			
$\mathbf{a}_1$	$\mathbf{v}_1^1$	$\mathbf{v}_2^1$	$\mathbf{v}_3^1$	$\mathbf{a}_2$	$\mathbf{w}_1^2$	$\mathbf{v}_1^2$	$\mathbf{v}_2^2$	$\mathbf{a}_3$	$\mathbf{w}_1^3$	$\mathbf{w}_2^3$	$\mathbf{v}_1^3$
−2	1	0	0	−1	1/2	1/2	0	0	1/3	1/3	0
1	0	1	0	−1	0	1	0	1	−1/3	2/3	0
0	0	0	1	0	0	0	1	−2	0	0	1
$\mathbf{t}^1$	−2	1	0	$\mathbf{t}^2$	−1/2	−3/2	0	$\mathbf{t}^3$	−1/3	2/3	−2

Iteration 4				Bidual			
$\mathbf{a}_4$	$\mathbf{w}_1^3$	$\mathbf{w}_2^3$	$\mathbf{w}_3^3$	$\mathbf{w}_1^4$	$\mathbf{w}_2^4$	$\mathbf{w}_3^4$	$\mathbf{w}_4^4$
2	1/3	1/3	0	1/3	0	4/3	1
−1	−1/3	2/3	0	2/3	0	2/3	−1
−6	−1/6	1/3	1/2	1/3	1/2	1/3	1/2
$\mathbf{t}^4$	2	−2	−3				

The bidual $\Omega(\Omega(\mathbf{A}_\pi))$ of $\mathbf{A}_\pi$, according to Table 7.9, is $\mathbf{W}_\pi$, with

$$\mathbf{W} = \begin{pmatrix} 1/3 & 0 & 4/3 & 1 \\ 2/3 & 0 & 2/3 & -1 \\ 1/3 & 1/2 & 1/3 & 1/2 \end{pmatrix}, \tag{7.14}$$

which, changing scales, is equivalent to

$$\mathbf{W} = \begin{pmatrix} 1 & 0 & 4 & 2 \\ 2 & 0 & 2 & -2 \\ 1 & 1 & 1 & 1 \end{pmatrix}, \tag{7.15}$$

from which we can conclude that the set of vertices of $\mathbf{A}_\lambda$ is

$$\{(1,2)^T, (0,0)^T, (4,2)^T, (2,-2)^T\}.$$

■

Definition 7.5 (Edge). *Let $\mathbf{A}_\pi$ be a polyhedron, the edges of which are defined as the set of single generators that cannot be written as linear combinations of the rest. In other words,*

$$\mathbf{a}_i \textit{ is an edge} \Leftrightarrow \mathbf{A}^{\mathbf{a}_i} \equiv \{\mathbf{a}_i\}. \tag{7.16}$$

Exercises

7.1 Given a three-dimensional polytope such that every two vertices are adjacent, show that it is a tetrahedron.

7.2 Consider a polytope $\mathbf{P}_\lambda$ in E^3 generated by

$$\mathbf{P} \equiv \{\mathbf{p}_1, \mathbf{p}_2, \mathbf{p}_3, \mathbf{p}_4\},$$

where

$$\mathbf{p}_1 = (0,0,0)^T;\ \mathbf{p}_2 = (0,1,0)^T;\ \mathbf{p}_3 = (1,1,0)^T;\ \mathbf{p}_4 = (1,1,1)^T.$$

If we apply an affine map π to a polytope $\mathbf{P}_\lambda$, then we get a new polytope $\pi(\mathbf{P}_\lambda)$. Calculate the projection of $\mathbf{P}_\lambda$ on E^2 and prove that it is a polytope too.

7.3 Obtain the intersection of the polyhedra generated by

$$\mathbf{A}_1 = \begin{pmatrix} -1 & 0 & 1 & 0 & 2 \\ 0 & -1 & 1 & 0 & 2 \\ 0 & 0 & 0 & -2 & 2 \end{pmatrix},$$

with the cone generated by

$$\mathbf{A}_2 = \begin{pmatrix} 1 & 0 & 0 \\ 0 & 1 & 0 \\ 0 & 0 & 1 \end{pmatrix}.$$

7.4 Let $\mathbf{A}_\pi$ be the cone generated by

$$\begin{aligned} \mathbf{a}_1 &= (2,1,0)^T, \\ \mathbf{a}_2 &= (2,-1,0)^T, \\ \mathbf{a}_3 &= (1,0,2)^T, \\ \mathbf{a}_4 &= (1,0,-2)^T. \end{aligned}$$

Obtain the intersection of $\mathbf{A}_\pi$ with the negative of $\Omega(\mathbf{A}_\pi)$.

7.5 Show that if $\mathbf{A}_\pi$ is any cone in E^{n+1} generated by arbitrary vectors $\mathbf{a}_i$, then $\{\mathbf{x} \in E^n \mid \begin{pmatrix} 1 \\ \mathbf{x} \end{pmatrix} \in \mathbf{A}_\pi\}$ is a polyhedron.

7.6 Consider the polytope generated by the set

$$\{(0,0)^T, (-2,3)^T, (0,2)^T, (2,1)^T, (4,0)^T,$$
$$(3,4)^T, (-4,2)^T, (-2,-2)^T, (-4,1)^T\},$$

and obtain its vertices.

8

Cones and Systems of Inequalities

8.1 INTRODUCTION

In this chapter we apply the theory of orthogonal sets, cones, polytopes, and polyhedra, developed in previous chapters, to analyze and solve systems of inequalities. A powerful method is proposed in which both the polyhedral convex cone $\mathbf{A}_\pi$ and the Farkas lemma play an important role. The latter is now understood not as a simple alternative theorem but as a mutual primal–dual interdependence of a deep content: "*Those questions stated in* $\mathbf{A}_\pi$ *are only solved by means of a call to the polar cone* $\Omega(\mathbf{A}_\pi)$ *and vice versa.*"

In order to make this interpretation we must first know $\Omega(\mathbf{A}_\pi)$. Then it is possible to answer the following questions with elegance and efficiency:

1. Is the system consistent?

2. Does the system have a unique solution?

3. What is the set of all solutions?

These are explained in the following sections. Section 8.2 discusses systems of linear equations with unrestricted and restricted unknowns. We discuss the problems of compatibility and existence of a unique solution. In both cases, obtaining the polar cone is the first step in answering these questions.

In Section 8.3 systems of linear equations and inequalities are solved and an algorithm is given. Several examples are used to illustrate the methods. Section 8.4 gives some applications in linear programming problems.

8.2 A DISCUSSION OF LINEAR SYSTEMS

In this section we discuss linear systems. More precisely, we deal with the first two problems above: compatibility and uniqueness. We start by analyzing the simplest linear system of equations and then we restrict variables gradually until the most general system of equations is achieved, which is the main object of our study.

8.2.1 Systems with Unrestricted Variables

In this section we revisit the problem of compatibility and uniqueness in the case of systems of linear equations with unrestricted variables, which was already treated in Section 5.2, but now we treat the linear spaces as cones (a different and more powerful point of view). Let

$$\mathbf{B}\mathbf{x}_\rho = \mathbf{a} \tag{8.1}$$

be a system of equations, where $\mathbf{b}_i, i = 1, 2, \ldots, n$ are the m-column vectors of $\mathbf{B} = (\mathbf{b}_1, \ldots, \mathbf{b}_n)$, $\mathbf{x}_\rho = (x_1, \ldots, x_n)^T$ is the vector of the n unknowns, $\mathbf{a} = (a_1, \ldots, a_m)^T$ is a column vector of real numbers, and the subindex ρ is used to indicate unrestricted variables. In this chapter we use the $\mathbf{B}$ matrix to refer to the coefficients of unrestricted variables.

The feasibility or consistency condition for system (8.1) can be stated as follows:

$$\mathbf{B}\mathbf{x}_\rho = \mathbf{a} \text{ has solution } \Leftrightarrow \mathbf{a} \in \mathbf{B}_\rho \tag{8.2}$$

That is, system (8.1) has solution iff the vector $\mathbf{a}$ is a linear combination of the columns of $\mathbf{B}$.

According to Corollary 6.1, Condition (8.2) can be written in the form

$$\boxed{\mathbf{B}\mathbf{x}_\rho = \mathbf{a} \text{ has solution } \Leftrightarrow \mathbf{V}^T\mathbf{a} = \mathbf{0},} \tag{8.3}$$

where $\mathbf{V}_\rho \equiv \Omega(\mathbf{B}_\rho)$.

Thus, checking compatibility reduces to obtaining the dual $\Omega(\mathbf{B}_\rho)$ and checking (8.3).

Remark 8.1 *Note that the homogeneous system* $\mathbf{B}\mathbf{x}_\rho = \mathbf{a} = \mathbf{0}$ *always has a solution, because* $\mathbf{a} = \mathbf{0} \in \mathbf{B}_\rho$ *and* $\mathbf{V}^T\mathbf{a} = \mathbf{V}^T\mathbf{0} = \mathbf{0}$.

The uniqueness of solution condition can be written as

$$\mathbf{B}\mathbf{x}_\rho = \mathbf{a} \text{ has unique solution } \Leftrightarrow \mathbf{a} \in \mathbf{B}_\rho \wedge \mathbf{0} \notin \mathbf{B}_\xi, \tag{8.4}$$

where $\mathbf{B}_\xi$ is the set of all non-null linear combinations of the columns of $\mathbf{B}$.

According to Corollary 6.2, Condition (8.4) can be written in the form

$$\boxed{\mathbf{B}\mathbf{x}_\rho = \mathbf{a} \text{ has unique solution } \Leftrightarrow \mathbf{V}^T\mathbf{a} = 0 \wedge {\mathbf{V}^{(i)}}^T\mathbf{b}_{i+1} \neq \mathbf{0}, \forall i.} \tag{8.5}$$

Conditions (8.2) and (8.4) are only of theoretical value. Their practical equivalent expressions are (8.3) and (8.5).

8.2.2 Systems with Restricted Variables

Let

$$\begin{aligned} \mathbf{C}\mathbf{x}_\pi &= \mathbf{a} \\ \mathbf{x}_\pi &\geq \mathbf{0} \end{aligned} \tag{8.6}$$

be a system of equations with restricted variables, where now the subindex π is used to refer to non-negative variables. In this chapter we use the $\mathbf{C}$ matrix to refer to the coefficients of non-negative variables.

It is worthwhile mentioning that other important cases can be reduced to this one. For example:

- *The case of inequalities*: In this case we use non-negative slack variables to reduce the inequalities to equalities (we add one new variable per inequality).

- *The mixed case of equalities and inequalities*: We reduce the inequalities as in the previous case.

- *The mixed case of unrestricted and restricted variables*: We replace the unrestricted variables by differences between two non-negative variables (in fact we add only one more variable).

Then we have

$$\mathbf{C}\mathbf{x}_\pi = \mathbf{a} \text{ has solution } \Leftrightarrow \mathbf{a} \in \mathbf{C}_\pi. \tag{8.7}$$

That is, system (8.6) has solution iff the vector $\mathbf{a}$ is a linear non-negative combination of the columns of $\mathbf{C}$.

According to Theorem 6.2, Condition (8.7) can be written in the form

$$\boxed{\mathbf{C}\mathbf{x}_\pi = \mathbf{a} \text{ has solution } \Leftrightarrow \mathbf{V}^T\mathbf{a} = \mathbf{0} \wedge \mathbf{W}^T\mathbf{a} \leq \mathbf{0},} \tag{8.8}$$

where $\Omega(\mathbf{C}_\pi) \equiv \mathbf{U}_\pi \equiv \mathbf{V}_\rho + \mathbf{W}_\pi$.

As in the case of unrestricted variables, checking the compatibility reduces to obtaining the dual $\Omega(\mathbf{C}_\pi) \equiv \mathbf{U}_\pi \equiv \mathbf{V}_\rho + \mathbf{W}_\pi$ and checking (8.8).

Remark 8.2 *Note that the homogeneous system* $\mathbf{C}\mathbf{x}_\pi = \mathbf{a} = \mathbf{0}$ *always has a solution, because* $\mathbf{a} = \mathbf{0} \in \mathbf{C}_\pi$ *and* $\mathbf{V}^T\mathbf{a} = \mathbf{V}^T\mathbf{0} = \mathbf{0} \wedge \mathbf{W}^T\mathbf{a} = \mathbf{W}^T\mathbf{0} \leq \mathbf{0}$.

The conditions for the uniqueness of the solution of systems with restricted variables requires the concept of binding sets (see Definition 6.13) and are given by the following theorem.

Theorem 8.1 *System (8.6) has a unique solution iff*

$$\begin{array}{ll} \mathbf{U}^T\mathbf{a} \le \mathbf{0} \wedge \mathbf{0} \notin \mathbf{C}^a_\xi \Leftrightarrow \mathbf{V}^T\mathbf{a} \le \mathbf{0} \wedge \mathbf{c}^a_{i+1} \notin \mathbf{C}^{a(i)}_\rho, \forall i, & if \quad \mathbf{a} \ne \mathbf{0}, \\ \mathbf{C}^a \equiv \emptyset & if \quad \mathbf{a} = \mathbf{0}, \end{array} \tag{8.9}$$

where $\mathbf{C}_\xi$ *is the set of all non-null linear combinations of the columns of* $\mathbf{C}$.

Proof: Assume that $\mathbf{a} \ne \mathbf{0}$ and that system (8.6) has solution. Then $\mathbf{a} \in \mathbf{C}_\pi$, and by Proposition 6.4, $\mathbf{a} = \mathbf{C}^a\boldsymbol{\sigma}^a$ with $\boldsymbol{\sigma}^a > \mathbf{0}$. The solution is unique iff

$$\mathbf{C}^a\boldsymbol{\sigma}^a \ne \mathbf{C}^a\boldsymbol{\sigma}, \forall \boldsymbol{\sigma} > \mathbf{0} \Leftrightarrow \mathbf{C}^a(\boldsymbol{\sigma} - \boldsymbol{\sigma}^a) \ne \mathbf{0}, \forall \boldsymbol{\sigma} > \mathbf{0} \Leftrightarrow \mathbf{0} \notin \mathbf{C}^a_\xi.$$

Obviously, if $\mathbf{a} = \mathbf{0}$ then $\mathbf{C}^a \equiv \emptyset$. ∎

8.2.3 Systems with Strictly Restricted Variables

Finally, let us consider the system

$$\begin{array}{r} \mathbf{F}\mathbf{x} = \mathbf{B}\mathbf{x}_\rho + \mathbf{C}\mathbf{x}_\pi + \mathbf{D}\mathbf{x}_\sigma = \mathbf{a} \\ \mathbf{x}_\pi \ge \mathbf{0}, \mathbf{x}_\sigma > \mathbf{0}, \end{array} \tag{8.10}$$

where now

$$\mathbf{A} = (\mathbf{B} : -\mathbf{B} : \mathbf{C} : \mathbf{D}), \;\; \Omega(\mathbf{A}_\pi) \equiv \mathbf{U}_\pi \equiv \mathbf{V}_\rho + \mathbf{W}_\pi \tag{8.11}$$

and the subindex σ refers to positive variables.

The consistency and uniqueness of solutions of system (8.10) are given by the following theorems.

Theorem 8.2 *System (8.10) has solution iff*

$$\mathbf{U}^T\mathbf{a} \le \mathbf{0} \wedge \mathbf{W}(\mathbf{a}) \subseteq \mathbf{W}(\mathbf{d}_i), \forall i, \tag{8.12}$$

where $\mathbf{W}(\mathbf{a})$ *and* $\mathbf{W}(\mathbf{d}_i)$ *are the subsets of* $\mathbf{W}$ *orthogonal to* $\mathbf{a}$ *and* $\mathbf{d}_i$, *respectively.*

Proof: $\mathbf{F}\mathbf{x} = \mathbf{a}$ has solution $\Leftrightarrow \mathbf{a} \in \mathbf{B}_\rho + \mathbf{C}_\pi + \mathbf{D}_\sigma \Leftrightarrow \mathbf{a} \in \mathbf{B}_\rho + \mathbf{C}_\pi + \mathbf{D}_\pi + \sigma\mathbf{d}_i, \forall i$ and from the definition of binding set, $\mathbf{d}_i \in \mathbf{A}^{\mathbf{a}}, \forall i$ and then the result holds. ∎

Theorem 8.3 *The necessary and sufficient conditions for system (8.10),* $\mathbf{F}\mathbf{x} = \mathbf{a}$, *to have a unique solution are:*

$$\begin{array}{ll} \mathbf{U}^T\mathbf{a} \le \mathbf{0} \wedge \mathbf{W}(\mathbf{a}) \subseteq \mathbf{W}(\mathbf{d}_i), \forall i \wedge \mathbf{0} \notin \mathbf{F}^a_\xi, & if \quad \mathbf{a} \ne \mathbf{0}, \\ \mathbf{C}^a \equiv \mathbf{D} \equiv \emptyset \wedge \mathbf{0} \notin \mathbf{B}_\xi, & if \quad \mathbf{a} = \mathbf{0}. \end{array} \tag{8.13}$$

Proof: [$\Leftrightarrow$] If $\mathbf{0} \neq \mathbf{a} \in \mathbf{A}_\pi$, taking into account Proposition 6.4, we get $\mathbf{a} \in \mathbf{A}^a_\sigma$, and by (8.12), $\mathbf{D} \subseteq \mathbf{A}^a$. In addition, we have

$$\mathbf{W}(\mathbf{b}_i) \equiv \mathbf{W}, \forall i \Rightarrow \mathbf{B}, -\mathbf{B} \subseteq \mathbf{A^a}.$$

Then

$$\mathbf{a} \in \mathbf{B}_\sigma - \mathbf{B}_\sigma + \mathbf{C}^a_\sigma + \mathbf{D}_\sigma \Rightarrow \mathbf{a} = \mathbf{B}\rho^a + (\mathbf{C}^a, \mathbf{D})\sigma^a = \mathbf{F}^a \begin{pmatrix} \rho^a \\ \sigma^a \end{pmatrix},$$

and if the solution is unique we get

$$\mathbf{0} \neq \mathbf{F}^a \sigma \begin{pmatrix} \rho - \rho^a \\ \sigma - \sigma^a \end{pmatrix}, \forall \begin{pmatrix} \rho \\ \sigma \end{pmatrix} \neq \begin{pmatrix} \rho^a \\ \sigma^a \end{pmatrix}.$$

Now it is easy to prove that

$$\left\{ \begin{pmatrix} \rho - \rho^a \\ \sigma - \sigma^a \end{pmatrix} \mid \begin{pmatrix} \rho \\ \sigma \end{pmatrix} \neq \begin{pmatrix} \rho^a \\ \sigma^a \end{pmatrix} \right\} \equiv \boldsymbol{\xi}.$$

In the homogeneous case we have:

- [$\Rightarrow$] If $\mathbf{C}^a \vee \mathbf{D} \not\equiv \emptyset$ with $\mathbf{Fx} = \mathbf{0}$ has solution, then $\mathbf{0} = \mathbf{B}\rho^a + (\mathbf{C}^a, \mathbf{D})\boldsymbol{\sigma}^a$ and also $\mathbf{0} = \mathbf{0}\sigma = \mathbf{B}\sigma\boldsymbol{\rho}^a + (\mathbf{C}^a, \mathbf{D})\sigma\boldsymbol{\sigma}^a$. Thus, the solution is not unique. Consequently, $\mathbf{C}^a \equiv \mathbf{D} \equiv \emptyset$.
- [$\Leftarrow$] $\mathbf{D} \equiv \emptyset$ and if $\mathbf{C}^a \equiv \emptyset$, then $\mathbf{0} \notin \mathbf{B}_\rho + \mathbf{C}_\pi + \sigma\mathbf{c}_i, \forall i \Rightarrow \mathbf{0} \notin \mathbf{B}_\rho + \mathbf{C}_\pi, \forall \boldsymbol{\pi} \neq \mathbf{0}$ and because $\mathbf{0} \notin \mathbf{B}_\xi \Rightarrow \mathbf{0} \notin \mathbf{B}_\rho + \mathbf{C}_\pi, \forall (\boldsymbol{\rho}, \boldsymbol{\pi}) \neq (\mathbf{0}, \mathbf{0})$ we have $\mathbf{Fx} = \mathbf{0}$ has unique solution.

■

Note that Conditions (8.13) are equivalent to

$$\boxed{\begin{array}{l} \mathbf{U}^T\mathbf{a} \leq \mathbf{0} \wedge \mathbf{W}(\mathbf{a}) \subseteq \mathbf{W}(\mathbf{d}_i), \forall i \wedge \mathbf{f}_{i+1} \notin \mathbf{F}^{a(i)}_\rho, \forall i \\ \mathbf{C}^a \equiv \mathbf{D} \equiv \emptyset \wedge \mathbf{b}^{i+1} \notin \mathbf{B}^{(i)}_\rho, \forall i, \end{array}} \tag{8.14}$$

respectively.

Example 8.1 (Point in a polygon). Suppose that we are interested in knowing whether or not a point lies inside a polygon defined by a set of points $\{\mathbf{x}_1, \ldots, \mathbf{x}_n\}$. Let $\mathbf{x}_i = (u_i, v_i)^T; i = 1, \ldots, n$. Then the point $(u, v)^T$ belongs to the polygon generated by $\{\mathbf{x}_1, \ldots, \mathbf{x}_n\}$, iff

$$\begin{pmatrix} u_1 & u_2 & \ldots & u_n \\ v_1 & v_2 & \ldots & v_n \\ 1 & 1 & \ldots & 1 \end{pmatrix} \begin{pmatrix} \pi_1 \\ \pi_2 \\ \vdots \\ \pi_n \end{pmatrix} = \begin{pmatrix} u \\ v \\ 1 \end{pmatrix}. \tag{8.15}$$

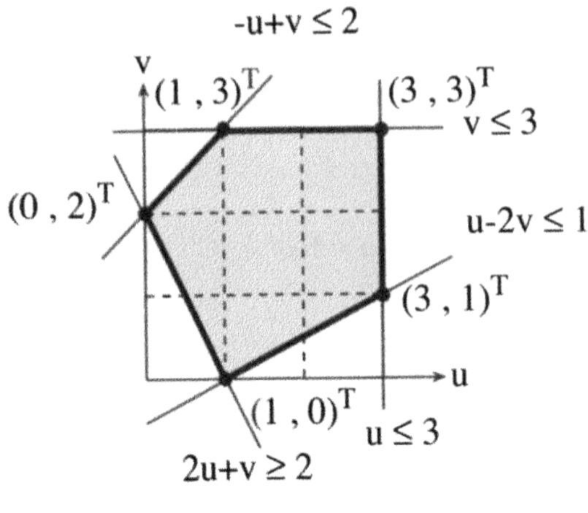

Fig. 8.1 Points and constraints defining a polygon.

Note that (8.15) expresses the point $(u, v)^T$ as a linear convex combination of the points in $\{\mathbf{x}_1, \ldots, \mathbf{x}_n\}$. For this to be possible, that is, for the point $(u, v)^T$ to belong to the polygon defined by $\{\mathbf{x}_1, \ldots, \mathbf{x}_n\}$, system (8.15) must be compatible.

Now we apply this to the case of the polygon generated by the points $\{(1,0)^T, (3,1)^T, (3,3)^T, (1,3)^T, (0,2)^T\}$ (see Fig. 8.1).

According to (8.8),

$$\mathbf{A}\mathbf{x}_\pi = \mathbf{a} \text{ has solution } \Leftrightarrow \mathbf{V}^T\mathbf{a} = \mathbf{0} \wedge \mathbf{W}^T\mathbf{a} \leq \mathbf{0}, \tag{8.16}$$

where $\Omega(\mathbf{A}_\pi) \equiv \mathbf{V}_\rho + \mathbf{W}_\pi$. Thus, we need the dual of the cone generated by the columns of $\mathbf{A}$, that is, the matrix in (8.15), which in this case becomes

$$\mathbf{A} = \begin{pmatrix} 1 & 3 & 3 & 1 & 0 \\ 0 & 1 & 3 & 3 & 2 \\ 1 & 1 & 1 & 1 & 1 \end{pmatrix}. \tag{8.17}$$

To obtain $\Omega(\mathbf{A}_\pi)$, we use the Γ-algorithm (see Chapter 6), as follows:

Step 1: Vector 1 is introduced, and the dot products are calculated, as shown in Table 8.1.

Step 2: We update the orthogonal sets $A(\mathbf{U})$:

$$A_{\mathbf{v}_1} \equiv \{\}, A_{\mathbf{v}_2} \equiv \{1\}, A_{\mathbf{v}_3} \equiv \{\}.$$

Step 3: The pivot is column vector 1.

Step 4: Since $s < 3$ and *pivot* > 0, go to the Γ_I-process.

Step 5: The pivot column is normalized and the pivoting process performed as shown in Table 8.2. Index h is increased to 2 and we go to Step 1.

Table 8.1 Added constraint, **U** matrix, and associated dot products.

Iteration 1			
$\mathbf{a}_1$	$\mathbf{v}_1^1$	$\mathbf{v}_2^1$	$\mathbf{v}_3^1$
1	**1**	0	0
0	**0**	1	0
1	**0**	0	1
$\mathbf{t}^1$	**1**	0	1

Table 8.2 Added constraint, **U** matrix, and associated dot products.

Iteration 2			
$\mathbf{a}_2$	$\mathbf{w}_1^2$	$\mathbf{v}_1^2$	$\mathbf{v}_2^2$
3	–1	**0**	–1
1	0	**1**	0
1	0	**0**	1
$\mathbf{t}^2$	–3	**1**	–2

Step 1: Constraint 2 is introduced, and the dot products are calculated, as shown in Table 8.2.

Step 2: We update the orthogonal sets $A(\mathbf{U})$:

$$A_{\mathbf{w}_1} \equiv \{\}, A_{\mathbf{v}_1} \equiv \{1\}, A_{\mathbf{v}_2} \equiv \{1\}.$$

Step 3: The pivot is column vector 2.

Step 4: Since $s < 3$ and *pivot* > 0, go to the Γ_I-process.

Step 5: The pivot column is normalized and the pivoting process performed as shown in Table 8.3. Index h is increased to 3 and we go to Step 1.

Step 1: Constraint 3 is introduced, and the dot products are calculated, as shown in Table 8.3.

Step 2: We update the orthogonal sets $A(\mathbf{U})$:

$$A_{\mathbf{w}_1} \equiv \{2\}, A_{\mathbf{w}_2} \equiv \{1\}, A_{\mathbf{v}_1} \equiv \{1, 2\}.$$

Step 3: The pivot is column vector 3.

Step 4: Since $s < 3$ and *pivot* > 0, go to the Γ_I-process.

Step 5: The pivot column is normalized and the pivoting process performed as shown in Table 8.4. Index h is increased to 4 and we go to Step 1.

Step 1: Constraint 4 is introduced, and the dot products are calculated, as shown in Table 8.4.

Step 2: We update the orthogonal sets $A(\mathbf{U})$:

$$A_{\mathbf{w}_1} \equiv \{2, 3\}, A_{\mathbf{w}_2} \equiv \{1, 3\}, A_{\mathbf{w}_3} \equiv \{1, 2\}.$$

Table 8.3 Added constraint, **U** matrix, and associated dot products.

Iteration 3			
$\mathbf{a}_3$	$\mathbf{w}_1^3$	$\mathbf{w}_2^3$	$\mathbf{v}_1^3$
3	-1	0	**-1**
3	3	-1	**2**
1	0	0	**1**
$\mathbf{t}^3$	6	-3	**4**

Table 8.4 Added constraint, **U** matrix, and associated dot products.

Iteration 4			
$\mathbf{a}_4$	$\mathbf{w}_1^4$	$\mathbf{w}_2^4$	$\mathbf{w}_3^4$
1	1/2	-3/4	1/4
3	0	1/2	-1/2
1	-3/2	3/4	-1/4
$\mathbf{t}^4$	-1	3/2	-3/2

Step 3: No pivot column can be obtained.
Step 4: Go to the Γ_{II}-process.
Step 6: $I(\mathbf{W}^-) \equiv \{1,3\}$; $I(\mathbf{W}^+) \equiv \{2\}$; $I(\mathbf{W}^0) \equiv \{\}$.
The initial list $A(\mathbf{W}_c^{-+})$ of vectors is:

$$\{\{3\},\{1\}\}.$$

The corresponding list of $\mathbf{W}_c^{-+}$ vectors is

$$\left\{\left(0,\frac{1}{2},-\frac{3}{2}\right)^T,\left(-\frac{3}{4},0,\frac{3}{4}\right)^T\right\}.$$

Step 7: The elimination of redundant vectors in $\mathbf{W}_c^{-+}$ is performed, as shown in Table 8.5. A new **U** matrix including $\mathbf{W}^0, \mathbf{W}^-, \mathbf{W}^{-+}$ is built.

Step 8: Since the dual is not empty, we go to Step 1.

Table 8.5 Removing redundant vectors from the $\mathbf{W}_c^{-+}$ list.

i	$A_{\mathbf{w}_i}$	$A(\mathbf{W}_c^{-+})$	Removed by
1	{3}	{{3},{1}}	None
2	{1}	{{3},{1}}	None

Table 8.6 Added constraint, **U** matrix, and associated dot products.

Iteration 5				
$\mathbf{a}_5$	$\mathbf{w}_1^5$	$\mathbf{w}_2^5$	$\mathbf{w}_3^5$	$\mathbf{w}_4^5$
0	1/2	1/4	0	–3/4
2	0	–1/2	1/2	0
1	–3/2	–1/4	–3/2	3/4
$\mathbf{t}^5$	–3/2	–5/4	–1/2	3/4

Table 8.7 Removing redundant vectors from the $\mathbf{W}_c^{-+}$ list.

i	$A_{\mathbf{w}_i}$	$A(\mathbf{W}_c^{-+})$	Removed by
1	{}	{{1},{4}}	$A_{\mathbf{w}_2}$
2	{1}	{{1},{4}}	None
3	{4}	{{1},{4}}	None

Step 1: Constraint 5 is introduced, and the dot products are calculated, as shown in Table 8.6.

Step 2: We update the orthogonal sets $A(\mathbf{U})$:

$$A_{\mathbf{w}_1} \equiv \{2,3\}.A_{\mathbf{w}_2} \equiv \{1,2\}, A_{\mathbf{w}_3} \equiv \{3,4\}, A_{\mathbf{w}_4} \equiv \{1,4\}.$$

Step 3: No pivot column can be obtained.

Step 4: Go to the Γ_{II}-process.

Step 6: $I(\mathbf{W}^-) \equiv \{1,2,3\}$; $I(\mathbf{W}^+) \equiv \{4\}$, $I(\mathbf{W}^0) \equiv \{\}$.

The initial list $A(\mathbf{W}_c^{-+})$ of vectors is

$$\{\{\},\{1\},\{4\}\}.$$

The corresponding list of $\mathbf{W}_c^{-+}$ vectors is

$$\left\{\left(-\frac{3}{4},0,0\right)^T,\left(-\frac{3}{4},-\frac{3}{8},\frac{3}{4}\right)^T,\left(-\frac{3}{8},\frac{3}{8},-\frac{3}{4}\right)^T\right\}.$$

Step 7: The elimination of redundant vectors in $\mathbf{W}_c^{-+}$ is performed, as shown in Table 8.7.

The set {} is excluded from the $A(\mathbf{W}_c^{-+})$ list because it is contained in the set $A_{\mathbf{w}_2} \equiv \{1\}$.

Step 7.5: A new **U** matrix including $\mathbf{W}^0, \mathbf{W}^-, \mathbf{W}^{-+}$ is built.

Step 8: Since the dual is not empty, we go to Step 1.

Thus, we have $\Omega(\mathbf{C}_\rho) \equiv \mathbf{W}_\pi$, where the vectors of **W** are

$$\left\{\left(\frac{1}{2},0,-\frac{3}{2}\right)^T,\left(\frac{1}{4},-\frac{1}{2},-\frac{1}{4}\right)^T,\left(0,\frac{1}{2},-\frac{3}{2}\right)^T,\left(-\frac{3}{4},-\frac{3}{8},\frac{3}{4}\right)^T,\left(-\frac{3}{8},\frac{3}{8},-\frac{3}{4}\right)^T\right\},$$

Table 8.8 Added constraint, **U** matrix, and associated dot products.

Dual				
$\mathbf{w}_1$	$\mathbf{w}_2$	$\mathbf{w}_3$	$\mathbf{w}_4$	$\mathbf{w}_5$
1/2	1/4	0	-3/4	-3/8
0	-1/2	1/2	-3/8	3/8
-3/2	-1/4	-3/2	3/4	-3/4

and then Condition (8.16) becomes

$$\mathbf{W}^T \begin{pmatrix} u \\ v \\ 1 \end{pmatrix} = \begin{pmatrix} 1/2 & 0 & -3/2 \\ 1/4 & -1/2 & -1/4 \\ 0 & 1/2 & -3/2 \\ -3/4 & -3/8 & 3/4 \\ -3/8 & 3/8 & -3/4 \end{pmatrix} \begin{pmatrix} u \\ v \\ 1 \end{pmatrix} = \begin{pmatrix} u/2 - 3/2 \\ u/4 - v/2 - 1/4 \\ v/2 - 3/2 \\ -3u/4 - 3v/8 + 3/4 \\ -3u/8 + 3v/8 - 3/4 \end{pmatrix} \le \mathbf{0}, \tag{8.18}$$

which is equivalent to (see Fig. 8.1)

$$\begin{array}{rrcl} u & & \le & 3 \\ u & -2v & \le & 1 \\ & v & \le & 3 \\ 2u & +v & \ge & 2 \\ -u & +v & \le & 2 \end{array} \tag{8.19}$$

■

Example 8.2 (Compatibility). In this example we analyze the system

$$\begin{array}{rrrcl} x_1 & +x_2 & +2x_3 & = & 2t^2 - 1 \\ 2x_1 & -x_2 & +x_3 & \ge & t^2 \\ -x_1 & +3x_2 & -2x_3 & < & t \\ & & x_2,\ x_3 & \ge & 0 \end{array}$$

By introducing the slack variables x_4 and x_5 we get the equivalent system

$$\begin{array}{rrrrrcl} x_1 & +x_2 & +2x_3 & & & = & 2t^2 - 1 \\ 2x_1 & -x_2 & +x_3 & -x_4 & & = & t^2 \\ -x_1 & +3x_2 & -2x_3 & & +x_5 & = & t \\ & & & & x_2,\ x_3,\ x_4 & \ge & 0 \\ & & & & x_5 & > & 0 \end{array}$$

From (8.10) we have

$$\begin{aligned}
\mathbf{B} &\equiv \{\mathbf{b}_1\} \equiv \{(1,2,-1)^T\},\\
\mathbf{C} &\equiv \{\mathbf{c}_1,\mathbf{c}_2,\mathbf{c}_3\} \equiv \{(1,-1,3)^T,(2,1,-2)^T,(0,-1,0)^T\},\\
\mathbf{D} &\equiv \{\mathbf{d}_1\} \equiv \{(0,0,1)^T\}.
\end{aligned}$$

To analyze the compatibility of this system, the dual

$$\mathbf{U}_\pi = \Omega(\mathbf{B}_\rho + (\mathbf{C},\mathbf{D})_\pi)$$

must be calculated. However, this dual has already been obtained in Example 6.10, Expression (6.43), and we have

$$\Omega(\mathbf{B}_\rho + \mathbf{C}_\pi + \mathbf{D}_\pi) \equiv \mathbf{U}_\pi^5 \equiv \pi_1(-1,0,-1)^T + \pi_2(-2,1,0)^T \equiv \pi_1\mathbf{w}_1 + \pi_2\mathbf{w}_2. \tag{8.20}$$

With $\mathbf{a} = (2t^2-1,t^2,t)^T$, the first constraint $\mathbf{U}^T\mathbf{a} \le \mathbf{0}$ in (8.12) becomes

$$\mathbf{U}^T\mathbf{a} \le \mathbf{0} \Leftrightarrow \begin{pmatrix} -1 & 0 & -1\\ -2 & 1 & 0\end{pmatrix}\begin{pmatrix}2t^2-1\\ t^2\\ t\end{pmatrix} \le \begin{pmatrix}0\\0\end{pmatrix}, \tag{8.21}$$

which leads to $-2t^2-t+1 \le 0$ and $-3t^2+2 \le 0$, which is equivalent to $t \ge \sqrt{2/3}$ or $t \le -1$.

The second condition in (8.12) requires the sets $\mathbf{W}(\mathbf{d}_1)$ and $\mathbf{W}(\mathbf{a})$. The first is $\mathbf{W}(\mathbf{d}_1) \equiv \{\mathbf{w}_2\}$, which is independent of the value t. For the second, we have three cases:

- $t = \sqrt{2/3}$: $\mathbf{a} = (1/3, 2/3, \sqrt{2/3})^T \Rightarrow \mathbf{W}(\mathbf{a}) = \{\mathbf{w}_2\} \Rightarrow \mathbf{W}(\mathbf{a}) \subseteq \mathbf{W}(\mathbf{d}_1)$ and the system has solution. To determine if the solution is unique, according to (8.13), we need to obtain $\mathbf{F}^a$. First we obtain $\mathbf{W}(\mathbf{b}_1) \equiv \mathbf{W}$, $\mathbf{W}(\mathbf{c}_1) \equiv \emptyset$, $\mathbf{W}(\mathbf{c}_2) \equiv \{\mathbf{w}_1\}$, and $\mathbf{W}(\mathbf{c}_3) \equiv \{\mathbf{w}_1\}$. Since only $\mathbf{W}(\mathbf{b}_1)$ and $\mathbf{W}(\mathbf{d}_1)$ contain $\mathbf{W}(\mathbf{a})$, we have $\mathbf{F}^a \equiv \{\mathbf{b}_1, \mathbf{d}_1\}$, and $\mathbf{0} \notin \mathbf{F}^a_\xi$ because the vectors in $\mathbf{F}^a$ are linearly independent. Consequently, the solution is unique.

- $t = -1$: $\mathbf{a} = (1,1,1)^T \Rightarrow \mathbf{W}(\mathbf{a}) = \{\mathbf{w}_1\}$, which implies $\mathbf{W}(\mathbf{a}) \not\subseteq \mathbf{W}(\mathbf{d}_1)$ and because of (8.12) the system has no solution.

- $t > \sqrt{2/3}$ *or* $t < -1$: In this case, $\mathbf{W}(\mathbf{a}) \equiv \emptyset \subseteq \mathbf{W}(\mathbf{d}_1)$ and the system has solution. However, $\mathbf{F}^a \equiv \{\mathbf{B}, \mathbf{C}, \mathbf{D}\}$, which obviously are not linearly independent. Thus, the solution is not unique.

Summarizing, we have:

- If $-1 \le t < \sqrt{2/3}$ the system is incompatible.

- If $t > \sqrt{2/3}$ *or* $t < -1$ the system is consistent but indeterminate.

- If $t = \sqrt{2/3}$ the system has unique solution.

■

8.3 SOLVING LINEAR SYSTEMS

In this section we show how the previous theory can be used for solving linear systems. Let us start by writing the system in the more convenient form

$$\begin{aligned} \mathbf{b}_i \bullet \mathbf{x} &= \mathbf{0} \\ \mathbf{c}_i \bullet \mathbf{x} &\leq \mathbf{0} \\ \mathbf{d}_i \bullet \mathbf{x} &\leq \mathbf{0} \\ \mathbf{d}_i \bullet \mathbf{x} &< \mathbf{0} \\ x_n &= 1 \end{aligned} \tag{8.22}$$

It is always possible to write a given system in this form just by transforming

- All relations of the type $\mathbf{b}_i \bullet \mathbf{x} = \mathbf{a}_i, \mathbf{c}_i \bullet \mathbf{x} \leq \mathbf{a}_i$ by means of

$$\mathbf{b}_i \bullet \mathbf{x} - x_n\mathbf{a}_i = \mathbf{0}, \mathbf{c}_i \bullet \mathbf{x} - x_n\mathbf{a}_i \leq \mathbf{0} \wedge x_n \geq 0 \wedge x_n = 1.$$

- The strict inequalities $\mathbf{d}_i \bullet \mathbf{x} < \mathbf{a}_i$ by the double constraint

$$\mathbf{d}_i \bullet \mathbf{x} - x_n\mathbf{a}_i \leq \mathbf{0} \wedge \mathbf{d}_i \bullet \mathbf{x} - x_n\mathbf{a}_i < \mathbf{0} \wedge x_n \geq 0 \wedge x_n = 1.$$

- The variables $x_i \geq 0$ by means of $\mathbf{c}_i \bullet \mathbf{x} \leq 0$ with $\mathbf{c}_i = (0, \ldots, -1, \ldots, 0)^T$ with the -1 in the i*th* position.

- The variables $x_i > 0$ by the double restriction

$$\mathbf{d}_i \bullet \mathbf{x} \leq 0 \wedge \mathbf{d}_i \bullet \mathbf{x} < 0,$$

  where, as before, $\mathbf{d}_i = (0, \ldots, 0, -1, 0, \ldots, 0)^T$.

If $\mathbf{X}$ is the set of solutions of (8.22), the definition of polar cone leads to

$$\mathbf{X} \equiv \{\mathbf{x} \in \Omega(\mathbf{B}_\rho + \mathbf{C}_\pi + \mathbf{D}_\pi) | \mathbf{d}_i \bullet \mathbf{x} < 0, \forall i \wedge x_n = 1\}. \tag{8.23}$$

This implies that we can find first the polar cone $\Omega(\mathbf{B}_\rho + \mathbf{C}_\pi + \mathbf{D}_\pi)$ associated with the first three sets of constraints of (8.22) and then add the remaining conditions.

Let $\Omega(\mathbf{B}_\rho + \mathbf{C}_\pi + \mathbf{D}_\pi) \equiv \mathbf{V}_\rho + \mathbf{W}_\pi$. Without loss of generality, we can assume that v_{nk} and w_{nk} are 0 or 1, because if they are not, we can devide by v_{nk} and w_{nk}, respectively.

Condition $x_n = 1$ leads to

$$\sum_{k \in K_V} \rho_k + \sum_{k \in K_W} \pi_k = 1, \tag{8.24}$$

where

$$K_V \equiv \{k | v_{nk} \neq 0\} \tag{8.25}$$

and

$$K_W \equiv \{k | w_{nk} \neq 0\}. \tag{8.26}$$

We have the following three cases:

1. $K_V \equiv K_W \equiv \emptyset$: In this case the problem (8.22) has no solution, since this implies $x_n = 0 \neq 1$.

2. $K_V \equiv \emptyset$ and $K_W \neq \emptyset$: Letting $\lambda_k = \pi_k$, (8.24) becomes

$$\sum_{k \in K_W} \lambda_k = 1, \tag{8.27}$$

and then the cone $\mathbf{V}_\rho + \mathbf{W}_\pi$ becomes $\mathbf{V}_\rho + \mathbf{W}^*_\pi + \mathbf{U}^*_\lambda$, where $\mathbf{W}^*$ and $\mathbf{U}^*$ are the matrices whose columns have null and non-null last component, respectively.

3. $K_V \neq \emptyset$: With $k_0 \in K_V$, from (8.24) we get

$$\rho_{k_0} = 1 - \sum_{k \in K_V \setminus \{k_0\}} \rho_k - \sum_{k \in K_W} \pi_k = 1, \tag{8.28}$$

and then the solution $\mathbf{V}_\rho + \mathbf{W}_\pi$ becomes

$$\begin{aligned} \mathbf{v}_{k_0} + \sum_{k \in K_V \setminus \{k_0\}} \rho_k(\mathbf{v}_k - \mathbf{v}_{k_0}) + \sum_{k \in K_W} \pi_k(\mathbf{w}_k - \mathbf{v}_{k_0}) \\ + \sum_{k \notin K_V} \rho_k \mathbf{v}_k + \sum_{k \notin K_W} \pi_k \mathbf{w}_k \end{aligned} \tag{8.29}$$

where $\mathbf{V}^*$ and $\mathbf{W}^*$ are the matrices from whose columns we have subtracted the column $\mathbf{v}_{k_0}$, respectively.

Finally, on top of this we impose conditions $\mathbf{d}_i \bullet \mathbf{x} < 0, \forall i$.

For the previous theory to have a practical application, an efficient algorithm for obtaining $\mathbf{U}_\pi \equiv \Omega(\mathbf{A}_\pi)$ becomes necessary. The advances of Minkowski (1911) were clearly insufficient. Only after 1953, when Motzkin, Raiffa, Thompson, and Thrall (1966) published the method of double description, which was later improved by Greenberg (1975) and others, were important advances made. However, some difficulties, associated with the range of $\mathbf{A}$, slowed down the development of this theory.

Algorithm 8.1 Solution of a system of linear equations and inequalities.

- **Input:** Coefficients $b_{ij}; i = 1, \ldots, n_b + n_c; j = 1, \ldots, n$ defining a system of n_b linear equations (the first n_b **b** vectors) and n_c linear inequalities of the form $\leq$ in n unknowns (the last n_c **b** vectors).

- **Output:** The general solution of the linear system of equations and inequalities, given as three lists of vectors, corresponding to the unrestricted **V**, non-negative **W**, and convex linear combinations **U**, respectively.

Step I: Use the Γ-algorithm for obtaining the dual

$$\mathbf{V}_\rho + \mathbf{W}_\pi \equiv \Omega\left[\mathbf{B}_\rho + (\mathbf{C}:\mathbf{D})_\pi\right].$$

Step II: Normalize generators

Normalize the **V** and **W** matrices such that the elements in their last rows are either zeroes or ones. To this end we divide each column by its last element if it is non-null.

Step III: Build the solution generators

Case 1: If $K_V \equiv K_W \equiv \emptyset$, there is no solution.

Case 2: If $K_V \equiv \emptyset$, return $\mathbf{V}_\rho + \mathbf{W}^*_\pi + \mathbf{Q}^*_\lambda$ as the solution, where $\mathbf{W}^*$ and $\mathbf{Q}^*$ are the submatrices of **W**, whose columns have null and non-null last element in **W**, respectively.

Case 3: If $K_V \neq \emptyset$, return (8.29):

$$\mathbf{v}_{k_0} + \sum_{k \in K_V \setminus \{k_0\}} \rho_k(\mathbf{v}_k - \mathbf{v}_{k_0}) + \sum_{k \in K_W} \pi_k(\mathbf{w}_k - \mathbf{v}_{k_0}) + \sum_{k \notin K_V} \rho_k \mathbf{v}_k + \sum_{k \notin K_W} \pi_k \mathbf{w}_k$$

Example 8.3 (The cube example). Consider the following system of inequalities:

$$\begin{array}{rrrcl} & & -x_3 & \leq & 0 \\ & & +x_3 & \leq & 1 \\ -x_1 & & & \leq & 0 \\ & -x_2 & & \leq & 0 \\ +x_1 & & & \leq & 1 \\ & +x_2 & & \leq & 1 \\ +x_1 & +x_2 & +x_3 & \leq & 5/2 \end{array} \tag{8.30}$$

which, using an auxiliary unknown, can be written as

$$\begin{array}{rrrrcl} & & -x_3 & & \leq & 0 \\ & & +x_3 & -x_4 & \leq & 0 \\ -x_1 & & & & \leq & 0 \\ & -x_2 & & & \leq & 0 \\ +x_1 & & & -x_4 & \leq & 0 \\ & +x_2 & & -x_4 & \leq & 0 \\ +x_1 & +x_2 & +x_3 & -5/2x_4 & \leq & 0 \\ & & & -x_4 & \leq & 0 \\ & & & x_4 & = & 1 \end{array} \tag{8.31}$$

If we consider the sequential incorporation of the inequalities in (8.31) to a system of inequalities, the sequential solutions are illustrated in Fig. 8.2 and 8.3. To solve this system the above algorithm is applied as follows.

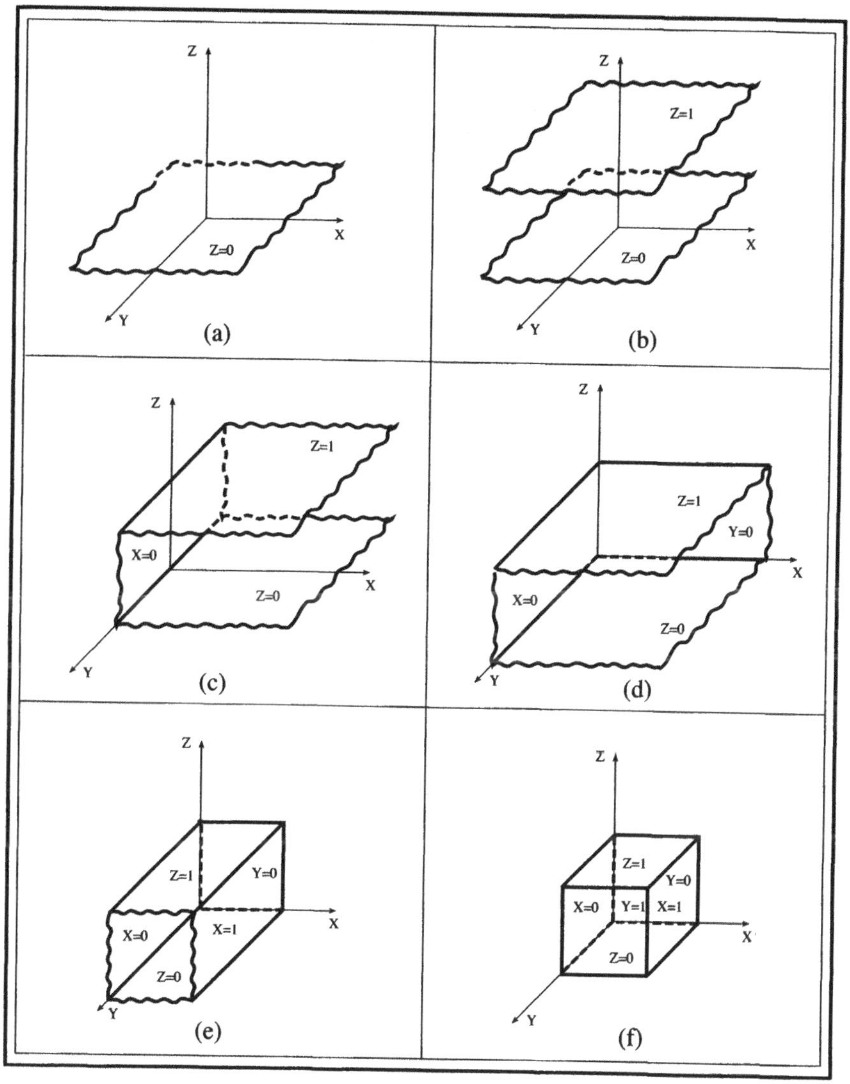

Fig. 8.2 Illustration of the six different solutions, after the sequential incorporation of the inequalities in (8.31), for the cube example.

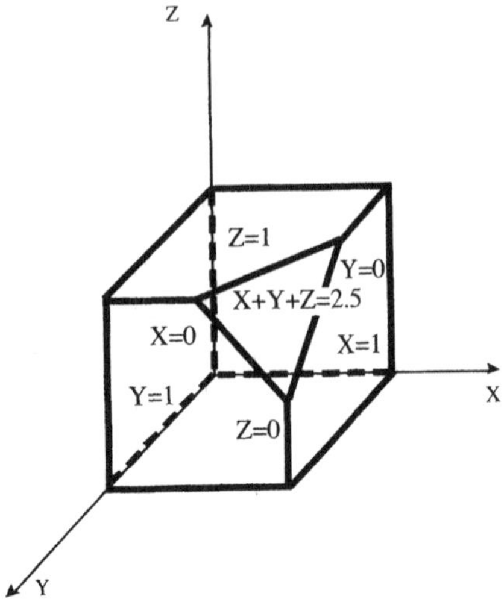

Fig. 8.3 Illustration of the new solution for the cube example.

Table 8.9 Added constraint, **U** matrix, and associated dot products.

Iteration 1				
$\mathbf{a}_1$	$\mathbf{v}_1^1$	$\mathbf{v}_2^1$	$\mathbf{v}_3^1$	$\mathbf{v}_4^1$
0	*1*	0	**0**	0
0	*0*	1	**0**	0
−1	*0*	0	**1**	0
0	*0*	0	**0**	1
$\mathbf{t}^1$	0	0	−1	0

Step I: We use the Γ-algorithm to obtain the dual

$$\mathbf{V}_\rho + \mathbf{W}_\pi \equiv \Omega(\mathbf{C}_\pi),$$

where **C** is the transpose matrix associated with the (8.31) system coefficients, as follows:

Step 1: Constraint 1 is introduced, and the dot products are calculated, as shown in Table 8.9.

Step 2: The orthogonal sets $A(\mathbf{U})$ are updated:

$$A_{\mathbf{v}_1} \equiv \{1\}, A_{\mathbf{v}_2} \equiv \{1\}, A_{\mathbf{v}_3} \equiv \{\}, A_{\mathbf{v}_4} \equiv \{1\}.$$

Step 3: The pivot is column vector 3.

Step 4: Since $s < 4$ and *pivot* > 0, we go to the Γ_I-process.

Table 8.10 Added constraint, **U** matrix, and associated dot products.

Iteration 2				
$\mathbf{a}_2$	$\mathbf{w}_1^2$	$\mathbf{v}_1^2$	$\mathbf{v}_2^2$	$\mathbf{v}_3^2$
0	0	*0*	1	**0**
0	0	*1*	0	**0**
1	1	*0*	0	**0**
–1	0	*0*	0	**1**
$\mathbf{t}^2$	1	0	0	**–1**

Step 5: The column vectors 1 (italic) and 3 (boldface) in Table 8.9 are permuted, because *pivot* = 3 is different from 1.

The pivot column is normalized and the pivoting process performed as shown in Table 8.10. Index h is increased to 2 and we return to Step 1.

Step 1: Constraint 2 is introduced, and the dot products are calculated, as shown in Table 8.10.

Step 2: The orthogonal sets $A(\mathbf{U})$ are updated:

$$A_{\mathbf{w}_1} \equiv \{\}, A_{\mathbf{v}_1} \equiv \{1,2\}, A_{\mathbf{v}_2} \equiv \{1,2\}, A_{\mathbf{v}_3} \equiv \{1\}.$$

Step 3: The pivot is column vector 4.

Step 4: Since $s < 4$ and $pivot > 0$, we go to the Γ_I-process.

Step 5: The column vectors 2 (italic) and 4 (boldface) in Table 8.10 are permuted, because *pivot* = 4 is different from 2.

The pivot column is normalized and the pivoting process performed as shown in Table 8.11. Index h is increased to 3 and we return to Step 1.

Step 1: Constraint 3 is introduced, and the dot products are calculated, as shown in Table 8.11.

Step 2: The orthogonal sets $A(\mathbf{U})$ are updated:

$$A_{\mathbf{w}_1} \equiv \{2,3\}, A_{\mathbf{w}_2} \equiv \{1,3\}, A_{\mathbf{v}_1} \equiv \{1,2\}, A_{\mathbf{v}_2} \equiv \{1,2,3\}.$$

Step 3: The pivot is column vector 3.

Step 4: Since $s < 4$ and $pivot > 0$, we go to the Γ_I-process.

The pivot column is normalized and the pivoting process performed as shown in Table 8.12. Index h is increased to 4 and we return to Step 1.

Step 1: Constraint 4 is introduced, and the dot products are calculated, as shown in Table 8.12.

Step 2: The orthogonal sets $A(\mathbf{U})$ are updated:

$$A_{\mathbf{w}_1} \equiv \{2,3,4\}, A_{\mathbf{w}_2} \equiv \{1,3,4\}, A_{\mathbf{w}_3} \equiv \{1,2,4\}, A_{\mathbf{v}_1} \equiv \{1,2,3\}.$$

Step 3: The pivot is column vector 4.

Step 4: Since $s < 4$ and $pivot > 0$, we go to the Γ_I-process.

Table 8.11 Added constraint, **U** matrix, and associated dot products.

Iteration 3				
$\mathbf{a}_3$	$\mathbf{w}_1^3$	$\mathbf{w}_2^3$	$\mathbf{v}_1^3$	$\mathbf{v}_2^3$
−1	0	0	**1**	0
0	0	0	**0**	1
0	1	0	**0**	0
0	1	1	**0**	0
$\mathbf{t}^3$	0	0	**−1**	0

Table 8.12 Added constraint, **U** matrix, and associated dot products.

Iteration 4				
$\mathbf{a}_4$	$\mathbf{w}_1^4$	$\mathbf{w}_2^4$	$\mathbf{w}_3^4$	$\mathbf{v}_1^4$
0	0	0	1	**0**
−1	0	0	0	**1**
0	1	0	0	**0**
0	1	1	0	**0**
$\mathbf{t}^4$	0	0	0	**−1**

The pivot column is normalized and the pivoting process performed as shown in Table 8.13. Index h is increased to 5 and we return to Step 1.

Step 1: Constraint 5 is introduced, and the dot products are calculated, as shown in Table 8.13.

Step 2: The orthogonal sets $A(\mathbf{U})$ are updated:

$$A_{\mathbf{w}_1} \equiv \{2,3,4\}, A_{\mathbf{w}_2} \equiv \{1,3,4\}, A_{\mathbf{w}_3} \equiv \{1,2,4\}, A_{\mathbf{w}_4} \equiv \{1,2,3,5\}.$$

Step 3: No pivot column can be obtained.

Step 4: Go to the Γ_{II}-process.

Table 8.13 Added constraint, **U** matrix, and associated dot products.

Iteration 5				
$\mathbf{a}_5$	$\mathbf{w}_1^5$	$\mathbf{w}_2^5$	$\mathbf{w}_3^5$	$\mathbf{w}_4^5$
1	0	0	1	0
0	0	0	0	1
0	1	0	0	0
−1	1	1	0	0
$\mathbf{t}^5$	−1	−1	1	0

Table 8.14 Removing redundant vectors from the $\mathbf{W}_c^{-+}$ list.

i	$A_{\mathbf{w}_i}$	$A(\mathbf{W}_c^{-+})$	Removed by
1	{2,4}	{{2,4},{1,4}}	None
2	{1,4}	{{2,4},{1,4}}	None

Table 8.15 Added constraint, **U** matrix, and associated dot products.

Iteration 6					
$\mathbf{a}_6$	$\mathbf{w}_1^6$	$\mathbf{w}_2^6$	$\mathbf{w}_3^6$	$\mathbf{w}_4^6$	$\mathbf{w}_5^6$
0	0	0	0	1	1
1	1	0	0	0	0
0	0	1	0	1	0
−1	0	1	1	1	1
$\mathbf{t}^6$	1	−1	−1	−1	−1

Step 6: $I(\mathbf{W}^-) \equiv \{1,2\}$; $I(\mathbf{W}^+) \equiv \{3\}$; $I(\mathbf{W}^0) \equiv \{4\}$.
The initial list $A(\mathbf{W}_c^{-+})$ of vectors is

$$\{\{2,4\},\{1,4\}\}.$$

The corresponding list of $\mathbf{W}_c^{-+}$ vectors is

$$\{(1,0,1,1)^T, (1,0,0,1)^T\}.$$

Step 7: The elimination of redundant vectors in $\mathbf{W}_c^{-+}$ is performed, as shown in Table 8.14.

A new **U** matrix including $\mathbf{W}^0, \mathbf{W}^-, \mathbf{W}^{-+}$ is built.

Step 8: Since the dual is not empty, we go to Step 1.

Step 1: Constraint 6 is introduced, and the dot products are calculated, as shown in Table 8.15.

Step 2: The orthogonal sets $A(\mathbf{U})$ are updated:

$$A_{\mathbf{w}_1} \equiv \{1,2,3,5\}, A_{\mathbf{w}_2} \equiv \{2,3,4\}, A_{\mathbf{w}_3} \equiv \{1,3,4\},$$

$$A_{\mathbf{w}_4} \equiv \{2,4,5\}, A_{\mathbf{w}_5} \equiv \{1,4,5\}.$$

Step 3: No pivot column can be obtained.

Step 4: Go to the Γ_{II}-process.

Step 6: $I(\mathbf{W}^-) \equiv \{2,3,4,5\}$; $I(\mathbf{W}^+) \equiv \{1\}$; $I(\mathbf{W}^0) \equiv \{\}$.
The initial list $A(\mathbf{W}_c^{-+})$ of vectors is

$$\{\{2,3\},\{1,3\},\{2,5\},\{1,5\}\}.$$

Table 8.16 Removing redundant vectors from the $\mathbf{W}_c^{-+}$ list.

i	$A_{\mathbf{w}_i}$	$A(\mathbf{W}_c^{-+})$	Removed by
1	{2,3}	{{2,3},{1,3},{2,5},{1,5}}	None
2	{1,3}	{{2,3},{1,3},{2,5},{1,5}}	None
3	{2,5}	{{2,3},{1,3},{2,5},{1,5}}	None
4	{1,5}	{{2,3},{1,3},{2,5},{1,5}}	None

Table 8.17 Added constraint, **U** matrix, and associated dot products.

Iteration 7								
$\mathbf{a}_7$	$\mathbf{w}_1^7$	$\mathbf{w}_2^7$	$\mathbf{w}_3^7$	$\mathbf{w}_4^7$	$\mathbf{w}_5^7$	$\mathbf{w}_6^7$	$\mathbf{w}_7^7$	$\mathbf{w}_8^7$
1	0	0	1	1	0	0	1	1
1	0	0	0	0	1	1	1	1
1	1	0	1	0	1	0	1	0
–5/2	1	1	1	1	1	1	1	1
$\mathbf{t}^7$	–3/2	–5/2	–1/2	–3/2	–1/2	–3/2	1/2	–1/2

The corresponding list of $\mathbf{W}_c^{-+}$ vectors is

$$\{(0,1,1,1)^T, (0,1,0,1)^T, (1,1,1,1)^T, (1,1,0,1)^T\}.$$

Step 7: The elimination of redundant vectors in $\mathbf{W}_c^{-+}$ is performed, as shown in Table 8.16.

A new **U** matrix including $\mathbf{W}^0, \mathbf{W}^-, \mathbf{W}^{-+}$ is built.

Step 8: Since the dual is not empty, we go to Step 1.

Step 1: Constraint 7 is introduced, and the dot products are calculated, as shown in Table 8.17.

Step 2: The orthogonal sets $A(\mathbf{U})$ are updated:

$$A_{\mathbf{w}_1} \equiv \{2,3,4\}, A_{\mathbf{w}_2} \equiv \{1,3,4\}, A_{\mathbf{w}_3} \equiv \{2,4,5\}, A_{\mathbf{w}_4} \equiv \{1,4,5\},$$

$$A_{\mathbf{w}_5} \equiv \{2,3,6\}, A_{\mathbf{w}_6} \equiv \{1,3,6\}, A_{\mathbf{w}_7} \equiv \{2,5,6\}, A_{\mathbf{w}_8} \equiv \{1,5,6\}.$$

Step 3: No pivot column can be obtained.

Step 4: Go to the Γ_{II}-process.

Step 6: $I(\mathbf{W}^-) \equiv \{1,2,3,4,5,6,8\}$; $I(\mathbf{W}^+) \equiv \{7\}$; $I(\mathbf{W}^0) \equiv \{\}$.

The initial list $A(\mathbf{W}_c^{-+})$ of vectors is

$$\{\{2\}, \{\}, \{2,5\}, \{5\}, \{2,6\}, \{6\}, \{5,6\}\}.$$

The corresponding list of $\mathbf{W}_c^{-+}$ vectors is

$$\left\{\left(\tfrac{3}{2}, \tfrac{3}{2}, 2, 2\right)^T, \left(\tfrac{5}{2}, \tfrac{5}{2}, \tfrac{5}{2}, 3\right)^T, \left(1, \tfrac{1}{2}, 1, 1\right)^T, \left(2, \tfrac{3}{2}, \tfrac{3}{2}, 2\right)^T, \left(\tfrac{1}{2}, 1, 1, 1\right)^T, \left(\tfrac{3}{2}, 2, \tfrac{3}{2}, 2\right)^T, \left(1, 1, \tfrac{1}{2}, 1\right)^T\right\}.$$

Table 8.18 Removing redundant vectors from the $\mathbf{W}_c^{-+}$ list.

i	$A_{\mathbf{w}_i}$	$A(\mathbf{W}_c^{-+})$	Removed by
1	{2}	{{},{2,5},{5},{2,6},{6},{5,6}}	$A_{\mathbf{w}_3}$
2	{}	{{2,5},{5},{2,6},{6},{5,6}}	$A_{\mathbf{w}_1}$
3	{2,5}	{{2,5},{5},{2,6},{6},{5,6}}	None
4	{5}	{{2,5},{2,6},{6},{5,6}}	$A_{\mathbf{w}_3}$
5	{2,6}	{{2,5},{2,6},{6},{5,6}}	None
6	{6}	{{2,5},{2,6},{5,6}}	$A_{\mathbf{w}_5}$
7	{5,6}	{{2,5},{2,6},{5,6}}	None

Table 8.19 Added constraint, **U** matrix, and associated dot products.

Iteration 8										
$\mathbf{a}_8$	$\mathbf{w}_1^8$	$\mathbf{w}_2^8$	$\mathbf{w}_3^8$	$\mathbf{w}_4^8$	$\mathbf{w}_5^8$	$\mathbf{w}_6^8$	$\mathbf{w}_7^8$	$\mathbf{w}_8^8$	$\mathbf{w}_9^8$	$\mathbf{w}_{10}^8$
0	0	0	1	1	0	0	1	1	1/2	1
0	0	0	0	0	1	1	1	1/2	1	1
0	1	0	1	0	1	0	0	1	1	1/2
–1	1	1	1	1	1	1	1	1	1	1
$\mathbf{t}^8$	–1	–1	–1	–1	–1	–1	–1	–1	–1	–1

Step 7: The elimination of redundant vectors in $\mathbf{W}_c^{-+}$ is performed, as shown in Table 8.18.

The set {2} is excluded from the $A(\mathbf{W}_c^{-+})$ list because it is contained in the set $A_{\mathbf{w}_3} \equiv \{2,5\}$.

The set {} is excluded from the $A(\mathbf{W}_c^{-+})$ list because it is contained in the set $A_{\mathbf{w}_1} \equiv \{-1\}$.

The set {5} is excluded from the $A(\mathbf{W}_c^{-+})$ list because it is contained in the set $A_{\mathbf{w}_3} \equiv \{2,5\}$.

The set {6} is excluded from the $A(\mathbf{W}_c^{-+})$ list because it is contained in the set $A_{\mathbf{w}_5} \equiv \{2,6\}$.

A new **U** matrix including $\mathbf{W}^0, \mathbf{W}^-, \mathbf{W}^{-+}$ is built.

Step 8: Since the dual is not empty, we go to Step 1.

Step 1: Constraint 8 is introduced and the dot products are calculated, as shown in Table 8.19.

Step 2: The orthogonal sets $A(\mathbf{U})$ are updated:

$$A_{\mathbf{w}_1} \equiv \{2,3,4\}, A_{\mathbf{w}_2} \equiv \{1,3,4\}, A_{\mathbf{w}_3} \equiv \{2,4,5\}, A_{\mathbf{w}_4} \equiv \{1,4,5\},$$

$$A_{\mathbf{w}_5} \equiv \{2,3,6\}, A_{\mathbf{w}_6} \equiv \{1,3,6\}, A_{\mathbf{w}_7} \equiv \{1,5,6\}, A_{\mathbf{w}_8} \equiv \{2,5,7\},$$

$$A_{\mathbf{w}_9} \equiv \{2,6,7\}, A_{\mathbf{w}_{10}} \equiv \{5,6,7\}.$$

Step 3: No pivot column can be obtained.

Step 4: Go to the Γ_{II}-process.

Step 6.1: $I(\mathbf{W}^-) = \{1,2,3,4,5,6,7,8,9,10\}$; $I(\mathbf{W}^+) = \{\}$; $I(\mathbf{W}^0) = \{\}$.

Table 8.20 Normalized dual generators.

Normalized dual									
$\mathbf{w}_1$	$\mathbf{w}_2$	$\mathbf{w}_3$	$\mathbf{w}_4$	$\mathbf{w}_5$	$\mathbf{w}_6$	$\mathbf{w}_7$	$\mathbf{w}_8$	$\mathbf{w}_9$	$\mathbf{w}_{10}$
0	0	1	1	0	0	1	1	1/2	1
0	0	0	0	1	1	1	1/2	1	1
1	0	1	0	1	0	0	1	1	1/2
1	1	1	1	1	1	1	1	1	1

Since there are no more constraints we go to Step 9.

Step 9: Since there are no more columns, the dual is found (see Table 8.19).

Step II (Normalization process): Since all column vectors in Table 8.19 have unit last component, we get the dual generators in Table 8.20.

Step III (Build the solution): We have

$$\mathbf{V} \equiv \{\},$$

$$\mathbf{W} \equiv \{\},$$

$$\mathbf{Q} = \{(0,0,1)^T, (0,0,0)^T, (1,0,1)^T, (1,0,0)^T, (0,1,1)^T, (0,1,0)^T, \\ (1,1,0)^T, (1,1/2,1)^T, (1/2,1,1)^T, (1,1,1/2)^T\}.$$

Thus, the general solution is

$$\begin{pmatrix} x_1 \\ x_2 \\ x_3 \end{pmatrix} = \begin{pmatrix} 0 & 0 & 1 & 1 & 0 & 0 & 1 & 1 & 1/2 & 1 \\ 0 & 0 & 0 & 0 & 1 & 1 & 1 & 1/2 & 1 & 1 \\ 1 & 0 & 1 & 0 & 1 & 0 & 0 & 1 & 1 & 1/2 \end{pmatrix}_\lambda .$$

It is worthwhile noting that during the above process the solutions corresponding to some subsystems of the system of inequalities (8.30) can be obtained in an incremental form, that is, those corresponding to Fig. 8.2. In fact, from Tables 8.10, 8.11, 8.12, 8.13, 8.15, and 8.17, using Step III of Algorithm 8.1, we find the following solutions for cases (a) – (f) of Fig. 8.2.

- *Case of Fig. 8.2(a):* Since it is Case 3 in Step III, we get

$$\mathbf{V} \equiv \{(1,0,0)^T, (0,1,0)^T\};\ \mathbf{W} \equiv \{(0,0,1)^T\};\ \mathbf{Q} = \{(0,0,0)^T\}.$$

- *Case of Fig. 8.2(b):* Since it is Case 2 in Step III, we get

$$\mathbf{V} \equiv \{(1,0,0)^T, (0,1,0)^T\};\ \mathbf{W} \equiv \{\};\ \mathbf{Q} = \{(0,0,0)^T, (0,0,1)^T\}.$$

- *Case of Fig. 8.2(c):* Since it is Case 2 in Step III, we get

$$\mathbf{V} \equiv \{(0,1,0)^T\};\ \mathbf{W} \equiv \{(1,0,0)^T\};\ \mathbf{Q} = \{(0,0,0)^T, (0,0,1)^T\}.$$

- *Case of Fig. 8.2(d):* Since it is Case 2 in Step III, we get

$$\mathbf{V} \equiv \{\}; \ \mathbf{W} \equiv \{(1,0,0)^T, (0,1,0)^T\}; \ \mathbf{Q} = \{(0,0,0)^T, (0,0,1)^T\}.$$

- *Case of Fig. 8.2(e):* Since it is Case 2 in Step III, we get

$$\mathbf{V} \equiv \{\}; \ \mathbf{W} \equiv \{(0,1,0)^T\};$$
$$\mathbf{Q} = \{(0,0,0)^T, (0,0,1)^T, (1,0,0)^T, (1,0,1)^T\}.$$

- *Case of Fig. 8.2(f):* Since it is Case 2 in Step III, we get

$$\mathbf{V} \equiv \{\}; \ \mathbf{W} \equiv \{\};$$
$$\begin{array}{rcl}\mathbf{Q} & = & \{(0,0,1)^T, (0,0,0)^T, (1,0,1)^T, (1,0,0)^T, \\ & & (0,1,1)^T, (0,1,0)^T, (1,1,1)^T, (1,1,0)^T\}.\end{array}$$

■

Example 8.4 (Solving a system of linear inequalities). Consider the following system of inequalities:

$$\begin{array}{cccccrr} +x_1 & & -x_3 & & & \leq & -1 \\ +4x_1 & & +2x_3 & & -6x_5 & \leq & -1 \\ & +x_2 & +2x_3 & +3x_4 & +3x_5 & \leq & -1 \\ -x_1 & & & & +x_5 & \leq & -1 \\ & +x_2 & +2x_3 & +2x_4 & +2x_5 & \leq & -1 \end{array} \tag{8.32}$$

which, using the auxiliary variable x_6, can be written as

$$\begin{array}{ccccccrr} +x_1 & & -x_3 & & & +x_6 & \leq & 0 \\ +4x_1 & & +2x_3 & & -6x_5 & +x_6 & \leq & 0 \\ & +x_2 & +2x_3 & +3x_4 & +3x_5 & +x_6 & \leq & 0 \\ -x_1 & & & & +x_5 & +x_6 & \leq & 0 \\ & +x_2 & +2x_3 & +2x_4 & +2x_5 & +x_6 & \leq & 0 \\ & & & & & -x_6 & \leq & 0 \\ & & & & & x_6 & = & 1 \end{array} \tag{8.33}$$

To solve this system the above algorithm is applied as follows.

Step I: We use the Γ-algorithm to obtain the dual

$$\mathbf{V}_\rho + \mathbf{W}_\pi \equiv \Omega(\mathbf{C}_\pi)$$

as follows:

Step 1: Constraint 1 is introduced, and the dot products are calculated, as shown in Table 8.21.

Step 2: The orthogonal sets $A(\mathbf{U})$ are updated:

$$A_{\mathbf{v}_1} \equiv \{\}, A_{\mathbf{v}_2} \equiv \{1\}, A_{\mathbf{v}_3} \equiv \{\},$$

Table 8.21 Added constraint, **U** matrix, and associated dot products.

Iteration 1						
$\mathbf{a}_1$	$\mathbf{v}_1^1$	$\mathbf{v}_2^1$	$\mathbf{v}_3^1$	$\mathbf{v}_4^1$	$\mathbf{v}_5^1$	$\mathbf{v}_6^1$
1	**1**	0	0	0	0	0
0	**0**	1	0	0	0	0
–1	**0**	0	1	0	0	0
0	**0**	0	0	1	0	0
0	**0**	0	0	0	1	0
1	**0**	0	0	0	0	1
$\mathbf{t}^1$	**1**	0	–1	0	0	1

Table 8.22 Added constraint, **U** matrix, and associated dot products.

Iteration 2						
$\mathbf{a}_2$	$\mathbf{w}_1^2$	$\mathbf{v}_1^2$	$\mathbf{v}_2^2$	$\mathbf{v}_3^2$	$\mathbf{v}_4^2$	$\mathbf{v}_5^2$
4	–1	*0*	**1**	0	0	–1
0	0	*1*	**0**	0	0	0
2	0	*0*	**1**	0	0	0
0	0	*0*	**0**	1	0	0
–6	0	*0*	**0**	0	1	0
1	0	*0*	**0**	0	0	1
$\mathbf{t}^2$	–4	0	**6**	0	–6	–3

$$A_{\mathbf{v}_4} \equiv \{1\}, A_{\mathbf{v}_5} \equiv \{1\}, A_{\mathbf{v}_6} \equiv \{\}.$$

Step 3: The pivot is column vector 1.

Step 4: Since $s < 6$ and *pivot* > 0, we go to the Γ_I-process.

Step 5: The pivot column is normalized and the pivoting process performed as shown in Table 8.22. Index h is increased to 2 and we return to Step 1.

Step 1: Constraint 2 is introduced, and the dot products are calculated, as shown in Table 8.22.

Step 2: The orthogonal sets $A(\mathbf{U})$ are updated:

$$A_{\mathbf{w}_1} \equiv \{\}, A_{\mathbf{v}_1} \equiv \{1,2\}, A_{\mathbf{v}_2} \equiv \{1\},$$

$$A_{\mathbf{v}_3} \equiv \{1,2\}, A_{\mathbf{v}_4} \equiv \{1\}, A_{\mathbf{v}_5} \equiv \{1\}.$$

Step 3: The pivot is column vector 3.

Step 4: Since $s < 6$ and *pivot* > 0, we go to the Γ_I-process.

Step 5: The column vectors 2 (italic) and 3 (boldface) in Table 8.22 are permuted, because *pivot* $= 3$ is different from 2.

Table 8.23 Added constraint, **U** matrix, and associated dot products.

	Iteration 3					
$\mathbf{a}_3$	$\mathbf{w}_1^3$	$\mathbf{w}_2^3$	$\mathbf{v}_1^3$	$\mathbf{v}_2^3$	$\mathbf{v}_3^3$	$\mathbf{v}_4^3$
0	-1/3	-1/6	**0**	0	1	-1/2
1	0	0	**1**	0	0	0
2	2/3	-1/6	**0**	0	1	1/2
3	0	0	**0**	1	0	0
3	0	0	**0**	0	1	0
1	0	0	**0**	0	0	1
$\mathbf{t}^3$	4/3	-1/3	**1**	3	5	2

The pivot column is normalized and the pivoting process performed as shown in Table 8.23. Index h is increased to 3 and we return to Step 1.

Step 1: Constraint 3 is introduced, and the dot products are calculated, as shown in Table 8.23.

Step 2: The orthogonal sets $A(\mathbf{U})$ are updated:

$$A_{\mathbf{w}_1} \equiv \{2\}, A_{\mathbf{w}_2} \equiv \{1\}, A_{\mathbf{v}_1} \equiv \{1,2\},$$

$$A_{\mathbf{v}_2} \equiv \{1,2\}, A_{\mathbf{v}_3} \equiv \{1,2\}, A_{\mathbf{v}_4} \equiv \{1,2\}.$$

Step 3: The pivot is column vector 3.

Step 4: Since $s < 6$ and *pivot* > 0, we go to the Γ_I-process.

The pivot column is normalized and the pivoting process performed as shown in Table 8.24. Index h is increased to 4 and we return to Step 1.

Step 1: Constraint 4 is introduced, and the dot products are calculated, as shown in Table 8.24.

Step 2: The orthogonal sets $A(\mathbf{U})$ are updated:

$$A_{\mathbf{w}_1} \equiv \{2,3\}, A_{\mathbf{w}_2} \equiv \{1,3\}, A_{\mathbf{w}_3} \equiv \{1,2,4\}$$

$$A_{\mathbf{v}_1} \equiv \{1,2,3,4\}, A_{\mathbf{v}_2} \equiv \{1,2,3,4\}, A_{\mathbf{v}_3} \equiv \{1,2,3\}.$$

Step 3: The pivot is column vector 6.

Step 4: Since $s < 6$ and *pivot* > 0, we go to the Γ_I-process.

Step 5: The column vectors 4 (italic) and 6 (boldface) in Table 8.24 are permuted, because *pivot* $= 6$ is different from 4.

The pivot column is normalized and the pivoting process performed as shown in Table 8.25. Index h is increased to 5 and we return to Step 1.

Step 1: Constraint 5 is introduced, and the dot products are calculated, as shown in Table 8.25.

Step 2: The orthogonal sets $A(\mathbf{U})$ are updated:

$$A_{\mathbf{w}_1} \equiv \{2,3,4,5\}, A_{\mathbf{w}_2} \equiv \{1,3,4,5\}, A_{\mathbf{w}_3} \equiv \{1,2,4\},$$

Table 8.24 Added constraint, **U** matrix, and associated dot products.

Iteration 4						
$\mathbf{a}_4$	$\mathbf{w}_1^4$	$\mathbf{w}_2^4$	$\mathbf{w}_3^4$	$\mathbf{v}_1^4$	$\mathbf{v}_2^4$	$\mathbf{v}_3^4$
−1	−1/3	−1/6	0	*0*	1	**−1/2**
0	−4/3	1/3	−1	*−3*	−5	**−2**
0	2/3	−1/6	0	*0*	1	**1/2**
0	0	0	0	*1*	0	**0**
1	0	0	0	*0*	1	**0**
1	0	0	0	*0*	0	**1**
$\mathbf{t}^4$	1/3	1/6	0	0	0	**3/2**

Table 8.25 Added constraint, **U** matrix, and associated dot products.

Iteration 5						
$\mathbf{a}_5$	$\mathbf{w}_1^5$	$\mathbf{w}_2^5$	$\mathbf{w}_3^5$	$\mathbf{w}_4^5$	$\mathbf{v}_1^5$	$\mathbf{v}_2^5$
0	−2/9	−1/9	0	1/3	**1**	0
1	−8/9	5/9	−1	4/3	**−5**	−3
2	5/9	−2/9	0	−1/3	**1**	0
2	0	0	0	0	**0**	1
2	0	0	0	0	**1**	0
1	−2/9	−1/9	0	−2/3	**0**	0
$\mathbf{t}^5$	0	0	−1	0	**−1**	−1

$$A_{\mathbf{w}_4} \equiv \{1,2,3,5\}, A_{\mathbf{v}_1} \equiv \{1,2,3,4\}, A_{\mathbf{v}_2} \equiv \{1,2,3,4\}.$$

Step 3: The pivot is column vector 5.

Step 4: Since we have found a pivot, we go to the Γ_I-process.

The pivot column is normalized and the pivoting process performed as shown in Table 8.26. Index h is increased to 6 and we return to Step 1.

Step 1: Constraint 6 is introduced, and the dot products are calculated, as shown in Table 8.26.

Step 2: The orthogonal sets $A(\mathbf{U})$ are updated:

$$A_{\mathbf{w}_1} \equiv \{2,3,4,5\}, A_{\mathbf{w}_2} \equiv \{1,3,4,5\}, A_{\mathbf{w}_3} \equiv \{1,2,4,5,6\},$$

$$A_{\mathbf{w}_4} \equiv \{1,2,3,5\}, A_{\mathbf{w}_5} \equiv \{1,2,3,4,6\}, A_{\mathbf{v}_1} \equiv \{1,2,3,4,5,6\}.$$

Step 3: No pivot column can be obtained.

Step 4: Go to the Γ_{II}-process.

Step 6.1: $I(\mathbf{W}^-) = \{\}$; $I(\mathbf{W}^+) = \{1,2,4\}$; $I(\mathbf{W}^0) = \{3,5\}$.

Step 6.2: Since $\mathbf{W}^- = \emptyset$, a new matrix **U** including $\mathbf{W}^0$ and column 6 is built.

Table 8.26 Added constraint, **U** matrix, and associated dot products.

Iteration 6						
$\mathbf{a}_6$	$\mathbf{w}_1^6$	$\mathbf{w}_2^6$	$\mathbf{w}_3^6$	$\mathbf{w}_4^6$	$\mathbf{w}_5^6$	$\mathbf{v}_1^6$
0	-2/9	-1/9	-1	1/3	1	*-1*
0	-8/9	5/9	4	4/3	-5	*2*
0	5/9	-2/9	-1	-1/3	1	*-1*
0	0	0	0	0	0	*1*
0	0	0	-1	0	1	*-1*
-1	-2/9	-1/9	0	-2/3	0	*0*
$\mathbf{t}^6$	2/9	1/9	0	2/3	0	0

Table 8.27 Dual generators.

Dual		
$\mathbf{w}_1$	$\mathbf{w}_2$	$\mathbf{v}_1$
-1	1	-1
4	-5	2
-1	1	-1
0	0	1
-1	1	1
0	0	0

Since there are no more vectors we have obtained the dual, as shown in Table 8.27.

Since there are no vectors with non-null last component, that is, $K_V \equiv K_W \equiv \emptyset$, we are in Case 1, and then no solution exists. ■

Example 8.5 (Solving a system of equalities). Consider now the system

$$\begin{array}{rrrrrrl}
+x_1 & & -x_3 & & & +x_6 & \le 0 \\
+4x_1 & & +2x_3 & & -6x_5 & +x_6 & \le 0 \\
& +x_2 & +2x_3 & +3x_4 & +3x_5 & +x_6 & \le 0 \\
-x_1 & & & & +x_5 & +x_6 & \le 0 \\
& +x_2 & +2x_3 & +2x_4 & +2x_5 & +x_6 & \le 0 \\
& & & & & -x_6 & \le 0
\end{array} \tag{8.34}$$

Since this system coincides with the system in (8.33), but with the constraint $x_6 = 1$ removed, we can use the calculations in Example 8.4 and get, from Table 8.27, its solution:

$$\begin{array}{rcl}
\mathbf{V} & \equiv & \{(-1,2,-1,1,-1,0)^T\}, \\
\mathbf{W} & \equiv & \{(-1,4,-1,0,-1,0)^T, (1,-5,1,0,1,0)^T\}, \\
\mathbf{Q} & \equiv & \{\}.
\end{array}$$

Table 8.28 Added constraint, **U** matrix, and associated dot products.

Iteration 1						
$\mathbf{a}_1$	$\mathbf{v}_1^1$	$\mathbf{v}_2^1$	$\mathbf{v}_3^1$	$\mathbf{v}_4^1$	$\mathbf{v}_5^1$	$\mathbf{v}_6^1$
1	**1**	0	0	0	0	0
2	**0**	1	0	0	0	0
0	**0**	0	1	0	0	0
1	**0**	0	0	1	0	0
-1	**0**	0	0	0	1	0
-4	**0**	0	0	0	0	1
$\mathbf{t}^1$	**1**	2	0	1	-1	-4

■

Example 8.6 (Solving a system of equalities). Consider the following system of equalities:

$$\begin{array}{rrrrrcr} +x_1 & +2x_2 & & +x_4 & -x_5 & = & 4 \\ +2x_1 & +x_2 & -x_3 & & +3x_5 & = & 0 \\ -x_1 & & +2x_3 & +x_4 & & = & -1 \end{array} \tag{8.35}$$

which, using the auxiliary unknown x_6, can be written as

$$\begin{array}{rrrrrrcr} +x_1 & +2x_2 & & +x_4 & -x_5 & -4x_6 & = & 0 \\ +2x_1 & +x_2 & -x_3 & & +3x_5 & & = & 0 \\ -x_1 & & +2x_3 & +x_4 & & +x_6 & = & 0 \\ & & & & & -x_6 & \le & 0 \\ & & & & & x_6 & = & 1 \end{array} \tag{8.36}$$

To solve this system the above algorithm is applied as follows

Step I: We use the Γ-algorithm to obtain the dual

$$\mathbf{V}_\rho + \mathbf{W}_\pi \equiv \Omega(\mathbf{B}_\rho + \mathbf{C}_\pi)$$

as follows:

Step 1: Constraint 1 is introduced, and the dot products are calculated, as shown in Table 8.28.

Step 2: The orthogonal sets $A(\mathbf{U})$ are updated:

$$A_{\mathbf{v}_1} \equiv \{\}, A_{\mathbf{v}_2} \equiv \{\}, A_{\mathbf{v}_3} \equiv \{1\},$$

$$A_{\mathbf{v}_4} \equiv \{\}, A_{\mathbf{v}_5} \equiv \{\}, A_{\mathbf{v}_6} \equiv \{\}.$$

Step 3: The pivot is column vector 1.

Step 4: Since $s < 6$ and $pivot > 0$, we go to the Γ_I-process.

Table 8.29 Added constraint, **U** matrix, and associated dot products.

Iteration 2					
$\mathbf{a}_2$	$\mathbf{v}_1^2$	$\mathbf{v}_2^2$	$\mathbf{v}_3^2$	$\mathbf{v}_4^2$	$\mathbf{v}_5^2$
2	**–2**	0	–1	1	4
1	**1**	0	0	0	0
–1	**0**	1	0	0	0
0	**0**	0	1	0	0
3	**0**	0	0	1	0
0	**0**	0	0	0	1
$\mathbf{t}^2$	**–3**	–1	–2	5	8

Step 5: The pivot column is normalized and the pivoting process performed as shown in Table 8.29. Index h is increased to 2 and we return to Step 1.

Step 1: Constraint 2 is introduced, and the dot products are calculated, as shown in Table 8.29.

Step 2: The orthogonal sets $A(\mathbf{U})$ are updated:

$$A_{\mathbf{v}_1} \equiv \{1\}, A_{\mathbf{v}_2} \equiv \{1\}, A_{\mathbf{v}_3} \equiv \{1\},$$

$$A_{\mathbf{v}_4} \equiv \{1\}, A_{\mathbf{v}_5} \equiv \{1\}.$$

Step 3: The pivot is column vector 1.

Step 4: Since $s < 5$ and $pivot > 0$, we go to the Γ_I-process.

Step 5: The pivot column is normalized and the pivoting process performed as shown in Table 8.30. Index h is increased to 3 and we return to Step 1.

Step 1: Constraint 3 is introduced, and the dot products are calculated, as shown in Table 8.30.

Step 2: The orthogonal sets $A(\mathbf{U})$ are updated:

$$A_{\mathbf{v}_1} \equiv \{1,2\}, A_{\mathbf{v}_2} \equiv \{1,2\}, A_{\mathbf{v}_3} \equiv \{1,2\}, A_{\mathbf{v}_4} \equiv \{1,2\}.$$

Step 3: The pivot is column vector 1.

Step 4: Since $s < 4$ and $pivot > 0$, we go to the Γ_I-process.

Step 5: The pivot column is normalized and the pivoting process performed as shown in Table 8.31. Index h is increased to 4 and we return to Step 1.

Step 1: Constraint 4 is introduced, and the dot products are calculated, as shown in Table 8.31.

Step 2: The orthogonal sets $A(\mathbf{U})$ are updated:

$$A_{\mathbf{v}_1} \equiv \{1,2,3,4\}, A_{\mathbf{v}_2} \equiv \{1,2,3,4\}, A_{\mathbf{v}_3} \equiv \{1,2,3\}.$$

Table 8.30 Added constraint, **U** matrix, and associated dot products.

Iteration 3				
$\mathbf{a}_3$	$\mathbf{v}_1^3$	$\mathbf{v}_2^3$	$\mathbf{v}_3^3$	$\mathbf{v}_4^3$
–1	**2/3**	1/3	–7/3	–4/3
0	**–1/3**	–2/3	5/3	8/3
2	**1**	0	0	0
1	**0**	1	0	0
0	**0**	0	1	0
1	**0**	0	0	1
$\mathbf{t}^3$	**4/3**	2/3	7/3	7/3

Table 8.31 Added constraint, **U** matrix, and associated dot products.

Iteration 4			
$\mathbf{a}_4$	$\mathbf{v}_1^4$	$\mathbf{v}_2^4$	$\mathbf{v}_3^4$
0	*0*	–7/2	**–5/2**
0	*–1/2*	9/4	**13/4**
0	*–1/2*	–7/4	**–7/4**
0	*1*	0	**0**
0	*0*	1	**0**
–1	*0*	0	**1**
$\mathbf{t}^4$	0	0	–1

Step 3: The pivot is column vector 3.

Step 4: Since $s < 3$ and *pivot* > 0, we go to the Γ_I-process.

Step 5.2: Columns 1 (italic) and 3 (boldfaced) in Table 8.31 are permuted, since the *pivot* $= 3$ is different from 1.

The pivot column is normalized and the pivoting process performed as shown in Table 8.32. Index h is increased to 5 and we return to Step 1.

Since there are no more constraints we go to Step 9.

Step 9.2: Since the unique vector with non-null component in Table 8.32 has such component equal to 1, we obtain Table 8.33.

Step III: Then the solution is $\mathbf{V}_\rho + \mathbf{Q}_\lambda$, where

$$\mathbf{V} = \left\{\left(-\frac{7}{2}, \frac{9}{4}, -\frac{7}{4}, 0, 1\right), \left(0, -\frac{1}{2}, -\frac{1}{2}, 1, 0\right)\right\},$$

$$\mathbf{W} = \{\},$$

$$\mathbf{Q} = \left\{\left(-\frac{5}{2}, \frac{13}{4}, -\frac{7}{4}, 0, 0\right)\right\}.$$

Table 8.32 Dual generators.

Dual		
$\mathbf{w}_1$	$\mathbf{v}_1$	$\mathbf{v}_2$
-5/2	-7/2	0
13/4	9/4	-1/2
-7/4	-7/4	-1/2
0	0	1
0	1	0
1	0	0

Table 8.33 Normalized dual generators.

Normalized dual		
$\mathbf{w}_1$	$\mathbf{v}_1$	$\mathbf{v}_2$
-5/2	-7/2	0
13/4	9/4	-1/2
-7/4	-7/4	-1/2
0	0	1
0	1	0
1	0	0

■

Example 8.7 (Set of solutions). In this example we determine the set of solutions of the system

$$\begin{array}{rrrcl} x_1 & +2x_2 & -x_3 & = & 4 \\ x_1 & -x_2 & +3x_3 & \leq & 3 \\ 2x_1 & +x_2 & +x_3 & < & 6 \\ & x_2 & & \geq & 0 \\ & & x_3 & > & 0 \end{array} \tag{8.37}$$

which, with the auxiliary variable x_4, becomes

$$\begin{array}{rrrrcl} x_1 & +2x_2 & -x_3 & -4x_4 & = & 0 \\ x_1 & -x_2 & +3x_3 & -3x_4 & \leq & 0 \\ 2x_1 & +x_2 & +x_3 & -6x_4 & < & 0 \\ & -x_2 & & & \leq & 0 \\ & & -x_3 & & < & 0 \\ & & & -x_4 & \leq & 0 \\ & & & x_4 & = & 1 \end{array} \tag{8.38}$$

According to (8.22) we have

$$\begin{aligned} \mathbf{B} &\equiv \{(1,2,-1,-4)^T\}, \\ \mathbf{C} &\equiv \{(1,-1,3,-3)^T, (0,-1,0,0)^T, (0,0,0,-1)^T\}, \\ \mathbf{D} &\equiv \{(2,1,1,-6)^T, (0,0,-1,0)^T\}, \end{aligned}$$

and proceeding with the Γ-algorithm we obtain

$$\begin{array}{rcl} \mathbf{U}_\pi & \equiv & \Omega(\mathbf{B}_\rho + \mathbf{C}_\pi + \mathbf{D}_\pi) \\ & \equiv & \pi_1(-2,1,0,0)^T + \pi_2(-5,4,3,0)^T + \pi_3(8,2,0,3)^T + \pi_4(5,5,3,3)^T, \end{array} \tag{8.39}$$

and by (8.23) we have

$$\mathbf{U}_\pi \bullet \mathbf{d}_1 = -3\pi_1 - 3\pi_2 < 0 \Rightarrow \pi_1 + \pi_2 > 0,$$

$$\mathbf{U}_\pi \bullet \mathbf{d}_2 = -3\pi_2 - 3\pi_4 < 0 \Rightarrow \pi_2 + \pi_4 > 0.$$

In addition, we must have $x_4 = 3\pi_3 + 3\pi_4 = 1$, and if we let $\lambda_1 = 3\pi_3$ and $\lambda_2 = 3\pi_4$, we finally get

$$X \equiv \pi_1(-2,1,0)^T + \pi_2(-5,4,3)^T + \lambda_1(8/3,2/3,0)^T + \lambda_2(5/3,5/3,1)^T \tag{8.40}$$

with $\pi_1 + \pi_2 > 0$, $\pi_2 + \lambda_2/3 > 0, \lambda_1, \lambda_2 \geq 0; \lambda_1 + \lambda_2 = 1$.

Note that the fourth component has been removed because it corresponds to the auxiliary variable. ■

Example 8.8 (Parametric and Cartesian representations). In this example we show how the parametric representation given by (8.40) coincides with the Cartesian representation given by (8.37).

From (8.39) we get

$$\begin{array}{rrrrrrcl} -2\pi_1 & -5\pi_2 & +8\pi_3 & +5\pi_4 & & & = & x_1 \\ \pi_1 & +4\pi_2 & +2\pi_3 & +5\pi_4 & & & = & x_2 \\ & 3\pi_2 & & +3\pi_4 & & & = & x_3 \\ & & +3\pi_3 & +3\pi_4 & & & = & 1 \\ \pi_1 & +\pi_2 & & & -\pi_5 & & = & 0 \\ & \pi_2 & & +\pi_4 & & -\pi_6 & = & 0 \end{array}$$

where $\pi_1, \pi_2, \pi_3, \pi_4 \geq 0, \pi_5, \pi_6 > 0$.

We analyze the compatibility conditions of the above system. A comparison with (8.10) leads to

$$\begin{array}{l} \mathbf{B} \equiv \emptyset, \\ \mathbf{C} \equiv \{(-2,1,0,0,1,0)^T, (-5,4,3,0,1,1)^T, (8,2,0,3,0,0)^T, (5,5,3,3,0,1)^T\}, \\ \mathbf{D} \equiv \{\mathbf{d}_1, \mathbf{d}_2\} \equiv \{(0,0,0,0,-1,0)^T, (0,0,0,0,0,-1)^T\}, \end{array}$$

and proceeding with the Γ-algorithm we obtain

$$\begin{array}{rcl} \Omega(\mathbf{C}_\pi + \mathbf{D}_\pi) & \equiv & \mathbf{U}_\pi \equiv \mathbf{V}_\rho + \mathbf{W}_\pi \\ & \equiv & \rho_1(1,2,-1,-4,0,0)^T + \pi_1(0,-3,3,2,3,0)^T \\ & & +\pi_2(0,0,-1,0,0,3)^T + \pi_3(0,0,-1,0,0,0)^T \\ & & +\pi_4(0,0,0,-1,0,0)^T + \pi_5(0,-3,3,2,0,0)^T \\ & & +\pi_6(0,-3,4,1,0,0)^T. \end{array}$$

The first condition in (8.12), $\mathbf{U}^T\mathbf{a} \leq \mathbf{0}$, with $\mathbf{a} = (x_1, x_2, x_3, 1)^T$ and (6.17) leads to (note that we have removed the last two components, because they correspond to auxiliary slack variables)

$$\begin{array}{rrrrcl} x_1 & +2x_2 & -x_3 & -4 & = & 0 \\ & -3x_2 & +3x_3 & +2 & \leq & 0 \\ & & -x_3 & & \leq & 0 \\ & -3x_2 & +4x_3 & +1 & \leq & 0 \end{array}$$

and the second condition in (8.12) gives

$$\begin{array}{l} \mathbf{W}(\mathbf{a}) \subseteq \mathbf{W}(\mathbf{d}_1) \equiv \{\mathbf{w}_2, \mathbf{w}_3, \mathbf{w}_4, \mathbf{w}_5, \mathbf{w}_6\} \Leftrightarrow \mathbf{w}_1 \bullet \mathbf{a} < 0, \\ \mathbf{W}(\mathbf{a}) \subseteq \mathbf{W}(\mathbf{d}_2) \equiv \{\mathbf{w}_1, \mathbf{w}_3, \mathbf{w}_4, \mathbf{w}_5, \mathbf{w}_6\} \Leftrightarrow \mathbf{w}_2 \bullet \mathbf{a} < 0, \end{array}$$

because, otherwise, $\mathbf{W}(\mathbf{a}) \not\subseteq \mathbf{W}(\mathbf{d}_1)$ and $\mathbf{W}(\mathbf{a}) \not\subseteq \mathbf{W}(\mathbf{d}_2)$. This leads to the system

$$\begin{array}{rcrrrrcl} (i) & \equiv & x_1 & +2x_2 & -x_3 & -4 & = & 0 \\ (ii) & \equiv & & -3x_2 & +3x_3 & +2 & < & 0 \\ (iii) & \equiv & & -3x_2 & +4x_3 & +1 & \leq & 0 \\ & & & & x_3 & & > & 0 \end{array}$$

which taking into account that

$$\begin{array}{l} (i) + (iii) \equiv x_1 - x_2 + 3x_3 \leq 3 \\ 2(i) + (ii) \equiv 2x_1 + x_2 + x_3 < 6 \\ (ii) \wedge x_3 > 0 \Rightarrow x_2 > 0 \Rightarrow x_2 \geq 0 \end{array}$$

is a compact form of system (8.37). ■

Examples 8.7 and 8.8 show a very simple methodology used to reduce a given system to its most compact form or, in other words, to eliminate its superfluous constraints.

8.4 APPLICATIONS TO LINEAR PROGRAMMING

In this section we illustrate some applications of the method for solving systems of inequalities to linear programming.

Example 8.9 (Solution of a linear programming problem). Consider the following linear programming problem:

$$\begin{array}{lrrcl} \text{Maximize} & x_1 & +x_2 & & \\ \text{Subject to} & -x_1 & +x_2 & \leq & 2 \\ & x_1 & +2x_2 & \leq & 6 \\ & 2x_1 & +x_2 & \leq & 6 \\ & & x_1, x_2 & \geq & 0 \end{array} \qquad (8.41)$$

The maximum value of the objective function can be found as follows. Consider the constraints in (8.41), using the slack variables x_3, x_4, and x_5 and adding the objective function as a new constraint $x_1 + x_2 = d$. The following system of linear inequalities is obtained:

$$\begin{array}{rrrrrcl} -x_1 & +x_2 & +x_3 & & & = & 2 \\ x_1 & +2x_2 & & +x_4 & & = & 6 \\ 2x_1 & +x_2 & & & +x_5 & = & 6 \\ x_1 & +x_2 & & & & = & d \\ & & & & x_1, x_2, x_3, x_4, x_5 & \geq & 0 \end{array} \tag{8.42}$$

The optimal value of (8.41) is the maximum value of d that makes system (8.42) compatible. So we consider

$$\mathbf{C} = \begin{pmatrix} -1 & 1 & 1 & 0 & 0 \\ 1 & 2 & 0 & 1 & 0 \\ 2 & 1 & 0 & 0 & 1 \\ 1 & 1 & 0 & 0 & 0 \end{pmatrix} \quad \text{and } \mathbf{a} = \begin{pmatrix} 2 \\ 6 \\ 6 \\ d \end{pmatrix},$$

and then (8.42) has solution iff $\mathbf{V}^T\mathbf{a} = 0$ and $\mathbf{W}^T\mathbf{a} \leq 0$, where $\Omega(\mathbf{C}_\pi) \equiv \mathbf{V}_\rho + \mathbf{W}_\pi$ is the dual cone of $\mathbf{C}_\pi$. It can be calculated with the Γ-algorithm:

$$\mathbf{V} \equiv \emptyset \tag{8.43}$$

and

$$\mathbf{W} = (\mathbf{w}_1, \mathbf{w}_2, \mathbf{w}_3, \mathbf{w}_4, \mathbf{w}_5, \mathbf{w}_6) = \begin{pmatrix} 0 & -1 & 0 & 0 & -1 & 0 \\ 0 & 0 & -1 & 0 & 0 & -1 \\ -1 & -2 & -1 & 0 & 0 & 0 \\ 1 & 3 & 3 & -1 & -1 & 1 \end{pmatrix}. \tag{8.44}$$

Then compatibility conditions can be stated as follows:

$$\begin{array}{llcl} \mathbf{w}_1 \bullet \mathbf{a} = -6 + d & \leq 0 & \Leftrightarrow & d \leq 6, \\ \mathbf{w}_2 \bullet \mathbf{a} = -14 + 3d & \leq 0 & \Leftrightarrow & d \leq 14/3, \\ \mathbf{w}_3 \bullet \mathbf{a} = -12 + 3d & \leq 0 & \Leftrightarrow & d \leq 4, \\ \mathbf{w}_4 \bullet \mathbf{a} = -d & \leq 0 & \Leftrightarrow & d \geq 0, \\ \mathbf{w}_5 \bullet \mathbf{a} = -2 - d & \leq 0 & \Leftrightarrow & d \geq -2, \\ \mathbf{w}_6 \bullet \mathbf{a} = -6 + d & \leq 0 & \Leftrightarrow & d \leq 6. \end{array} \tag{8.45}$$

The maximum value of d that satisfies (8.45) is $d = 4$. This is the optimal value of the linear programming problem (8.41). ■

Exercises

8.1 We are interested in obtaining the convex polygon (convex hull) generated by a set of points $\{\mathbf{x}_1, \ldots, \mathbf{x}_p\}$.

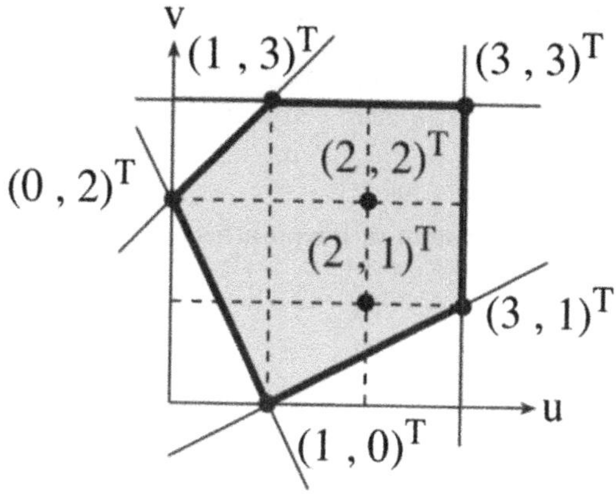

Fig. 8.4 Points defining a polygon.

(a) Obtain a set of rules or conditions to that aim.

(b) Apply them to the case of the polygon generated by the points $\{(1,0)^T, (3,1)^T, (3,3)^T, (1,3)^T, (0,2)^T, (2,2)^T, (2,1)^T\}$ (see Fig. 8.4).

(c) Comment on the result.

Hint: Use the Γ-algorithm to obtain the dual of the dual of $\mathbf{C}_\rho$, where

$$\mathbf{C} = \begin{pmatrix} 1 & 3 & 3 & 1 & 0 & 2 & 2 \\ 0 & 1 & 3 & 3 & 2 & 2 & 1 \\ 1 & 1 & 1 & 1 & 1 & 1 & 1 \end{pmatrix}. \tag{8.46}$$

8.2 Solve the following system of inequalities:

$$\begin{array}{cccl} x_1 & -x_2 & & \leq 0 \\ x_1 & & -x_3 & \leq 0 \\ & +x_2 & & \leq 1 \\ & & x_3 & \leq 1 \end{array} \tag{8.47}$$

8.3 Solve the following system of inequalities:

$$\begin{array}{ccl} x_1 & -x_2 & \leq 0 \\ x_1 & & \leq 1 \end{array} \tag{8.48}$$

8.4 Solve the following system of inequalities:

$$\begin{array}{ccrl} x_1 & +x_2 & = & -1 \\ -x_1 & +2x_2 & \leq & 1 \\ -x_1 & -x_2 & \leq & -2 \end{array} \tag{8.49}$$

8.5 Solve the following system of inequalities:

$$\begin{array}{rrcr} x_1 & +x_2 & = & -1 \\ -x_1 & -x_2 & \leq & 0 \\ & x_2 & \leq & 0 \end{array} \tag{8.50}$$

8.6 Solve the following system of inequalities:

$$\begin{array}{rrcr} -x_1 & & = & -1 \\ x_1 & -x_2 & \leq & 0 \\ & x_2 & \leq & 1 \end{array} \tag{8.51}$$

8.7 Solve the following system of inequalities:

$$\begin{array}{rrrcr} -x_1 & & & \leq & 0 \\ & -x_2 & & \leq & 0 \\ & & -x_3 & \leq & 0 \\ & & & \leq & 1 \\ -x_1 & +2x_2 & -x_3 & \leq & 0 \\ & -x_2 & +x_3 & \leq & 1 \\ 2x_1 & & -x_3 & \leq & 1 \end{array} \tag{8.52}$$

8.8 Solve the following system of inequalities:

$$\begin{array}{rrrcr} -x_1 & & & \leq & 0 \\ & -x_2 & & \leq & 0 \\ & & -x_3 & \leq & 0 \\ & & & \leq & 1 \\ -x_1 & & & \leq & 1 \\ -x_1 & -x_2 & -x_3 & \leq & -1 \\ -x_1 & & +x_3 & \leq & 0 \\ x_1 & -x_2 & -x_3 & \leq & -2 \\ & -2x_2 & +x_3 & \leq & 0 \end{array} \tag{8.53}$$

Part III

Linear Programming

Introduction to Part III

Linear programming is one of the most important areas of applied mathematics. It is used in economics, business administration, engineering, and many other areas of science, industry, and technology.

Economists of the former Soviet Union were the first in applying linear programming techniques in the organization and planning of production. However, it was during WorldWar II that linear programming gained prominence. The U.S.A. Air Force created the SCOOP project (Scientific Computation of Optima Programs) directed by G. B. Dantzig. The best known method for solving linear programming problems, *the simplex method*, is due to Dantzig, who presented the method in 1947 (see also Dantzig (1963)). Fortunately, the growth of large and fast computing facilities has aided in the use of the developed techniques.

During the last decades very intensive work and research has been devoted to the linear programming problem (LPP). While implementation of the simplex method algorithm has undergone major modification, the underlying concept has withstood the test of time. A full description of the computer implementation of this method can be seen, for example, in the IBM Optimization Subroutine Library (OSL) (1992) or in Kuenzi, Tzchach, and Zehnder (1971).

For many linear programming problems, the simplex method is still the best method. However, several improvements have been introduced, such as the revised simplex, the dual, or the primal–dual methods. The primal–dual methods work simultaneously with both the primal problem and its dual. This can have advantages because we can exploit the well–known relationships between primal and dual problems (see Forrest and Tomlin (1992)). However, as indicated, for example, by Press, Teukolsky, Vetterling, and Flannery (1985):

> There seems to be no clearcut evidence that these methods are superior to the usual method by any factor substantially larger than the "tender-loving-care factor", which reflects the programming effort of the proponents.

For most cases the primal is the preferred method. However, in some special cases the other methods give superior performance.

Recently, the interior point method was introduced as an alternative to the simplex (see, for example, Gill et al. (1986)). The primal–dual method has been superseded by the predictor–corrector method, which is a modified version of the primal-dual originated by Mehrotra (1990). This method usually requires less iterations and is the interior point method of choice.

Since there are many books dealing with these methods, we shall not treat them here. The interested reader can consult any of the references.

In this part we study new programming methods, which arise from the concept of the convex polyhedral cone. We start with the ideas of nineteenth century mathematicians, such as Weyl, Gordan, Farkas, and Minkowski, and the duality concept, which allows incredible possibilities to be shown.

We develop the exterior point method, which is more efficient and elegant than the simplex method and can greatly be simplified by updating only the pivot row and column.

In addition, we show that the exterior point method is adequate for finding the new solution of a linear programming problem after adding or removing constraints and/or variables without the need of redoing all the computations. Thus, it is useful for sensitivity analysis.

9

An Introduction to Linear Programming

9.1 INTRODUCTION

This chapter is devoted to the linear programming problem, which has many practical applications, as it is illustrated with the collection of examples in Chapter 11.

In Section 9.2 we introduce the linear programming problem and illustrate the cases of unique and multiple solutions, and the cases of unbounded and nonfeasible solutions.

Section 9.3 describes the standard form of a linear programming problem, while Section 9.4 introduces the concept of basic solutions, which are the candidate solutions for the linear programming problem.

Finally, in Section 9.5, we show how every linear programming problem has a dual and we give some examples with their corresponding interpretations of the primal and dual approaches.

9.2 PROBLEM STATEMENT AND BASIC DEFINITIONS

Linear programming deals with the problem of optimizing (minimizing or maximizing) a linear function of n variables subject to equality and/or inequality linear constraints. In other words, a linear programming problem consists of finding the optimum (maximum or minimum) of a linear function in a set that is expressed as the intersection of a finite number of hyperplanes and halfspaces in $\mathbb{R}^n$.

Thus, we have the following definitions.

Definition 9.1 (Linear programming problem). *A linear programming problem (LPP) consists of:*

Minimizing

$$f(\mathbf{x})$$

subject to

$$\begin{array}{rcl} g_i(\mathbf{x}) & \leq & d_i \text{ for } i = 1, \ldots, r, \\ h_i(\mathbf{x}) & = & e_i \text{ for } i = 1, \ldots, s, \end{array} \tag{9.1}$$

where $f, g_1, \ldots, g_r, h_1, \ldots, h_s$ are linear functions defined on $\mathbb{R}^n$ and $\mathbf{x}$ is a vector of n components $x_1, \ldots, x_n$. The function $f(\mathbf{x})$ is known as the objective or cost function.

Obviously, a linear programming problem can be stated as a maximization problem because maximizing $f(\mathbf{x})$ is equivalent to minimizing $-f(\mathbf{x})$.

Definition 9.2 (Feasible solution). *A vector satisfying all the constraints in (9.1) is called a feasible solution. The collection of all such solutions forms the feasible region.*

In a linear programming problem, the feasible region is a polytope or a polyhedron.

Definition 9.3 (Optimal solution). *A feasible point $\tilde{\mathbf{x}}$ such that $f(\mathbf{x}) \geq f(\tilde{\mathbf{x}})$ for any other feasible point $\mathbf{x}$ is called an optimal solution, or simply a solution, of the problem.*

A linear programming problem has the following interesting properties:

- If the feasible region is bounded, the problem always has a solution (it is a sufficient but not necessary condition for existence of solution).
- The optimum of a linear programming problem is always a global optimum.
- If $\mathbf{x}$ and $\mathbf{y}$ are optima of a linear programming problem, then any convex linear combination of them is also an optimum.
- The optimum is always attained at least on a vertex of the feasible region.

Example 9.1 (Unique solution). Consider the following linear programming problem:

Maximize

$$x_1 + x_2$$

subject to

$$\begin{array}{rcrcl} 5x_1 & + & 2x_2 & \leq & 10 \\ 3x_1 & + & 5x_2 & \leq & 15 \\ & & x_1, x_2 & \geq & 0. \end{array} \tag{9.2}$$

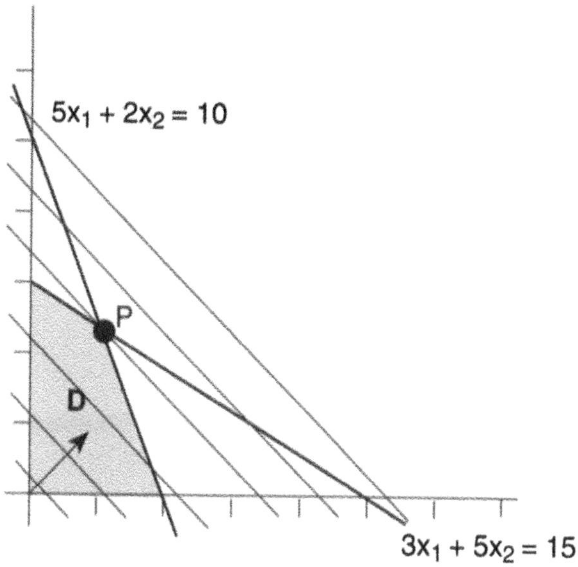

Fig. 9.1 Graphical illustration of a linear programming problem with a unique solution.

Figure 9.1 shows the feasible region, the contour plots of the objective function, and its gradient vector indicating the increasing direction. The solution is reached at point P, because it is on the last contour plot with feasible points in the indicated direction. P is the intersection of the lines

$$5x_1 + 2x_2 = 10,$$
$$3x_1 + 5x_2 = 15.$$

So the maximum is attained at the point

$$P = \left(\frac{20}{19}, \frac{45}{19}\right)^T.$$

■

Example 9.2 (Multiple solutions). If the objective function of Example 9.1 is replaced by

$$f(x_1, x_2) = 6x_1 + 10x_2,$$

the resulting contour plots are parallel to one of the constraint lines (the second). In this case, the optimum is reached at the points of a polytope's edge, as shown in Fig. 9.2.

Any point on the line segment between the points $(0,3)^T$ and $(\frac{20}{19}, \frac{45}{19})^T$ is a global maximum of the problem. ■

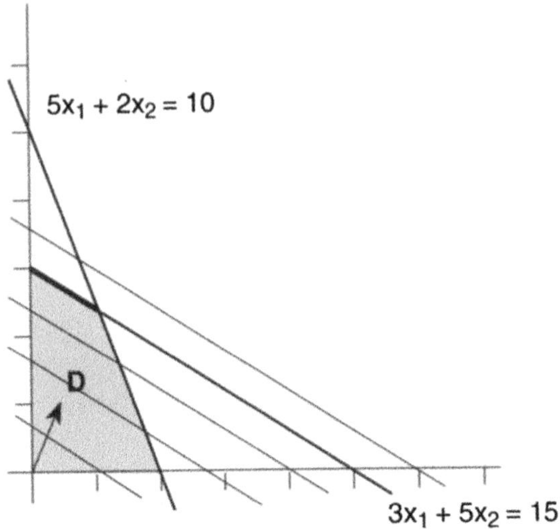

Fig. 9.2 Graphical illustration of a linear programming problem with multiple solutions.

Example 9.3 (No solution). The linear programming problem:

Maximize

$$2x_1 + 3x_2$$

subject to

$$\begin{array}{rrcl} x_1 & +x_2 & \geq & 4 \\ x_1 & & \geq & 2 \\ & x_2 & \geq & 0, \end{array} \tag{9.3}$$

has unbounded solution, because, as shown in Fig. 9.3, the feasible region is unbounded in the increasing direction of the objective function.

■

Example 9.4 (Bounded solution). Consider the following linear programming problem:

Minimize

$$2x_1 - x_2 - x_3$$

subject to

$$\begin{array}{rrrcl} x_1 & & +x_3 & \leq & 1 \\ & x_2 & +x_3 & \leq & 1 \\ & & x_1, x_2, x_3 & \geq & 0 \end{array} \tag{9.4}$$

The feasible region is the polytope in $\mathbb{R}^3$ shown in Fig. 9.4. This set is bounded, so the problem has solution. It is well known that the solution is

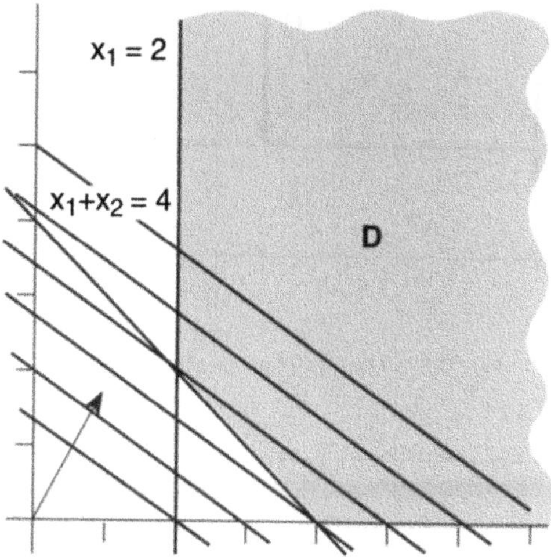

Fig. 9.3 Graphical illustration of a linear programming problem with unbounded feasible region.

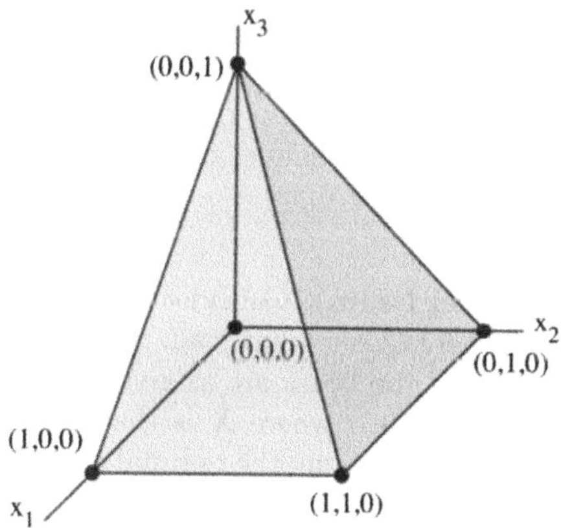

Fig. 9.4 The three-dimensional feasible region for the constraints in (9.4).

reached, at least, on a vertex of the feasible region. Fig. 9.4 shows all these vertices; evaluating the objective function in them, the optimum can be found:

$$
\begin{aligned}
f(0,0,0) &= 0, \\
f(1,0,0) &= 2, \\
f(1,1,0) &= 1, \\
f(0,1,0) &= -1, \\
f(0,0,1) &= -1.
\end{aligned}
$$

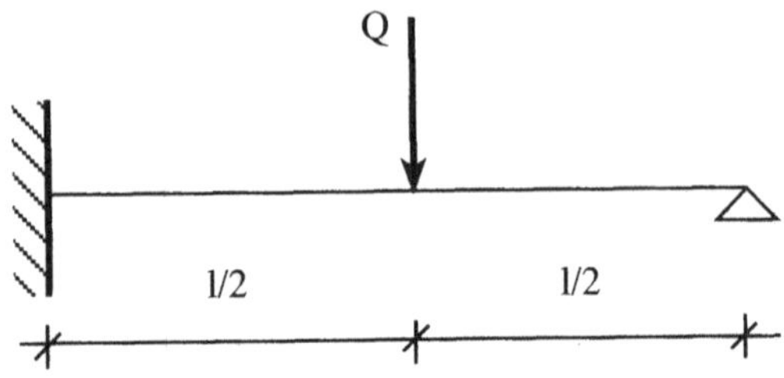

Fig. 9.5 A supported cantilever beam.

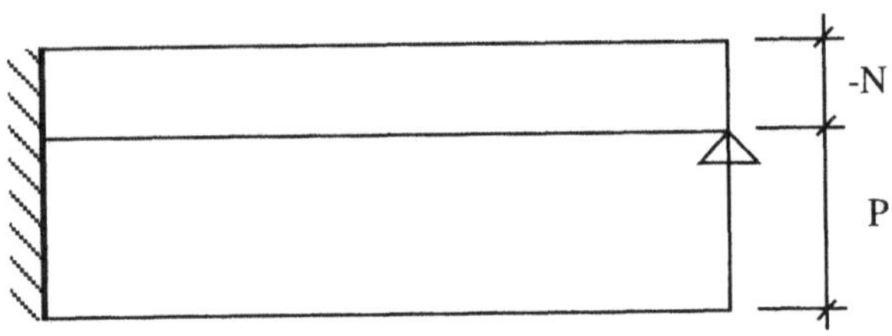

Fig. 9.6 Plastic moments (strength) at all points of the supported cantilever beam.

Then the problem has an infinite number of minima: the points on the line segment between $(0,1,0)^T$ and $(0,0,1)^T$. ■

Example 9.5 (Cantilever beam). Consider the supported cantilever beam of length ℓ in Fig. 9.5, where Q is a point load. Assume that the positive and negative plastic moments of the beam are as given in Fig. 9.6. The bending moment law is given in Fig. 9.7, where X is the moment at the supported end. We are interested in determining the optimal clamping degree (the value of X) to maximize the load Q.

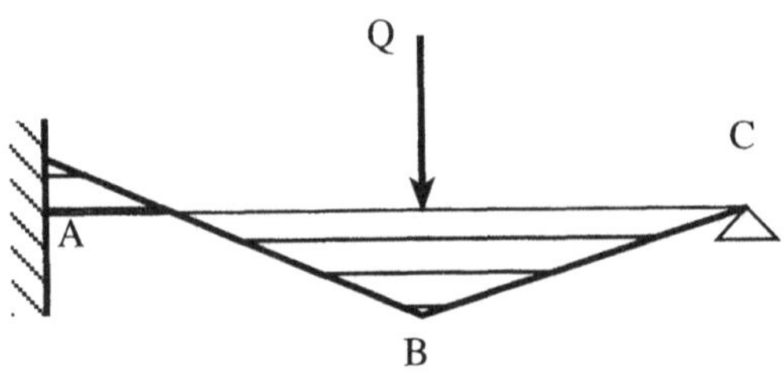

Fig. 9.7 Bending moment law when the moment at the supported end is X.

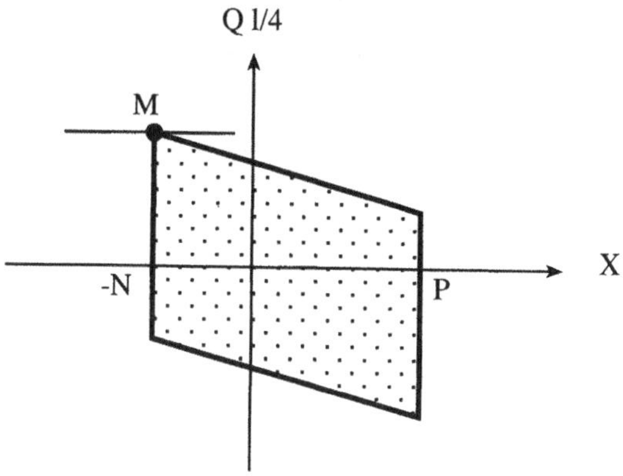

Fig. 9.8 Feasible region in terms of X and Q.

The moments at the three critical points (those with relative maxima) A, B, and C are

$$\begin{aligned} M_A &= X, \\ M_B &= \frac{Q\ell}{4} + \frac{X}{2}, \\ M_C &= 0. \end{aligned} \tag{9.5}$$

Thus, failure does not occur if

$$\begin{aligned} -N &\leq X \leq P, \\ -N &\leq \frac{Q\ell}{4} + \frac{X}{2} \leq P, \\ -N &\leq 0 \leq P. \end{aligned} \tag{9.6}$$

These conditions are illustrated in Fig. 9.8, where the safe region has been shadowed.

The maximum feasible load Q is attained at point M, that is, when $X = -N$ and $Q = (4P + 2N)/\ell$.

Fig. 9.9 illustrates the optimal solution, which clearly corresponds to the case of points A and B satisfying the plasticity condition.

■

9.3 LINEAR PROGRAMMING PROBLEM IN STANDARD FORM

Definition 9.4 (Standard form). *A linear programming problem is said to be in standard form if it is expressed as follows:*

Minimize

$$c_1x_1 + c_2x_2 + \cdots + c_nx_n$$

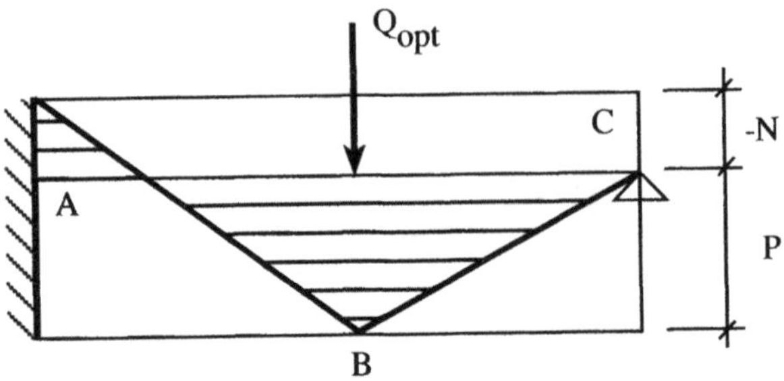

Fig. 9.9 Illustration of the optimal solution (points A and B satisfy plasticity condition).

subject to

$$\begin{array}{ccccccccc} a_{11}x_1 & + & a_{12}x_2 & + & \cdots & + & a_{1n}x_n & = & b_1 \\ a_{21}x_1 & + & a_{22}x_2 & + & \cdots & + & a_{2n}x_n & = & b_2 \\ \cdots & \cdots & \cdots & \cdots & \cdots & \cdots & \cdots & \cdots & \cdots \\ a_{m1}x_1 & + & a_{m2}x_2 & + & \cdots & + & a_{mn}x_n & = & b_m \\ & & & & & & \multicolumn{3}{r}{x_i \geq 0; i = 1, 2, \ldots, n,} \end{array} \tag{9.7}$$

where $n \geq m$ *and* $b_i \geq 0;\ i = 1, 2, \ldots, m.$

Defining the following matrices

$$\mathbf{c} = (c_1, c_2, \ldots, c_n); \quad \mathbf{x} = \begin{pmatrix} x_1 \\ x_2 \\ \vdots \\ x_n \end{pmatrix}; \quad \mathbf{b} = \begin{pmatrix} b_1 \\ b_2 \\ \vdots \\ b_n \end{pmatrix};$$

$$\mathbf{A} = \begin{pmatrix} a_{11} & a_{12} & \cdots & a_{1n} \\ a_{21} & a_{22} & \cdots & a_{2n} \\ \cdots & \cdots & \cdots & \cdots \\ a_{m1} & a_{m2} & \cdots & a_{mn} \end{pmatrix},$$

the problem (9.7) can be formulated as:
Minimize

$$\mathbf{cx}$$

subject to

$$\begin{array}{rcl} \mathbf{Ax} & = & \mathbf{b} \\ \mathbf{x} & \geq & 0, \end{array} \tag{9.8}$$

and is said to be in *standard matrix form.*

9.3.1 Transformation to Standard Form

A linear programming problem can always be formulated in a standard form by performing some manipulations:

1. A maximization problem is equivalent to a minimization one.
2. Inequality constraints can be converted to equality constraints by adding new variables known as *slack variables*:
   - If $a_{i1}x_1 + a_{i2}x_2 + \cdots + a_{in}x_n \leq b_i$, then there exists a variable $x_{n+1} \geq 0$ such that
   $$a_{i1}x_1 + a_{i2}x_2 + \cdots + a_{in}x_n + x_{n+1} = b_i.$$
   - If $a_{i1}x_1 + a_{i2}x_2 + \cdots + a_{in}x_n \geq b_i$, then there exists a variable $x_{n+1} \geq 0$ such that
   $$a_{i1}x_1 + a_{i2}x_2 + \cdots + a_{in}x_n - x_{n+1} = b_i.$$
3. Unrestricted variables can be replaced by differences between two non-negative variables.
4. Finally, if some $b_i < 0$, the corresponding constraint can be multiplied by -1.

Example 9.6 (Standard form). The standard form of the linear programming problem:

Maximize
$$2x_1 - 3x_2 + 5x_3$$
subject to
$$\begin{array}{lllcl} x_1 & +x_2 & & \leq & 2 \\ 3x_1 & +x_2 & -x_3 & \geq & 3 \\ & & x_1, x_2 & \geq & 0 \end{array} \quad (9.9)$$
is

Minimize
$$-2x_1 + 3x_2 + 5(x_6 - x_7)$$
subject to
$$\begin{array}{llllllcl} x_1 & +x_2 & & +x_4 & & = & 2 \\ 3x_1 & +x_2 & -(x_6 - x_7) & & -x_5 & = & 3 \\ & & & & x_1, x_2, x_4, x_5, x_6, x_7 & \geq & 0 \end{array} \quad (9.10)$$

■

9.4 BASIC SOLUTIONS

Consider a linear programming problem in its standard matrix form (9.8), where $\mathbf{A}$ is a full-rank $m \times n$ matrix.

Definition 9.5 (Basic matrix). *A nonsingular $m \times m$ submatrix $\mathbf{B}$ of $\mathbf{A}$ is called a basic matrix. It is also called a feasible basic matrix iff $\mathbf{B}^{-1}\mathbf{b} \geq 0$.*

Each basic matrix has an associated vector called *basic solution*. The process for calculating this solution is the following.

Let $\mathbf{x}_B$ be the vector of variables associated with the columns of $\mathbf{A}$ used to build $\mathbf{B}$. Variables in $\mathbf{x}_B$ are known as *basic variables* and the remaining ones are called *nonbasic variables*. Giving zero values to nonbasic variables and letting

$$\mathbf{x}_B = \mathbf{B}^{-1}\mathbf{b},$$

the basic solution associated with $\mathbf{B}$ can be obtained. If $\mathbf{B}$ is a feasible basic matrix, its basic solution is said to be feasible.

The number of feasible basic solutions of a linear programming problem is always finite and each of them corresponds to a vertex of its feasible region. Thus, if a linear programming problem has a solution, it is reached by at least one of its feasible basic solutions.

9.5 DUALITY

In this section the concept and significance of duality in linear programming are explained. First, the dual problem of a given linear programming problem is formulated and next, the mathematical relationship between the two problems is studied. We provide simple examples that lead to a better understanding of this important concept.

Given a linear programming problem, called *primal problem*, there is another linear programming problem, called *dual problem*, closely associated with it. Both problems are said to be in duality. Under certain assumptions, the primal and dual problems have the same optimal objective value, and hence, it is possible to solve the primal problem indirectly by solving the corresponding dual problem.

Definition 9.6 (Dual problem). *Given the linear programming problem:*

Minimize

$$\mathbf{c}\mathbf{x}$$

subject to

$$\begin{array}{rcl} \mathbf{A}\mathbf{x} & \geq & \mathbf{b} \\ \mathbf{x} & \geq & 0, \end{array} \tag{9.11}$$

its dual problem is

Maximize

$$\mathbf{b}^T\mathbf{y}$$

subject to

$$\begin{array}{rcl} \mathbf{A}^T\mathbf{y} & \leq & \mathbf{c}^T \\ \mathbf{y} & \geq & 0, \end{array} \tag{9.12}$$

where $\mathbf{y} = (y_1, \ldots, y_m)^T$ *are called dual variables.*

Remark 9.1 *As can be observed, every constraint of the primal problem has a dual variable associated with it, the coefficients of the objective function of the primal problem are the right-hand-side terms of the dual constraints and vice versa, and the coefficient matrix of the dual constraints is the transpose of the primal's one. The primal is a minimization problem but its dual is a maximization problem.*

A linear programming problem whose formulation is not in the form (9.11) has also a dual problem. It can be constructed using the following rules:

- An equality constraint in the primal implies that the corresponding dual variable is unrestricted.
- A $\geq$ inequality constraint in the primal implies that its dual variable is restricted (non-negative).
- A primal restricted variable has a $\leq$ inequality constraint in the dual problem.
- A primal unrestricted variable has an equality constraint in the dual problem.

Remark 9.2 *The duality is a symmetric relationship, that is, if problem D is the dual of problem P, then P is also the dual of D.*

Example 9.7 (Dual problem). The dual of the linear programming problem

Minimize

$$x_1 + x_2 - x_3$$

subject to

$$\begin{array}{rrrcl} 2x_1 & +x_2 & & \geq & 3 \\ x_1 & & -x_3 & = & 2 \\ & & x_3 & \geq & 0 \end{array} \tag{9.13}$$

is

Maximize

$$3y_1 + 2y_2$$

subject to

$$\begin{array}{rrcr} 2y_1 & +y_2 & = & 1 \\ y_1 & & = & 1 \\ & -y_2 & \leq & -1 \\ & y_1 & \geq & 0. \end{array} \tag{9.14}$$

■

The importance of the dual problem is established in the following theorems

Theorem 9.1 (Weak duality theorem). *Let P and D be two feasible problems in duality and let $\mathbf{x}$ be a feasible solution to P and $\mathbf{y}$ a feasible solution to D. Then*

$$f(\mathbf{x}) \geq g(\mathbf{y}),$$

where f and g are the objective functions of P and D, respectively.

Theorem 9.2 (Duality theorem). *Let P be a linear programming problem (primal) and D be its dual problem. Then one of the following statements is true:*

1. *Both problems have optimal solution and their optimal values coincide.*

2. *One problem has infinite optimum and the other one has an empty feasible region.*

3. *Both problems are infeasible (empty feasible regions).*

Example 9.8 (Dual of a planing mill problem). A small planing mill manufactures two types of wood tables. Each table of the first type requires 4 machine-hours for cutting from the wood stock to size and 4 machine-hours finishing time (assembly, painting, etc.). Similarly, each second type table requires 3 machine-hours cutting time and 7 machine-hours finishing time. The cutting and finishing equipment capacities available per day are 40 and 56 machine-hours, respectively. Finally, each first type table provides \$70 profit, while each second type table provides \$90 profit. These data are summarized in Table 9.1.

We wish to know the quantity of tables produced per day of each type, in order to maximize the daily total profit. The linear programming model of the problem is:

Maximize

$$70x_1 + 90x_2$$

subject to

$$\begin{array}{rrcr} 4x_1 & +3x_2 & \leq & 40 \\ 4x_1 & +7x_2 & \leq & 56 \\ & x_1, x_2 & \geq & 0, \end{array} \tag{9.15}$$

Table 9.1 Data of the planing mill problem.

	First type table	Second type table	Available machine-hours per day
Cutting	4	3	40
Finishing	4	7	56
Profit	\$70	\$90	

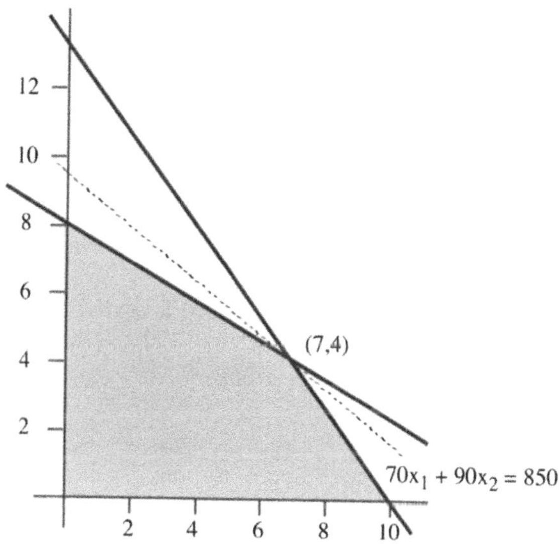

Fig. 9.10 Graphical analysis of the planing mill problem.

where x_1 and x_2 are the quantities of the first and second types of table, respectively, to be produced per day.

The optimal solution of this model, as can be seen in Fig. 9.10, indicates a daily production of 7 tables of the first type and 4 of the second type and a daily maximum profit of \$850. This result indicates that both cutting and finishing resources are fully utilized, because both constraints are binding.

Now suppose we wish to obtain higher daily profits. Then an expansion of the available capacities is necessary. Suppose that the available finishing capacity can be expanded from 56 machine-hours to 72 machine-hours per day. How does this expansion affect the total daily profit? The answer is given by the solution of the following linear programming problem:

Maximize

$$70x_1 + 90x_2$$

subject to

$$\begin{array}{rrl} 4x_1 & +3x_2 & \leq \ 40 \\ 4x_1 & +7x_2 & \leq \ 72 \\ & x_1, x_2 & \geq \ 0. \end{array} \tag{9.16}$$

In this case the optimal solution would be $x_1 = 4$ and $x_2 = 8$ with a maximum daily profit of \$1000. This solution indicates that we can increase daily profit by \$150 if we increase the daily finishing capacity by 16 machine-hours. The rate \$150/16=\$75/8, at which the objective function value will increase as the finishing capacity is increased by 1 hour, is called the *shadow price* (also *dual price*) of the finishing resource.

In general, the shadow price of a constraint is the change in the optimal value of the objective function, per unit change in the right-hand-side value of the constraint, assuming that all other parameter values of the problem remain unchanged. For many linear programming applications the shadow prices are at least as important as the solution of the problem itself; they indicate whether or not certain changes in the use of productive capacities affect the objective function value. Shadow prices can be determined by formulating and solving the *dual problem.*

The dual of the planing mill problem (9.15) is stated as follows:

Minimize

$$40y_1 + 56y_2$$

subject to

$$\begin{array}{rrl} 4y_1 & +4y_2 & \geq \ 70 \\ 3y_1 & +7y_2 & \geq \ 90 \\ & y_1, y_2 & \geq \ 0. \end{array} \tag{9.17}$$

The optimal solution of this problem is $y_1 = 65/8$, $y_2 = 75/8$ and the optimal value is 850. Note that y_1 and y_2 values are the shadow prices of cutting and finishing resources, respectively, and optimal values of problems (9.15) and (9.17) coincide.

The dual problem (9.17) can be interpreted as follows. Assume that we are willing to sell machine time (rather than produce and sell tables), provided that we can obtain at least the same level of total profit. In this case, producing tables and selling machine time would be two equivalent alternatives; the y_1 and y_2 variables would represent the sale prices of one hour on the cutting and finishing machine, respectively. In order to maintain a competitive business we should set the total daily profit as low as possible, that is, minimize the function $40y_1 + 56y_2$, where the 40 and 56 coefficients represent the machine-hours available per day for sale of cutting and finishing machines. The constraints in (9.17) state that the total prices of the cutting and finishing hours required for producing one unit of each type of table must not be less than the profit expected from the sale of one unit of each type of table. ■

Table 9.2 Vitamin content per kilogram of the two nutrients P_1 and P_2.

	P_1	P_2
Vitamin A	5	3
Vitamin B	1.5	3
Vitamin C	1	1.3

Table 9.3 Minimum daily required vitamin amount per animal.

Vitamin A	20
Vitamin B	15
Vitamin C	8

Example 9.9 (Diet problem (primal)). A rancher uses two types of nutrients (P_1 and P_2) for cattle breeding. The unit cost per kilogram of each of those items is \$2 and \$3, respectively. The vitamin content of each kilogram is shown in Table 9.2.

The animal nutrition experts recommend each animal to receive at least the daily vitamin amounts shown in Table 9.3.

The aim of the rancher is to decide the daily amounts of each of the two nutrients to be given to each animal, such that, on the one hand, the recommendations are satisfied and, on the other hand, the cost of the cattle breeding is minimal. To this aim we must solve the following LPP:

Minimize

$$2x_1 + 3x_2$$

subject to

$$\begin{array}{rrcl} 5x_1 & +3x_2 & \geq & 20 \\ 1.5x_1 & +3x_2 & \geq & 15 \\ x_1 & +1.3x_2 & \geq & 8 \\ & x_1, x_2 & \geq & 0, \end{array} \tag{9.18}$$

where x_1 and x_2 are the daily amounts, in kilograms, of the two nutrients P_1 and P_2, respectively, given to each animal.

The minimum of the above problem is attained at the point $x_1 = 4.286$, $x_2 = 2.857$ and the minimum cost becomes \$17.143 per animal and day. ∎

Example 9.10 (Diet problem (dual)). In this example we state and give a physical interpretation to the dual of the problem in Example 9.9. Think now of a company that manufactures cattle breeding nutrients. Assume that it sells three types of vitamin pills to the ranch in Example 9.9. This firm must convince the rancher to use these pills as a better alternative to other items. Thus, the pill prices must be competitive when compared to the nutrients P_1 and P_2.

Let y_1, y_2, and y_3 be the unit prices of vitamins A, B, and C, respectively. The aim of the company is to fix some prices such that the profit is maximized and, at the same time, the rancher is satisfied.

Each kilogram of nutrient P_1 includes 5 units of vitamin A, 1.5 units of vitamin B, and 1 unit of vitamin C. The total price to be paid by the rancher to get the same amount of vitamins is

$$5y_1 + 1.5y_2 + y_3.$$

The rancher is interested in the pills only if

$$5y_1 + 1.5y_2 + y_3 \leq 2. \tag{9.19}$$

Similarly, the company must satisfy the constraint

$$3y_1 + 3y_2 + 1.3y_3 \leq 3. \tag{9.20}$$

It is clear that the prices must be positive. Thus, we have the constraints

$$y_1, y_2, y_3 \geq 0. \tag{9.21}$$

Assuming the rancher decides to buy the minimum amount required to satisfy the constraints for each vitamin. This means that the rancher will buy 20 unit of vitamin A, 15 of vitamin B, and 8 of vitamin C, per animal per day. Consequently, the gross income of the company per animal per day will be

$$V(y_1, y_2, y_3) = 20y_1 + 15y_2 + 8y_3.$$

To fix the prices, the company must solve the following LPP:
Maximize

$$20y_1 + 15y_2 + 8y_3$$

subject to

$$\begin{array}{rrrcl} 5y_1 & +1.5y_2 & +y_3 & \leq & 2 \\ 3y_1 & +3y_2 & +1.3y_3 & \leq & 3 \\ & & y_1, y_2, y_3 & \geq & 0. \end{array} \tag{9.22}$$

The solution of this problem is attained at the point $y_1 = 0$, $y_2 = 0.381$, $y_2 = 1.428$, which corresponds to a maximum of $V(0, 0.381, 1.428) = 17.143$.

Note that one of the prices is null. This means that the rancher with nutrients P_1 and P_2 only has to worry about giving the cattle the required amounts for vitamins B and C, because with these conditions, the required amount for vitamin A is automatically satisfied. This is the reason why the rancher does not need to buy additional units of vitamin A. ■

Exercises

9.1 A furniture manufacturing company wishes to determine how many tables, chairs, desks, and bookcases it should make to optimize the use of its available resources. These products utilize two different types of lumber, and the company has on hand 1500 board-feet of the first type and 1000 board-feet of the second type. There are 800 labor-hours available for the job. Sales forecasts plus back orders require the company to make at least 40 tables, 130 chairs, 30 desks, and no more than 10 bookcases. Each table, chair, desk, and bookcase requires $5, 1, 9$, and 12 board-feet, respectively, of the first type of lumber and $2, 3, 4$, and 1 board-feet of the second type. A table requires 3 labor-hours to make, a chair 2 hours, a desk 5 hours, and a bookcase 10 hours. The company makes a total of \$12 profit on a table, \$5 on a chair, \$15 on a desk, and \$10 on a bookcase. Write out the linear programming model of this problem in terms of maximizing profit. Modify the problem to impose that four chairs be made for every table.

9.2 Using the graphical method, solve the following LPP:

Minimize

$$2x_1 + x_2$$

subject to

$$\begin{array}{rrcl} 3x_1 & +x_2 & \geq & 3 \\ 4x_1 & +3x_2 & \geq & 6 \\ x_1 & +2x_2 & \geq & 2 \\ & x_1, x_2 & \geq & 0. \end{array}$$

9.3 Write and solve the dual of the problem:

Maximize

$$3x_1 - 4x_2 + 9x_3 + x_4$$

subject to

$$\begin{array}{rrrcl} x_1 & -5x_2 & +x_3 & \geq & 0 \\ 3x_1 & -5x_2 & +x_3 & \geq & 10 \\ & & x_2, x_3 & \geq & 0 \\ & & x_4 & \leq & 0. \end{array}$$

9.4 A refrigerator manufacturer has three factories that supply four retail stores. Consider the transportation problem shown in Fig. 9.11; that is, one is interested in shipping 10 units from F_1, 15 units from F_2, and 17 units from F_3, and receiving 10 units in S_1, 9 units in S_2, 11 units in S_3, and 12 units in S_4. Let x_{ij} be the unknown number of refrigerators to be shipped from factory i to store j.

(a) Obtain the set of all feasible solutions.

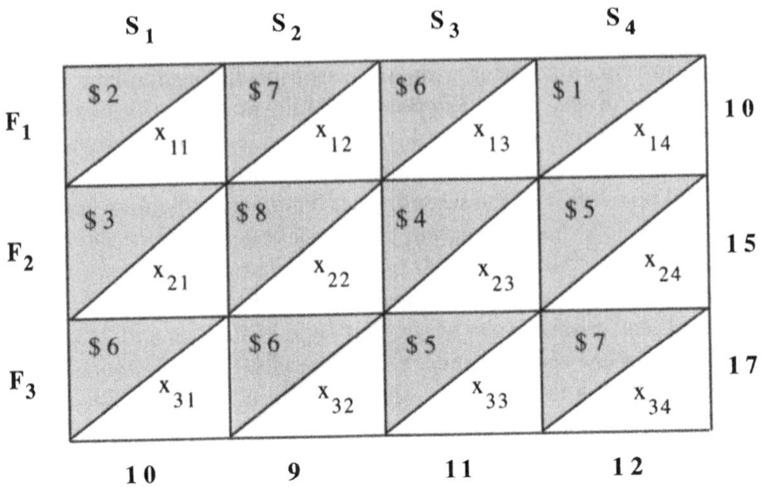

Fig. 9.11 Transportation problem showing the origins and destinations.

(b) State the associated linear programming problem.

9.5 Consider the problem:

Maximize

$$8x_1 - 4x_2$$

subject to

$$\begin{array}{rrcr} 4x_1 & -5x_2 & \leq & 2 \\ 2x_1 & +4x_2 & \geq & 6 \\ x_1 & -6x_2 & \leq & -14. \end{array}$$

(a) Draw the feasible region and identify the vertices of the feasible region.

(b) Find a point in the feasible region at which the objective function attains the maximum.

9.6 Determine the feasible region and its vertices of the following linear programming problem:

Minimize

$$-4x_1 + 7x_2$$

subject to

$$\begin{array}{rrcl} x_1 & +x_2 & \geq & 3 \\ -x_1 & +x_2 & \leq & 3 \\ 2x_1 & +x_2 & \leq & 8 \\ & x_1, x_2 & \geq & 0. \end{array}$$

Obtain the optimal vertex.

9.7 Prove that the following problem does not have a finite optimum:

Minimize

$$-6x_1 + 2x_2$$

subject to

$$\begin{array}{rrrr} -3x_1 & +x_2 & \leq & 6 \\ 3x_1 & +5x_2 & \geq & 15 \\ & x_1, x_2 & \geq & 0. \end{array}$$

9.8 Prove that the problem

Maximize

$$-6x_1 + 2x_2$$

subject to

$$\begin{array}{rrrr} -3x_1 & +x_2 & \leq & 6 \\ 3x_1 & +5x_2 & \geq & 15 \\ & x_1, x_2 & \geq & 0 \end{array}$$

has an infinite maxima.

9.9 Formulate the dual problem of the following:

Minimize

$$5x_1 - 6x_2 + 7x_3 + x_4$$

subject to

$$\begin{array}{rrrrrr} x_1 & +2x_2 & -x_3 & -x_4 & = & -7 \\ 6x_1 & -3x_2 & +x_3 & +7x_4 & \geq & 14 \\ -2x_1 & -7x_2 & +4x_3 & +2x_4 & \leq & -3 \\ & & & x_1, x_2 & \geq & 0. \end{array}$$

9.10 A manufacturer produces two types of articles using a unique raw material. Suppose that 100 kg of this material is available for production and each article of the first type requires 2 kg while each article of the second type requires 4 kg. The manufacturer makes a profit of \$3 and \$2, respectively, on each of the first and second type of article. A market forecast predicts that at least 40 first-type articles and not more than 20 second-type articles can be sold during the month. The manufacturer wants to determine how many of each article should be produced in order to maximize profits. Formulate the problem as a linear programming problem.

9.11 A firm producing a product P has two plants. Each plant produces 90 tons of P monthly, and the product is distributed in three different markets. Table 9.4 shows the unit costs of shipping 1 ton of product P

Table 9.4 Distribution problem.

	Markets		
Plants	$Market_1$	$Market_2$	$Market_3$
$Plant_1$	1	3	5
$Plant_2$	2	5	4

Table 9.5 Working hours for the production of one sofa and one chair.

	Wood-work section	Manufacturing section	Finishing section
Sofa	6	2	1
Chair	3	6	1

from a given plant to a given market. The firm wants to send the same number of tons to each market and minimize the total cost. Formulate the corresponding linear programming problem.

9.12 A firm can produce 120 items of type A and 360 items of type B on a daily basis. The items are subject to a quality control process that can control 200 items per day. Items of type A are sold in the market at a price four times that of type B. Make a decision about the production in order to maximize profits.

9.13 A furniture manufacturing company produces sofas and chairs that are sold at \$160 and \$140 per unit, respectively. The production of a single unit of each requires the working hours shown in Table 9.5. The firm has available 240 working hours per week in the wood-work section, 180 hours in the manufacturing section, and 45 hours in the finishing section. Make a decision about the weekly plan to maximize profits.

10

The Exterior Point Method

10.1 INTRODUCTION

In this chapter we introduce the exterior point method (EPM) that, unlike the simplex and the interior point methods, works outside the feasible region.

In Section 10.2 we describe the EPM in its simpler form and we give an algorithm with a detailed explanation of the method. We also give conditions for detecting the cases of finite solution, unbounded solution, or infeasibility.

In Section 10.3 we simplify the method by drastically reducing the working tableau. In particular, we show that some rows and columns can be removed.

In Section 10.4 we analyze its complexity and make a comparison with the simplex method.

In Section 10.5 we show how the final tableau can be constructed based on the initial tableau and the list of basic variables.

Finally, in Section 10.6 we show how the solution of a linear programming problem (LPP) resulting from adding or removing constraints and/or variables can be obtained without the need of redoing all the computations.

10.2 THE EXTERIOR POINT METHOD

To understand how the EPM works and its main differences from the well-known simplex method, we start this section with a simple example. Consider the following LPP:

Maximize

$$x_f = 2x_1 + x_2 \tag{10.1}$$

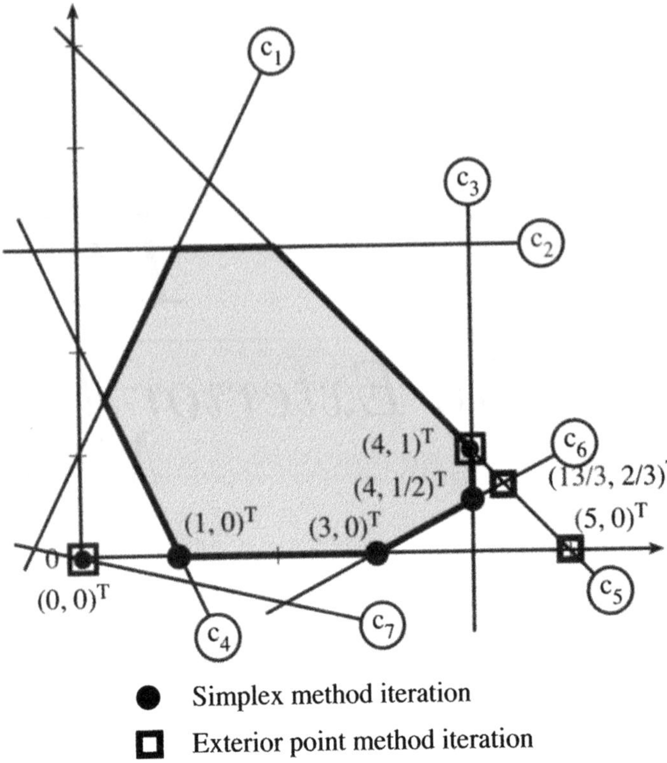

Fig. 10.1 Illustration of the differences between the EPM and the simplex method.

subject to

$$
\begin{array}{rrcr}
-2x_1 & +x_2 & \leq & 1 \\
 & +x_2 & \leq & 3 \\
x_1 & & \leq & 4 \\
-2x_1 & -x_2 & \leq & -2 \\
x_1 & +x_2 & \leq & 5 \\
x_1 & -2x_2 & \leq & 3 \\
-x_1 & -4x_2 & \leq & 0 \\
 & x_1, x_2 & \geq & 0.
\end{array}
\tag{10.2}
$$

The set of points satisfying the above constraints defines the feasible region. The remaining points form the infeasible region. See Fig. 10.1.

In this case, the simplex method begins at the origin and moves, in the following iterations, to vertices $(1,0)^T, (3,0)^T$, and $(4,1/2)^T$ before reaching the optimum at vertex $(4,1)^T$. Observe that all points, but the initial (the origin), belong to the feasible region of the problem.

On the contrary, the EPM works outside the feasible region. It also starts at the origin but moves to the points $(5,0)^T$ and $(13/3, 2/3)^T$ before reaching the optimum at the vertex $(4,1)^T$.

This geometric interpretation of both methods helps in understanding the EPM, whose theoretical foundations are given below.

The EPM will be applied to a LPP in the following form:

Maximize

$$f(\mathbf{x}) = c_1 x_1 + \cdots + c_n x_n \tag{10.3}$$

subject to

$$a_{i1}x_1 + \cdots + a_{in}x_n \le b_i; \; i = 1, \ldots, m \tag{10.4}$$

and

$$x_1, \ldots, x_n \ge 0. \tag{10.5}$$

Constraints have to appear in the convenient form of inequalities. This is not a problem, since m equalities can always be written as $m+1$ inequalities. More precisely, the set of equalities

$$a_{i1}x_1 + \cdots + a_{in}x_n = b_i; \; i = 1, \ldots, m, \tag{10.6}$$

can be replaced by the set of inequalities

$$\begin{array}{lllll} a_{i1}x_1 & + \cdots & + a_{in}x_n & \le & b_i; \; i = 1, \ldots, m, \\ \left(\sum\limits_{i=1}^{m} a_{i1}\right) x_1 & + \cdots & + \left(\sum\limits_{i=1}^{m} a_{in}\right) x_n & \ge & \sum\limits_{i=1}^{m} b_i, \end{array} \tag{10.7}$$

as the following proposition shows.

Proposition 10.1 *Let* $\mathbf{a}_1, \ldots, \mathbf{a}_m$ *be vectors in* $\mathbb{R}^n$ *and* $b_1, \ldots, b_m \in \mathbb{R}$. *Then*

$$\mathbf{a}_i \bullet \mathbf{x} = b_i, \; i = 1, \ldots, m \Leftrightarrow \left\{ \begin{array}{l} \mathbf{a}_i \bullet \mathbf{x} \le b_i, \; i = 1, \ldots, m, \\ \left(\sum\limits_{i=1}^{m} \mathbf{a}_i\right) \bullet \mathbf{x} \ge \sum\limits_{i=1}^{m} b_i. \end{array} \right. \tag{10.8}$$

Proof: [$\Rightarrow$] This implication is obvious.
[$\Leftarrow$] Conversely, consider $k \in \{1, \ldots, m\}$. Then

$$\mathbf{a}_k \bullet \mathbf{x} = \left(\sum_{i=1}^{m} \mathbf{a}_i\right) \bullet \mathbf{x} - \left(\sum_{i \ne k} \mathbf{a}_i\right) \bullet \mathbf{x} \ge \sum_{i=1}^{m} b_i - \sum_{i \ne k} b_i = b_k,$$

and, taking into account that $\mathbf{a}_k \bullet \mathbf{x} \le b_k$, then

$$\mathbf{a}_k \bullet \mathbf{x} = b_k.$$

■

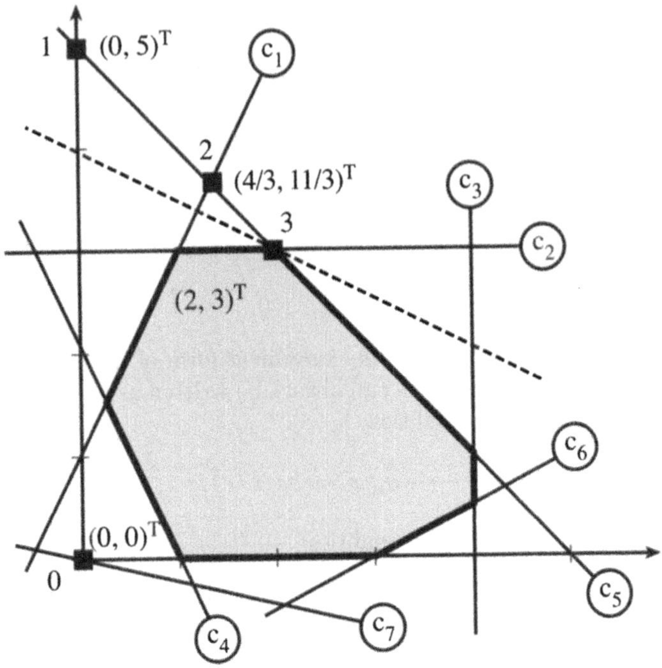

Fig. 10.2 Illustration of the different steps followed by the EPM in Example 10.2.

and we perform the pivoting, obtaining Table 10.10.

Step 6: Since the solution condition does not hold and

$$\min_{u_{i0}^{(t)}<0} u_{i0}^{(t)} = \min(-4, -2) = -4,$$

the variable entering J_t is x_3.

Step 7: Since $\max_{u_{ij}^{(t)}>0}\left(u_{fj}/u_{ij}^{(t)}\right) = \max\left(\frac{-1}{3}\right) = -1/3$, the pivot is 3, which corresponds to the column of x_1 and the row of x_3. Next, we return to Step 5.

Step 5: We update I_t and J_t and get

$$I_2 \equiv \{1, 4, 5, 6, 2, 8, 9\}; J_2 \equiv \{3, 7\},$$

and we perform the pivoting, obtaining Table 10.11.

Step 6: Since the solution condition does not hold and

$$\min_{u_{i0}^{(t)}<0} u_{i0}^{(t)} = \min(-2/3) = -2/3,$$

In other words, J_t is the set of indices of the x-variables such that the associated coefficient row matrix in the t-state is $\mathbf{e}_j$ for some j, and I_t is the set of indices of the x-variables such that they do not belong to the nonbasic set. Then J_t is the set of *nonbasic* variables, and I_t is the set of *basic variables.*

Obviously, in the initial state ($t = 0$), $J_0 \equiv \mathcal{P}$. The essence of the exterior point method consists of adequately changing the sets of basic and nonbasic variables in the successive steps of the method. In each iteration, a basic variable and a nonbasic variable interchange their roles.

The linear programming problem in the t step can be stated as follows: Maximize

$$x_f$$

subject to

$$\mathbf{x} = \begin{pmatrix} x_f \\ \mathbf{x}_{\mathcal{P}} \\ \mathbf{x}_{\mathcal{S}} \end{pmatrix} = \begin{pmatrix} x_f \\ \mathbf{x}_{\mathcal{R}} \end{pmatrix} = \begin{pmatrix} \mathbf{u}_f^{(t)} \\ \mathbf{U}_{\mathcal{P}}^{(t)} \\ \mathbf{U}_{\mathcal{S}}^{(t)} \end{pmatrix} \mathbf{x}_{J_t} = \begin{pmatrix} \mathbf{u}_f^{(t)} \\ \mathbf{U}_{\mathcal{R}}^{(t)} \end{pmatrix} \mathbf{x}_{J_t} = \mathbf{U}^{(t)}\, \mathbf{x}_{J_t} \tag{10.16}$$

with

$$x_0 = 1, \mathbf{x}_{\mathcal{R}} \geq \mathbf{0}. \tag{10.17}$$

In the following, we shall call $\mathbf{u}^{(t)}{}_j$, $\mathbf{u}_{\mathcal{R}}^{(t)}{}_j$, $\mathbf{u}_{\mathcal{P}}^{(t)}{}_j$, and $\mathbf{u}_{\mathcal{S}}^{(t)}{}_j$ the j-columns of matrices $\mathbf{U}^{(t)}$, $\mathbf{U}_{\mathcal{R}}^{(t)}$, $\mathbf{U}_{\mathcal{P}}^{(t)}$, and $\mathbf{U}_{\mathcal{S}}^{(t)}$, respectively.

The linear programming problem can also be stated as:

$$\textit{Maximize } x_f \textit{ subject to } \mathbf{x} \in X^{(t)}, \mathbf{x}_{I_t} \geq \mathbf{0}, \tag{10.18}$$

or

$$\textit{Maximize } x_f \textit{ subject to } \mathbf{x} \in X, \tag{10.19}$$

where

$$X^{(t)} \equiv \{\mathbf{x} \in \mathbf{U}^{(t)}\, \mathbf{x}_{J_t} \mid x_0 = 1 \wedge \mathbf{x}_{J_t} \geq \mathbf{0}\} \tag{10.20}$$

and

$$X \equiv \{\mathbf{x} \in X^{(t)} | \mathbf{x}_{I_t} \geq \mathbf{0}\}. \tag{10.21}$$

The LPP of (10.1) and (10.2) stated in the form (10.16) becomes

$$\begin{pmatrix} x_f \\ x_0 \\ x_1 \\ x_2 \\ x_3 \\ x_4 \\ x_5 \\ x_6 \\ x_7 \\ x_8 \\ x_9 \end{pmatrix} = \begin{pmatrix} 0 & 2 & 1 \\ 1 & 0 & 0 \\ 0 & 1 & 0 \\ 0 & 0 & 1 \\ 1 & 2 & -1 \\ 3 & 0 & -1 \\ 4 & -1 & 0 \\ -2 & 2 & 1 \\ 5 & -1 & -1 \\ 3 & -1 & 2 \\ 0 & 1 & 4 \end{pmatrix} \begin{pmatrix} x_0 \\ x_1 \\ x_2 \end{pmatrix} \tag{10.22}$$

Table 10.1 Initial tableau.

	$\mathbf{U}^{(0)}$		
	x_0	x_1	x_2
x_f	0	2	1
x_3	1	2	–1
x_4	3	0	–1
x_5	4	–1	0
x_6	–2	2	1
x_7	5	–1	–1
x_8	3	–1	2
x_9	0	1	4

This information can be organized in tabular form, and the rows associated with the main variables $\{x_0, x_1, x_2\}$ can be eliminated, because they always lead to a unit matrix. This leads to the tableau in Table 10.1. Each table of this form has an associated point that can be calculated using the following rules:

1. The basic variables, which appear in the first column, take the values in the second column.

2. The nonbasic variables, which appear in the first row, take null values.

For example, Table 10.1 has the associated point

$$(x_1, x_2, x_3, x_4, x_5, x_6, x_7, x_8, x_9)^T = (0, 0, 1, 3, 4, -2, 5, 3, 0)^T,$$

which corresponds to the point $(x_1, x_2)^T = (0, 0)^T$ in Fig. 10.1.

The most important transformation used in the EPM is the elemental pivoting operation, which is described next.

10.2.1 Elemental Pivoting Operation

The exterior point method reaches the solution of the LPP by means of elemental transformations, which act initially on $\mathbf{U}^{(0)}$, to obtain $\mathbf{U}^{(1)}$ and sequentially $\mathbf{U}^{(2)}, \ldots, \mathbf{U}^{(t)}$, until the solution is obtained or no solution can be found. The theoretical basis of this transformation is given in the following theorem.

Theorem 10.1 *Let $u_{\alpha\beta}^{(t)}$, with $\alpha \in I_t$, $\beta \in J_t - \{0\}$, be a non-null element of $\mathbf{U}^{(t)}$, which is called a pivot. If we transform the system of equations $\mathbf{x} = \mathbf{U}^{(t)}\mathbf{x}_{J_t}$ into the system $\mathbf{x} = \mathbf{U}^{(t+1)}\mathbf{x}_{J_{t+1}}$, by the following column trans-*

formation

$$u_{ij}^{(t+1)} = \left\{ \begin{array}{ll} \dfrac{u_{ij}^{(t)}}{u_{\alpha\beta}^{(t)}}, & if \quad j = \beta \\ u_{ij}^{(t)} - \dfrac{u_{\alpha j}^{(t)}}{u_{\alpha\beta}^{(t)}} u_{i\beta}^{(t)}, & if \quad j \neq \beta \end{array} \right\} \forall i \in \mathcal{R}, \tag{10.23}$$

and

$$\begin{array}{l} J_{t+1} \equiv J_t - \{\beta\} \cup \{\alpha\}, \\ I_{t+1} \equiv I_t - \{\alpha\} \cup \{\beta\}, \end{array} \tag{10.24}$$

then both systems of equations are equivalent.

Proof: If we obtain x_β from the equation associated with row α of $\mathbf{U}^{(t)}$,

$$x_\alpha = \mathbf{u}_\alpha^{(t)} \, \mathbf{x}_{J_t} = \sum_{j \in J_t} u_{\alpha j}^{(t)} x_j,$$

then we find

$$x_\beta = \sum_{j \neq \beta} \left(\frac{-u_{\alpha j}^{(t)}}{u_{\alpha\beta}^{(t)}} x_j \right) + \frac{1}{u_{\alpha\beta}^{(t)}} x_\alpha = \mathbf{u}_\beta^{(t+1)} \, \mathbf{x}_{J_{t+1}}. \tag{10.25}$$

Replacing x_β in the identity $x_\beta = x_\beta$, we reproduce the initial constraint, which, because of its redundancy, can be eliminated and replaced by the identity $x_\alpha = x_\alpha$.

Finally, replacing x_β in the remaining equations associated with I_t, we get

$$x_i = \sum_{j \neq \beta} \left(u_{ij}^{(t)} - \frac{u_{\alpha j}^{(t)}}{u_{\alpha\beta}^{(t)}} u_{i\beta}^{(t)} \right) x_j + \frac{u_{i\beta}^{(t)}}{u_{\alpha\beta}^{(t)}} x_\alpha; \quad i \in I_t - \{\alpha\}, \tag{10.26}$$

and then

$$x_i = \sum_{j \neq \beta} u_{ij}^{(t+1)} \, x_j + u_{i\alpha}^{(t+1)} \, x_\alpha = \mathbf{u}_i^{(t+1)} \, \mathbf{x}_{J_{t+1}} \Rightarrow \mathbf{x} = \mathbf{U}^{(t+1)} \, \mathbf{x}_{J_{t+1}}. \tag{10.27}$$

■

Remark 10.1 *In the transformation above the pivot row α transforms to a β unit row $\mathbf{e}_\beta$.*

The exterior point method consists of two stages:

1. **First Stage: Regulating iteration.** We use a regulating iteration to get $u_{fj}^{(1)} \leq 0; \forall j \neq 0$; that is, we look for a cost function with all nonpositive cost coefficients.

2. **Second Stage: Standard iterations.** In this stage, we perform elemental transformations such that we have

$$u_{fj}^{(t)} \leq 0; \forall j \neq 0, j \in J_t \text{ and } t \geq 1, \tag{10.28}$$

until we have

$$u_{i0}^{(t)} \geq 0; \forall i \in I_t, \tag{10.29}$$

which is the bounded solution, or we detect unboundedness or no solution.

10.2.2 First Stage: Regulating Iteration

The regulating iteration is one in which we transform the cost function to a function with all nonpositive coefficients; that is, we transform $\mathbf{U}^{(0)}$ in $\mathbf{U}^{(1)}$, such that

$$u_{fj}^{(1)} \leq 0; \forall j \neq 0. \tag{10.30}$$

We can have one of two situations:

1. There exists a constraint $x_i = \mathbf{u}_i^{(0)} \mathbf{x}_{J_0}$ such that

$$\left.\begin{array}{l} u_{ij}^{(0)} < 0 \text{ if } u_{fj}^{(0)} \geq 0 \\ u_{ij}^{(0)} \leq 0 \text{ if } u_{fj}^{(0)} < 0 \end{array}\right\} \forall j \neq 0. \tag{10.31}$$

2. There is no such constraint. In this case we introduce the additional constraint

$$x_1 + \cdots + x_n \leq M \Rightarrow x_{n+m+1} = Mx_0 - x_1 - \cdots - x_n = \mathbf{u}_{n+m+1}^{(0)} \mathbf{x}_{J_0}, \tag{10.32}$$

where x_{n+m+1} is the new slack variable, M is a number much larger than any other in $\mathbf{U}^{(0)}$, which obviously satisfies (10.31), and we let $i = n + m + 3$.

In either case, we satisfy (10.30) by using the transformation (10.23) with the pivot $u_{i\beta}^{(0)}$ given by

$$\frac{u_{f\beta}^{(0)}}{u_{i\beta}^{(0)}} = \min_{u_{fj}^{(0)} > 0} \frac{u_{fj}^{(0)}}{u_{ij}^{(0)}}. \tag{10.33}$$

Remark 10.2 *With the convenient selection of M, we have that*

$$LPP \textit{ has bounded solution } \Rightarrow x_{n+m+1} > 0.$$

Our illustrative example requires the regulating iteration. The constraint associated with variable x_7 satisfies the regulating condition (10.31). To use the pivoting operation (10.23), the pivot element must be selected in the row

Table 10.2 Tableau at the end of the regulating iteration. Pivot element is boldfaced.

$\mathbf{U}^{(1)}$			
1	x_0	x_7	x_2
x_f	10	–2	–1
x_3	11	–2	–3
x_4	3	0	–1
x_5	–1	1	1
x_6	8	–2	–1
x_1	5	–1	–1
x_8	–2	1	**3**
x_9	5	–1	3

of variable $\mathbf{x}_7$. Since $\min\left(\frac{2}{-1}, \frac{1}{-1}\right) = -2$, the pivot is -1, which corresponds to the column of x_1 and the row of x_7. Performing the pivoting, Table 10.2 is obtained, whose associated point is

$$(x_1, x_2, x_3, x_4, x_5, x_6, x_7, x_8, x_9)^T = (5, 0, 11, 3, -1, 8, 0, -2, 5)^T,$$

which corresponds to the point $(x_1, x_2)^T = (5, 0)^T$ in Fig. 10.1.

10.2.3 Second Stage: Standard Iterations

In this stage, transformation (10.23) is used in obtaining $\mathbf{U}^{(2)}, \ldots, \mathbf{U}^{(t)}$, such that:

1. We obtain an equivalent system of constraints:

$$\mathbf{x} = \mathbf{U}^{(0)}\, \mathbf{x}_{J_0} \Leftrightarrow \mathbf{x} = \mathbf{U}^{(t)}\, \mathbf{x}_{J_t}, \forall t \geq 1. \tag{10.34}$$

2. The condition

$$u_{fj}^{(t)} \leq 0, \forall j \neq 0 \tag{10.35}$$

still holds.

3. The value of the objective function does not increase:

$$u_{f0}^{(2)} \geq \cdots \geq u_{f0}^{(t)}. \tag{10.36}$$

4. The pivot elements are selected to maximize the decrease of $x_f^{(t)} = u_{f0}^{(t)}$.

Thus, in each iteration the LPP is equivalent to maximizing x_f subject to

$$\mathbf{x} = \mathbf{U}^{(t)}\, \mathbf{x}_{J_t};\; x_0 = 1, \mathbf{x}_{J_t} \geq \mathbf{0}, \mathbf{x}_{I_t} \geq \mathbf{0}, \tag{10.37}$$

or to the equivalent condition

$$\mathbf{x} \in X^{(t)}, \mathbf{x}_{I_t} \geq \mathbf{0}. \tag{10.38}$$

To perform the transformation (10.23), we choose a pivot element $u_{\alpha\beta}$ such that $\alpha \in I_t$ and $\beta \in J_t$. In other words, we replace one variable from J_t by another variable from I_t. To this end, we must:

1. Determine which variable must leave the set J_t and enter the set I_t.
2. Determine which variable must enter the set J_t and leave the set I_t.

10.2.4 Selecting the Variable Leaving $\mathbf{J_t}$

Given a row i, the pivot $u_{i\beta}^{(t)}$ of the transformation (10.23) is selected for $\mathbf{U}^{(t)}$ to satisfy (10.35). Thus, the pivot column β and the pivot value $u_{i\beta}^{(t)}$ are obtained by means of

$$\Lambda_i^{(t)} = \frac{u_{f\beta}^{(t)}}{u_{i\beta}^{(t)}} = \max_{u_{ij}^{(t)}>0} \frac{u_{fj}^{(t)}}{u_{ij}^{(t)}}, \tag{10.39}$$

where $\Lambda_i^{(t)}$ is the *function-pivot ratio*. Thus, a variable associated with column β leaves the set J_t.

10.2.5 Solution, Feasibility, and Unboundedness

Before selecting the variable to enter J_t, we need to discuss some important points. During the iteration process we can achieve one of the following states:

1. **Solution state:** This occurs when we have

$$u_{fj}^{(t)} \leq 0, \forall j \neq 0 \text{ and } j \in J_t, u_{i0}^{(t)} \geq 0, \forall i \in I_t. \tag{10.40}$$

In this case, the solution of the LPP is

$$\mathbf{x}^* = \mathbf{u}_0^{(t)}; \tag{10.41}$$

that is, the solution appears in the first column of the matrix $\mathbf{U}^{(t)}$ for the variables in the set I_t, and the variables in the set J_t take null values. This is proved below.

Proof: In fact, we have

$$x_f = \mathbf{u}_f^{(t)} \mathbf{x}_{J_t} = u_{f0}^{(t)} + \sum_{j\neq 0} u_{fj}^{(t)} x_j. \tag{10.42}$$

Since $\mathbf{x} \in X^{(t)}$, (10.20) implies $x_j \geq 0, \forall j \neq 0$ and $j \in J_t$; the maximum of the sum on the right-hand side of (10.42) is zero, because the coefficient $u_{fj} \leq 0$, and thus,

$$\max_{X^{(t)}} x_f = u_{f0}^{(t)}.$$

In addition, $X \subseteq X^{(t)}$, which implies

$$\max_{X} x_f \leq u_{f0}^{(t)},$$

but $\mathbf{u}_0^{(t)} \in X$, which proves the result. ■

2. **Unbounded solution state:** This occurs when we have $x_{n+m+1}^* = 0$ and (10.40).

   Proof: It is obvious, since $\mathbf{x}^* = \mathbf{u}_0^{(t)}$ is the solution of the LPP and because of Remark 10.2, that it is unbounded. ■

3. **Infeasible solution state:** This occurs when, for any t, there exists a row $i \in I_t$ such that

$$u_{i0}^{(t)} < 0, u_{ij}^{(t)} \leq 0, \forall j. \tag{10.43}$$

   Proof: We have

$$\max_{X^{(t)}} x_i = \max_{X^{(t)}} \sum u_{ij}^{(t)} x_j < 0 \Rightarrow x_i < 0 \Rightarrow X \equiv \emptyset. \tag{10.44}$$

■

10.2.6 Selecting the Variable Entering $\mathbf{J_t}$

According to the previous results, if when we are in state t, the solution has not been attained and the infeasibility and unboundedness conditions have not appeared yet, a certain i and j must exist such that $u_{i0}^{(t)} < 0$ and $u_{ij}^{(t)} > 0$.

Operating with the corresponding function-pivot ratio $\Lambda_i^{(t)}$, given in (10.39), we get

$$u_{f0}^{(t+1)} = u_{f0}^{(t)} - \Delta_i^{(t)}, \tag{10.45}$$

where

$$\Delta_i^{(t)} = \Lambda_i^{(t)} u_{i0}^{(t)}. \tag{10.46}$$

We give two options for selecting the pivot row:

1. **Exhaustive criterion:** The largest variation of $x_f^{(t)} = u_{f0}^{(t)}$ can be obtained by selecting the pivot $u_{\alpha\beta}^{(t)}$, by means of

$$\Delta_\alpha^{(t)} = \max_{u_{i0}^{(t)} < 0} \Delta_i^{(t)}. \tag{10.47}$$

Table 10.3 Tableau after the first standard iteration. Pivot element is boldfaced and the common denominator of all elements is given in the upper left corner of the table.

$\mathbf{U}^{(2)}$			
3	x_0	x_7	x_8
x_f	28	–5	–1
x_3	27	–3	–3
x_4	7	1	–1
x_5	–1	2	**1**
x_6	22	–5	–1
x_1	13	–2	–1
x_2	2	–1	1
x_9	21	–6	3

This is equivalent to obtaining the maximum local variation in the simplex method.

2. **Reduced criterion:** Since $\Delta_i^{(t)}$ is proportional to $u_{i0}^{(t)}$ (see (10.46)), the following alternative and simpler rule for obtaining the pivot can be used

$$u_{\alpha 0}^{(t)} = \min_{i \in I_t} u_{i0}^{(t)} . \tag{10.48}$$

This is similar to the simplex method's reduced cost criteria of obtaining $\max(z_j - c_j)$.

Like in the simplex method, the most efficient criterion corresponds to (10.48). In this way, according to the previous assumptions, we get

$$u_{f0}^{(0)} \geq \cdots \geq u_{f0}^{(t)} \Rightarrow f(\mathbf{x}^{(0)}) \geq \cdots \geq f(\mathbf{x}^{(t)}),$$

a relation that, interpreted in a geometric sense, gives a sequence of points $\mathbf{x}^{(0)}, \ldots, \mathbf{x}^{(t)}$, which approach the hyperplanes $f(\mathbf{x}) = f(\mathbf{x}^{(t)})$ to the contact point $\mathbf{x}^* \in X$. This justifies the use of the *exterior point* name, which has been selected for the proposed method.

In our illustrative example we need to perform two standard iterations to reach the optimum. In the first iteration we use the element in the x_8 row and x_2 column as pivot and, after pivoting, we get Table 10.3, whose associated point is

$$(x_1, x_2, x_3, x_4, x_5, x_6, x_7, x_8, x_9)^T = (13/3, 2/3, 9, 7/3, -1/3, 22/3, 0, 0, 7)^T,$$

which corresponds to the point $(x_1, x_2)^T = (13/3, 2/3)^T$ in Fig. 10.1.

In the second iteration we use the element in the x_5 row and x_8 column as pivot and, after pivoting, we get Table 10.4, whose associated point is

$$(x_1, x_2, x_3, x_4, x_5, x_6, x_7, x_8, x_9)^T = (4, 1, 8, 2, 0, 7, 0, 1, 8)^T,$$

Table 10.4 Tableau after the second standard iteration. Pivot element is boldfaced.

$\mathbf{U}^{(3)}$			
	x_0	x_7	x_5
x_f	9	-1	-1
x_3	8	1	-3
x_4	2	1	-1
x_8	1	-2	3
x_6	7	-1	-1
x_1	4	0	-1
x_2	1	-1	1
x_9	8	-4	3

which corresponds to the point $(x_1, x_2)^T = (4, 1)^T$ in Fig. 10.1. Since all values in the second column are non-negative, this is the solution of the problem.

Next, we give the EPM algorithm, which works only with the $\mathbf{U}_S^{(t)}$ matrix, because $\mathbf{U}_\mathcal{P}$ is always the identity matrix.

Algorithm 10.1 The EPM algorithm.

- **Input:** The starting tableau $\mathbf{U}_S^{(0)}$ of the LPP, including the objective function, and the m problem constraints.
- **Output:** The solution of the maximization problem or a message indicating infeasibility or unbounded solution.

Step 1: *Initialization.* Let $columns = n + 1$, $rows = m + 1$, $J_t \equiv \{0, \ldots, n\}$, $I_t \equiv \{n + 1, \ldots, n + m\}$, and $t = 0$.

Step 2: *Testing the objective function.* If $u_{fj}^{(0)} > 0$ for some j, go to Step 3. Otherwise, go to Step 6.

Regulating Iteration

Step 3: *Transform cost function to negative coefficients.* Check if there exists a constraint i satisfying condition (10.31). If the answer is positive, let $\alpha = i$ and continue with Step 4. Otherwise, introduce the artificial constraint (10.32) and let $rows = rows + 1$, $m = m + 1$, and $\alpha = m + 1$.

Step 4: *Find pivot.* Find the pivot column β of row α using (10.33) and go to Step 5.

Standard Iterations

Step 5: *Pivoting.* Let $t = t + 1$, update I_t and J_t using (10.24), and perform the pivoting transformation (10.23).

Step 6: *Select the entering variable.* The variable x_α entering J_t is selected using

$$u_{\alpha 0}^{(t)} = \min_{u_{i0}^{(t)}<0} u_{i0}^{(t)}. \tag{10.49}$$

If the solution condition

$$u_{fj}^{(t)} \leq 0, \forall j \neq 0 \text{ and } j \in J_t, u_{i0}^{(t)} \geq 0, \forall i \in I_t \tag{10.50}$$

and the unbounded condition $x_{n+m+1}^* = 0$ hold, print the unbounded condition message; otherwise, print the solution

$$\mathbf{x}^* = \mathbf{u}_0^{(t)} \tag{10.51}$$

and quit. If none of them hold, continue with Step 7.

Step 7: *Select the leaving variable.* The variable x_β leaving J_t is selected using

$$\Lambda_i^{(t)} = \frac{u_{f\beta}^{(t)}}{u_{i\beta}^{(t)}} = \max_{u_{ij}^{(t)}>0} \frac{u_{fj}^{(t)}}{u_{ij}^{(t)}}. \tag{10.52}$$

If the infeasible solution conditions

$$u_{i0}^{(t)} < 0, u_{ij}^{(t)} \leq 0, \forall j \tag{10.53}$$

are found, send the corresponding message and quit. Otherwise, return to Step 5. ■

Example 10.1 (Linear programming problem). Consider again the illustrative example of the introduction:

Maximize

$$x_f = 2x_1 + x_2 \tag{10.54}$$

subject to

$$\begin{array}{rrcr} -2x_1 & +x_2 & \leq & 1 \\ & +x_2 & \leq & 3 \\ x_1 & & \leq & 4 \\ -2x_1 & -x_2 & \leq & -2 \\ x_1 & +x_2 & \leq & 5 \\ x_1 & -2x_2 & \leq & 3 \\ -x_1 & -4x_2 & \leq & 0 \end{array} \tag{10.55}$$

with

$$x_1 \geq 0, x_2 \geq 0. \tag{10.56}$$

Table 10.5 Initial tableau.

	$\mathbf{U}^{(0)}$		
	x_0	x_1	x_2
x_f	0	2	1
x_3	1	2	–1
x_4	3	0	–1
x_5	4	–1	0
x_6	–2	2	1
x_7	5	**–1**	–1
x_8	3	–1	2
x_9	0	1	4

Using the EPM algorithm the following results are obtained:

Step 1: Table 10.5 shows the initial tableau.

Step 2: Since some of the elements in the first row are positive, we go to Step 3.

Step 3: Since constraint 5, associated with variable x_7, satisfies the regulating condition, we let $\alpha = 5$ and continue with Step 4.

Step 4: Since $\min\left(\frac{2}{-1}, \frac{1}{-1}\right) = -2$, the pivot is -1, which corresponds to the column of x_1 and the row of x_7.

Step 5: We update I_t and J_t and get

$$I_1 \equiv \{3,4,5,6,1,8,9\}; J_1 \equiv \{7,2\},$$

and we perform the pivoting, obtaining Table 10.6.

Step 6: Since the solution condition does not hold and

$$\min_{u_{i0}^{(t)}<0} u_{i0}^{(t)} = \min(-1,-2) = -2,$$

the variable entering J_t is x_8.

Step 7: Since $\max_{u_{ij}^{(t)}>0}\left(u_{fj}/u_{ij}^{(t)}\right) = \max\left(\frac{-2}{1}, \frac{-1}{3}\right) = -1/3$, the pivot is 3, which corresponds to the column of x_2 and the row of x_8. Next, we return to Step 5.

Step 5: We update I_t and J_t and get

$$I_2 \equiv \{3,4,5;6,1,2,9\}; J_2 \equiv \{7,8\},$$

and we perform the pivoting, obtaining Table 10.7.

Step 6: Since the solution condition does not hold and

$$\min_{u_{i0}^{(t)}<0} u_{i0}^{(t)} = \min(-1/3) = -1/3,$$

the variable entering J_t is x_5.

Table 10.6 Tableau at the end of Step 7 for iteration 1. Pivot element is boldfaced.

$\mathbf{U}^{(1)}$			
1	x_0	x_7	x_2
x_f	10	-2	-1
x_3	11	-2	-3
x_4	3	0	-1
x_5	-1	1	1
x_6	8	-2	-1
x_1	5	-1	-1
x_8	-2	1	**3**
x_9	5	-1	3

Table 10.7 Tableau at the end of Step 7 for iteration 2. Pivot element is boldfaced and the common denominator of all elements is given in the upper left corner of the table.

$\mathbf{U}^{(2)}$			
3	x_0	x_7	x_8
x_f	28	-5	-1
x_3	27	-3	-3
x_4	7	1	-1
x_5	-1	2	**1**
x_6	22	-5	-1
x_1	13	-2	-1
x_2	2	-1	1
x_9	21	-6	3

Step 7: Since $\max_{u_{ij}^{(t)}>0}\left(u_{fj}/u_{ij}^{(t)}\right) = \max\left(\frac{-5/3}{2/3}, \frac{-1/3}{1/3}\right) = -1$, the pivot is 1/3, which corresponds to the column of x_8 and the row of x_5. Next, we return to Step 5.

Step 5: We update I_t and J_t and get

$$I_3 \equiv \{3, 4, 8, 6, 1, 2, 9\}; J_3 \equiv \{7, 5\},$$

and we perform the pivoting, obtaining Table 10.8.

Step 6: Since all values in the x_0 column are positive, the solution is found. The value of the objective function is 9, which is attained at

$$x_1 = 4, x_2 = 1.$$

Table 10.8 Tableau at the end of Step 7 for iteration 3. Pivot element is boldfaced.

	$\mathbf{U}^{(3)}$		
	x_0	x_7	x_5
x_f	9	-1	-1
x_3	8	1	-3
x_4	2	1	-1
x_8	1	-2	3
x_6	7	-1	-1
x_1	4	0	-1
x_2	1	-1	1
x_9	8	-4	3

Table 10.9 Initial tableau.

	$\mathbf{U}^{(0)}$		
	x_0	x_1	x_2
x_f	0	1	2
x_3	1	2	-1
x_4	3	0	1
x_5	4	-1	0
x_6	-2	2	1
x_7	5	-1	**-1**
x_8	3	-1	2
x_9	0	1	4

In Fig. 10.1 it can be seen that the algorithm obtains a sequence of infeasible points until reaching the optimum. ∎

Example 10.2 (Linear programming problem revisited). Suppose that in Example 10.1 the objective function is replaced by $x_f = x_1 + 2x_2$.

The iterations required to reach the optimum solution are also given in Fig. 10.2. Using the EPM algorithm the following results are obtained:

Step 1: Table 10.9 shows the initial tableau.

Step 2: Since some of the elements in the first row are positive, we go to Step 3.

Step 3: Since constraint 5, associated with variable x_7, satisfies the regulating condition, we let $\alpha = 5$ and continue with Step 4.

Step 4: Since $\min\left(\frac{1}{-1}, \frac{2}{-1}\right) = -2$, the pivot is -1, which corresponds to the column of x_2 and the row of x_7.

Step 5: We update I_t and J_t and get

$$I_1 \equiv \{3,4,5,6,2,8,9\}; J_1 \equiv \{1,7\},$$

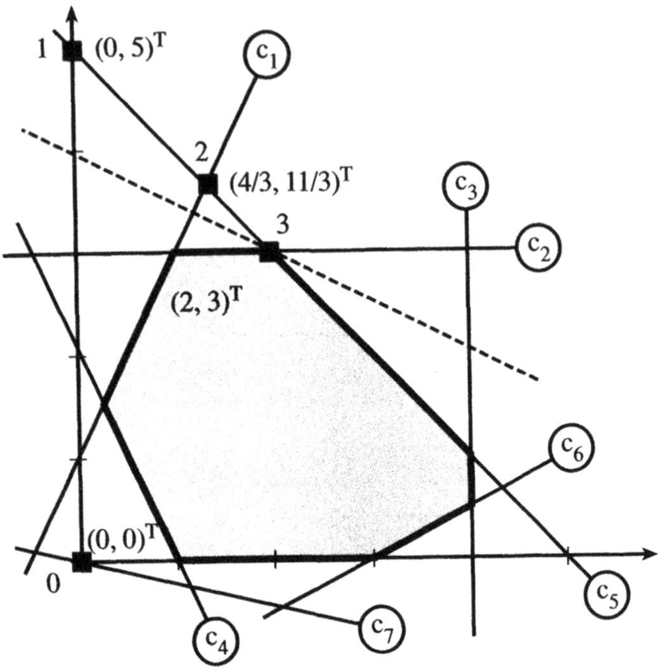

Fig. 10.2 Illustration of the different steps followed by the EPM in Example 10.2.

and we perform the pivoting, obtaining Table 10.10.

Step 6: Since the solution condition does not hold and

$$\min_{u_{i0}^{(t)}<0} u_{i0}^{(t)} = \min(-4, -2) = -4,$$

the variable entering J_t is x_3.

Step 7: Since $\max_{u_{ij}^{(t)}>0} \left(u_{fj}/u_{ij}^{(t)}\right) = \max\left(\frac{-1}{3}\right) = -1/3$, the pivot is 3, which corresponds to the column of x_1 and the row of x_3. Next, we return to Step 5.

Step 5: We update I_t and J_t and get

$$I_2 \equiv \{1, 4, 5, 6, 2, 8, 9\}; J_2 \equiv \{3, 7\},$$

and we perform the pivoting, obtaining Table 10.11.

Step 6: Since the solution condition does not hold and

$$\min_{u_{i0}^{(t)}<0} u_{i0}^{(t)} = \min(-2/3) = -2/3,$$

Table 10.10 Tableau at the end of Step 7 for iteration 1. Pivot element is boldfaced.

$\mathbf{U}^{(1)}$			
1	x_0	x_1	x_7
x_f	10	–1	-2
x_3	–4	**3**	1
x_4	–2	1	1
x_5	4	–1	0
x_6	3	1	–1
x_2	5	–1	–1
x_8	13	–3	–2
x_9	20	–3	4

Table 10.11 Tableau at the end of Step 7 for iteration 2. Pivot element is boldfaced and the common denominator of all elements is given in the upper left corner of the table.

$\mathbf{U}^{(2)}$			
3	x_0	x_3	x_7
x_f	26	–1	–5
x_1	4	1	-1
x_4	–2	**1**	2
x_5	8	–1	1
x_6	13	1	–4
x_2	11	–1	-2
x_8	27	–3	–3
x_9	48	–3	–9

the variable entering J_t is x_4.

Step 7: Since $\max_{u_{ij}^{(t)}>0}\left(u_{fj}/u_{ij}^{(t)}\right) = \max\left(\frac{-1/3}{1/3}\right) = -1$, the pivot is 1/3, which corresponds to the column of x_3 and the row of x_4. Next, we return to Step 5.

Step 5: We update I_t and J_t and get

$$I_3 \equiv \{1, 3, 5, 6, 2, 8, 9\}; J_3 \equiv \{4, 7\},$$

and we perform the pivoting, obtaining Table 10.12.

Step 6: Since all values in the x_0 column are positive, the solution is found. The value of the objective function is 8, which is attained at $x_1 = 2$ and $x_2 = 3$.

■

Table 10.12 Tableau at the end of Step 7 for iteration 3.

$\mathbf{U}^{(3)}$			
1	x_0	x_4	x_7
x_f	8	-1	-1
x_1	2	1	-1
x_3	2	3	-2
x_5	2	-1	1
x_6	5	1	-2
x_2	3	-1	0
x_8	7	-3	1
x_9	14	-3	-1

Example 10.3 (Linear programming problem). Consider the following LPP:
Maximize

$$x_f = 7x_1 + 5x_2 + 5x_3 - 3x_4 + 0x_5 + x_6 + 2x_7 + 3x_8 \tag{10.57}$$

subject to

$$\begin{array}{rrrrrrrrcr} 5x_1 & +4x_2 & +3x_3 & +2x_4 & +x_5 & +2x_6 & +3x_7 & +4x_8 & \leq & 17 \\ 2x_1 & -x_2 & +4x_3 & -2x_4 & +3x_5 & +3x_6 & +x_7 & -x_8 & \leq & 11 \\ 3x_1 & +2x_2 & +3x_3 & -2x_4 & +x_5 & -x_6 & +2x_7 & -x_8 & \leq & 8 \\ x_1 & -x_2 & +2x_3 & +2x_4 & -x_5 & & +x_7 & +x_8 & \leq & 4 \\ 4x_1 & +3x_2 & +x_3 & -x_4 & -2x_5 & +x_6 & +4x_7 & +3x_8 & \geq & 14 \end{array} \tag{10.58}$$

with

$$x_1 \geq 0, \ldots, x_8 \geq 0. \tag{10.59}$$

Table 10.13 Initial tableau.

	$\mathbf{U}^{(0)}$								
	x_0	x_1	x_2	x_3	x_4	x_5	x_6	x_7	x_8
x_f	0	7	5	5	–3	0	1	2	3
x_9	17	–5	–4	**–3**	–2	–1	–2	–3	–4
x_{10}	11	–2	1	–4	2	-3	–3	–1	1
x_{11}	8	–3	–2	–3	2	–1	1	–2	1
x_{12}	4	–1	1	–2	–2	1	0	–1	–1
x_{13}	–14	4	3	1	–1	–2	1	4	3

The corresponding form (10.16) is

$$\begin{pmatrix} x_f \\ x_0 \\ x_1 \\ x_2 \\ x_3 \\ x_4 \\ x_5 \\ x_6 \\ x_7 \\ x_8 \\ x_9 \\ x_{10} \\ x_{11} \\ x_{12} \\ x_{13} \end{pmatrix} = \begin{pmatrix} 0 & 7 & 5 & 5 & -3 & 0 & 1 & 2 & 3 \\ 1 & 0 & 0 & 0 & 0 & 0 & 0 & 0 & 0 \\ 0 & 1 & 0 & 0 & 0 & 0 & 0 & 0 & 0 \\ 0 & 0 & 1 & 0 & 0 & 0 & 0 & 0 & 0 \\ 0 & 0 & 0 & 1 & 0 & 0 & 0 & 0 & 0 \\ 0 & 0 & 0 & 0 & 1 & 0 & 0 & 0 & 0 \\ 0 & 0 & 0 & 0 & 0 & 1 & 0 & 0 & 0 \\ 0 & 0 & 0 & 0 & 0 & 0 & 1 & 0 & 0 \\ 0 & 0 & 0 & 0 & 0 & 0 & 0 & 1 & 0 \\ 0 & 0 & 0 & 0 & 0 & 0 & 0 & 0 & 1 \\ 17 & -5 & -4 & -3 & -2 & -1 & -2 & -3 & -4 \\ 11 & -2 & 1 & -4 & 2 & -3 & -3 & -1 & 1 \\ 8 & -3 & -2 & -3 & 2 & -1 & 1 & -2 & 1 \\ 4 & -1 & 1 & -2 & -2 & 1 & 0 & -1 & -1 \\ -14 & 4 & 3 & 1 & -1 & -2 & 1 & 4 & 3 \end{pmatrix} \begin{pmatrix} x_0 \\ x_1 \\ x_2 \\ x_3 \\ x_4 \\ x_5 \\ x_6 \\ x_7 \\ x_8 \end{pmatrix} \tag{10.60}$$

with

$$x_1 \geq 0, \ldots, x_{13} \geq 0.$$

We illustrate the EPM by giving in detail all the steps corresponding to Algorithm 10.1

Step 1: Table 10.13 shows the initial tableau.

Step 2: Since some of the elements in the first row are positive, we go to Step 3.

Step 3: Since constraint 1, associated with variable x_9, satisfies the regulating condition, we let $\alpha = 1$ and continue with Step 4.

Step 4: Since $\min\left(\frac{7}{-5}, \frac{5}{-4}, \frac{5}{-3}, \frac{1}{-2}, \frac{2}{-3}, \frac{3}{-4}\right) = -5/3$, the pivot is -3, which corresponds to the column of x_3 and the row of x_9.

Step 5: We update I_t and J_t and get

$$I_1 \equiv \{3, 10, 11, 12, 13\}; J_1 \equiv \{1, 2, 9, 4, 5, 6, 7, 8\},$$

Table 10.14 Tableau at the end of Step 7 for iteration 1. Pivot element is boldfaced.

$\mathbf{U}^{(1)}$									
	x_0	x_1	x_2	x_9	x_4	x_5	x_6	x_7	x_8
x_f	85/3	-4/3	-5/3	-5/3	-19/3	-5/3	-7/3	-3	-11/3
x_3	17/3	-5/3	-4/3	-1/3	-2/3	-1/3	-2/3	-1	-4/3
x_{10}	-35/3	14/3	**19/3**	4/3	14/3	-5/3	-1/3	3	19/3
x_{11}	-9	2	2	1	4	0	3	1	5
x_{12}	-22/3	7/3	11/3	2/3	-2/3	5/3	4/3	1	5/3
x_{13}	-25/3	7/3	5/3	-1/3	-5/3	-7/3	1/3	3	5/3

and we perform the pivoting, obtaining Table 10.14.

Step 6: Since the solution condition does not hold and

$$\min_{u_{i0}^{(t)}<0} u_{i0}^{(t)} = \min\left(-35/3, -9, -22/3, -25/3\right) = -35/3,$$

the variable entering J_t is x_{10}.

Step 7: Since $\max_{u_{ij}^{(t)}>0}\left(u_{fj}/u_{ij}^{(t)}\right) = \max\left(\frac{-4/3}{14/3}, \frac{-5/3}{19/3}\right) = -5/19$, the pivot is 19/3, which corresponds to the column of x_2 and the row of x_{10}. Next, we return to Step 5.

Step 5: We update I_t and J_t and get

$$I_2 \equiv \{3, 2, 11, 12, 13\}; J_2 \equiv \{1, 10, 9, 4, 5, 6, 7, 8\},$$

and we perform the pivoting, obtaining Table 10.15.

Step 6: Since the solution condition does not hold and

$$\min_{u_{i0}^{(t)}<0} u_{i0}^{(t)} = \min\left(-101/19, -11/19, -100/19\right) = -101/19,$$

the variable entering J_t is x_{11}.

Step 7: Since $\max_{u_{ij}^{(t)}>0}\left(u_{fj}/u_{ij}^{(t)}\right) = \max\left(\frac{-2/19}{10/19}\right) = -1/5$, the pivot is 10/19, which corresponds to the column of x_1 and the row of x_{11}. Next, we return to Step 5.

Step 5: We update I_t and J_t and get

$$I_3 \equiv \{3, 2, 1, 12, 13\}; J_3 \equiv \{11, 10, 9, 4, 5, 6, 7, 8\},$$

and we perform the pivoting, obtaining Table 10.16.

Step 6: Since the solution condition does not hold and

$$\min_{u_{i0}^{(t)}<0} u_{i0}^{(t)} = \min\left(-37/10, -28/5, -43/10\right) = -28/5,$$

the variable entering J_t is x_2.

Table 10.15 Tableau at the end of Step 7 for iteration 2. Pivot element is boldfaced and the common denominator of all elements is given in the upper left corner of the table.

$\mathbf{U}^{(2)}$									
19	x_0	x_1	x_{10}	x_9	x_4	x_5	x_6	x_7	x_8
x_f	480	−2	−5	−25	−97	−40	−46	−42	−2
x_3	61	−13	−4	−1	6	−13	−14	−7	0
x_2	35	−14	3	−4	−14	5	1	−9	−1
x_{11}	−101	**10**	6	11	48	10	59	1	3
x_{12}	−11	−7	11	−2	−64	50	29	−14	−2
x_{13}	−100	21	5	−13	−55	−36	8	42	0

Table 10.16 Tableau at the end of Step 7 for iteration 3. Pivot element is boldfaced and the common denominator of all elements is given in the upper left corner of the table.

$\mathbf{U}^{(3)}$									
10	x_0	x_{11}	x_{10}	x_9	x_4	x_5	x_6	x_7	x_8
x_f	242	−2	−2	−12	−46	−20	−18	−22	−14
x_3	−37	−13	2	7	36	0	33	−3	39
x_2	−56	−14	**6**	6	28	1	44	−4	32
x_1	101	19	−6	−11	−48	−1	−59	−1	−57
x_{12}	−43	−7	8	3	−16	30	37	−7	1
x_{13}	59	21	−4	−19	−82	−30	−61	21	−63

Step 7: Since $\max_{u_{ij}^{(t)}>0}\left(u_{fj}/u_{ij}^{(t)}\right) = \max\left(\frac{-1/5}{3/5}\right) = -1/3$, the pivot is 3/5, which corresponds to the column of x_{10} and the row of x_2. Next, we return to Step 5.

Step 5: We update I_t and J_t and get

$$I_4 \equiv \{3, 10, 1, 12, 13\}; J_4 \equiv \{11, 2, 9, 4, 5, 6, 7, 8\},$$

and we perform the pivoting, obtaining Table 10.17.

Step 6: Since the solution condition does not hold and

$$\min_{u_{i0}^{(t)}<0} u_{i0}^{(t)} = \min(-11/6) = -11/6,$$

the variable entering J_t is x_3.

Step 7: Since $\max_{u_{ij}^{(t)}>0}\left(u_{fj}/u_{ij}^{(t)}\right) = \max\left(\frac{-1/3}{1/3}, \frac{-1/3}{11/6}, \frac{-1/3}{17/6}\right) = -2/17$, the pivot is 17/6, which corresponds to the column of x_8 and the row of x_3. Next, we return to Step 5.

Table 10.17 Tableau at the end of Step 7 for iteration 4. Pivot element is boldfaced and the common denominator of all elements is given in the upper left corner of the table.

$\mathbf{U}^{(4)}$									
6	x_0	x_{11}	x_2	x_9	x_4	x_5	x_6	x_7	x_8
x_f	134	–4	–2	–6	–22	–10	–2	–14	–2
x_3	–11	–5	2	3	16	–2	11	–1	**17**
x_{10}	56	14	10	–6	–28	–10	–44	4	–32
x_1	27	3	–6	–3	–12	0	–9	–3	–15
x_{12}	19	7	8	–3	–32	10	–13	–1	–25
x_{13}	13	7	–4	–9	–38	–14	–19	11	–25

Table 10.18 Tableau at the end of Step 7 for iteration 5. Pivot element is boldfaced and the common denominator of all elements is given in the upper left corner of the table.

$\mathbf{U}^{(5)}$									
17	x_0	x_{11}	x_2	x_9	x_4	x_5	x_6	x_7	x_3
x_f	376	–13	–5	–16	–57	–29	–2	–40	–2
x_8	11	5	–2	–3	–16	2	–11	1	6
x_{10}	100	13	39	–1	6	–39	–66	6	–32
x_1	49	–4	–12	–1	6	–5	2	–11	–15
x_{12}	8	–1	31	4	–24	20	9	–7	–25
x_{13}	–9	–1	–3	–13	–41	–48	–8	**27**	–25

Step 5: We update I_t and J_t and get

$$I_5 \equiv \{8, 10, 1, 12, 13\}; J_5 \equiv \{11, 2, 9, 4, 5, 6, 7, 3\},$$

and we perform the pivoting, obtaining Table 10.18.

Step 6: Since the solution condition does not hold and

$$\min_{u_{i0}^{(t)}<0} u_{i0}^{(t)} = \min(-9/17) = -9/17,$$

the variable entering J_t is x_{13}.

Step 7: Since $\max_{u_{ij}^{(t)}>0} \left(u_{fj}/u_{ij}^{(t)}\right) = \max\left(\frac{-40/17}{27/17}\right) = -40/27$, the pivot is 27/17, which corresponds to the column of x_7 and the row of x_{13}. Next, we return to Step 5.

Step 5: We update I_t and J_t and get

$$I_6 \equiv \{8, 10, 1, 12, 7\}; J_6 \equiv \{11, 2, 9, 4, 5, 6, 13, 3\},$$

and we perform the pivoting, obtaining Table 10.19.

Table 10.19 Tableau at the end of Step 7 for iteration 6.

$\mathbf{U}^{(6)}$									
27	x_0	x_{11}	x_2	x_9	x_4	x_5	x_6	x_{13}	x_3
x_f	576	-23	-15	-56	-187	-159	-22	-40	-62
x_8	18	8	-3	-4	-23	6	-17	1	11
x_{10}	162	21	63	3	24	-45	-102	6	-42
x_1	72	-7	-21	-10	-17	-39	-2	-11	-40
x_{12}	9	-2	48	1	-55	12	11	-7	-50
x_7	9	1	3	13	41	48	8	17	25

Step 6: Since all values in the x_0 column are positive, the solution is found. The value of the objective function is 64/3, which is attained at

$$x_1 = 8/3, x_2 = 0, x_3 = 0, x_4 = 0, x_5 = 0, x_6 = 0, x_7 = 1/3, x_8 = 2/3.$$

■

10.3 MAKING THE EPM MORE EFFICIENT

The above algorithm allows working with the EPM. However, the main efficiency of the EPM lies in the simplifications we can do in the previous formulas. We have already ignored the rows corresponding to the nonbasic variables because they give an identity matrix. Next, we shall prove that we can also ignore:

1. All the rows associated with the secondary variables except for the pivot row.

2. All the columns corresponding to the main variables except for the pivot column.

To this end, we have the following theorem.

Theorem 10.2 *Let $I_t(\mathcal{P})$ and $I_t(\mathcal{S})$ be the set of indices of I_t associated with the basic main and secondary variables, respectively. Let $J_t(\mathcal{P})$ and $J_t(\mathcal{S})$ be the set of indices of J_t associated with the nonbasic main and secondary variables, respectively. Consider the following partitions of the systems of equations $\mathbf{x} = \mathbf{U}^{(t)}\,\mathbf{x}_{J_t}$ and $\mathbf{x} = \mathbf{U}^{(0)}\,\mathbf{x}_{J_0}$:*

	$\mathbf{x}_{I_t(\mathcal{P})}$	$\mathbf{x}_{J_t(\mathcal{P})}$
$\mathbf{x}_{J_t(\mathcal{S})}$	$\mathbf{A}^{(0)}$	$\mathbf{B}^{(0)}$
$\mathbf{x}_{I_t(\mathcal{S})}$	$\mathbf{C}^{(0)}$	$\mathbf{D}^{(0)}$

	$\mathbf{x}_{J_t(\mathcal{S})}$	$\mathbf{x}_{J_t(\mathcal{P})}$
$\mathbf{x}_{I_t(\mathcal{P})}$	$\mathbf{A}^{(t)}$	$\mathbf{B}^{(t)}$
$\mathbf{x}_{I_t(\mathcal{S})}$	$\mathbf{C}^{(t)}$	$\mathbf{D}^{(t)}$

associated with the partitions of $\mathbf{U}^{(0)}$ *and* $\mathbf{U}^{(t)}$*:*

$$\mathbf{U}^{(0)} = \left(\begin{array}{c|c} \mathbf{A}^{(0)} & \mathbf{B}^{(0)} \\ \hline \mathbf{C}^{(0)} & \mathbf{D}^{(0)} \end{array}\right); \quad \mathbf{U}^{(t)} = \left(\begin{array}{c|c} \mathbf{A}^{(t)} & \mathbf{B}^{(t)} \\ \hline \mathbf{C}^{(t)} & \mathbf{D}^{(t)} \end{array}\right). \tag{10.61}$$

Then the information in Table $\mathbf{U}^{(t)}$ *can be obtained from* $\mathbf{U}^{(0)}$ *and* $\mathbf{A}^{(t)}$ *by means of*

	$\mathbf{x}_{J_t(\mathcal{S})}$	$\mathbf{x}_{J_t(\mathcal{P})}$
$\mathbf{x}_{I_t(\mathcal{P})}$	$\mathbf{A}^{(t)}$	$\mathbf{B}^{(t)}$
$\mathbf{x}_{I_t(\mathcal{S})}$	$\mathbf{C}^{(t)}$	$\mathbf{D}^{(t)}$

$\equiv$

	$\mathbf{x}_{J_t(\mathcal{S})}$	$\mathbf{x}_{J_t(\mathcal{P})}$
$\mathbf{x}_{I_t(\mathcal{P})}$	$\mathbf{A}^{(t)}$	$-\mathbf{A}^{(t)}\mathbf{B}^{(0)}$
$\mathbf{x}_{I_t(\mathcal{S})}$	$\mathbf{C}^{(0)}\mathbf{A}^{(t)}$	$\mathbf{D}^{(0)} - \mathbf{C}^{(0)}\mathbf{A}^{(t)}\mathbf{B}^{(0)}$

Proof: With the above partition we can write:

- **For the main variables:**

$$\begin{aligned} \mathbf{x}_{J_t(\mathcal{S})} &= \mathbf{A}^{(0)}\, \mathbf{x}_{I_t(\mathcal{P})} + \mathbf{B}^{(0)}\, \mathbf{x}_{J_t(\mathcal{P})}, \\ \mathbf{x}_{I_t(\mathcal{P})} &= \mathbf{A}^{(t)}\, \mathbf{x}_{J_t(\mathcal{S})} + \mathbf{B}^{(t)}\, \mathbf{x}_{J_t(\mathcal{P})}, \end{aligned} \tag{10.62}$$

from which we get

$$\mathbf{A}^{(0)}\, \mathbf{x}_{I_t(\mathcal{P})} = \mathbf{I}\, \mathbf{x}_{J_t(\mathcal{S})} - \mathbf{B}^{(0)}\, \mathbf{x}_{J_t(\mathcal{P})}, \tag{10.63}$$

where $\mathbf{I}$ is the identity matrix of dimension $|J_t(\mathcal{S})|$. Multiplying the second equation in (10.62) by $\mathbf{A}^{(0)}$, we obtain

$$\mathbf{A}^{(0)}\, \mathbf{x}_{I_t(\mathcal{P})} = \mathbf{A}^{(0)}\, \mathbf{A}^{(t)} \mathbf{x}_{J_t(\mathcal{S})} + \mathbf{A}^{(0)}\, \mathbf{B}^{(t)}\, \mathbf{x}_{J_t(\mathcal{P})}. \tag{10.64}$$

Since the systems of equations corresponding to the initial and the tth steps are equivalent, Expressions (10.63) and (10.64) are identical, which implies

$$\begin{aligned} \mathbf{A}^{(0)}\, \mathbf{A}^{(t)} &= \mathbf{I} \Rightarrow \mathbf{A}^{(0)^{-1}} = \mathbf{A}^{(t)}, \\ \mathbf{A}^{(0)}\, \mathbf{B}^{(t)} &= -\mathbf{B}^{(0)} \Rightarrow \mathbf{B}^{(t)} = -\mathbf{A}^{(t)}\, \mathbf{B}^{(0)}. \end{aligned} \tag{10.65}$$

The last equation in (10.65) proves that we can determine all elements of $\mathbf{U}^{(t)}$ related to the main variables using $\mathbf{A}^{(t)}$ and the original matrix $\mathbf{B}^{(0)}$. Consequently, the knowledge we need in the EPM of elements $b_{ij}^{(t)}$, according to (10.65), can be obtained as follows:

$$b_{ij}^{(t)} = -\sum_{k \in J_t(\mathcal{S})} a_{ik}^{(t)} b_{kj}^{(0)}; \quad i \in I_t(\mathcal{P}), j \in J_t(\mathcal{P}). \tag{10.66}$$

- **For the secondary variables:**

$$\begin{aligned} \mathbf{x}_{I_t(\mathcal{S})} &= \mathbf{C}^{(t)}\, \mathbf{x}_{J_t(\mathcal{S})} + \mathbf{D}^{(t)}\, \mathbf{x}_{J_t(\mathcal{P})}, \\ \mathbf{x}_{I_t(\mathcal{S})} &= \mathbf{C}^{(0)}\, \mathbf{x}_{I_t(\mathcal{P})} + \mathbf{D}^{(0)}\, \mathbf{x}_{J_t(\mathcal{P})}. \end{aligned} \tag{10.67}$$

Replacing $\mathbf{x}_{J_t(\mathcal{S})}$ obtained from (10.62) in the first equation of (10.67), we get

$$\mathbf{x}_{I_t(\mathcal{S})} = \mathbf{C}^{(t)}\, \mathbf{A}^{(0)} \mathbf{x}_{I_t(\mathcal{P})} + \left(\mathbf{C}^{(t)} \mathbf{B}^{(0)} + \mathbf{D}^{(t)}\right) \mathbf{x}_{J_t(\mathcal{P})}. \tag{10.68}$$

Again, since system (10.68) and the second system of (10.67) are equivalent, we have

$$\mathbf{C}^{(t)}\, \mathbf{A}^{(0)} = \mathbf{C}^{(0)} \quad \Rightarrow \quad \mathbf{C}^{(t)} = \mathbf{C}^{(0)}\, \mathbf{A}^{(t)}, \tag{10.69}$$

$$\mathbf{C}^{(t)}\, \mathbf{B}^{(0)} + \mathbf{D}^{(t)} = \mathbf{D}^{(0)} \quad \Rightarrow \quad \mathbf{D}^{(t)} = \mathbf{D}^{(0)} - \mathbf{C}^{(t)} \mathbf{B}^{(0)}. \tag{10.70}$$

Then we obtain

$$c_{ij}^{(t)} = \sum_{k \in I_t(\mathcal{P})} c_{ik}^{(0)} a_{kj}^{(t)}; i \in I_t(\mathcal{S}), j \in J_t(\mathcal{S}), \tag{10.71}$$

$$d_{ij}^{(t)} = d_{ij}^{(0)} - \sum_{k \in J_t(\mathcal{S})} c_{ik}^{(t)} b_{kj}^{(0)}; i \in I_t(\mathcal{S}), j \in J_t(\mathcal{P}). \tag{10.72}$$

Considering (10.71) and (10.65), Equation (10.72) gives

$$\mathbf{D}^{(t)} = \mathbf{D}^{(0)} - \mathbf{C}^{(0)} \mathbf{A}^{(t)} \mathbf{B}^{(0)} = \mathbf{D}^{(0)} + \mathbf{C}^{(0)} \mathbf{B}^{(t)}, \tag{10.73}$$

and then

$$d_{ij}^{(t)} = d_{ij}^{(0)} + \sum_{i \in I_t(\mathcal{P})} c_{ik}^{(0)} b_{kj}^{(t)}; i \in I_t(\mathcal{S}), j \in J_t(\mathcal{P}). \tag{10.74}$$

The above expressions prove that we can determine all elements of $\mathbf{U}^{(t)}$ related to the secondary variables. Thus, the statement of the theorem is proved.

■

The above theorem allows a substantial simplification of the EPM and leads to the following algorithm.

Algorithm 10.2 The Simplified EPM Algorithm.

- **Input:** The starting tableau $\mathbf{U}^{(0)}$ of the LPP, including the objective function and the m problem constraints.

- **Output:** The solution of the maximization problem or a message indicating infeasibility or unbounded solution.

Step 1: *Initialization.* Let $columns = n + 1$, $rows = m + 1$, $J_t \equiv \{0, \ldots, n\}$, $I_t \equiv \{n + 1, \ldots, n + m\}$, and $t = 0$.

Step 2: *Testing the objective function.* If $u_{fj}^{(0)} > 0$ for some j, go to Step 3. Otherwise, go to Step 7.

Regulating Iteration

Step 3: *Transform cost function to negative coefficients.* Check if there exists a constraint i satisfying condition (10.31). If the answer is positive, let $\alpha = i$ and continue with Step 4. Otherwise, introduce the artificial constraint (10.32) and let $rows = rows + 1$, $m = m + 1$, and $\alpha = m + 1$.

Step 4: *Find pivot.* Find the pivot column β of row α using (10.33) and go to Step 5.

Standard Iterations

Step 5: *Pivoting.* Let $t = t + 1$, update I_t and J_t using (10.24), and perform the pivoting transformation (10.23) only with the $u_{ij}^{(t)}$, such that $i \in I_t(\mathcal{P})$ and $j \in \{0\} \cup J_t(\mathcal{S})$, and the x_f row, and go to Step 6.

Step 6: *Complete the* $\mathbf{u}_0^{(t)}$ *column.* This is done by using Equation (10.73) with $j = 0$, that is, with

$$u_{i0}^{(t)} = u_{i0}^{(0)} + \sum_{k \in I_t(\mathcal{P})} u_{ik}^{(0)} u_{k0}^{(t)}; \quad i \in I_t(\mathcal{S}). \tag{10.75}$$

Step 7: *Select the entering variable.* The variable x_α entering J_t is selected using

$$u_{\alpha 0}^{(t)} = \min_{u_{i0}^{(t)} < 0} u_{i0}^{(t)}. \tag{10.76}$$

If the solution condition

$$u_{fj}^{(t)} \le 0, \forall j \ne 0 \text{ and } j \in J_t, u_{i0}^{(t)} \ge 0, \forall i \in I_t. \tag{10.77}$$

and the unbounded condition $x^*_{n+m+1} = 0$ hold, print the unbounded condition message; otherwise, print the solution

$$\mathbf{x}^* = \mathbf{u}_0^{(t)}, \tag{10.78}$$

and quit. If none of them hold, continue with Step 8.

Step 8: *Complete the row associated with the entering variable.* If the entering variable is a main variable, the row associated with x_α is completed using Equation (10.66) with $i = \alpha$, that is,

$$u_{\alpha j}^{(t)} = -\sum_{k \in J_t(\mathcal{S})} u_{\alpha k}^{(t)} u_{kj}^{(0)}; \quad j \in J_t(\mathcal{P}). \tag{10.79}$$

Otherwise, we use the two equations in (10.71) and (10.72) with $i = \alpha$:

$$u_{\alpha j}^{(t)} = \sum_{k \in I_t(\mathcal{P})} u_{\alpha k}^{(0)} u_{kj}^{(t)}; j \in J_t(\mathcal{S}), \tag{10.80}$$

$$u_{\alpha j}^{(t)} = u_{\alpha j}^{(0)} - \sum_{k \in J_t(\mathcal{S})} u_{\alpha k}^{(t)} u_{kj}^{(0)}; j \in J_t(\mathcal{P}). \tag{10.81}$$

Step 9: *Select the leaving variable.* The variable x_β leaving J_t is selected using

$$\Lambda_i^{(t)} = \frac{u_{f\beta}^{(t)}}{u_{i\beta}^{(t)}} = \max_{u_{ij}^{(t)}>0} \frac{u_{fj}^{(t)}}{u_{ij}^{(t)}}. \tag{10.82}$$

If the infeasible solution conditions

$$u_{i0}^{(t)} < 0, u_{ij}^{(t)} \leq 0, \forall j \tag{10.83}$$

are found, send the corresponding message and quit.

Step 10: *Complete the pivot column.* If the pivot column β corresponds to a main variable, use Equation (10.66) with $j = \beta$:

$$u_{i\beta}^{(t)} = -\sum_{k \in J_t(\mathcal{S})} u_{ik}^{(t)} u_{k\beta}^{(0)}; \quad i \in I_t(\mathcal{P}), \tag{10.84}$$

to complete the pivot column. Next, go to Step 5. ■

Example 10.4 (Simplified EPM algorithm). Consider again the LPP in Example 10.1. Using the simplified EPM algorithm the following results are obtained:

Step 1: Table 10.20 shows the initial tableau.

Step 2: Since some of the elements in the first row are positive, we go to Step 3.

Step 3: Since constraint 5, associated with variable x_7, satisfies the regulating condition, we let $\alpha = 5$ and continue with Step 4.

Table 10.20 Initial tableau.

$\mathbf{U}^{(0)}$			
	x_0	x_1	x_2
x_f	0	2	1
x_3	1	2	-1
x_4	3	0	-1
x_5	4	-1	0
x_6	-2	2	1
x_7	5	-1	-1
x_8	3	-1	2
x_9	0	1	4

Step 4: Since $\min\left(\frac{2}{-1}, \frac{1}{-1}\right) = -2$, the pivot is -1, which corresponds to the column of x_1 and the row of x_7.

Step 5: We update I_t and J_t and get

$$I_1 \equiv \{3,4,5,6,1,8,9\};\, J_1 \equiv \{7,2\}.$$

We perform the pivoting for the pivot column, the x_f row, and the rows associated with the set $I(\mathcal{P}) \equiv \{x_1\}$ and the columns of the set $\{x_0\} \cup J(\mathcal{S}) \equiv \{x_0\}$.

Step 6: Next, we complete the first column $\mathbf{u}_0$, using

$$\mathbf{d}_0^{(t)} = \mathbf{d}_0^{(0)} + \mathbf{C}^{(0)}\mathbf{b}_0^{(t)} = \begin{pmatrix} 1 \\ 3 \\ 4 \\ -2 \\ 3 \\ 0 \end{pmatrix} + \begin{pmatrix} 2 \\ 0 \\ -1 \\ 2 \\ -1 \\ 1 \end{pmatrix} \times (5) = \begin{pmatrix} 11 \\ 3 \\ -1 \\ 8 \\ -2 \\ 5 \end{pmatrix}. \qquad (10.85)$$

Step 7: Since the solution condition does not hold and

$$\min_{u_{i0}^{(t)}<0} u_{i0}^{(t)} = \min(-1,-2) = -2,$$

the variable entering J_t is x_8.

Step 8: Next, we complete the row associated with the entering variable. Since this is a secondary variable, we use

$$\mathbf{c}_\alpha^{(t)T} = \left(\mathbf{c}_\alpha^{(0)}\mathbf{C}^{(t)}\right)^T = (-1) \times (-1) = (1) \qquad (10.86)$$

and

$$\mathbf{d}_\alpha^{(t)T} = \left(\mathbf{d}_\alpha^{(0)} - \mathbf{c}_\alpha^{(t)}\mathbf{B}^{(0)}\right)^T = (2) - (-1) \times (1) = (3). \qquad (10.87)$$

Table 10.21 Tableau at the end of iteration 1. Pivot element is boldfaced.

$\mathbf{U}^{(1)}$			
	x_0	x_7	x_2
x_f	10	-2	-1
x_3	11		
x_4	3		
x_5	-1		
x_6	8		
x_1	5	-1	-1
x_8	-2	1	**3**
x_9	5		

Step 9: Since $\max_{u_{ij}^{(t)}>0}\left(u_{fj}/u_{ij}^{(t)}\right) = \max\left(\frac{-2}{1}, \frac{-1}{3}\right) = -1/3$, the pivot is 3, which corresponds to the column of x_2 and the row of x_8.

Step 10: We complete the pivot column. Since the pivot column corresponds to a main variable, we use

$$\mathbf{b}_{\beta}^{(t)} = -\mathbf{A}_{\alpha}^{(t)}\mathbf{b}_{\beta}^{(0)} = -(-1) \times (-1) = (-1). \tag{10.88}$$

The resulting $\mathbf{U}^{(t)}$ matrix is shown in Table 10.21.

Step 5: We update I_t and J_t and get

$$I_2 \equiv \{3,4,5,6,1,2,9\}; J_2 \equiv \{7,8\}.$$

We perform the pivoting for the pivot column, the x_f row, and the rows associated with the set $I(\mathcal{P}) \equiv \{x_1, x_2\}$ and the columns of the set $\{x_0\} \cup J(\mathcal{S}) \equiv \{x_0, x_7\}$.

Step 6: Next, we complete the first column $\mathbf{u}_0$, using

$$\mathbf{d}_0^{(t)} = \mathbf{d}_0^{(0)} + \mathbf{C}^{(0)}\mathbf{b}_0^{(t)} = \begin{pmatrix} 1 \\ 3 \\ 4 \\ -2 \\ 0 \end{pmatrix} + \begin{pmatrix} 2 & -1 \\ 0 & -1 \\ -1 & 0 \\ 2 & 1 \\ 1 & 4 \end{pmatrix} \times \begin{pmatrix} 13/3 \\ 2/3 \end{pmatrix} = \begin{pmatrix} 9 \\ 7/3 \\ -1/3 \\ 22/3 \\ 7 \end{pmatrix}. \tag{10.89}$$

Step 7: Since the solution condition does not hold and

$$\min_{u_{i0}^{(t)}<0} u_{i0}^{(t)} = \min(-1/3) = -1/3,$$

the variable entering J_t is x_5.

Step 8: Next, we complete the row associated with the entering variable. Since this is a secondary variable, we use

$$\mathbf{c}_{\alpha}^{(t)T} = \left(\mathbf{c}_{\alpha}^{(0)}\mathbf{C}^{(t)}\right)^T = \begin{pmatrix} -2/3 & -1/3 \\ -1/3 & 1/3 \end{pmatrix} \times \begin{pmatrix} -1 \\ 0 \end{pmatrix} = \begin{pmatrix} 2/3 \\ 1/3 \end{pmatrix} \tag{10.90}$$

Table 10.22 Tableau at the end of iteration 2. Pivot element is boldfaced.

$\mathbf{U}^{(2)}$			
	x_0	x_7	x_8
x_f	28/3	-5/3	-1/3
x_3	9		
x_4	7/3		
x_5	-1/3	2/3	**1/3**
x_6	22/3		
x_1	13/3	-2/3	-1/3
x_2	2/3	-1/3	1/3
x_9	7		

Step 9: Since $\max_{u_{ij}^{(t)}>0}\left(u_{fj}/u_{ij}^{(t)}\right) = \max\left(\frac{-5/3}{2/3}, \frac{-1/3}{1/3}\right) = -1$, the pivot is 1/3, which corresponds to the column of x_8 and the row of x_5.

Step 10: Since the pivot column does not correspond to a main variable, we do nothing.

The resulting $\mathbf{U}^{(t)}$ matrix is shown in Table 10.22.

Step 5: We update I_t and J_t and get

$$I_3 \equiv \{3,4,8,6,1,2,9\};\ J_3 \equiv \{7,5\}.$$

We perform the pivoting for the pivot column, the x_f row, and the rows associated with the set $I(\mathcal{P}) \equiv \{x_1, x_2\}$ and the columns of the set $\{x_0\} \cup J(\mathcal{S}) \equiv \{x_0\}$.

Step 6: Next, we complete the first column $\mathbf{u}_0$, using

$$\mathbf{d}_0^{(t)} = \mathbf{d}_0^{(0)} + \mathbf{C}^{(0)}\mathbf{b}_0^{(t)} = \begin{pmatrix} 1 \\ 3 \\ 3 \\ -2 \\ 0 \end{pmatrix} + \begin{pmatrix} 2 & -1 \\ 0 & -1 \\ -1 & 2 \\ 2 & 1 \\ 1 & 4 \end{pmatrix} \times \begin{pmatrix} 4 \\ 1 \end{pmatrix} = \begin{pmatrix} 8 \\ 2 \\ 1 \\ 7 \\ 8 \end{pmatrix}. \quad (10.91)$$

Step 7: Since all values in the x_0 column are positive, the solution is found. The value of the objective function is 9, which is attained at (see Table 10.23) $x_1 = 4$ and $x_2 = 1$.

■

Example 10.5 (Simplified EPM algorithm). Consider again the LPP in Example 10.2. Using the simplified EPM algorithm the following results are obtained:

Step 1: Table 10.24 shows the initial tableau.

Step 2: Since some of the elements in the first row are positive, we go to Step 3.

Table 10.23 Tableau at the end of iteration 3.

$\mathbf{U}^{(3)}$			
	x_0	x_7	x_5
x_f	9	–1	–1
x_3	8		
x_4	2		
x_8	1		
x_6	7		
x_1	4	0	–1
x_2	1	–1	1
x_9	8		

Table 10.24 Initial tableau.

$\mathbf{U}^{(0)}$			
	x_0	x_1	x_2
x_f	0	1	2
x_3	1	2	–1
x_4	3	0	–1
x_5	4	–1	0
x_6	–2	2	1
x_7	5	–1	–1
x_8	3	–1	2
x_9	0	1	4

Step 3: Since constraint 5, associated with variable x_7, satisfies the regulating condition, we let $\alpha = 5$ and continue with Step 4.

Step 4: Since $\min\left(\frac{1}{-1}, \frac{2}{-1}\right) = -2$, the pivot is -1, which corresponds to the column of x_2 and the row of x_7.

Step 5: We update I_t and J_t and get

$$I_1 \equiv \{3, 4, 5, 6, 2, 8, 9\}; J_1 \equiv \{1, 7\}.$$

We perform the pivoting for the pivot column, the x_f row, and the rows associated with the set $I(\mathcal{P}) \equiv \{x_2\}$ and the columns associated with the set $\{x_0\} \cup J(\mathcal{S}) \equiv \{x_0\}$.

Step 6: Next, we complete the first column $\mathbf{u}_0$, using

$$\mathbf{d}_0^{(t)} = \mathbf{d}_0^{(0)} + \mathbf{C}^{(0)}\mathbf{b}_0^{(t)} = \begin{pmatrix} 1 \\ 3 \\ 4 \\ -2 \\ 3 \\ 0 \end{pmatrix} + \begin{pmatrix} -1 \\ -1 \\ 0 \\ 1 \\ 2 \\ 4 \end{pmatrix} \times (5) = \begin{pmatrix} -4 \\ -2 \\ 4 \\ 3 \\ 13 \\ 20 \end{pmatrix}. \quad (10.92)$$

Step 7: Since the solution condition does not hold and

$$\min_{u_{i0}^{(t)}<0} u_{i0}^{(t)} = \min(-4, -2) = -4,$$

the variable entering J_t is x_3.

Step 8: Next, we complete the row associated with the entering variable. Since this is a secondary variable, we use

$$\mathbf{c}_\alpha^{(t)T} = \left(\mathbf{c}_\alpha^{(0)}\mathbf{C}^{(t)}\right)^T = (-1) \times (-1) = (1) \quad (10.93)$$

and

$$\mathbf{d}_\alpha^{(t)T} = \left(\mathbf{d}_\alpha^{(0)} - \mathbf{c}_\alpha^{(t)}\mathbf{B}^{(0)}\right)^T = (2) - (-1) \times (1) = (3). \quad (10.94)$$

Step 9: Since $\max_{u_{ij}^{(t)}>0}\left(u_{fj}/u_{ij}^{(t)}\right) = \max\left(\frac{-1}{3}\right) = -1/3$, the pivot is 3, which corresponds to the column of x_1 and the row of x_3.

Step 10: We complete the pivot column. Since the pivot column corresponds to a main variable, we use

$$\mathbf{b}_\beta^{(t)} = -\mathbf{A}_\alpha^{(t)}\mathbf{b}_\beta^{(0)} = -(-1) \times (-1) = (-1). \quad (10.95)$$

The resulting $\mathbf{U}^{(t)}$ matrix is shown in Table 10.25.

Step 5: We update I_t and J_t and get

$$I_2 \equiv \{1, 4, 5, 6, 2, 8, 9\}; J_2 \equiv \{3, 7\}.$$

We perform the pivoting for the pivot column, the x_f row, and the rows associated with the set $I(\mathcal{P}) \equiv \{x_1, x_2\}$ and the columns of the set $\{x_0\} \cup J(\mathcal{S}) \equiv \{x_0, x_7\}$.

Step 6: Next, we complete the first column $\mathbf{u}_0$, using

$$\mathbf{d}_0^{(t)} = \mathbf{d}_0^{(0)} + \mathbf{C}^{(0)}\mathbf{b}_0^{(t)} = \begin{pmatrix} 3 \\ 4 \\ -2 \\ 3 \\ 0 \end{pmatrix} + \begin{pmatrix} 0 & -1 \\ -1 & 0 \\ 2 & 1 \\ -1 & 2 \\ 1 & 4 \end{pmatrix} \times \begin{pmatrix} 4/3 \\ 11/3 \end{pmatrix} = \begin{pmatrix} -2/3 \\ 8/3 \\ 13/3 \\ 9 \\ 16 \end{pmatrix}. \quad (10.96)$$

Table 10.25 Tableau at the end of iteration 1. Pivot element is boldfaced.

$\mathbf{U}^{(1)}$			
	x_0	x_1	x_7
x_f	10	-1	-2
x_3	-4	**3**	1
x_4	-2		
x_5	4		
x_6	3		
x_2	5	-1	-1
x_8	13		
x_9	20		

Step 7: Since the solution condition does not hold and

$$\min_{u_{i0}^{(t)}<0} u_{i0}^{(t)} = \min(-2/3) = -2/3,$$

the variable entering J_t is x_4.

Step 8: Next, we complete the row associated with the entering variable. Since this is a secondary variable, we use

$$\mathbf{c}_\alpha^{(t)T} = \left(\mathbf{c}_\alpha^{(0)}\mathbf{C}^{(t)}\right)^T = \begin{pmatrix} 1/3 & -1/3 \\ -1/3 & -2/3 \end{pmatrix} \times \begin{pmatrix} 0 \\ -1 \end{pmatrix} = \begin{pmatrix} 1/3 \\ 2/3 \end{pmatrix}. \qquad (10.97)$$

Step 9: Since $\max_{u_{ij}^{(t)}>0}\left(u_{fj}/u_{ij}^{(t)}\right) = \max\left(\frac{-1/3}{1/3}\right) = -1$, the pivot is 1/3, which corresponds to the column of x_3 and the row of x_4.

Step 10: Since the pivot column does not correspond to a main variable, we do nothing.

The resulting $\mathbf{U}^{(t)}$ matrix is shown in Table 10.26.

Step 5: We update I_t and J_t and get

$$I_3 \equiv \{1,3,5,6,2,8,9\};\, J_3 \equiv \{4,7\}.$$

We perform the pivoting for the pivot column, the x_f row, and the rows associated with the set $I(\mathcal{P}) \equiv \{x_1, x_2\}$ and the columns of the set $\{x_0\} \cup J(\mathcal{S}) \equiv \{x_0\}$.

Step 6: Next, we complete the first column $\mathbf{u}_0$, using

$$\mathbf{d}_0^{(t)} = \mathbf{d}_0^{(0)} + \mathbf{C}^{(0)}\mathbf{b}_0^{(t)} = \begin{pmatrix} 1 \\ 4 \\ -2 \\ 3 \\ 0 \end{pmatrix} + \begin{pmatrix} 2 & -1 \\ -1 & 0 \\ 2 & 1 \\ -1 & 2 \\ 1 & 4 \end{pmatrix} \times \begin{pmatrix} 2 \\ 3 \end{pmatrix} = \begin{pmatrix} 2 \\ 2 \\ 5 \\ 7 \\ 14 \end{pmatrix}. \qquad (10.98)$$

Table 10.26 Tableau at the end of iteration 2. Pivot element is boldfaced.

	$\mathbf{U}^{(2)}$		
	x_0	x_3	x_7
x_f	26/3	-1/3	-5/3
x_1	4/3	1/3	-1/3
x_4	-2/3	**1/3**	2/3
x_5	8/3		
x_6	13/3		
x_2	11/3	-1/3	-2/3
x_8	9		
x_9	16		

Table 10.27 Tableau at the end of iteration 3.

	$\mathbf{U}^{(3)}$		
	x_0	x_4	x_7
x_f	8	-1	-1
x_1	2	1	-1
x_3	2		
x_5	2		
x_6	5		
x_2	3	-1	0
x_8	7		
x_9	14		

Step 7: Since all values in the x_0 column are positive, the solution is found. The value of the objective function is 8, which is attained at (see Table 10.27) $x_1 = 2$ and $x_2 = 3$.

■

Example 10.6 (Detailed simplified EPM algorithm). Consider again the LPP in Example 10.3. We illustrate the simplified EPM by giving in detail all the steps corresponding to Algorithm 10.2

Step 1: Table 10.28 shows the initial tableau.

Step 2: Since some of the elements in the first row are positive, we go to Step 3.

Step 3: Since constraint 1, associated with variable x_9, satisfies the regulating condition, we let $\alpha = 1$ and continue with Step 4.

Step 4: Since $\min\left(\frac{7}{-5}, \frac{5}{-4}, \frac{5}{-3}, \frac{1}{-2}, \frac{2}{-3}, \frac{3}{-4}\right) = -5/3$, the pivot is -3, which corresponds to the column of x_3 and the row of x_9.

Table 10.28 Initial Tableau.

$\mathbf{U}^{(0)}$									
	x_0	x_1	x_2	x_3	x_4	x_5	x_6	x_7	x_8
x_f	0	7	5	5	-3	0	1	2	3
x_9	17	-5	-4	-3	-2	-1	-2	-3	-4
x_{10}	11	-2	1	-4	2	-3	-3	-1	1
x_{11}	8	-3	-2	-3	2	-1	1	-2	1
x_{12}	4	-1	1	-2	-2	1	0	-1	-1
x_{13}	-14	4	3	1	-1	-2	1	4	3

Step 5: We update I_t and J_t and get

$$I_1 \equiv \{3, 10, 11, 12, 13\}; J_1 \equiv \{1, 2, 9, 4, 5, 6, 7, 8\}.$$

We perform the pivoting for the pivot column, the x_f row, and the rows associated with the set $I(\mathcal{P}) \equiv \{x_3\}$ and the columns of the set $\{x_0\} \cup J(\mathcal{S}) \equiv \{x_0\}$.

Step 6: Next, we complete the first column $\mathbf{u}_0$, using

$$\mathbf{d}_0^{(t)} = \mathbf{d}_0^{(0)} + \mathbf{C}^{(0)}\mathbf{b}_0^{(t)} = \begin{pmatrix} 11 \\ 8 \\ 4 \\ -14 \end{pmatrix} + \begin{pmatrix} -4 \\ -3 \\ -2 \\ 1 \end{pmatrix} \times (17/3) = \begin{pmatrix} -35/3 \\ -9 \\ -22/3 \\ -25/3 \end{pmatrix}. \quad (10.99)$$

Step 7: Since the solution condition does not hold and

$$\min_{u_{i0}^{(t)}<0} u_{i0}^{(t)} = \min(-35/3, -9, -22/3, -25/3) = -35/3,$$

the variable entering J_t is x_{10}.

Step 8: Next, we complete the row associated with the entering variable. Since this is a secondary variable, we use

$$\mathbf{c}_\alpha^{(t)T} = \left(\mathbf{c}_\alpha^{(0)}\mathbf{C}^{(t)}\right)^T = (-1/3) \times (-4) = (4/3) \quad (10.100)$$

and

$$\mathbf{d}_\alpha^{(t)T} = \left(\mathbf{d}_\alpha^{(0)} - \mathbf{c}_\alpha^{(t)}\mathbf{B}^{(0)}\right)^T = \begin{pmatrix} -2 \\ 1 \\ 2 \\ -3 \\ -3 \\ -1 \\ 1 \end{pmatrix} - \begin{pmatrix} -5 \\ -4 \\ -2 \\ -1 \\ -2 \\ -3 \\ -4 \end{pmatrix} \times (4/3) = \begin{pmatrix} 14/3 \\ 19/3 \\ 14/3 \\ -5/3 \\ -1/3 \\ 3 \\ 19/3 \end{pmatrix}. \quad (10.101)$$

Table 10.29 Tableau at the end of Step 10 for iteration 1. Pivot element is boldfaced and the common denominator of all elements is given in the upper left corner of the table.

$\mathbf{U}^{(1)}$									
3	x_0	x_1	x_2	x_9	x_4	x_5	x_6	x_7	x_8
x_f	85	-4	-5	-5	-19	-5	-7	-9	-11
x_3	17		-4	-1					
x_{10}	-35	14	**19**	4	14	-5	-1	9	19
x_{11}	-27								
x_{12}	-22								
x_{13}	-25								

Step 9: Since $\max_{u_{ij}^{(t)}>0}\left(u_{fj}/u_{ij}^{(t)}\right) = \max\left(\frac{-4/3}{14/3}, \frac{-5/3}{19/3}\right) = -5/19$, the pivot is 19/3, which corresponds to the column of x_2 and the row of x_{10}.

Step 10: Since the pivot column corresponds to a main variable, we use

$$\mathbf{b}_\beta^{(t)} = -\mathbf{A}_\alpha^{(t)}\mathbf{b}_\beta^{(0)} = -(-1/3)\times(-4) = (-4/3)\,. \tag{10.102}$$

The resulting $\mathbf{U}^{(t)}$ matrix is shown in Table 10.29.

Step 5: We update I_t and J_t and get

$$I_2 \equiv \{3,2,11,12,13\}; J_2 \equiv \{1,10,9,4,5,6,7,8\}.$$

We perform the pivoting for the pivot column, the x_f row, and the rows associated with the set $I(\mathcal{P}) \equiv \{x_3, x_2\}$ and the columns of the set $\{x_0\} \cup J(\mathcal{S}) \equiv \{x_0, x_9\}$.

Step 6: Next, we complete the first column $\mathbf{u}_0$, using

$$\begin{aligned}\mathbf{d}_0^{(t)} &= \mathbf{d}_0^{(0)} + \mathbf{C}^{(0)}\mathbf{b}_0^{(t)} \\ &= \begin{pmatrix} 8 \\ 4 \\ -14 \end{pmatrix} + \begin{pmatrix} -3 & -2 \\ -2 & 1 \\ 1 & 3 \end{pmatrix} \times \begin{pmatrix} 61/19 \\ 35/19 \end{pmatrix} = \begin{pmatrix} -101/19 \\ -11/19 \\ -100/19 \end{pmatrix}.\end{aligned} \tag{10.103}$$

Step 7: Since the solution condition does not hold and

$$\min_{u_{i0}^{(t)}<0} u_{i0}^{(t)} = \min(-101/19, -11/19, -100/19) = -101/19,$$

the variable entering J_t is x_{11}.

Step 8: Next, we complete the row associated with the entering variable. Since this is a secondary variable, we use

$$\mathbf{c}_\alpha^{(t)^T} = \left(\mathbf{c}_\alpha^{(0)}\mathbf{C}^{(t)}\right)^T = \begin{pmatrix} -4/19 & 3/19 \\ -1/19 & -4/19 \end{pmatrix} \times \begin{pmatrix} -3 \\ -2 \end{pmatrix} = \begin{pmatrix} 6/19 \\ 11/19 \end{pmatrix} \tag{10.104}$$

Table 10.30 Tableau at the end of Step 10 for iteration 2. Pivot element is boldfaced and the common denominator of all elements is given in the upper left corner of the table.

$\mathbf{U}^{(2)}$									
19	x_0	x_1	x_{10}	x_9	x_4	x_5	x_6	x_7	x_8
x_f	480	-2	-5	-25	-97	-40	-46	-42	-38
x_3	61	-13	-4	-1					
x_2	35	-14	3	-4					
x_{11}	-101	**10**	6	11	48	10	59	1	57
x_{12}	-11								
x_{13}	-100								

and

$$\begin{aligned} \mathbf{d}_\alpha^{(t)T} &= \left(\mathbf{d}_\alpha^{(0)} - \mathbf{c}_\alpha^{(t)}\mathbf{B}^{(0)}\right)^T \\ &= \begin{pmatrix} -3 \\ 2 \\ -1 \\ 1 \\ -2 \\ 1 \end{pmatrix} - \begin{pmatrix} -2 & -5 \\ 2 & -2 \\ -3 & -1 \\ -3 & -2 \\ -1 & -3 \\ 1 & -4 \end{pmatrix} \times \begin{pmatrix} 6/19 \\ 11/19 \end{pmatrix} = \begin{pmatrix} 10/19 \\ 48/19 \\ 10/19 \\ 59/19 \\ 1/19 \\ 3 \end{pmatrix}. \end{aligned} \tag{10.105}$$

Step 9: Since $\max_{u_{ij}^{(t)}>0}\left(u_{fj}/u_{ij}^{(t)}\right) = \max\left(\frac{-2/19}{10/19}\right) = -1/5$, the pivot is 10/19, which corresponds to the column of x_1 and the row of x_{11}.

Step 10: Since the pivot column corresponds to a main variable, we use

$$\mathbf{b}_\beta^{(t)} = -\mathbf{A}_\alpha^{(t)}\mathbf{b}_\beta^{(0)} = -\begin{pmatrix} -4/19 & -1/19 \\ 3/19 & -4/19 \end{pmatrix} \times \begin{pmatrix} -2 \\ -5 \end{pmatrix} = \begin{pmatrix} -13/19 \\ -14/19 \end{pmatrix}. \tag{10.106}$$

The resulting $\mathbf{U}^{(t)}$ matrix is shown in Table 10.30.

Step 5: We update I_t and J_t and get

$$I_3 \equiv \{3, 2, 1, 12, 13\};\ J_3 \equiv \{11, 10, 9, 4, 5, 6, 7, 8\}.$$

We perform the pivoting for the pivot column, the x_f row, and the rows associated with the set $I(\mathcal{P}) \equiv \{x_3, x_2, x_1\}$ and the columns of the set $\{x_0\} \cup J(\mathcal{S}) \equiv \{x_0, x_{10}, x_9\}$.

Step 6: Next, we complete the first column $\mathbf{u}_0$, using

$$\begin{aligned} \mathbf{d}_0^{(t)} &= \mathbf{d}_0^{(0)} + \mathbf{C}^{(0)}\mathbf{b}_0^{(t)} \\ &= \begin{pmatrix} 4 \\ -14 \end{pmatrix} + \begin{pmatrix} -2 & 1 & -1 \\ 1 & 3 & 4 \end{pmatrix} \times \begin{pmatrix} -37/10 \\ -28/5 \\ 101/10 \end{pmatrix} = \begin{pmatrix} -43/10 \\ 59/10 \end{pmatrix}. \end{aligned} \tag{10.107}$$

Table 10.31 Tableau at the end of Step 10 for iteration 3. Pivot element is boldfaced and the common denominator of all elements is given in the upper left corner of the table.

$\mathbf{U}^{(3)}$									
10	x_0	x_{11}	x_{10}	x_9	x_4	x_5	x_6	x_7	x_8
x_f	242	-2	-2	-12	-46	-20	-18	-225	-14
x_3	-37	-13	2	7					
x_2	-56	-14	**6**	6	28	1	44	-4	32
x_1	101	19	-6	-11					
x_{12}	-43								
x_{13}	59								

Step 7: Since the solution condition does not hold and

$$\min_{u_{i0}^{(t)}<0} u_{i0}^{(t)} = \min(-37/10, -28/5, -43/10) = -28/5,$$

the variable entering J_t is x_2.

Step 8: Next, we complete the row associated with the entering variable. Since this is a main variable, we use

$$\mathbf{b}_\alpha^{(t)T} = -\left(\mathbf{a}_\alpha^{(t)}\mathbf{B}^{(0)}\right)^T = -\begin{pmatrix} 2 & 2 & -2 \\ -1 & -3 & -1 \\ 1 & -3 & -2 \\ -2 & -1 & -3 \\ 1 & 1 & -4 \end{pmatrix} \times \begin{pmatrix} -7/5 \\ 3/5 \\ 3/5 \end{pmatrix} = \begin{pmatrix} 14/5 \\ 1 \\ 22/5 \\ -2/5 \\ 16/5 \end{pmatrix}. \tag{10.108}$$

Step 9: Since $\max_{u_{ij}^{(t)}>0}\left(u_{fj}/u_{ij}^{(t)}\right) = \max\left(\frac{-1/5}{3/5}\right) = -1/3$, the pivot is 3/5, which corresponds to the column of x_{10} and the row of x_2.

Step 10: Since the pivot column does not correspond to a main variable, we do nothing.

The resulting $\mathbf{U}^{(t)}$ matrix is shown in Table 10.31.

Step 5: We update I_t and J_t and get

$$I_4 \equiv \{3, 10, 1, 12, 13\}; J_4 \equiv \{11, 2, 9, 4, 5, 6, 7, 8\}.$$

We perform the pivoting for the pivot column, the x_f row, and the rows associated with the set $I(\mathcal{P}) \equiv \{x_3, x_1\}$ and the columns of the set $\{x_0\} \cup J(\mathcal{S}) \equiv \{x_0\}$.

Step 6: Next, we complete the first column $\mathbf{u}_0$, using

$$\mathbf{d}_0^{(t)} = \mathbf{d}_0^{(0)} + \mathbf{C}^{(0)}\mathbf{b}_0^{(t)} = \begin{pmatrix} 11 \\ 4 \\ -14 \end{pmatrix} + \begin{pmatrix} -4 & -2 \\ -2 & -1 \\ 1 & 4 \end{pmatrix} \times \begin{pmatrix} -11/6 \\ 9/2 \end{pmatrix} = \begin{pmatrix} 28/3 \\ 19/6 \\ 13/6 \end{pmatrix}. \tag{10.109}$$

Table 10.32 Tableau at the end of Step 10 for iteration 4. Pivot element is boldfaced and the common denominator of all elements is given in the upper left corner of the table.

$\mathbf{U}^{(4)}$									
6	x_0	x_{11}	x_2	x_9	x_4	x_5	x_6	x_7	x_8
x_f	134	–4	–2	–16	–22	–10	–2	–14	–2
x_3	–11	–5	2	3	16	–2	11	–1	**17**
x_{10}	56								
x_1	27	3	–6	–3					–15
x_{12}	19								
x_{13}	13								

Step 7: Since the solution condition does not hold and

$$\min_{u_{i0}^{(t)}<0} u_{i0}^{(t)} = \min(-11/6) = -11/6,$$

the variable entering J_t is x_3.

Step 8: Next, we complete the row associated with the entering variable. Since this is a main variable, we use

$$\mathbf{b}_\alpha^{(t)T} = -\left(\mathbf{a}_\alpha^{(t)}\mathbf{B}^{(0)}\right)^T = -\begin{pmatrix} -2 & -4 \\ 2 & -2 \\ -1 & -1 \\ 1 & -2 \\ -2 & -3 \\ 1 & -4 \end{pmatrix} \times \begin{pmatrix} -5/6 \\ 1/2 \end{pmatrix} = \begin{pmatrix} 1/3 \\ 8/3 \\ -1/3 \\ 11/6 \\ -1/6 \\ 17/6 \end{pmatrix}. \tag{10.110}$$

Step 9: Since $\max_{u_{ij}^{(t)}>0}\left(u_{fj}/u_{ij}^{(t)}\right) = \max\left(\frac{-1/3}{1/3}, \frac{-1/3}{11/6}, \frac{-1/3}{17/6}\right) = -2/17$, the pivot is 17/6, which corresponds to the column of x_8 and the row of x_3.

Step 10: Since the pivot column corresponds to a main variable, we use

$$\mathbf{b}_\beta^{(t)} = -\mathbf{A}_\alpha^{(t)}\mathbf{b}_\beta^{(0)} = -\begin{pmatrix} -5/6 & 1/2 \\ 1/2 & -1/2 \end{pmatrix} \times \begin{pmatrix} 1 \\ -4 \end{pmatrix} = \begin{pmatrix} 17/6 \\ -5/2 \end{pmatrix}. \tag{10.111}$$

The resulting $\mathbf{U}^{(t)}$ matrix is shown in Table 10.32.

Step 5: We update I_t and J_t and get

$$I_5 \equiv \{8, 10, 1, 12, 13\}; J_5 \equiv \{11, 2, 9, 4, 5, 6, 7, 3\}.$$

We perform the pivoting for the pivot column, the x_f row, and the rows associated with the set $I(\mathcal{P}) \equiv \{x_8, x_1\}$ and the columns of the set $\{x_0\} \cup J(\mathcal{S}) \equiv \{x_0, x_{11}, x_9\}$.

Step 6: Next, we complete the first column $\mathbf{u}_0$, using

$$\begin{aligned} \mathbf{d}_0^{(t)} &= \mathbf{d}_0^{(0)} + \mathbf{C}^{(0)}\mathbf{b}_0^{(t)} \\ &= \begin{pmatrix} 11 \\ 4 \\ -14 \end{pmatrix} + \begin{pmatrix} 1 & -2 \\ -1 & -1 \\ 3 & 4 \end{pmatrix} \times \begin{pmatrix} 11/17 \\ 49/17 \end{pmatrix} = \begin{pmatrix} 100/17 \\ 8/17 \\ -9/17 \end{pmatrix}. \end{aligned} \tag{10.112}$$

Step 7: Since the solution condition does not hold and

$$\min_{u_{i0}^{(t)}<0} u_{i0}^{(t)} = \min(-9/17) = -9/17,$$

the variable entering J_t is x_{13}.

Step 8: Next, we complete the row associated with the entering variable. Since this is a secondary variable, we use

$$\mathbf{c}_\alpha^{(t)T} = \left(\mathbf{c}_\alpha^{(0)}\mathbf{C}^{(t)}\right)^T = \begin{pmatrix} 5/17 & -4/17 \\ -3/17 & -1/17 \end{pmatrix} \times \begin{pmatrix} 3 \\ 4 \end{pmatrix} = \begin{pmatrix} -1/17 \\ -13/17 \end{pmatrix}. \tag{10.113}$$

and

$$\begin{aligned} \mathbf{d}_\alpha^{(t)T} &= \left(\mathbf{d}_\alpha^{(0)} - \mathbf{c}_\alpha^{(t)}\mathbf{B}^{(0)}\right)^T \\ &= \begin{pmatrix} 3 \\ -1 \\ -2 \\ 1 \\ 4 \\ 1 \end{pmatrix} - \begin{pmatrix} -2 & -4 \\ 2 & -2 \\ -1 & -1 \\ 1 & -2 \\ -2 & -3 \\ -3 & -3 \end{pmatrix} \times \begin{pmatrix} -1/17 \\ -13/17 \end{pmatrix} = \begin{pmatrix} -3/17 \\ -41/17 \\ -48/17 \\ -8/17 \\ 27/17 \\ -25/17 \end{pmatrix}. \end{aligned} \tag{10.114}$$

Step 9: Since $\max_{u_{ij}^{(t)}>0}\left(u_{fj}/u_{ij}^{(t)}\right) = \max\left(\frac{-40/17}{27/17}\right) = -40/27$, the pivot is $27/17$, which corresponds to the column of x_7 and the row of x_{13}.

Step 10: Since the pivot column corresponds to a main variable, we use

$$\mathbf{b}_\beta^{(t)} = -\mathbf{A}_\alpha^{(t)}\mathbf{b}_\beta^{(0)} = -\begin{pmatrix} 5/17 & -3/17 \\ -4/17 & -1/17 \end{pmatrix} \times \begin{pmatrix} -2 \\ -3 \end{pmatrix} = \begin{pmatrix} 1/17 \\ -11/17 \end{pmatrix}. \tag{10.115}$$

The resulting $\mathbf{U}^{(t)}$ matrix is shown in Table 10.33.

Step 5: We update I_t and J_t and get

$$I_6 \equiv \{8, 10, 1, 12, 7\};\ J_6 \equiv \{11, 2, 9, 4, 5, 6, 13, 3\}.$$

We perform the pivoting for the pivot column, the x_f row, and the rows associated with the set $I(\mathcal{P}) \equiv \{x_8, x_1, x_7\}$ and the columns of the set $\{x_0\} \cup J(\mathcal{S}) \equiv \{x_0, x_{11}, x_9\}$.

Table 10.33 Tableau at the end of Step 10 for iteration 5. Pivot element is boldfaced and the common denominator of all elements is given in the upper left corner of the table.

				$\mathbf{U}^{(5)}$					
17	x_0	x_{11}	x_2	x_9	x_4	x_5	x_6	x_7	x_3
x_f	376	-13	-5	-16	-57	-29	-2	-40	-2
x_8	11	5		-3				1	6
x_{10}	100								
x_1	49	-4		-1				-11	-15
x_{12}	8								
x_{13}	-9	-1	-3	-13	-41	-48	-8	**27**	-25

Table 10.34 Tableau at the end of Step 10 for iteration 6. Pivot element is boldfaced.

				$\mathbf{U}^{(6)}$					
27	x_0	x_{11}	x_2	x_9	x_4	x_5	x_6	x_{13}	x_3
x_f	576	-23	-15	-56	-187	-159	-22	-40	-62
x_8	18	8		-4				1	
x_{10}	162								
x_1	72	-7		-10				-11	
x_{12}	9								
x_7	9	1		13				17	

Step 6: Next, we complete the first column $\mathbf{u}_0$, using

$$\mathbf{d}_0^{(t)} = \mathbf{d}_0^{(0)} + \mathbf{C}^{(0)}\mathbf{b}_0^{(t)} = \begin{pmatrix} 11 \\ 4 \end{pmatrix} + \begin{pmatrix} 1 & -2 & -1 \\ -1 & -1 & -1 \end{pmatrix} \times \begin{pmatrix} 2/3 \\ 8/3 \\ 1/3 \end{pmatrix} = \begin{pmatrix} 6 \\ 1/3 \end{pmatrix}. \tag{10.116}$$

Step 7: Since all values in the x_0 column are positive, the solution is found. The value of the objective function is 64/3, which is attained at (see Table 10.34).

$$x_1 = 8/3, x_2 = 0, x_3 = 0, x_4 = 0, x_5 = 0, x_6 = 0, x_7 = 1/3, x_8 = 2/3.$$

■

10.4 COMPLEXITY

In this section we analyze the complexity of the EPM. We calculate the number of products or divisions required to reach the solution of the problem. To this end we define the following parameters:

- $\alpha_1 = m\alpha$: The number of constraints that are strictly satisfied. This value coincides with the number of nonbasic secondary variables and the number of main basic variables.
- $\beta_1 = m\beta$: The number of iterations required to reach the solution.
- $\gamma = n/m$: The ratio of the number of main variables to number of constraints.

Next, we analyze the complexity of the EPM and the simplex method.

10.4.1 Complexity of the EPM

We divide the calculation into several parts:

- The number of operations required for selecting the entering variable is

$$\begin{aligned} N_p^1 &= 1 \times (m-1) + \cdots + \alpha_1 \times (m - \alpha_1) + (\beta_1 - \alpha_1)\alpha_1(m - \alpha_1) \\ &= \frac{m^3\alpha}{6}[4\alpha^2 - 3\alpha(2\beta + 1) + 6\beta], \end{aligned} \tag{10.117}$$

  where we have assumed that the α_1 nonbasic secondary variables are in J_t after the first α_1 iterations and that later the J_t cardinal remains constant. We note that the nonbasic secondary variables row products are null, and thus they are not counted.

- The number of operations required for calculating the row of the entering variable x_α is

$$N_p^2 = 1 \times n + \cdots + \alpha_1 \times n + (\beta_1 - \alpha_1)\alpha_1 n = \frac{m^3\alpha\gamma}{2}(-\alpha + 2\beta)\,. \tag{10.118}$$

- The number of operations required for calculating the pivot column is

$$\begin{aligned} N_p^3 &= [1 \times 1 + \cdots + \alpha_1 \times \alpha_1 + (\beta_1 - \alpha_1)\alpha_1\alpha_1] \\ &= \frac{m^3\alpha^2}{3}(3\beta - 2\alpha)\,. \end{aligned} \tag{10.119}$$

- The number of operations required for pivoting is

$$N_p^4 = N_p^3. \tag{10.120}$$

Thus, the total number of operations is

$$N_p = N_p^1 + N_p^2 + N_p^3 + N_p^4 = \frac{m^3\alpha}{6}[3(2\beta - \alpha)(\gamma + 1) + 2\alpha(3\beta - 2\alpha)]\,. \tag{10.121}$$

Table 10.35 Relative complexity of the EPM with respect to the simplex method.

α	β	γ	$\frac{N_P}{N_S}$
0.6	1.3	10	0.3644
0.1	1.3	10	0.0974
0.6	2.0	10	0.3906
0.1	2.0	10	0.0986
0.6	1.3	2	0.5589
0.1	1.3	2	0.1359

10.4.2 Complexity of the Simplex

Considering the simplex in its basic form, we divide, as before, the calculation into several parts (see Grossman (1994)):

- The number of operations N_S^1 required for obtaining z_j under the same hypotheses above is

$$N_S^1 = 1 \times n + \cdots + \alpha_1 \times n + (\beta_1 - \alpha_1)\alpha_1 n = N_p^2. \tag{10.122}$$

- The number of operations N_S^2 required for pivoting, working with $m \times n$ matrices, is

$$N_S^2 = \beta_1 mn. \tag{10.123}$$

Thus, the total number of operations is

$$N_S = N_S^1 + N_S^2 = \frac{m^3\gamma}{6}\left[3\alpha(2\beta - \alpha) + 6\beta\right]. \tag{10.124}$$

Example 10.7 (Complexity of the EPM algorithm). In Table 10.35 the relative complexity of the EPM with respect to the simplex method is illustrated by giving the ratio of the number of operations required for both methods. As can be seen, the EPM is very efficient relative to the simplex method.

■

As we have seen, the EPM partitions the operating tableau in four submatrices, that is, it distinguishes among four different variables, while the simplex and its variants methods tableaux consider only two: basic and nonbasic. This explains the superiority of the EPM. In fact, the simplex method operates with the well known expression $\mathbf{x}_I = \mathbf{B}^{-1}\mathbf{b} - \mathbf{B}^{-1}\mathbf{N}\mathbf{x}_J$, where $\mathbf{B}^{-1}$ is of order $m \times m$, that is, the tableau has m rows. The EPM tableau requires only knowledge of the matrix $\mathbf{A}^{(t)} = \left(\mathbf{A}^{(0)}\right)^{-1}$, which order coincides with the number of main basic variables, which in the initial step is 0×0 and in the final step oscilates between $0.1m$ and $0.6m$. This leads to a considerable smaller number of computations as shown in Table 10.35.

10.5 RECOVERING THE FINAL TABLEAU FROM THE SOLUTION

In this section we show how to recover the final tableau $\mathbf{U}^{final}$ from the initial tableau $\mathbf{U}^{(0)}$, when we know the list of basic variables. This is a very important step when we want to find the solution of a new LPP coming from some simple modifications of a given LPP with known solution. We consider three parts:

Part 1. Recovering $\mathbf{A}^{(t)}$: We just use Equation (10.65), which in this case becomes

$$\mathbf{A}^{(t)} = \mathbf{A}^{(0)^{-1}}. \tag{10.125}$$

Part 2. Recovering the first row: Since x_f is a secondary variable, we must use Expressions (10.71) and (10.72):

$$\mathbf{c}_f^{(t)} = \mathbf{c}_f^{(0)} \mathbf{A}^{(0)^{-1}} \Leftrightarrow u_{fj}^{(t)} = \sum_{k \in I_t(\mathcal{P})} u_{fk}^{(0)} u_{kj}^{(t)}; j \in J_t(\mathcal{S}) \tag{10.126}$$

and

$$\mathbf{d}_f^{(t)} = \mathbf{d}_f^{(0)} - \mathbf{c}_f^{(t)} \mathbf{B}^{(0)} \Leftrightarrow u_{fj}^{(t)} = u_{fj}^{(0)} - \sum_{k \in J_t(\mathcal{S})} u_{fk}^{(t)} u_{kj}^{(0)}; j \in J_t(\mathcal{P}). \tag{10.127}$$

Part 3. Recovering the x_0 column: Since x_0 is a main variable, we use (10.66):

$$\mathbf{b}_0^{(t)} = -\mathbf{A}^{(t)}\, \mathbf{b}_0^{(0)} \Leftrightarrow u_{i0}^{(t)} = -\sum_{k \in J_t(\mathcal{S})} u_{ik}^{(t)} u_{k0}^{(0)}; i \in I_t(\mathcal{P}) \tag{10.128}$$

and

$$\mathbf{d}_0^{(t)} = \mathbf{d}_0^{(0)} - \mathbf{C}^{(0)} \mathbf{A}^{(t)}\, \mathbf{b}_0^{(0)} \Leftrightarrow u_{i0}^{(t)} = u_{i0}^{(0)} - \sum_{j \in J_t(\mathcal{S})} \sum_{k \in I_t(\mathcal{P})} u_{ik}^{(0)} u_{kj}^{(t)} u_{j0}^{(0)}, \tag{10.129}$$

which holds for all $i \in I_t(\mathcal{P})$.

Example 10.8 (Recovering the final tableau). In this example, we show how to recover the final tableau $\mathbf{U}^{(6)}$ in Table 10.34. We consider the above three parts:

Part 1. Recovering $\mathbf{A}^{(6)}$: We use Equation (10.125), which in this case becomes

$$\mathbf{A}^{(6)} = \mathbf{A}^{(0)^{-1}} = \begin{pmatrix} -5 & -3 & -4 \\ -3 & -2 & 1 \\ 4 & 4 & 3 \end{pmatrix}^{-1} = \frac{1}{27} \begin{pmatrix} -10 & -7 & -11 \\ 13 & 1 & 17 \\ -4 & 8 & 1 \end{pmatrix}. \tag{10.130}$$

Part 2. Recovering the x_f row: We use Expressions (10.126) and (10.127)

$$\begin{aligned} \mathbf{c}_f^{(t)} &= \mathbf{c}_f^{(0)} \mathbf{A}^{(0)^{-1}} \\ &= (7 \quad 2 \quad 3) \begin{pmatrix} -10 & -7 & -11 \\ 13 & 1 & 17 \\ -4 & 8 & 1 \end{pmatrix} \frac{1}{27} = \frac{1}{27}(-56 \quad -23 \quad -40), \end{aligned} \tag{10.131}$$

$$\begin{aligned}
\mathbf{d}_f^{(t)} &= \mathbf{d}_f^{(0)} - \mathbf{c}_f^{(t)}\mathbf{B}^{(0)} = (5 \quad 5 \quad -3 \quad 0 \quad 1) \\
&\quad -\frac{1}{27}(-56 \quad -23 \quad -40)\begin{pmatrix} -4 & -3 & -2 & -1 & -2 \\ -2 & -3 & 2 & -1 & 1 \\ 3 & 1 & -1 & -2 & 1 \end{pmatrix} \\
&= \frac{1}{27}(-15 \quad -62 \quad -187 \quad -159 \quad -22).
\end{aligned} \tag{10.132}$$

Part 3. Recovering the x_0 column: We use (10.128) and get

$$\mathbf{b}_0^{(t)} = -\mathbf{A}^{(t)}\,\mathbf{b}_0^{(0)} = \frac{1}{27}\begin{pmatrix} -10 & -7 & -11 \\ 13 & 1 & 17 \\ -4 & 8 & 1 \end{pmatrix}\begin{pmatrix} 17 \\ 8 \\ -14 \end{pmatrix} = \frac{1}{27}\begin{pmatrix} 72 \\ 9 \\ 18 \end{pmatrix} \tag{10.133}$$

■

10.6 MODIFYING A LINEAR PROGRAMMING PROBLEM

Assume that we know the solution of a given linear programming problem. In this section we discuss how to solve a new problem, which comes from the initial problem by making some small changes, without the need of redoing the computations. If the final tableau of the initial LPP is known, we start from it; otherwise, it can be obtained, using the method described in the previous section.

In particular, we consider the following four cases:

1. Eliminating one or several constraints.

2. Adding new constraints.

3. Removing one or several variables.

4. Adding one variable.

10.6.1 Eliminating Constraints

Assume that we eliminate the constraints

$$x_{q_k} = \mathbf{u}_{q_k}^{(0)}\mathbf{x}_{J_0};\ k = 1, \ldots, s.$$

We have two cases:

1. All the $x_{q_k};\ k = 1, \ldots, s$ are basic variables in the final tableau, that is, $q_k \in I_t;\ k = 1, \ldots, s$. Then the new solution is obtained by removing the $q_k \in I_t;\ k = 1, \ldots, s$ rows from the table $\mathbf{U}^{(t)}$. In other words, the optimal value of the objective function and the values of the main variables remain the same.

2. Some of the x_{q_k}; $k = 1, \ldots, s$ are nonbasic variables in the final tableau, that is, $q_k \in J_t$ for some k. Then we start by removing the set of rows $\{q_k \in I_t | k = 1, \ldots, s\}$ from the table $\mathbf{U}^{(t)}$ and correct the s value accordingly. Since the remaining constraints cannot be removed because they are in J_t, we need first to move them to I_t, one by one.

   To this end, we can use different methods. We include here three options:

   First option: We first relax the condition associated with x_{q_j}, using the change of variable

$$x_{q_j} = x'_{q_j} - m; x'_{q_j} \geq 0; \tag{10.134}$$

   where m is a non-negative constant that is determined by

$$m = \min_{u_{iq_j} > 0} \frac{u_{i0}}{u_{iq_j}} = \frac{u_{\alpha 0}}{u_{\alpha q_j}}. \tag{10.135}$$

   Note that m is non-negative because $u_{iq_j} > 0$ and $u_{i0} \geq 0$. Note also that (10.134) is a relaxing condition because $x'_{q_j} \geq 0$ is equivalent to $x_{q_j} \geq -m$, and previously it was $x_{q_j} \geq 0$.

   Note that this change of variable leads only to modifying the x_0 column by

$$\mathbf{u}_0 = \mathbf{u}_0 - m\mathbf{u}_\beta, \tag{10.136}$$

   where β is the column of the constraint being removed.

   Next, we perform the pivoting process for x_{q_j} to leave J_t and enter I_t. At this point we can remove the constraint associated with x_{q_j} and repeat the same process with all the remaining constraints.

   Note that the first column elements do not change, because the first column value associated with the α row is null.

   Finally, we get a $\mathbf{U}$ table with non-negative elements in its first column, and we use the dual EPM process, described below, until a solution is found.

   The idea behind this method consists of translating the hyperplane associated with each constraint until it becomes non active, that is, its column element is null. In that case we can exchange it with some variables in I_t, with no change in the value of the optimization function.

   Second option: Since the LPP is now not subject to the set of constraints $\{x_{q_k} | k = 1, \ldots, s\}$, no sign constraints in $x_{q_k}; k = 1, \ldots, s$, must be satisfied. Thus, we can write

$$x_{q_k} = \begin{cases} x'_h - x'_{q_k}, & \text{if} \quad u_{1q_k} < 0, \\ x'_{q_k} - x'_h, & \text{if} \quad u_{1q_k} > 0, \end{cases} \quad k = 1, \ldots, s, \tag{10.137}$$

where x'_h and $x'_{q_k}; k = 1, \ldots, s$, are new non-negative variables. Replacing (10.137) in $\mathbf{U}^{(t)}$ we obtain $\mathbf{U}^{(t+1)}$, which is identical to $\mathbf{U}^{(t)}$ but with the new columns

$$\mathbf{u}_h^{(t+1)} = \mathbf{u}_{q_k}^{(t)} \text{ and } \mathbf{u}_{q_k}^{(t+1)} = -\mathbf{u}_{q_k}^{(t)}$$

or their negatives.

Next, we perform the pivoting process with the entering variable α using

$$M = \frac{u_{\alpha 0}^{(t)}}{u_{\alpha\beta}^{(t)}} = \max_{\{u_{i\beta}^{(t)}<0;\, i>0\}} \frac{u_{i0}^{(t)}}{u_{i\beta}^{(t)}}, \tag{10.138}$$

and we repeat the process for each of the remaining variables. Finally, we use the dual EPM algorithm as before, to get the solution.

Note that the pivoting process using (10.138) leads to a null column for x'_h, and thus this variable can be removed from J_t (in fact, we never include it in J_t).

Third option: The last alternative consists of writing (10.137) as

$$x_{q_k} = x'_{q_k} - x'_h;\ k = 1, \ldots, s. \tag{10.139}$$

However, for convenience, in this book we use (10.137).

In this way, the EPM goes backward using this process, which we call the *dual EPM.*

Algorithm 10.3 The Dual EPM.

- **Input:** *The initial* $\mathbf{U}^{(0)}$ *and the final tableau* $\mathbf{U}^{final}$ *after transformation (10.137).*

- **Output:** *The solution of the maximization problem or a message indicating infeasibility or unbounded solution.*

Step 1: Select the leaving variable. *If* $u_{fj}^{(t)} \le 0; \forall j \ne 0$, *we have the optimal solution. We return this solution and quit. Otherwise, the variable* x_β *leaving* J_t *is selected using the reduced method*

$$u_{f\beta}^{(t)} = \max_{u_{fj}^{(t)}>0;\, j>0} u_{fj}^{(t)}. \tag{10.140}$$

Step 2: Complete the pivot column. *If the leaving variable* x_β *corresponds to a secondary variable, use (10.71):*

$$u_{i\beta}^{(t)} = \sum_{k \in I_t(\mathcal{P})} u_{ik}^{(0)} u_{k\beta}^{(t)};\quad i \in I_t(\mathcal{S}). \tag{10.141}$$

Otherwise, use Equations (10.66),

$$u_{i\beta}^{(t)} = - \sum_{k \in J_t(\mathcal{S})} u_{ik}^{(t)} u_{k\beta}^{(0)}; \quad i \in I_t(\mathcal{P}), \tag{10.142}$$

and (10.74),

$$u_{i\beta}^{(t)} = u_{i\beta}^{(0)} + \sum_{i \in I_t(\mathcal{P})} u_{ik}^{(0)} u_{k\beta}^{(t)}; i \in I_t(\mathcal{S}). \tag{10.143}$$

Step 3: Select the entering variable. *If some i such that $u_{i\beta}^{(t)} < 0$ does not exist, the LPP is unbounded. Otherwise, the variable x_α entering J_t is selected using by means of*

$$\frac{u_{\alpha 0}^{(t)}}{u_{\alpha\beta}^{(t)}} = \max_{u_{i\beta}^{(t)}<0;\, i>0} \frac{u_{i0}^{(t)}}{u_{i\beta}^{(t)}}. \tag{10.144}$$

Step 4: Complete the αth row associated with the entering variable. *If the entering variable is a main variable, the row associated with x_α is completed using Equation (10.66), that is,*

$$u_{\alpha j}^{(t)} = - \sum_{k \in J_t(\mathcal{S})} u_{\alpha k}^{(t)} u_{kj}^{(0)}; j \in J_t(\mathcal{P}). \tag{10.145}$$

Otherwise, we use the left two equations in (10.71) and (10.72):

$$\begin{aligned} u_{\alpha j}^{(t)} &= \sum_{k \in I_t(\mathcal{P})} u_{\alpha k}^{(0)} u_{kj}^{(t)}; j \in J_t(\mathcal{S}), \\ u_{\alpha j}^{(t)} &= u_{\alpha j}^{(0)} - \sum_{k \in J_t(\mathcal{S})} u_{\alpha k}^{(t)} u_{kj}^{(0)}; j \in J_t(\mathcal{P}). \end{aligned} \tag{10.146}$$

Step 5: Pivoting. *Let $t = t + 1$. Update I_t and J_t using (10.24) and perform the pivoting transformation (10.23) only with the $u_{ij}^{(t)}$, such that $i \in I_t(\mathcal{P})$ and $j \in \{0\} \cup J_t(\mathcal{S})$, the $\mathbf{u}^0$ column, and the x_f row, and go to Step 2.* ∎

Note that:

1. The objective function does not decrease, since

$$\mathbf{u}_R^{(t+1)} \geq \mathbf{0}; u_{f0}^{(t+1)} - u_{f0}^{(t)} = \frac{u_{\alpha 0}^{(t)}}{u_{\alpha q}^{(t)}} u_{fq}^{(t)} \geq 0. \tag{10.147}$$

2. If only one constraint has been removed, the x_h column can be omitted, since $\mathbf{u}_h^{(t+1)} = \mathbf{0}$.

Table 10.36 Tableau at the end of Step 7 for iteration 6. The common denominator of all elements is given in the upper left corner of the table.

$\mathbf{U}^{(6)}$									
27	x_0	x_{11}	x_2	x_9	x_4	x_5	x_6	x_{13}	x_3
x_f	576	–23	–15	–56	–187	–159	–22	–40	-62
x_8	18	8	-3	-4	–23	6	–17	1	11
x_{10}	162	21	63	3	24	–45	–102	6	-42
x_1	72	-7	–21	–10	–17	–39	–2	–11	–40
x_{12}	9	-2	48	1	–55	12	11	-7	-50
x_7	9	1	3	13	41	48	8	17	25

If now $u_{fj}^{(t+1)} \leq 0; \forall j \neq 0$, the solution of the new LPP has been obtained. Otherwise, we must repeat this process until this condition is satisfied.

Example 10.9 (Removing constraints). Consider again Example 10.3 and assume that we want to remove constraints associated with variables x_{11} and x_{13}. Using the first of the methods described above we proceed as follows:
1: We start by removing x_{11} from the final tableau shown in Table 10.19, which we reproduce in Table 10.36 for convenience. Thus, $\beta = 2$ (the column of x_{11}).
2: We select the row α using (10.135):

$$m = \min\left(\frac{18/27}{8/27}, \frac{6}{7/9}, \frac{1/3}{1/27}\right) = \min(9/4, 54/7, 9) = 9/4, \tag{10.148}$$

which implies $\alpha = 2$, that is, x_8 enters J_t.
3: We modify the x_0 column using $\mathbf{u}_0 = \mathbf{u}_0 - m\mathbf{u}_\beta$:

$$\begin{pmatrix} 93/4 \\ 0 \\ 17/4 \\ 13/4 \\ 1/2 \\ 1/4 \end{pmatrix} = \begin{pmatrix} 64/3 \\ 2/3 \\ 6 \\ 8/3 \\ 1/3 \\ 1/3 \end{pmatrix} - 9/4 \begin{pmatrix} -23/27 \\ 8/27 \\ 7/9 \\ -7/27 \\ -2/27 \\ 1/27 \end{pmatrix},$$

and get Table 10.37.
4: We perform the pivoting process and get Table 10.38.
5: We remove the row associated with x_{11}.
6: We start removing constraint x_{13}. Thus, $\beta = 8$ (the column of x_{13}).
7: We select the row α using (10.135):

$$m = \min\left(\frac{17/4}{1/8}, \frac{1/4}{5/8}\right) = \min(34, 2/5) = 2/5, \tag{10.149}$$

which implies $\alpha = 4$, that is, x_7 enters J_t.

Table 10.37 Tableau after modifying the first column. Pivot element is boldfaced and the common denominator of all elements is given in the upper left corner of the table.

108	x_0	x_{11}	x_2	x_9	x_4	x_5	x_6	x_{13}	x_3
x_f	2511	-92	-60	-224	-748	-636	-88	-160	-248
x_8	0	**32**	-12	-16	-92	24	-68	4	44
x_{10}	459	84	252	12	96	-180	-408	24	-168
x_1	351	-28	-84	-40	-68	-156	-8	-44	-160
x_{12}	78	-8	192	4	-220	48	44	-28	-200
x_7	27	4	12	52	164	192	32	68	100

Table 10.38 Tableau after pivoting. Pivot element is boldfaced and the common denominator of all elements is given in the upper left corner of the table.

8	x_0	x_8	x_2	x_9	x_4	x_5	x_6	x_{13}	x_3
x_f	186	-23	-7	-20	-75	-42	-21	-11/8	-9
x_{11}	0	27	3	4	23	-6	17	-1	-11
x_{10}	34	21	21	4	25	-18	-17	1	-21
x_1	26	-7	-7	-4	-11	-10	-5	-3	-9
x_{12}	4	-2	14	0	-18	4	2	-2	-14
x_7	2	1	1	4	13	14	3	**5**	7

8: We modify the x_0 column using $\mathbf{u}_0 = \mathbf{u}_0 - m\mathbf{u}_\beta$:

$$\begin{pmatrix} 119/5 \\ 21/5 \\ 17/5 \\ 3/5 \\ 0 \end{pmatrix} = \begin{pmatrix} 93/4 \\ 17/4 \\ 13/4 \\ 1/2 \\ 1/4 \end{pmatrix} - 2/5 \begin{pmatrix} -11/8 \\ 1/8 \\ -3/8 \\ -1/4 \\ 5/8 \end{pmatrix},$$

and get Table 10.39.
9: We perform the pivoting process and get Table 10.40.

Table 10.39 Tableau after modifying the first column. Pivot element is boldfaced and the common denominator of all elements is given in the upper left corner of the table.

40	x_0	x_8	x_2	x_9	x_4	x_5	x_6	x_{13}	x_3
x_f	952	-115	-35	-100	-375	-210	-105	-55	-45
x_{10}	168	105	105	20	125	-90	-85	5	-105
x_1	136	-35	-35	-20	-55	-50	-25	-15	-45
x_{12}	24	-10	70	0	-90	20	10	-10	-70
x_7	0	5	5	20	65	70	15	**25**	35

Table 10.40 Tableau after pivoting. The common denominator of all elements is given in the upper left corner of the table.

5	x_0	x_8	x_2	x_9	x_4	x_5	x_6	x_7	x_3
x_f	119	-13	-3	-7	-29	-7	-9	-11	4
x_{10}	21	13	13	2	14	-13	-11	1	-14
x_1	17	-4	-4	-1	-2	-1	-2	-3	-3
x_{12}	3	-1	9	1	-8	6	2	-2	-7
x_{13}	0	-1	-1	-4	-13	-14	-3	8	-7

Table 10.41 The EPM algorithm process. The common denominator of all elements is given in the upper left corner of the table.

5	x_0	x_8	x_2	x_9	x_4	x_5	x_6	x_7	x_3
x_f	119	-13	-3	-7	-29	-7	-9	-11	4
x_{10}	21	13	13	2	14	-13	-11	1	-14
x_1	17	-4	-4	-1	-2	-1	-2	-3	-3
x_{12}	3	-1	9	1	-8	6	2	-2	-7
7	x_0	x_8	x_2	x_9	x_4	x_5	x_6	x_7	x_{12}
x_f	169	-19	3	-9	-47	-5	-11	-17	-4
x_{10}	3	3	-1	0	6	-5	-3	1	2
x_1	22	-5	**-11**	-2	2	-5	-4	-3	3
x_3	3	-1	9	1	-8	6	2	-2	-5
11	x_0	x_8	x_1	x_9	x_4	x_5	x_6	x_7	x_{12}
x_f	25	-32	-3	-15	-73	-10	-19	-28	-5
x_{10}	1	38	7	2	64	-50	-29	14	19
x_2	2	-5	-7	-2	2	-5	-4	-3	3
x_3	3	-8	-9	-1	-10	3	-2	-7	-4

10: We remove the row associated with x_{13}.
11: We use the EPM algorithm and get Table 10.41, from which we find that the maximum value of the objective function is 25, which is obtained when

$$x_1 = 0; x_2 = 2; x_3 = 3; x_4 = x_5 = x_6 = x_7 = x_8 = 0.$$

Next, we solve the same problem using the second method described above.
1: First, we remove constraint x_{11}. Since $u_{12} = -23/27 < 0$ we make the change of variable $x'_{11} = x'_h - x_{11}$. Thus, $\beta = 2$ (the column of x_{11}) and we change sign to that column of Table 10.36 and get Table 10.42.
2: We select the row α using (10.138):

$$M = \max\left(\frac{2/3}{-8/27}, \frac{6}{-7/19}, \frac{1/3}{-1/27}\right) = \max\left(-9/4, -54/7, -9\right) = -9/4, \tag{10.150}$$

Table 10.42 Initial tableau. Pivot element is boldfaced and the common denominator of all elements is given in the upper left corner of the table.

27	x_0	x_{11}	x_2	x_9	x_4	x_5	x_6	x_{13}	x_3
x_f	576	23	–15	–56	–187	–159	–22	–40	–62
x_8	18	**–8**	–3	–4	–23	6	–17	1	11
x_{10}	162	–21	63	3	24	–45	–102	6	–42
x_1	72	7	–21	–10	–17	–39	–2	–11	–40
x_{12}	9	2	48	1	–55	12	11	–7	–50
x_7	9	–1	3	13	41	48	8	17	25

Table 10.43 Tableau after pivoting. Pivot element is boldfaced and the common denominator of all elements is given in the upper left corner of the table.

8	x_0	x_8	x_2	x_9	x_4	x_5	x_6	x_{13}	x_3
x_f	186	–23	–7	–20	–75	–42	–21	–11/8	–9
x_{11}	18	27	3	4	23	–6	17	–1	–11
x_{10}	34	21	21	4	25	–18	–17	1	–21
x_1	26	–7	–7	–4	–11	–10	–5	–3	–9
x_{12}	4	–2	14	0	–18	4	2	–2	–14
x_7	2	1	1	4	13	14	3	**5**	7

which implies $\alpha = 2$, that is, x_8 enters J_t.
3: We perform the pivoting process and get Table 10.43.
4: We remove the row associated with x_{11}.
5: Next, we start removing constraint x_{13}. Thus, $\beta = 8$ (the column of x_{13}). Since $u_{18} = -11/8 < 0$, we make the change of variable $x'_{13} = x'_h - x_{13}$. Thus, we change sign to that column and get Table 10.44.

Table 10.44 Tableau after changing sign to column x_{13}. Pivot element is boldfaced and the common denominator of all elements is given in the upper left corner of the table.

8	x_0	x_8	x_2	x_9	x_4	x_5	x_6	x_{13}	x_3
x_f	186	–23	–7	–20	–75	–42	–21	11	–9
x_{10}	34	21	21	4	25	–18	–17	–1	–21
x_1	26	–7	–7	–4	–11	–10	–5	3	–9
x_{12}	4	–2	14	0	–18	4	2	2	–14
x_7	2	1	1	4	13	14	3	**–5**	7

Table 10.45 Tableau after pivoting. The common denominator of all elements is given in the upper left corner of the table.

5	x_0	x_8	x_2	x_9	x_4	x_5	x_6	x_7	x_3
x_f	119	-13	-3	-7	-29	-7	-9	-11	4
x_{10}	21	13	13	2	14	-13	-11	1	-14
x_1	17	-4	-4	-1	-2	-1	-2	-3	-3
x_{12}	3	-1	9	1	-8	6	2	-2	-7
x_{13}	2	1	1	4	13	14	3	-8	7

Table 10.46 Final tableau with the solution.

5	x_0	x_8	x_2	x_9	x_4	x_5	x_6	x_7	x_3
x_f	119	-13	-3	-7	-29	-7	-9	-11	4
x_{10}	21	13	13	2	14	-13	-11	1	-14
x_1	17	-4	-4	-1	-2	-1	-2	-3	-3
x_{12}	3	-1	9	1	-8	6	2	-2	-7
7	x_0	x_8	x_2	x_9	x_4	x_5	x_6	x_7	x_{12}
x_f	169	-19	3	-9	-47	-5	-11	-17	-4
x_{10}	3	3	-1	0	6	-5	-3	1	2
x_1	22	-5	-11	-2	2	-5	-4	-3	3
x_3	3	-1	9	1	-8	6	2	-2	-5
11	x_0	x_8	x_1	x_9	x_4	x_5	x_6	x_7	x_{12}
x_f	25	-32	-3	-15	-73	-10	-19	-28	-5
x_{10}	1	38	7	2	64	-50	-29	14	19
x_2	2	-5	-7	-2	2	-5	-4	-3	3
x_3	3	-8	-9	-1	-10	3	-2	-7	-4

6: We select the row α using (10.138):

$$M = \max\left(\frac{17/4}{-1/8}, \frac{1/4}{-5/18}\right) = \max(-34, -2/5) = -2/5, \qquad (10.151)$$

which implies $\alpha = 4$, that is, x_7 enters J_t.
7: We perform the pivoting process and get Table 10.45.
8: We remove the row associated with x_{13}.
9: We use the EPM algorithm and get Table 10.46, from which we find that the maximum value of the objective function is 25, which is obtained when

$$x_1 = 0; x_2 = 2; x_3 = 3; x_4 = x_5 = x_6 = x_7 = x_8 = 0.$$

■

Example 10.10 (Removing constraints). Consider again the LPP (10.57) to (10.59) and assume that we know $\mathbf{U}^{(6)}$ in Table 10.34 with its solution. We consider two cases:

1. Assume that constraint x_{12} is to be removed. Since it is a basic variable, we know that the new solution is the same. Thus, we can remove the corresponding row and get Table 10.47 for the new LPP.

2. Assume now, as in the previous example, that constraints x_{11} and x_{13} are to be removed. Since both are nonbasic variables, we use the second option above, but now removing both constraints simultaneously. Thus, we define $x_{11} = x'_h - x'_{11}$ and $x_{13} = x'_h - x'_{13}$ and change signs in their corresponding columns, which now correspond to $x'_{11} \geq 0$ and $x'_{13} \geq 0$ (see $\mathbf{U}^{(7)}$ in Table 10.48). Note that now we need to incorporate the x'_h column.

   Since $40 = \max(23, 40)$, the leaving variable is x'_{13} and the entering variable is, as before, x_7. Pivoting with $u_{7,13} = -17$ we obtain $\mathbf{U}^{(8)}$ in Table 10.48.

   In $\mathbf{U}^{(8)}$ the leaving variable is x'_{11} and the entering variable is x_8, since

$$\frac{11}{-5} = \max\left(\frac{11}{-5}, \frac{9}{-1}, \frac{100}{-13}\right). \tag{10.152}$$

   Thus, pivoting with $u_{8,11} = -5$ we get $\mathbf{U}^{(9)}$ in Table 10.48.

   In $\mathbf{U}^{(9)}$ the leaving variable is x_3 and the entering variable is x_{12}, since

$$\frac{3}{-7} = \max\left(\frac{72}{-7}, \frac{21}{-14}, \frac{3}{-7}\right). \tag{10.153}$$

   Thus, pivoting with $u_{12,3} = -7$ we get $\mathbf{U}^{(10)}$ in Table 10.48.

   In $\mathbf{U}^{(10)}$ the leaving variable is x_2 and the entering variable is x_{10}, since

$$\frac{22}{-11} = \max\left(\frac{22}{-11}, \frac{21}{-7}\right). \tag{10.154}$$

   Thus, pivoting with $u_{1,2} = -11$ we get $\mathbf{U}^{(11)}$ in Table 10.48. Since $u_{fj}^{(11)} \leq 0; \forall j \neq 0$, the following solution has been obtained:

$$x_2^* = 2;\ x_3^* = 3;\ x_1^* = x_4^* = x_5^* = x_6^* = x_7^* = x_8^* = 0;\ x_f^* = 25.$$

■

Table 10.47 Terms in matrix $\mathbf{U}^{(0)}$.

					$\mathbf{U}^{(0)}$					
1	x_0	x_1	x_2	x_3	x_4	x_5	x_6	x_7	x_8	x'_h
$\mathbf{u}_9^{(0)}$	17	-5	-4	-3	-2	-1	-2	-3	-4	0
$\mathbf{u}_{10}^{(0)}$	11	-2	1	-4	2	-3	-3	-1	1	0
$\mathbf{u}_{11}^{(0)\prime}$	-8	3	2	3	-2	1	-1	2	1	1
$\mathbf{u}_{12}^{(0)}$	4	-1	1	-2	-2	1	0	-1	-1	0
$\mathbf{u}_{13}^{(0)\prime}$	14	-4	-3	-1	1	2	-1	-4	-3	1

10.6.2 Adding New Constraints

Adding one or several constraints is not a problem since we can add more rows to the tableau and continue the normal process. Assume that we add the constraints $x_{q_k} = \mathbf{u}_{q_k}^{(0)}\mathbf{x}_{J_0};\ k = 1, \ldots, s$. Since x_{q_k} are secondary variables, we must use Expressions (10.71) and (10.72) to find the corresponding rows in the final tableau:

$$u_{qj}^{(t)} = \sum_{k \in I_t(\mathcal{P})} u_{qk}^{(0)} u_{kj}^{(t)}; \quad j \in J_t(\mathcal{S}) \tag{10.155}$$

$$u_{qj}^{(t)} = u_{qj}^{(0)} - \sum_{k \in J_t(\mathcal{S})} u_{qk}^{(t)} u_{kj}^{(0)}; \quad j \in J_t(\mathcal{P}). \tag{10.156}$$

Next, we proceed with the usual EPM algorithm.

Example 10.11 (Adding new constraints). In this example we consider the following LPP:

Maximize

$$x_f = 7x_1 + 5x_2 + 5x_3 - 3x_4 + 0x_5 + x_6 + 2x_7 + 3x_8 \tag{10.157}$$

subject to

$$\begin{array}{rrrrrrrrcr}
5x_1 & +4x_2 & +3x_3 & +2x_4 & +x_5 & +2x_6 & +3x_7 & +4x_8 & \leq & 17 \\
2x_1 & -x_2 & +4x_3 & -2x_4 & +3x_5 & +3x_6 & +x_7 & -x_8 & \leq & 11 \\
3x_1 & +2x_2 & +3x_3 & -2x_4 & +x_5 & -x_6 & +2x_7 & -x_8 & \leq & 8 \\
x_1 & -x_2 & +2x_3 & +2x_4 & -x_5 & & +x_7 & +x_8 & \leq & 4 \\
4x_1 & +3x_2 & +x_3 & -x_4 & -2x_5 & +x_6 & +4x_7 & +3x_8 & \geq & 14 \\
x_1 & & & & & & +x_7 & +x_8 & \leq & 3 \\
x_1 & +x_2 & & & & & & & \leq & 2 \\
 & & & & x_1, x_2, x_3, & x_4, x_5, & x_6, x_7, & x_8 & \geq & 0;
\end{array} \tag{10.158}$$

that is, we add the last two constraints to the LPP in Example 10.3. The initial tableau is shown in Table 10.49.

Table 10.48 Terms in matrices $\mathbf{U}^{(7)}$ to $\mathbf{U}^{(11)}$ that need to be calculated using the EPM. Pivot elements are boldfaced and the common denominator of all elements is given in the upper left corner of the table.

$\mathbf{U}^{(7)}$

27	x_0	x'_{11}	x_2	x_9	x_4	x_5	x_6	x'_{13}	x_3	x'_h
x_f	576	23	–15	–56	–187	–159	–22	40	–62	–63
x_1	72	7		–10				11		
x_7	9	–1	3	13	41	48	8	**–17**	25	18
x_8	18	–8		–4				–1		
x_{10}	162							–6		
x_{12}	9							7		

$\mathbf{U}^{(8)}$

17	x_0	x'_{11}	x_2	x_9	x_4	x_5	x_6	x_7	x_3	x'_h
x_f	376	13	–5	–16	–57	–29	–2	–40	–2	–13
x_1	49	4		–1						
x_8	11	**–5**	–2	–3	–16	2	–11	1	6	5
x_{10}	100	–13								
x_{12}	8	1								
x'_{13}	9	–1								

$\mathbf{U}^{(9)}$

5	x_0	x_8	x_2	x_9	x_4	x_5	x_6	x_7	x_3	x'_h
x_f	119	–13	–3	–7	–29	–7	–9	–11	4	0
x_1	17			–1					–3	
x_{10}	21								–14	
x'_{11}	11								6	
x_{12}	3	–1	9	1	–8	6	2	–2	**–7**	0
x'_{13}	2								7	

$\mathbf{U}^{(10)}$

7	x_0	x_8	x_2	x_9	x_4	x_5	x_6	x_7	x_{12}	x'_h
x_f	169	–19	3	–9	–47	–5	–11	–17	–4	0
x_1	22	–5	**–11**	–2	2	–5	–4	–3	3	0
x_3	3		9	1					–5	
x_{10}	21		–7							
x'_{11}	19		8							
x'_{13}	7		14							

$\mathbf{U}^{(11)}$

11	x_0	x_8	x_1	x_9	x_4	x_5	x_6	x_7	x_{12}	x'_h
x_f	275	–32	–3	–15	–73	–10	–19	–28	–5	0
x_2	22									
x_3	33									
x_{10}	11									
x'_{11}	55									
x'_{13}	55									

Table 10.49 Initial tableau.

	$\mathbf{U}^{(0)}$								
	x_0	x_1	x_2	x_3	x_4	x_5	x_6	x_7	x_8
x_f	0	7	5	5	–3	0	1	2	3
x_9	17	–5	–4	–3	–2	–1	–2	–3	–4
x_{10}	11	–2	1	–4	2	–3	–3	–1	1
x_{11}	8	–3	–2	–3	2	–1	1	–2	1
x_{12}	4	–1	1	–2	–2	1	0	–1	–1
x_{13}	–14	4	3	1	–1	–2	1	4	3
x_{14}	3	–1	0	0	0	0	0	–1	–1
x_{15}	2	–1	–1	0	0	0	0	0	0

First we recover the last tableau, using the results in Section 10.3. In other words, we use

$$\mathbf{U}^{(t)} = \left(\begin{array}{c|c} \mathbf{A}^{(t)} & -\mathbf{A}^{(t)}\ \mathbf{B}^{(0)} \\ \hdashline \mathbf{C}^{(0)}\ \mathbf{A}^{(t)} & \mathbf{D}^{(0)} - \mathbf{C}^{(0)}\ \mathbf{A}^{(t)}\ \mathbf{B}^{(0)} \end{array}\right), \tag{10.159}$$

where

$$\mathbf{U}^{(0)} = \left(\begin{array}{c|c} \mathbf{A}^{(0)} & \mathbf{B}^{(0)} \\ \hdashline \mathbf{C}^{(0)} & \mathbf{D}^{(0)} \end{array}\right); \ \mathbf{A}^{(t)} = \mathbf{A}^{(0)^{-1}}. \tag{10.160}$$

Taking into account that

$$\begin{aligned} I_t(\mathcal{P}) &\equiv \{x_1, x_7, x_8\}, \\ I_t(\mathcal{S}) &\equiv \{x_{10}, x_{12}, x_{15}, x_{16}\}, \\ J_t(\mathcal{P}) &\equiv \{x_2, x_3, x_4, x_5, x_6\}, \\ J_t(\mathcal{S}) &\equiv \{x_9, x_{11}, x_{13}\}, \end{aligned}$$

and using the numeric ordering of the variables, from Table 10.49 we get

$$\mathbf{A}^{(0)} = \begin{pmatrix} -5 & -3 & -4 \\ -3 & -2 & 1 \\ 4 & 4 & 3 \end{pmatrix}; \ \mathbf{B}^{(0)} = \begin{pmatrix} 17 & -4 & -3 & -2 & -1 & -2 \\ 8 & -2 & -3 & 2 & -1 & 1 \\ -14 & 3 & 1 & -1 & -2 & 1 \end{pmatrix};$$

$$\mathbf{C}^{(0)} = \begin{pmatrix} 7 & 2 & 3 \\ -2 & -1 & 1 \\ -1 & -1 & -1 \\ -1 & -1 & -1 \\ -1 & 0 & 0 \end{pmatrix}; \ \mathbf{D}^{(0)} = \begin{pmatrix} 0 & 5 & 5 & -3 & 0 & 1 \\ 11 & 1 & -4 & 2 & -3 & -3 \\ 4 & 1 & -2 & -2 & 1 & 0 \\ 3 & 0 & 0 & 0 & 0 & 0 \\ 2 & -1 & 0 & 0 & 0 & 0 \end{pmatrix};$$

$$-\mathbf{A}^{(t)}\ \mathbf{B}^{(0)} = \frac{1}{27}\begin{pmatrix} 72 & -21 & -40 & -17 & -39 & -2 \\ 9 & 3 & 25 & 41 & 48 & 8 \\ 18 & -3 & 11 & -23 & 6 & -17 \end{pmatrix};$$

Table 10.50 Recovered last tableau. The common denominator of all elements is given in the upper left corner of the table.

$\mathbf{U}^{(7)}$									
27	x_0	x_2	x_3	x_4	x_5	x_6	x_9	x_{11}	x_{13}
x_f	576	-15	-62	-187	-159	-22	-56	-23	-40
x_{10}	162	63	-42	24	-45	-102	3	21	6
x_{12}	9	48	-50	-55	12	11	1	-2	-7
x_{14}	-18	**14**	4	-1	-15	11	1	-2	-7
x_{15}	-18	-6	40	17	39	2	10	7	11
x_1	72	-21	-40	-17	-39	-2	-10	-7	-11
x_7	9	3	25	41	48	8	13	1	17
x_8	18	-3	11	-23	6	-17	-4	8	1

$$\mathbf{A}^{(t)} = \frac{1}{27}\begin{pmatrix} -10 & -7 & -11 \\ 13 & 1 & 17 \\ -4 & 8 & 1 \end{pmatrix};$$

$$\mathbf{D}^{(0)} - \mathbf{C}^{(0)}\mathbf{A}^{(t)}\mathbf{B}^{(0)} = \frac{1}{27}\begin{pmatrix} 576 & -15 & -62 & -187 & -159 & -22 \\ 162 & 63 & -42 & 24 & -45 & -102 \\ 9 & 48 & -50 & -55 & 12 & 11 \\ -18 & 21 & 4 & -1 & -15 & 11 \\ -18 & -6 & 40 & 17 & 39 & 2 \end{pmatrix};$$

$$\mathbf{C}^{(0)}\mathbf{A}^{(t)} = \frac{1}{27}\begin{pmatrix} -56 & -23 & -40 \\ 3 & 21 & 6 \\ 1 & -2 & -7 \\ 1 & -2 & -7 \\ 10 & 7 & 11 \end{pmatrix}.$$

Then we get the final tableau shown in Table 10.50.

Once we have recovered the last tableau we apply the EPM algorithm. Table 10.50 shows that the pivot is the element associated with the row of x_{14} and the column of x_2. Thus, we perform the pivoting process and get Table 10.51. The next pivot is the element associated with the row of x_{15} and the column of x_3. Thus, we perform the pivoting process and get Table 10.52, which shows that the new solution leads to a value of the objective function of 157/8, which is attained at

$$x_1 = 5/4, x_2 = 3/4, x_3 = 9/16, x_4 = x_5 = x_6 = 0, x_7 = 15/16, x_8 = 13/16.$$

■

Table 10.51 Intermediate tableau. The common denominator of all elements is given in the upper left corner of the table.

$\mathbf{U}^{(8)}$									
21	x_0	x_{14}	x_3	x_4	x_5	x_6	x_9	x_{11}	x_{13}
x_f	438	–15	–46	–146	–132	–11	–43	–19	–35
x_{10}	168	63	–42	21	0	–105	0	21	21
x_{12}	39	48	–46	–41	36	–11	–1	2	7
x_2	18	27	–4	1	15	–11	–1	2	7
x_{15}	–18	–6	**32**	13	27	4	8	5	7
x_1	42	–21	–28	–14	–42	7	–7	–7	–14
x_7	0	3	19	32	39	5	10	1	14
x_8	12	–13	9	–18	3	–12	–3	6	0

Table 10.52 Last tableau. The common denominator of all elements is given in the upper left corner of the table.

$\mathbf{U}^{(9)}$									
32	x_0	x_{14}	x_{15}	x_4	x_5	x_6	x_9	x_{11}	x_{13}
x_f	628	–36	–46	–194	–142	–8	–48	–18	–38
x_{10}	220	84	–42	58	54	–152	16	42	46
x_{12}	20	60	–46	–34	114	–8	16	14	26
x_2	24	40	–4	4	28	–16	0	4	12
x_3	18	6	21	–13	–27	–4	–8	–5	–7
x_1	40	–40	–35	–4	–28	16	0	–4	–12
x_7	30	·10	19	37	35	4	8	–3	15
x_8	26	–2	9	–33	–7	–20	–8	7	–3

10.6.3 Removing Variables

The problem of eliminating a single variable x_p is identical to the problem of making $x_p = 0$. We distinguish among three cases:

1. If x_p is a nonbasic variable, that is, $p \in J_t$, the new constraint $x_p = 0$ is equivalent to deleting the $\mathbf{u}_p^{(t)}$ column. Thus, the resulting tableau gives the optimal solution.

2. If x_p is a basic variable, that is, $p \in I_t$, and $u_{p0}^{(t)} = 0$, the x_p row is removed to obtain the final tableau of the new LPP.

3. If x_p is a basic variable, that is, $p \in I_t$, and $u_{p0}^{(t)} > 0$, we consider the new variable $x_p' \geq 0$ in the form $x_p + x_p{}' = 0$. Thus, the new constraint

Table 10.53 Terms in matrices $\mathbf{U}^{(8)}$ and $\mathbf{U}^{(9)}$ for Example 10.12. The common denominator of all elements is given in the left upper corner of the table.

$\mathbf{U}^{(8)}$									
27	x_0	x_{11}	x_2	x_9	x_4	x_5	x_6	x_{13}	x_3
x_f	576	-23	-15	-56	-187	-159	-22	-40	-62
x_1	72	-7		-10			-2	-11	
x_7	9	1		13			8	17	
x'_8	-18	-8	3	4	23	-6	**17**	-1	-11
x_{10}	162						-102		
x_{12}	9						11		

$\mathbf{U}^{(9)}$									
17	x_0	x_{11}	x_2	x_9	x_4	x_5	x_{13}	x_3	
x_f	348	-21	-7	-32	-99	-105	-26	-48	
x_1	44								
x_6	18								
x_7	11								
x_{10}	34								
x_{12}	13								

$x'_p = -x_p$ is incorporated into the tableau as

$$x'_p = \mathbf{u}'^{(t)}_p \mathbf{x}_{J_t} = -\mathbf{u}^{(t)}_p \mathbf{x}_{J_t}. \tag{10.161}$$

Since $u'^{(t)}_{p0} < 0$ the row x'_p is chosen as the entering variable. We select the leaving variable as usual, and then we perform the pivoting process. However, since x'_p becomes a nonbasic variable, $x'_p = 0$, which can be deleted. The same is true for the x_p row, since now $\mathbf{u}_p = \mathbf{0}$. After removing the row and column we continue with the standard process.

Example 10.12 (Removing variables). Consider again Example 10.3 and assume that we want to remove variable x_8. Thus, we start by introducing the constraint $x'_8 = -x_8$ (see Table 10.53). Using now the usual EPM, we obtain the pivot $u^{(8)}_{8',6} = 17$. After pivoting, we remove the x'_8 row to get Table 10.53. From it, the optimal solution is

$$x^*_1 = \frac{44}{17};\ x^*_6 = \frac{18}{17};\ x^*_7 = \frac{11}{17};\ x^*_f = \frac{348}{17}.$$

■

10.6.4 Adding Variables

Consider the LPP resulting from adding the new main variables $x_{p_1}, \ldots, x_{p_q}$ to the initial LPP:

Maximize

$$x_g = x_f + \sum_{j=1}^{q} c_j x_{p_j} \tag{10.162}$$

subject to

$$a_{i1}x_1 + \cdots + a_{in}x_n + \sum_{j=1}^{q} a_{n+i,j}x_{p_j} \le b_i;\ i = 1, \ldots, m, \tag{10.163}$$

and

$$x_1, \ldots, x_n, x_{p_1}, \ldots, x_{p_q} \ge 0. \tag{10.164}$$

Let $\mathbf{U}'^{(0)}$ be the initial state of the new LPP. Since $\mathbf{A}^{(t)'} = \mathbf{A}^{(t)}$ is known, we can obtain $\mathbf{U}^{(t)'}$using Expressions (10.66) and (10.72):

$$u_{ij}^{(t)} = -\sum_{k\in J_t(\mathcal{S})} u_{ik}^{(t)}u_{kj}^{(0)};\quad i \in I_t(\mathcal{P}), j \in \{x_{p_1}, \ldots, x_{p_q}\} \tag{10.165}$$

and

$$u_{ij}^{(t)} = u_{ij}^{(0)} - \sum_{k\in J_t(\mathcal{S})} u_{ik}^{(t)}u_{kj}^{(0)};\quad i \in I_t(\mathcal{S}), j \in \{x_{p_1}, \ldots, x_{p_q}\}, \tag{10.166}$$

and proceed with the dual EPM to solve the new problem.

Example 10.13 (Adding variables). Assume that we add two new variables x_{p_1} and x_{p_2} to the LPP in Example 10.3, that is, consider the following LPP:

Maximize

$$x_f = 7x_1 + 5x_2 + 5x_3 - 3x_4 + 0x_5 + x_6 + 2x_7 + 3x_8 - x_{p_1} \tag{10.167}$$

subject to

$$\begin{array}{rrrrrrrrrrl}
5x_1 & +4x_2 & +3x_3 & +2x_4 & +x_5 & +2x_6 & +3x_7 & +4x_8 & & +x_{p_2} & \le 17\\
2x_1 & -x_2 & +4x_3 & -2x_4 & +3x_5 & +3x_6 & +x_7 & -x_8 & -2x_{p_1} & -2x_{p_2} & \le 11\\
3x_1 & +2x_2 & +3x_3 & -2x_4 & +x_5 & -x_6 & +2x_7 & -x_8 & +x_{p_1} & -x_{p_2} & \le 8\\
x_1 & -x_2 & +2x_3 & +2x_4 & -x_5 & & +x_7 & +x_8 & -2x_{p_1} & +x_{p_2} & \le 4\\
4x_1 & +3x_2 & +x_3 & -x_4 & -2x_5 & +x_6 & +4x_7 & +3x_8 & & +x_{p_2} & \ge 14
\end{array} \tag{10.168}$$

with

$$x_1 \ge 0, \ldots, x_8, x_{p_1}, x_{p_2} \ge 0. \tag{10.169}$$

Using the method above, we obtain the $\mathbf{U}^{(0)}$ tableau in Table 10.54. Since $u_{f2}^{(8)} > 0$, we must proceed with the dual EPM. Choose the pivot $u_{82}^{(8)} = -25$ and obtain the $\mathbf{U}^{(9)}$ tableau, which gives the optimal solution:

$$\begin{array}{l}
x_1^* = \dfrac{71}{25};\ x_2^* = \dfrac{7}{25};\ x_7^* = \dfrac{3}{25};\ x_{p_2}^* = \dfrac{33}{25};\\
x_3^* = x_4^* = x_5^* = x_6^* = x_8^* = x_{p_1}^* = 0;\ x_g^* = \dfrac{538}{25}.
\end{array} \tag{10.170}$$

■

Table 10.54 Terms in matrices $\mathbf{U}^{(0)'}$,$\mathbf{U}^{(7)'}$, $\mathbf{U}^{(8)'}$ and $\mathbf{U}^{(9)'}$ in Example 10.13. The common denominator of all elements is given in the upper left corner of the table.

$\mathbf{U}^{(0)'}$

1	x_0	x_1	x_2	x_3	x_4	x_5	x_6	x_7	x_8	x_{p_1}	x_{p_2}
x_f	0	7	5	5	–3	0	1	2	3	–1	0
x_9	17	–5	–4	**–3**	–2	–1	–2	–3	–4	0	–1
x_{10}	11	–2	1	–4	2	–3	–3	–1	1	2	2
x_{11}	8	–3	–2	–3	2	–1	1	–2	1	–1	1
x_{12}	4	–1	1	–2	–2	1	0	–1	–1	2	–1
x_{13}	–14	4	3	1	–1	–2	1	4	3	0	1

$\mathbf{U}^{(7)'}$

27	x_0	x_{11}	x_2	x_9	x_4	x_5	x_6	x_{13}	x_3	x_{p_1}	x_{p_2}
x_f	576	–23	–15	–56	–187	–159	–22	–40	–62	–50	7
x_1	72	–7		–10				–11		–7	8
x_7	9	1		13				17		1	–5
x_8	18	8		–4				1		8	–13
x_{10}	162									75	30
x_{12}	9	–2	48	1	–55	12	11	–7	–50	52	**–17**

$\mathbf{U}^{(8)'}$

17	x_0	x_{11}	x_2	x_9	x_4	x_5	x_6	x_{13}	x_3	x_{p_1}	x_{12}
x_f	365	–15	3	–35	–132	–97	–11	–27	–52	–18	–7
x_1	48	–5	1	–6				–9			–8
x_7	4	1	–7	8				12			5
x_8	7	6	**–25**	–3	12	–2	–16	4	31	–20	13
x_{p_2}	9	–2	48	1				–7			–27
x_{10}	112		93								

$\mathbf{U}^{(9)'}$

25	x_0	x_{11}	x_8	x_9	x_4	x_5	x_6	x_{13}	x_3	x_{p_1}	x_{12}
x_f	538	–21	–3	–52	–192	–143	–19	–39	–71	–30	–8
x_1	71										
x_2	7										
x_7	3										
x_{p_2}	33										
x_{10}	203										

Table 10.55 Amounts of nutrients for thiamine, niacin, and calories.

Nutrients	Amount of each nutrient (mg) in 1 ounce of Krunchies (K)	Amount of each nutrient (mg) in 1 ounce of Crispies (C)	Requirement of each nutrient (in mg)
Thiamine	0.10	0.25	1
Niacin	1.00	0.25	5
Calories	110.00	120.00	400
Cost	3.8 cents per ounce	4.2 cents per ounce	

Exercises

10.1 Consider the following LPP:

Maximize

$$3x_1 + x_2 \tag{10.171}$$

subject to

$$\begin{array}{rrcr} x_1 & +2x_2 & \leq & 8 \\ 3x_1 & -4x_2 & \leq & 12 \\ x_1 & & \geq & 0 \\ & x_2 & \geq & 0 \end{array} \tag{10.172}$$

(a) Solve this problem using the EPM algorithm with complete tables.

(b) Solve this problem using the simplified EPM algorithm.

(c) Solve the same problem after removing the second constraint.

(d) Solve the previous problems graphically.

10.2 Solve the previous exercise after the following modifications:

(a) Adding the new constraints

$$\begin{array}{rrcr} x_1 & +x_2 & \leq & 4 \\ & x_2 & \leq & 3 \end{array} \tag{10.173}$$

(b) Removing the first constraint.

(c) Removing the first constraint after adding the above constraints.

10.3 Solve the diet problem in Table 10.55.

10.4 Consider the following LPP:

Maximize

$$x_1 + 2x_2 + 3x_3 - x_4 \tag{10.174}$$

subject to

$$\begin{array}{rrrrcl} x_1 & +2x_2 & +3x_3 & & = & 15 \\ 2x_1 & +x_2 & +5x_3 & & = & 20 \\ x_1 & +2x_2 & +x_3 & +x_4 & = & 10 \\ x_1 & & & & \geq & 0 \\ & x_2 & & & \geq & 0. \end{array} \tag{10.175}$$

(a) Solve this problem using the EPM algorithm with complete tables.

(b) Solve this problem using the simplified EPM algorithm.

(c) Solve the same problem after removing the second constraint.

10.5 Solve the following LPP:

Minimize

$$2x_1 - x_2 + x_3 - 5x_4 + 22x_5$$

subject to

$$\begin{array}{rrrrrcl} x_1 & & & +x_4 & +x_5 & = & 6 \\ & x_2 & & +x_4 & -4x_5 & = & 3 \\ & & x_3 & +3x_4 & +2x_5 & = & 10 \\ & & x_j & \geq & 0; (j = 1, \ldots, 5). & & \end{array}$$

10.6 Figure 10.3 is a PERT (program evaluation and review) network with the distances interpreted as the time to complete an activity. Node 0 is the start of the project and node m is the completion of the project. The intermediate nodes are specific events. Variable x_i represents the occurrence time of event (node) i. Let $t_{ij} \geq 0$ be the assumed duration of activity (arc) (i, j). The aim is to determine the earliest start time for each activity that will result in the shortest total project time; that is, we want to minimize the time difference between event 0 and event m that is given by $x_m - x_0$. The problem is then:

Minimize

$$x_m - x_0$$

subject to

$$x_j - x_i \geq t_{ij}; \forall (i, j).$$

In other words, solve the following LPP:

Minimize

$$x_m$$

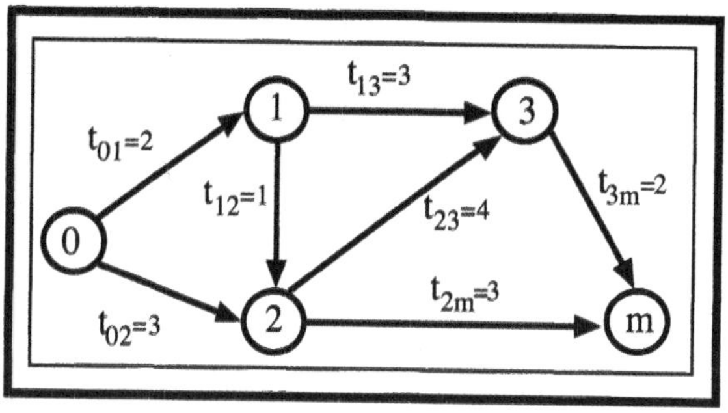

Fig. 10.3 Diagram for the PERT problem.

subject to

$$\begin{array}{rrrrl} x_1 & & & & \geq 2 \\ & +x_2 & & & \geq 3 \\ -x_1 & +x_2 & & & \geq 1 \\ -x_1 & & +x_3 & & \geq 3 \\ & -x_2 & +x_3 & & \geq 4 \\ & -x_2 & & +x_m & \geq 3 \\ & & -x_3 & +x_m & \geq 2 \\ & & & & x_j \geq 0; \forall j. \end{array}$$

Part IV

Applications

Introduction to Part IV

While a mathematician is mainly interested in theoretical problems, an engineer normally looks for real–life applications. In this part of the book, a collection of practical examples are given to illustrate the power of the methods described in previous chapters and their applications to real problems.

Five different analyses are performed in each example:

- **The physical interpretation of the problem:** The problem is described in terms of its physical meaning. It is the first step of the process and building the model of the problem.

- **The mathematical representation of the problem:** A mathematical model for the problem, in terms of a system of equations and/or inequalities, including the adequate constraints of the variables involved, is derived.

- **The compatibility problem:** The conditions for the mathematical problem to make sense and the corresponding physical implications of these mathematical conditions are given.

- **The solution problem:** The general solution of the problem, regardless of whether it is unique or includes several solutions, is obtained.

- **The physical interpretation of the solutions:** Physical interpretations of its three components – the linear space $\mathbf{V}$, the acute cone $\mathbf{W}$, and the polytope $\mathbf{U}$ summand – are given.

11

Applications

11.1 INTRODUCTION

In this chapter we describe several examples of applications. The first selected applications are taken from matrix analysis of engineering structures. They illustrate how the stiffness matrices of some structures can be obtained, leading to systems of equations that relate forces and moments with displacements and rotations. We solve different linear systems of equations associated with several structures, but we start from the solution of one of them to avoid redoing a great part of the computations.

It is shown that changing the geometric or structural characteristics of the structural pieces, or adding/removing pieces, leads to changes in the system of equations, allowing us to obtain new solutions from the previous ones.

We also consider other examples, such as the transportation problem, the production-scheduling problem, the input–output tables, the diet problem, or the network flow problems, which are classical problems in linear programming.

However, unlike the classical treatment of these problems, we do not limit our analysis to obtaining the optimal solution; instead, we obtain the feasible set of solutions, that is, the general solution of the system of equalities and inequalities, corresponding to the particular problem. In addition, we give the corresponding physical interpretations, which show the strong connection between the mathematical models and the reality behind them.

11.2 A MOTIVATING EXAMPLE FROM MATRIX ANALYSIS OF ENGINEERING STRUCTURES

Consider the structural piece in Fig. 11.1, where

$$H_i, V_i, M_i; i = 1, 2,$$

are the horizontal and vertical forces and the moments applied to nodes 1 and 2, and

$$h_i, v_i, \theta_i; i = 1, 2,$$

are the horizontal and vertical displacements and the rotations, respectively, of these nodes.

Using the well-known two cases of *clamped at both ends* and *clamped and supported* beams in Fig. 11.2(a) and (b), respectively, and the well-known simple axial compression law, the following system of equations establishes the relationships among these variables.

$$\begin{pmatrix} H_1 \\ V_1 \\ M_1 \\ -- \\ H_2 \\ V_2 \\ M_2 \end{pmatrix} = \left(\begin{array}{c|c} \mathbf{K}_1 & \mathbf{K}_2 \\ \hline \mathbf{K}_2^T & \mathbf{K}_3 \end{array}\right) \begin{pmatrix} h_1 \\ v_1 \\ \theta_1 \\ -- \\ h_2 \\ v_2 \\ \theta_2 \end{pmatrix} = \mathbf{K} \begin{pmatrix} h_1 \\ v_1 \\ \theta_1 \\ -- \\ h_2 \\ v_2 \\ \theta_2 \end{pmatrix}$$

$$= \left(\begin{array}{ccc|ccc} \frac{AE}{L} & 0 & 0 & -\frac{AE}{L} & 0 & 0 \\ 0 & \frac{12EI}{L^3} & \frac{6EI}{L^2} & 0 & -\frac{12EI}{L^3} & \frac{6EI}{L^2} \\ 0 & \frac{6EI}{L^2} & \frac{4EI}{L} & 0 & -\frac{6EI}{L^2} & \frac{2EI}{L} \\ \hline -\frac{AE}{L} & 0 & 0 & \frac{AE}{L} & 0 & 0 \\ 0 & -\frac{12EI}{L^3} & -\frac{6EI}{L^2} & 0 & \frac{12EI}{L^3} & -\frac{6EI}{L^2} \\ 0 & \frac{6EI}{L^2} & \frac{2EI}{L} & 0 & -\frac{6EI}{L^2} & \frac{4EI}{L} \end{array}\right) \begin{pmatrix} h_1 \\ v_1 \\ \theta_1 \\ -- \\ h_2 \\ v_2 \\ \theta_2 \end{pmatrix}, \tag{11.1}$$

where the matrix $\mathbf{K}$ of the system is known as the *stiffness matrix*, A is the cross section, E is Young's modulus, I is the moment of inertia of the cross section of the piece, L is the length, $\mathbf{K}_2^T$ is the transpose of $\mathbf{K}_2$, and the meaning of the block matrices $\mathbf{K}_1$, $\mathbf{K}_2$, and $\mathbf{K}_3$ becomes obvious from (11.1).

The compatibility of system (11.1) leads to (see Equation (8.3))

$$(H_1 + H_2)^2 + (M_1 + M_2 - LV_1)^2 + (V_1 + V_2)^2 = 0 \tag{11.2}$$

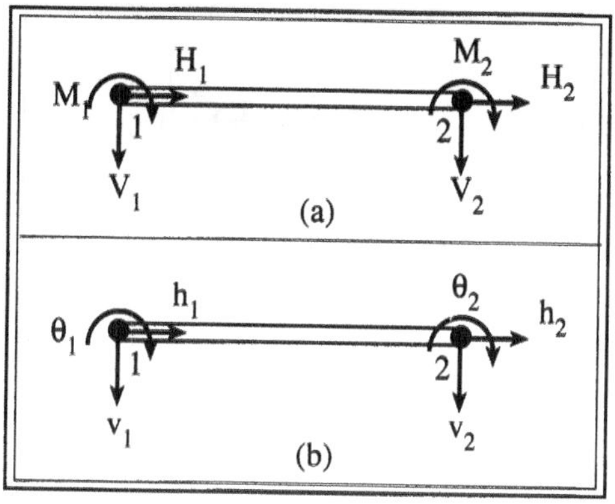

Fig. 11.1 (a) Node forces and moments, and (b) node displacements and rotations for an elementary horizontal piece.

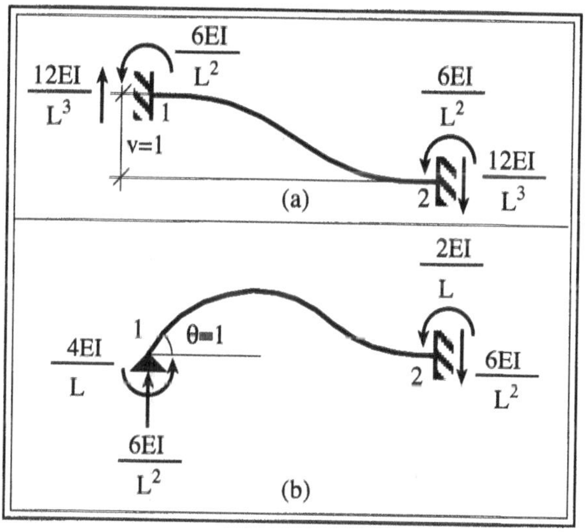

Fig. 11.2 Basic cases of clamped at both ends and clamped and supported beams used for building the stiffness matrix. The relative displacement in (a) and the rotation in (b) are unity.

or, equivalently, to

$$H_1 + H_2 = 0, \tag{11.3}$$
$$V_1 + V_2 = 0, \tag{11.4}$$
$$M_1 + M_2 - LV_1 = 0, \tag{11.5}$$

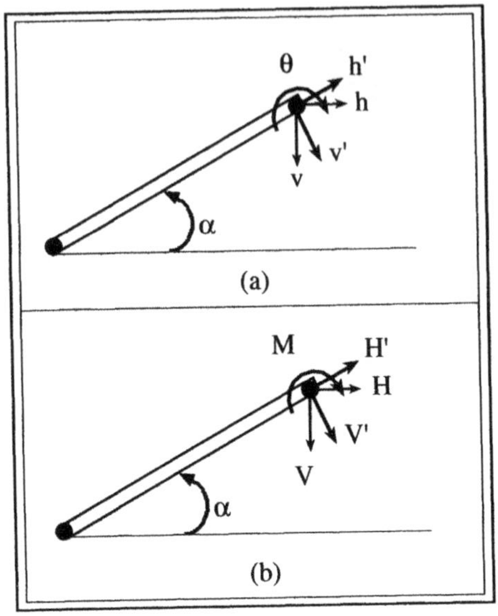

Fig. 11.3 Piece rotated an angle α with respect to the horizontal line.

which are the equations of equilibrium of horizontal and vertical forces and of moments, respectively. This is an important result: *compatibility* of the above system means *equilibrium.*

The above relationships among forces, moments, displacements, and rotations for a horizontal piece can be generalized to include pieces in any position by means of the following considerations.

Consider the piece in Fig. 11.3, where h' and v', and H' and V' are the displacements and forces in the piece and its orthogonal direction, respectively. Then we have

$$\begin{pmatrix} h \\ v \\ m \end{pmatrix} = \mathbf{G} \begin{pmatrix} h' \\ v' \\ m' \end{pmatrix}; \begin{pmatrix} H \\ V \\ M \end{pmatrix} = \mathbf{G} \begin{pmatrix} H' \\ V' \\ M' \end{pmatrix}, \tag{11.6}$$

where

$$\mathbf{G} = \begin{pmatrix} \cos\alpha & \sin\alpha & 0 \\ -\sin\alpha & \cos\alpha & 0 \\ 0 & 0 & 1 \end{pmatrix} \tag{11.7}$$

is the rotation matrix.

Thus, for the piece in Fig. 11.3, we have

$$\begin{pmatrix} H_1 \\ V_1 \\ M_1 \\ H_2 \\ V_2 \\ M_2 \end{pmatrix} = \begin{pmatrix} \mathbf{G} & 0 \\ 0 & \mathbf{G} \end{pmatrix} \mathbf{K}' \begin{pmatrix} \mathbf{G}^{-1} & 0 \\ 0 & \mathbf{G}^{-1} \end{pmatrix} \begin{pmatrix} h_1 \\ v_1 \\ \theta_1 \\ h_2 \\ v_2 \\ \theta_2 \end{pmatrix}. \tag{11.8}$$

When we have a structure composed of several pieces, its stiffness matrix can be obtained from the stiffness matrices of all its pieces. The following Mathematica program allows assembling stiffness matrices of different pieces in the stiffness matrix $\mathbf{K}$ of a composed structure.

Program 11.1 (Module for obtaining the assembled matrix)

```
AssembleMatrix[K_,i1_,j1_,L_,A_,II_,EE_,alfa_]:=
  Module[{K1=K,K3},
    K3=K2[L,A,II,EE,alfa];
    Do[K1[[3*i1+i-3,3*i1+j-3]]+=K3[[i,j]],{i,1,3},{j,1,3}];
    Do[K1[[3*i1+i-3,3*j1+j-6]]+=K3[[i,j]],{i,1,3},{j,4,6}];
    Do[K1[[3*j1+i-6,3*i1+j-3]]+=K3[[i,j]],{i,4,6},{j,1,3}];
    Do[K1[[3*j1+i-6,3*j1+j-6]]+=K3[[i,j]],{i,4,6},{j,4,6}];
    K1
]
```

In this program

$$K2[L, A, II, EE, alfa] = \begin{pmatrix} \mathbf{G} & 0 \\ 0 & \mathbf{G} \end{pmatrix} \mathbf{K}' \begin{pmatrix} \mathbf{G}^{-1} & 0 \\ 0 & \mathbf{G}^{-1} \end{pmatrix}, \tag{11.9}$$

that is, the stiffness matrix in (11.8).

For example, consider the structure in Fig. 11.4(a). The following Mathematica program allows obtaining its associated system of equations, which is shown in Fig. 11.5.

```
NumberUnknowns=12;
K=DiagonalMatrix[Table[0,{i,1,NumberUnknowns}]];
Bars={{1,2,1,1,1,1,Pi/2},{2,3,1,1,1,1,0},{3,4,1,1,1,1,-Pi/2}};
MatrixForm[Transpose[Bars]]
Do[K=AssembleMatrix[K,Bars[[jj,1]],Bars[[jj,2]],Bars[[jj,3]],
Bars[[jj,4]],Bars[[jj,5]],Bars[[jj,6]],Bars[[jj,7]]]
,{jj,1,Length[Bars]}]
MatrixForm[K]
```

In the upper part of Fig. 11.5, the building process of the stiffness matrix is illustrated. First, the stiffness matrices $\mathbf{K}^I, \mathbf{K}^{II}$, and $\mathbf{K}^{III}$ of the three pieces I, II, and III are calculated, and next, their corresponding blocks (see

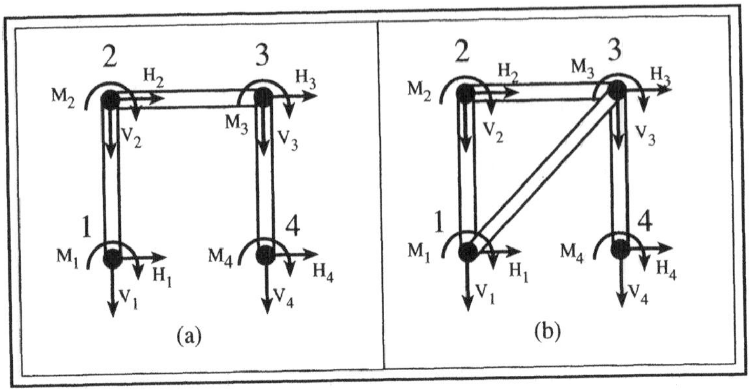

Fig. 11.4 Two one-level structures.

(11.1)) are assembled in the corresponding places (see the node numbers and the block matrices in the upper right matrix in Fig. 11.5).

It is worth mentioning that assembling the stiffness matrix by adding the block matrices corresponding to all pieces is equivalent to establishing the equilibrium of horizontal and vertical forces and of moments in all nodes.

The corresponding stiffness matrix for $A_i = E_i = I_i = L_i = 1; i = 1, 2, 3,$ becomes:

$$\begin{pmatrix} 12 & 0 & 6 & -12 & 0 & 6 & 0 & 0 & 0 & 0 & 0 & 0 \\ 0 & 1 & 0 & 0 & -1 & 0 & 0 & 0 & 0 & 0 & 0 & 0 \\ 6 & 0 & 4 & -6 & 0 & 2 & 0 & 0 & 0 & 0 & 0 & 0 \\ -12 & 0 & -6 & 13 & 0 & -6 & -1 & 0 & 0 & 0 & 0 & 0 \\ 0 & -1 & 0 & 0 & 13 & 6 & 0 & -12 & 6 & 0 & 0 & 0 \\ 6 & 0 & 2 & -6 & 6 & 8 & 0 & -6 & 2 & 0 & 0 & 0 \\ 0 & 0 & 0 & -1 & 0 & 0 & 13 & 0 & -6 & -12 & 0 & -6 \\ 0 & 0 & 0 & 0 & -12 & -6 & 0 & 13 & -6 & 0 & -1 & 0 \\ 0 & 0 & 0 & 0 & 6 & 2 & -6 & -6 & 8 & 6 & 0 & 2 \\ 0 & 0 & 0 & 0 & 0 & 0 & -12 & 0 & 6 & 12 & 0 & 6 \\ 0 & 0 & 0 & 0 & 0 & 0 & 0 & -1 & 0 & 0 & 1 & 0 \\ 0 & 0 & 0 & 0 & 0 & 0 & -6 & 0 & 2 & 6 & 0 & 4 \end{pmatrix} . \tag{11.10}$$

As can be shown, the compatibility of the above system leads to (see Equation (8.3))

$$\begin{aligned} H_1 + H_2 + H_3 + H_4 &= 0 \\ V_1 + V_2 + V_3 + V_4 &= 0 \\ H_2 + H_3 + M_1 + M_2 + M_3 + M_4 - V_1 - V_2 &= 0, \end{aligned} \tag{11.11}$$

which, again, are the equations of equilibrium of horizontal and vertical forces and of moments, respectively. Note that taking moments with respect to node 4, we obtain (11.11).

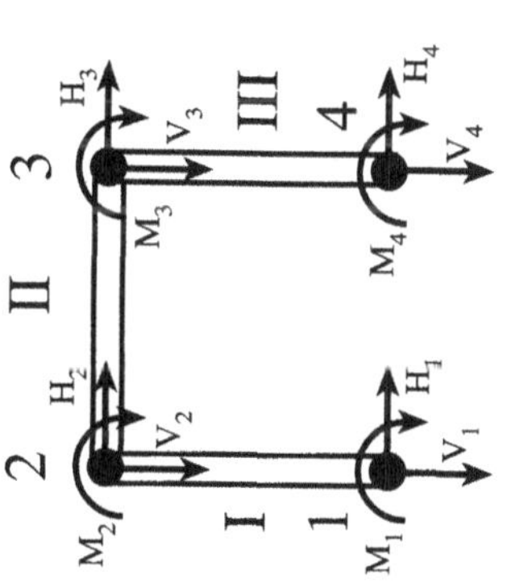

$$
\begin{array}{c|cccc}
 & 1 & 2 & 3 & 4 \\
\hline
1 & K_1^{I} & K_2^{I} & & \\
2 & {K_2^{I}}^{T} & K_3^{I}+K_1^{II} & K_2^{II} & \\
3 & & {K_2^{II}}^{T} & K_3^{II}+K_1^{III} & K_2^{III} \\
4 & & & {K_2^{III}}^{T} & K_3^{III}
\end{array}
$$

$$
\left(\begin{array}{cccccccccccc}
\frac{12E_1I_1}{L_1} & 0 & \frac{6E_1I_1}{L_1} & -\frac{12E_1I_1}{L_1} & 0 & \frac{6E_1I_1}{L_1} & 0 & 0 & 0 & 0 & 0 & 0 \\
0 & \frac{A_1E_1}{L_1} & 0 & 0 & -\frac{A_1E_1}{L_1} & 0 & 0 & 0 & 0 & 0 & 0 & 0 \\
\frac{6E_1I_1}{L_1} & 0 & \frac{4E_1I_1}{L_1} & -\frac{6E_1I_1}{L_1} & 0 & \frac{2E_1I_1}{L_1} & 0 & 0 & 0 & 0 & 0 & 0 \\
-\frac{12E_1I_1}{L_1} & 0 & -\frac{6E_1I_1}{L_1} & \frac{12E_1I_1}{L_1}+\frac{A_2E_2}{L_2} & 0 & -\frac{6E_1I_1}{L_1} & -\frac{A_2E_2}{L_2} & 0 & 0 & 0 & 0 & 0 \\
0 & -\frac{A_1E_1}{L_1} & 0 & 0 & \frac{A_1E_1}{L_1}+\frac{12E_2I_2}{L_2} & \frac{6E_2I_2}{L_2} & 0 & -\frac{12E_2I_2}{L_2} & \frac{6E_2I_2}{L_2} & 0 & 0 & 0 \\
\frac{6E_1I_1}{L_1} & 0 & \frac{2E_1I_1}{L_1} & -\frac{6E_1I_1}{L_1} & \frac{6E_2I_2}{L_2} & \frac{4E_1I_1}{L_1}+\frac{4E_2I_2}{L_2} & 0 & -\frac{6E_2I_2}{L_2} & \frac{2E_2I_2}{L_2} & 0 & 0 & 0 \\
0 & 0 & 0 & -\frac{A_2E_2}{L_2} & 0 & 0 & \frac{A_2E_2}{L_2}+\frac{12E_3I_3}{L_3} & 0 & -\frac{6E_3I_3}{L_3} & -\frac{12E_3I_3}{L_3} & 0 & -\frac{6E_3I_3}{L_3} \\
0 & 0 & 0 & 0 & -\frac{12E_2I_2}{L_2} & -\frac{6E_2I_2}{L_2} & 0 & \frac{12E_2I_2}{L_2}+\frac{A_3E_3}{L_3} & -\frac{6E_2I_2}{L_2} & 0 & -\frac{A_3E_3}{L_3} & 0 \\
0 & 0 & 0 & 0 & \frac{6E_2I_2}{L_2} & \frac{2E_2I_2}{L_2} & -\frac{6E_3I_3}{L_3} & -\frac{6E_2I_2}{L_2} & \frac{4E_2I_2}{L_2}+\frac{4E_3I_3}{L_3} & \frac{6E_3I_3}{L_3} & 0 & \frac{2E_3I_3}{L_3} \\
0 & 0 & 0 & 0 & 0 & 0 & -\frac{12E_3I_3}{L_3} & 0 & \frac{6E_3I_3}{L_3} & \frac{12E_3I_3}{L_3} & 0 & \frac{6E_3I_3}{L_3} \\
0 & 0 & 0 & 0 & 0 & 0 & 0 & -\frac{A_3E_3}{L_3} & 0 & 0 & \frac{A_3E_3}{L_3} & 0 \\
0 & 0 & 0 & 0 & 0 & 0 & -\frac{6E_3I_3}{L_3} & 0 & \frac{2E_3I_3}{L_3} & \frac{6E_3I_3}{L_3} & 0 & \frac{4E_3I_3}{L_3}
\end{array}\right)
$$

Fig. 11.5 Stiffness matrix for the structure in Figure 11.4(a), and illustration of how to build it.

Assuming the compatibility conditions hold, the general solution of the system with associated matrix in Fig. 11.5 is

$$
6\begin{pmatrix} h_1\\ v_1\\ \theta_1\\ h_2\\ v_2\\ \theta_2\\ h_3\\ v_3\\ \theta_3\\ h_4\\ v_4\\ \theta_4 \end{pmatrix}
= H_1\begin{pmatrix} 16\\ 1\\ -12\\ 5\\ 6\\ -9\\ -1\\ 0\\ -3\\ 0\\ 0\\ 0 \end{pmatrix}
+ H_2\begin{pmatrix} 6\\ -3\\ 3\\ 8\\ -3\\ 3\\ 2\\ 0\\ 3\\ 0\\ 0\\ 0 \end{pmatrix}
+ H_3\begin{pmatrix} -1\\ -3\\ 3\\ 2\\ -3\\ 3\\ 2\\ 0\\ 3\\ 0\\ 0\\ 0 \end{pmatrix}
+ M_1\begin{pmatrix} -12\\ -9\\ 18\\ 3\\ -9\\ 12\\ 3\\ 0\\ 6\\ 0\\ 0\\ 0 \end{pmatrix}
$$

$$
+M_2\begin{pmatrix} -9\\ -9\\ 12\\ 3\\ -9\\ 12\\ 3\\ 0\\ 6\\ 0\\ 0\\ 0 \end{pmatrix}
+ M_3\begin{pmatrix} -3\\ -6\\ 6\\ 3\\ -6\\ 6\\ 3\\ 0\\ 6\\ 0\\ 0\\ 0 \end{pmatrix}
+ V_1\begin{pmatrix} 6\\ 20\\ -9\\ -3\\ 14\\ -9\\ -3\\ 6\\ -6\\ 0\\ 0\\ 0 \end{pmatrix}
+ V_2\begin{pmatrix} 6\\ 14\\ -9\\ -3\\ 14\\ -9\\ -3\\ 6\\ -6\\ 0\\ 0\\ 0 \end{pmatrix}
$$

$$
+V_3\begin{pmatrix} 0\\ 6\\ 0\\ 0\\ 6\\ 0\\ 0\\ 0\\ 0\\ 0\\ 0\\ 0 \end{pmatrix}
+ \rho_1\begin{pmatrix} 1\\ 0\\ 0\\ 1\\ 0\\ 0\\ 1\\ 0\\ 0\\ 1\\ 0\\ 0 \end{pmatrix}
+ \rho_2\begin{pmatrix} 0\\ 1\\ 0\\ 0\\ 1\\ 0\\ 0\\ 1\\ 0\\ 0\\ 1\\ 0 \end{pmatrix}
+ \rho_3\begin{pmatrix} 0\\ -1\\ 1\\ 1\\ -1\\ 1\\ 1\\ 0\\ 1\\ 0\\ 0\\ 1 \end{pmatrix}, \tag{11.12}
$$

where ρ_1, ρ_2, and ρ_3 are arbitrary real constants.

Let us now consider several cases of structures with different boundary conditions.

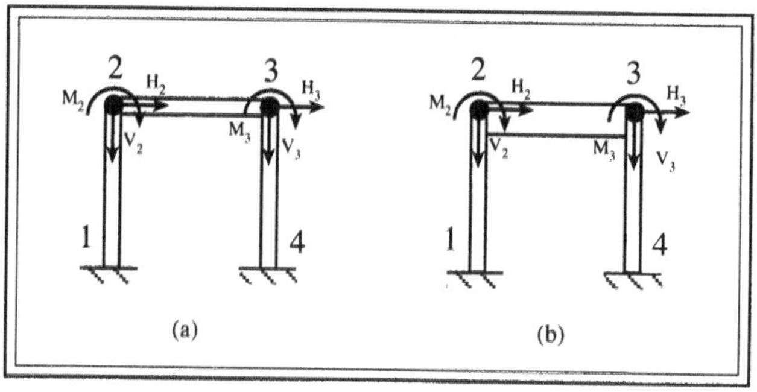

Fig. 11.6 (a) Initial structure S_1 and (b) modified structure S_2 obtained by increasing the cross section of the horizontal beam but keeping the rest of the characteristics identical to the initial structure.

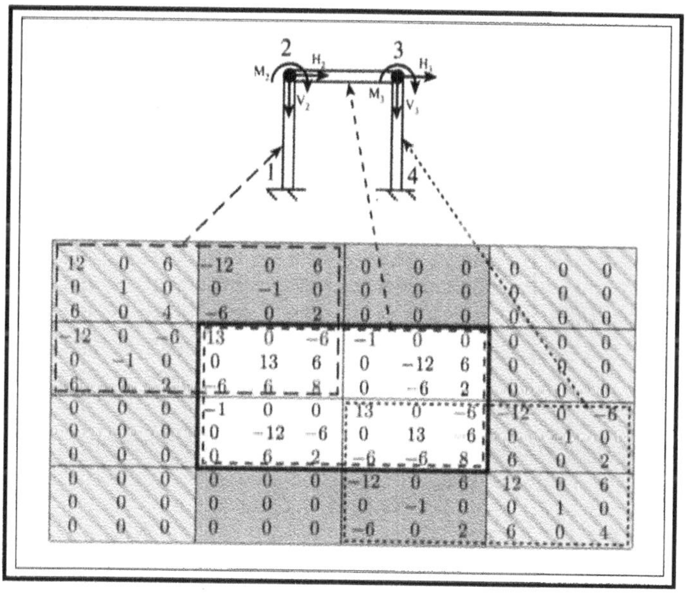

Fig. 11.7 Stiffness matrix for Case 1.

Case 1: Consider the structure in Fig. 11.6(a), where

$$h_1 = v_1 = \theta_1 = 0 \text{ and } h_4 = v_4 = \theta_4 = 0. \tag{11.13}$$

Since the corresponding displacement and rotations of nodes 1 and 4 are null, we can remove the first and the last three columns (see Fig. 11.7). In addition, to avoid compatibility problems, we remove the first and last three equations, corresponding to the reactions in nodes 1 and 4, and obtain the following system of equations (see Fig. 11.7):

$$\begin{pmatrix} H_2 \\ V_2 \\ M_2 \\ H_3 \\ V_3 \\ M_3 \end{pmatrix} = \begin{pmatrix} 13 & 0 & -6 & -1 & 0 & 0 \\ 0 & 13 & 6 & 0 & -12 & 6 \\ -6 & 6 & 8 & 0 & -6 & 2 \\ -1 & 0 & 0 & 13 & 0 & -6 \\ 0 & -12 & -6 & 0 & 13 & -6 \\ 0 & 6 & 2 & -6 & -6 & 8 \end{pmatrix} \begin{pmatrix} h_2 \\ v_2 \\ \theta_2 \\ h_3 \\ v_3 \\ \theta_3 \end{pmatrix}, \qquad (11.14)$$

from which we get

$$\begin{pmatrix} h_2 \\ v_2 \\ \theta_2 \\ h_3 \\ v_3 \\ \theta_3 \end{pmatrix} = \begin{pmatrix} 13 & 0 & -6 & -1 & 0 & 0 \\ 0 & 13 & 6 & 0 & -12 & 6 \\ -6 & 6 & 8 & 0 & -6 & 2 \\ -1 & 0 & 0 & 13 & 0 & -6 \\ 0 & -12 & -6 & 0 & 13 & -6 \\ 0 & 6 & 2 & -6 & -6 & 8 \end{pmatrix}^{-1} \begin{pmatrix} H_2 \\ V_2 \\ M_2 \\ H_3 \\ V_3 \\ M_3 \end{pmatrix}$$

$$= \frac{1}{1488} \begin{pmatrix} 305 & -144 & 393 & 119 & 144 & 207 \\ -144 & 912 & -288 & -144 & 576 & -288 \\ 393 & -288 & 817 & 207 & 288 & 383 \\ 119 & -144 & 207 & 305 & 144 & 393 \\ 144 & 576 & 288 & 144 & 912 & 288 \\ 207 & -288 & 383 & 393 & 288 & 817 \end{pmatrix} \begin{pmatrix} H_2 \\ V_2 \\ M_2 \\ H_3 \\ V_3 \\ M_3 \end{pmatrix}; \qquad (11.15)$$

and from the first and last three rows we obtain the reactions

$$\begin{pmatrix} H_1 \\ V_1 \\ M_1 \\ H_4 \\ V_4 \\ M_4 \end{pmatrix} = \begin{pmatrix} -12 & 0 & 6 & 0 & 0 & 0 \\ 0 & -1 & 0 & 0 & 0 & 0 \\ -6 & 0 & 2 & 0 & 0 & 0 \\ 0 & 0 & 0 & -12 & 0 & 6 \\ 0 & 0 & 0 & 0 & -1 & 0 \\ 0 & 0 & 0 & -6 & 0 & 2 \end{pmatrix} \begin{pmatrix} h_2 \\ v_2 \\ \theta_2 \\ h_3 \\ v_3 \\ \theta_3 \end{pmatrix}. \qquad (11.16)$$

Thus, first the unknown displacements are obtained from (11.14), and then the reactions are obtained by (11.15).

Case 2: Consider the structure in Fig. 11.6(b), which is exactly the same as the structure in Fig. 11.6(a), but obtained by increasing the cross section of the horizontal beam and keeping the rest of the characteristics unchanged.

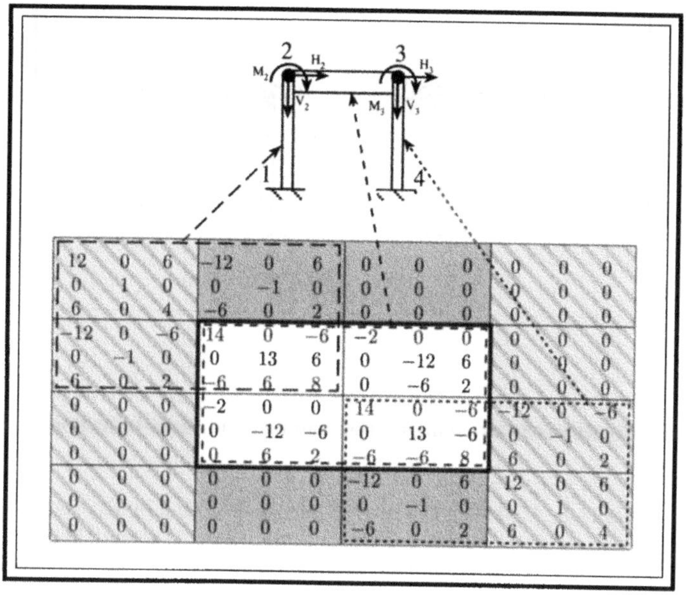

Fig. 11.8 Stiffness matrix for Case 2.

Thus, Equation (11.15) transforms to (see Fig. 11.8)

$$\begin{pmatrix} h_2 \\ v_2 \\ \theta_2 \\ h_3 \\ v_3 \\ \theta_3 \end{pmatrix} = \begin{pmatrix} \mathbf{14} & 0 & -6 & \mathbf{-2} & 0 & 0 \\ 0 & 13 & 6 & 0 & -12 & 6 \\ -6 & 6 & 8 & 0 & -6 & 2 \\ \mathbf{-2} & 0 & 0 & \mathbf{14} & 0 & -6 \\ 0 & -12 & -6 & 0 & 13 & -6 \\ 0 & 6 & 2 & -6 & -6 & 8 \end{pmatrix}^{-1} \begin{pmatrix} H_2 \\ V_2 \\ M_2 \\ H_3 \\ V_3 \\ M_3 \end{pmatrix}$$

$$= \frac{1}{930} \begin{pmatrix} 179 & -90 & 234 & 86 & 90 & 141 \\ -90 & 570 & -180 & -90 & 360 & -180 \\ 234 & -180 & 499 & 141 & 180 & 251 \\ 86 & -90 & 141 & 179 & 90 & 234 \\ 90 & 360 & 180 & 90 & 570 & 180 \\ 141 & -180 & 251 & 234 & 180 & 499 \end{pmatrix} \begin{pmatrix} H_2 \\ V_2 \\ M_2 \\ H_3 \\ V_3 \\ M_3 \end{pmatrix}. \tag{11.17}$$

Note that the only differences are elements $k_{1,1}, k_{1,4}, k_{4,1}$, and $k_{4,4}$, which are boldface in (11.17).

Since we know the inverse (11.15) of the previous matrix (11.14), the new inverse can be calculated based on the old one. The process is illustrated in Table 11.1. In the first step we incorporate the change of the first row, in the second step, we incorporate the changes of the fourth row, and finally, we get the desired inverse in Step 3.

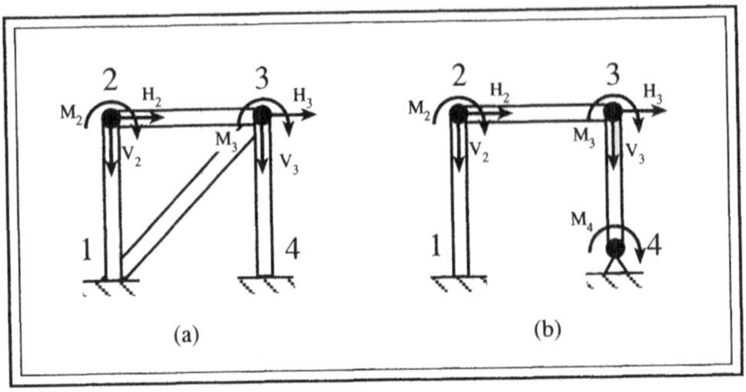

Fig. 11.9 (a) Modified structure S_3 resulting after adding an inclined piece to S_1, and (b) modified structure S_4 by changing the clamped end of S_1 to a support.

Table 11.1 Updating the inverse matrix in Case 2 by incorporating the changes of rows 1 (Step 1) and 4 (Step 2) to get the final inverse matrix (Step 3). The common denominator of all elements is given in the upper left corner of the table.

1488			**Step 1**			
14	305	−144	393	119	144	207
0	−144	912	−288	−144	576	−288
−6	393	−288	817	207	288	383
−2	119	−144	207	305	144	393
0	144	576	288	144	912	288
0	207	−288	383	393	288	817
	1674	0	186	−186	0	−186
1674			**Step 2**			
−2	305	−162	404	172	162	271
0	−144	1026	−306	−180	648	−342
0	393	−324	870	282	324	480
14	119	−162	218	358	162	457
0	144	648	306	180	1026	342
−6	207	−324	405	468	324	945
	−186	0	−186	1860	0	186
930			**Step 3**			
	179	−90	234	86	90	141
	−90	570	−180	−90	360	−180
	234	−180	499	141	180	251
	86	−90	141	179	90	234
	90	360	180	90	570	180
	141	−180	251	234	180	499

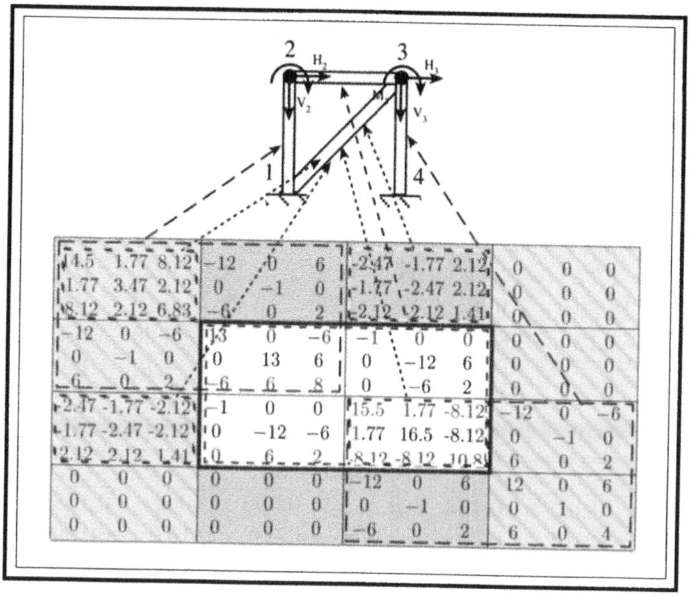

Fig. 11.10 Stiffness matrix for Case 3.

Case 3: Consider the structure in Fig. 11.9(a), which is the same as the structure in Fig. 11.6(a) after adding an inclined piece.

Thus, Equation (11.15) transforms to (see Fig. 11.10)

$$\begin{pmatrix} h_2 \\ v_2 \\ \theta_2 \\ h_3 \\ v_3 \\ \theta_3 \end{pmatrix} = \begin{pmatrix} 13 & 0 & -6 & -1 & 0 & 0 \\ 0 & 13 & 6 & 0 & -12 & 6 \\ -6 & 6 & 8 & 0 & -6 & 2 \\ -1 & 0 & 0 & \mathbf{15.50} & \mathbf{1.77} & \mathbf{-8.12} \\ 0 & -12 & -6 & \mathbf{1.77} & \mathbf{16.47} & \mathbf{-8.12} \\ 0 & 6 & 2 & \mathbf{-8.12} & \mathbf{-8.12} & \mathbf{10.80} \end{pmatrix}^{-1} \begin{pmatrix} H_2 \\ V_2 \\ M_2 \\ H_3 \\ V_3 \\ M_3 \end{pmatrix}$$

$$= \begin{pmatrix} 0.190 & -0.110 & 0.237 & 0.057 & 0.047 & 0.095 \\ -0.110 & 0.349 & -0.222 & -0.089 & 0.119 & -0.130 \\ 0.237 & -0.222 & 0.496 & 0.096 & 0.095 & 0.174 \\ 0.057 & -0.089 & 0.096 & 0.160 & 0.044 & 0.185 \\ 0.047 & 0.119 & 0.095 & 0.044 & 0.242 & 0.131 \\ 0.095 & -0.130 & 0.174 & 0.185 & 0.131 & 0.369 \end{pmatrix} \begin{pmatrix} H_2 \\ V_2 \\ M_2 \\ H_3 \\ V_3 \\ M_3 \end{pmatrix}. \tag{11.18}$$

Note that the only differences are in elements $k_{4,4}$, $k_{4,5}$, $k_{4,6}$, $k_{5,4}$, $k_{5,5}$, $k_{5,6}$, $k_{6,4}$, $k_{6,5}$, and $k_{6,6}$, which are boldface in (11.18).

Again, since the inverse of the previous matrix is known, the new inverse based on the old one (11.15) can be calculated, with important savings in computation. The process is illustrated in Table 11.2. In the first step we

Table 11.2 Updating the inverse matrix in Case 3 by incorporating the block formed by the last three rows (those with changes).

First step								
–1	0	0	0.21	–0.10	0.26	0.08	0.10	0.14
0	–12	–6	–0.10	0.61	–0.19	–0.10	0.39	–0.19
0	6	2	0.26	–0.19	0.55	0.14	0.19	0.26
15.50	1.77	–8.12	0.08	–0.10	0.14	0.21	0.10	0.26
1.77	15.50	–8.12	0.10	0.39	0.19	0.10	0.61	0.19
–8.12	–8.12	10.80	0.14	–0.19	0.26	0.26	0.19	0.55
			0.07	0.86	0.14	1.12	0.91	–0.17
			0.09	1.20	0.18	0.04	2.28	–0.22
			0.02	–1.16	0.02	0.11	–0.96	1.58

Second step					
0.193	–0.102	0.242	0.059	0.062	0.103
–0.102	0.368	–0.208	–0.082	0.157	–0.109
0.242	–0.208	0.508	0.101	0.124	0.190
0.059	–0.082	0.101	0.163	0.058	0.192
0.062	0.157	0.124	0.058	0.318	0.172
0.103	–0.109	0.190	0.192	0.172	0.391

incorporate the change of the block formed by the last three rows, and we get the desired inverse in Step 2.

Case 4: Consider the structure in Fig. 11.9(b), where the node 4 clamped end has been changed to a support. Then we have

$$h_1 = v_1 = \theta_1 = 0 \text{ and } h_4 = v_4 = 0. \tag{11.19}$$

Thus, columns $1, 2, 3, 10$, and 11 can be removed (see Fig. 11.11). In addition, to avoid compatibility problems, the same rows, corresponding to the associated reactions in nodes 1 and 4, are removed, and the following system of equations is obtained:

$$\begin{pmatrix} H_2 \\ V_2 \\ M_2 \\ H_3 \\ V_3 \\ M_3 \\ M_4 \end{pmatrix} = \begin{pmatrix} 13 & 0 & -6 & -1 & 0 & 0 & \mathbf{0} \\ 0 & 13 & 6 & 0 & -12 & 6 & \mathbf{0} \\ -6 & 6 & 8 & 0 & -6 & 2 & \mathbf{0} \\ -1 & 0 & 0 & 13 & 0 & -6 & \mathbf{-6} \\ 0 & -12 & -6 & 0 & 13 & -6 & \mathbf{0} \\ 0 & 6 & 2 & -6 & -6 & 8 & \mathbf{2} \\ \mathbf{0} & \mathbf{0} & \mathbf{0} & \mathbf{-6} & \mathbf{0} & \mathbf{2} & \mathbf{4} \end{pmatrix} \begin{pmatrix} h_2 \\ v_2 \\ \theta_2 \\ h_3 \\ v_3 \\ \theta_3 \\ \theta_4 \end{pmatrix}, \tag{11.20}$$

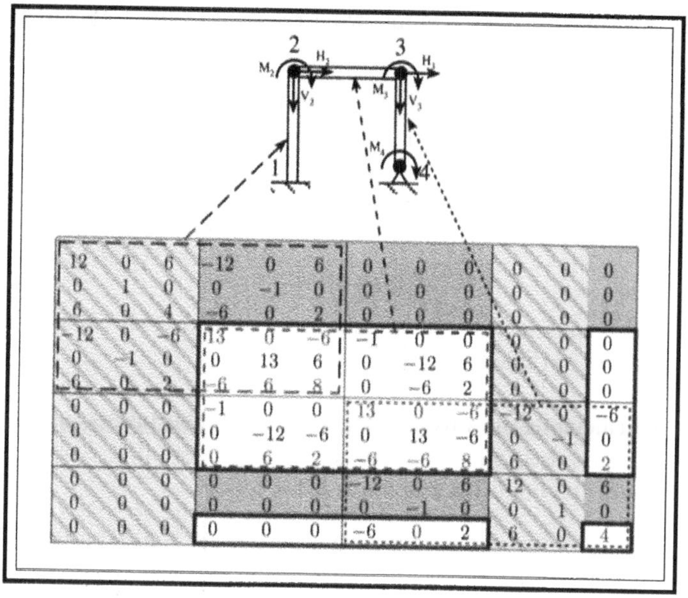

Fig. 11.11 Stiffness matrix for Case 4.

from which we find the displacements and rotations, where the matrix to be inverted is (11.15) with added one row and one column.

$$
\begin{pmatrix} h_2 \\ v_2 \\ \theta_2 \\ h_3 \\ v_3 \\ \theta_3 \\ \theta_4 \end{pmatrix} = \begin{pmatrix} 13 & 0 & -6 & -1 & 0 & 0 & \mathbf{0} \\ 0 & 13 & 6 & 0 & -12 & 6 & \mathbf{0} \\ -6 & 6 & 8 & 0 & -6 & 2 & \mathbf{0} \\ -1 & 0 & 0 & 13 & 0 & -6 & \mathbf{-6} \\ 0 & -12 & -6 & 0 & 13 & -6 & \mathbf{0} \\ 0 & 6 & 2 & -6 & -6 & 8 & \mathbf{2} \\ \mathbf{0} & \mathbf{0} & \mathbf{0} & \mathbf{-6} & \mathbf{0} & \mathbf{2} & \mathbf{4} \end{pmatrix}^{-1} \begin{pmatrix} H_2 \\ V_2 \\ M_2 \\ H_3 \\ V_3 \\ M_3 \\ M_4 \end{pmatrix}
$$

$$
= \begin{pmatrix} 0.26 & -0.15 & 0.35 & 0.27 & 0.15 & 0.27 & 0.26 \\ -0.15 & 0.66 & -0.27 & -0.27 & 0.34 & -0.32 & -0.25 \\ 0.35 & -0.27 & 0.68 & 0.43 & 0.27 & 0.46 & 0.42 \\ 0.27 & -0.27 & 0.43 & 0.85 & 0.27 & 0.71 & 0.92 \\ 0.15 & 0.34 & 0.27 & 0.27 & 0.66 & 0.32 & 0.25 \\ 0.27 & -0.32 & 0.46 & 0.71 & 0.32 & 0.86 & 0.64 \\ 0.26 & -0.25 & 0.42 & 0.92 & 0.25 & 0.64 & 1.30 \end{pmatrix} \begin{pmatrix} H_2 \\ V_2 \\ M_2 \\ H_3 \\ V_3 \\ M_3 \\ M_4 \end{pmatrix}; \tag{11.21}
$$

and from the removed rows, the following reaction expressions are obtained:

$$\begin{pmatrix} H_1 \\ V_1 \\ M_1 \\ H_4 \\ V_4 \end{pmatrix} = \begin{pmatrix} -12 & 0 & 6 & 0 & 0 & 0 & 0 \\ 0 & -1 & 0 & 0 & 0 & 0 & 0 \\ -6 & 0 & 2 & 0 & 0 & 0 & 0 \\ 0 & 0 & 0 & -12 & 0 & 6 & 6 \\ 0 & 0 & 0 & 0 & -1 & 0 & 0 \end{pmatrix} \begin{pmatrix} h_2 \\ v_2 \\ \theta_2 \\ h_3 \\ v_3 \\ \theta_3 \\ \theta_4 \end{pmatrix}. \tag{11.22}$$

Thus, first the unknown displacements are obtained from (11.21), and then the reactions are obtained by (11.22).

Next, we calculate the inverse in (11.21) from the inverse in (11.15). In order for the proposed algorithm to be applicable, we first reorder the rows and columns of the modified matrix to get

$$\begin{pmatrix} \mathbf{4} & \mathbf{0} & \mathbf{0} & \mathbf{0} & \mathbf{-6} & \mathbf{0} & \mathbf{2} \\ \mathbf{0} & 13 & 0 & -6 & -1 & 0 & 0 \\ \mathbf{0} & 0 & 13 & 6 & 0 & -12 & 6 \\ \mathbf{0} & -6 & 6 & 8 & 0 & -6 & 2 \\ \mathbf{-6} & -1 & 0 & 0 & 13 & 0 & -6 \\ \mathbf{0} & 0 & -12 & -6 & 0 & 13 & -6 \\ \mathbf{2} & 0 & 6 & 2 & -6 & -6 & 8 \end{pmatrix}. \tag{11.23}$$

We apply the proposed method to this matrix and modify its first row, except the first element, and obtain the results in Table 11.3, which is the same as Table 3.12. Then we transpose the inverse and modify the columns, as shown in Table 11.4.

Next, we obtain the transpose of the last matrix in Table 11.4, which is the inverse matrix of (11.23). Finally, if we undo the permutations of rows and columns over this matrix, we get the desired inverse matrix:

$$\begin{pmatrix} 0.258 & -0.148 & 0.349 & 0.265 & 0.148 & 0.268 & 0.264 \\ -0.148 & 0.662 & -0.275 & -0.275 & 0.338 & -0.317 & -0.254 \\ 0.349 & -0.275 & 0.683 & 0.433 & 0.275 & 0.461 & 0.419 \\ 0.265 & -0.275 & 0.433 & 0.850 & 0.275 & 0.711 & 0.919 \\ 0.148 & 0.338 & 0.275 & 0.275 & 0.662 & 0.317 & 0.254 \\ 0.268 & -0.317 & 0.461 & 0.711 & 0.317 & 0.859 & 0.637 \\ 0.264 & -0.254 & 0.419 & 0.919 & 0.254 & 0.637 & 1.310 \end{pmatrix}. \tag{11.24}$$

An alternative option consists of using Expression (3.26).

Case 5: Consider the structure in Fig. 11.12(a), where

$$h_1 = v_1 = \theta_1 = 0, \; h_4 = v_4 = \theta_4 = 0 \text{ and } h_3 = 0. \tag{11.25}$$

Table 11.3 Updating the inverse matrix in Case 4 (phase I).

Phase I							
1	1	0	0	0	0	0	0
0	0	0.205	–0.097	0.264	0.080	0.097	0.139
0	0	–0.097	0.613	–0.194	–0.097	0.387	–0.194
0	0	0.264	–0.194	0.549	0.139	0.194	0.257
–6	0	0.080	–0.097	0.139	0.205	0.097	0.264
0	0	0.097	0.387	0.194	0.097	0.613	0.194
2	0	0.139	–0.194	0.257	0.264	0.194	0.549
	1	–0.202	0.194	–0.320	–0.702	–0.194	–0.487

Output of phase I						
1	0.202	–0.194	0.320	0.702	0.194	0.487
0	0.205	–0.097	0.264	0.080	0.097	0.139
0	–0.097	0.613	–0.194	–0.097	0.387	–0.194
0	0.264	–0.194	0.549	0.139	0.194	0.257
0	0.080	–0.097	0.139	0.205	0.097	0.264
0	0.097	0.387	0.194	0.097	0.613	0.194
0	0.139	–0.194	0.257	0.264	0.194	0.549

Table 11.4 Updating the inverse matrix in Case 4 (phase II).

Phase II							
4	1	0	0	0	0	0	0
0	0.202	0.205	–0.097	0.264	0.080	0.097	0.139
0	–0.194	–0.097	0.613	–0.194	–0.097	0.387	–0.194
0	0.320	0.264	–0.194	0.549	0.139	0.194	0.257
–6	0.702	0.080	–0.097	0.139	0.205	0.097	0.264
0	0.194	0.097	0.387	0.194	0.097	0.613	0.194
2	0.487	0.139	–0.194	0.257	0.264	0.194	0.549
	0.763	–0.202	0.194	–0.320	–0.702	–0.194	–0.487

Output of phase II						
1.310	0.264	–0.254	0.419	0.919	0.254	0.637
0.264	0.258	–0.148	0.349	0.265	0.148	0.268
–0.254	–0.148	0.662	–0.275	–0.275	0.338	–0.317
0.419	0.349	–0.275	0.683	0.433	0.275	0.461
0.919	0.265	–0.275	0.433	0.850	0.275	0.711
0.254	0.148	0.338	0.275	0.275	0.662	0.317
0.637	0.268	–0.317	0.461	0.711	0.317	0.859

Thus, columns $1, 2, 3, 7, 10, 11$, and 12 can be removed (see Fig. 11.13). In addition, to avoid compatibility problems, the same columns, corresponding

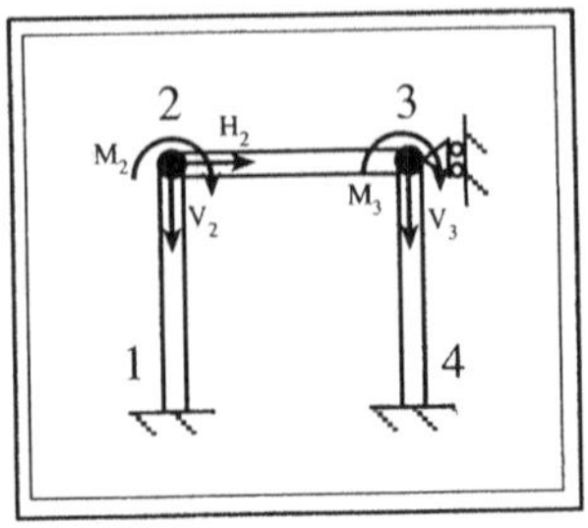

Fig. 11.12 Modified structure S_5 achieved by adding one horizontal support in the upper right corner of S_1.

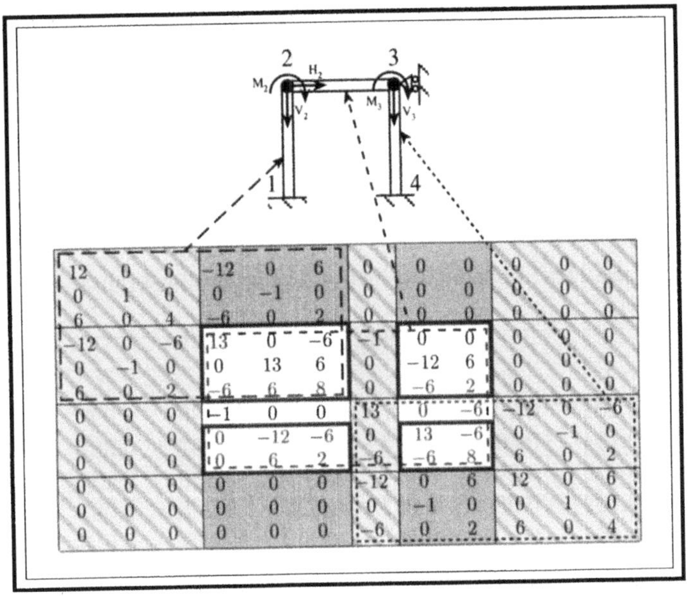

Fig. 11.13 Stiffness matrix for Case 5.

to the associated reactions in nodes 1, 3, and 4, are removed to obtain the following system of equations:

$$\begin{pmatrix} H_2 \\ V_2 \\ M_2 \\ V_3 \\ M_3 \end{pmatrix} = \begin{pmatrix} 13 & 0 & -6 & 0 & 0 \\ 0 & 13 & 6 & -12 & 6 \\ -6 & 6 & 8 & -6 & 2 \\ 0 & -12 & -6 & 13 & -6 \\ 0 & 6 & 2 & -6 & 8 \end{pmatrix} = \begin{pmatrix} h_2 \\ v_2 \\ \theta_2 \\ v_3 \\ \theta_3 \end{pmatrix}, \tag{11.26}$$

from which we find

$$\begin{pmatrix} h_2 \\ v_2 \\ \theta_2 \\ v_3 \\ \theta_3 \end{pmatrix} = \begin{pmatrix} 13 & 0 & -6 & 0 & 0 \\ 0 & 13 & 6 & -12 & 6 \\ -6 & 6 & 8 & -6 & 2 \\ 0 & -12 & -6 & 13 & -6 \\ 0 & 6 & 2 & -6 & 8 \end{pmatrix}^{-1} \begin{pmatrix} H_2 \\ V_2 \\ M_2 \\ V_3 \\ M_3 \end{pmatrix}$$

$$= \frac{1}{1830} \begin{pmatrix} 318 & -108 & 384 & 108 & 66 \\ -108 & 1038 & -234 & 792 & -126 \\ 384 & -234 & 832 & 234 & 143 \\ 108 & 792 & 234 & 1038 & 126 \\ 66 & -126 & 143 & 126 & 382 \end{pmatrix} \begin{pmatrix} H_2 \\ V_2 \\ M_2 \\ V_3 \\ M_3 \end{pmatrix} \tag{11.27}$$

and from the first and last three rows the reactions become

$$\begin{pmatrix} H_1 \\ V_1 \\ M_1 \\ H_3 \\ H_4 \\ V_4 \\ M_4 \end{pmatrix} = \begin{pmatrix} -12 & 0 & 6 & 0 & 0 \\ 0 & -1 & 0 & 0 & 0 \\ -6 & 0 & 2 & 0 & 0 \\ -1 & 0 & 0 & 0 & -6 \\ 0 & 0 & 0 & 0 & 6 \\ 0 & 0 & 0 & -1 & 0 \\ 0 & 0 & 0 & 0 & 2 \end{pmatrix} \begin{pmatrix} h_2 \\ v_2 \\ \theta_2 \\ v_3 \\ \theta_3 \end{pmatrix}. \tag{11.28}$$

Thus, first the unknown displacements are obtained from (11.27), and then the reactions are obtained by (11.28).

Note that the matrices in (11.17), (11.18), (11.21), and (11.27) differ only in some rows and/or columns from the matrix in (11.15). Thus, the methods developed in Chapter 3 can be used to obtain the inverses with important savings in calculations. The reason for using the inverses – a more costly process than that for solving the system of equations – is that several assumptions for the external forces can be used without the need to solve different systems of equations.

Next, we calculate the inverse in (11.27) from the inverse in (11.15). First, we permute the fourth row and column with the first row and column, to get

$$\begin{pmatrix} 0.205 & 0.080 & -0.097 & 0.139 & 0.097 & 0.264 \\ 0.080 & 0.205 & -0.097 & 0.264 & 0.097 & 0.139 \\ -0.097 & -0.097 & 0.613 & -0.194 & 0.387 & -0.194 \\ 0.139 & 0.264 & -0.194 & 0.549 & 0.194 & 0.257 \\ 0.097 & 0.097 & 0.387 & 0.194 & 0.613 & 0.194 \\ 0.264 & 0.139 & -0.194 & 0.257 & 0.194 & 0.549 \end{pmatrix}. \tag{11.29}$$

Table 11.5 Updating the inverse matrix.

First step						
1	0.205	0.080	-0.097	0.139	0.097	0.264
0	0.080	0.205	-0.097	0.264	0.097	0.139
0	-0.097	-0.097	0.613	-0.194	0.387	-0.194
0	0.139	0.264	-0.194	0.549	0.194	0.257
0	0.097	0.097	0.387	0.194	0.613	0.194
0	0.264	0.139	-0.194	0.257	0.194	0.549
	0.205	0.080	-0.097	0.139	0.097	0.264

Second step					
1.000	0.000	0.000	0.000	0.000	0.000
0.390	0.174	-0.059	0.210	0.059	0.036
-0.472	-0.059	0.567	-0.128	0.433	-0.069
0.679	0.210	-0.128	0.455	0.128	0.078
0.472	0.059	0.433	0.128	0.567	0.069
1.290	0.036	-0.069	0.078	0.069	0.209

Finally, removing the first row and column, we obtain the inverse (see Table 11.5):

$$\begin{pmatrix} 0.174 & -0.059 & 0.210 & 0.059 & 0.036 \\ -0.059 & 0.567 & -0.128 & 0.433 & -0.069 \\ 0.210 & -0.128 & 0.455 & 0.128 & 0.078 \\ 0.059 & 0.433 & 0.128 & 0.567 & 0.069 \\ 0.036 & -0.069 & 0.078 & 0.069 & 0.209 \end{pmatrix}. \tag{11.30}$$

An alternative option consists of using Expression (3.32), that is:

$$\begin{aligned} \mathbf{A}_{22}^{-1} &= \mathbf{A}^{22} - \mathbf{A}^{21}\mathbf{A}^{11^{-1}}\mathbf{A}^{12} \\ &= \begin{pmatrix} 0.205 & -0.097 & 0.264 & 0.097 & 0.139 \\ -0.097 & 0.613 & -0.194 & 0.387 & -0.194 \\ 0.264 & -0.194 & 0.549 & 0.194 & 0.257 \\ 0.097 & 0.387 & 0.194 & 0.613 & 0.194 \\ 0.139 & -0.194 & 0.257 & 0.194 & 0.549 \end{pmatrix} \\ &\quad - \frac{1}{0.205}\begin{pmatrix} 0.08 \\ -0.097 \\ 0.139 \\ 0.097 \\ 0.264 \end{pmatrix} (0.08 \quad -0.097 \quad 0.139 \quad 0.097 \quad 0.264). \end{aligned} \tag{11.31}$$

11.3 THE TRANSPORTATION PROBLEM

In this section we discuss the transportation problem. First, we obtain the set of feasible solutions in Section 11.3.1, and then we find the optimal solution in Section 11.3.2.

11.3.1 The Feasible Solution Set Problem

A homogeneous product is to be shipped in the amounts $a_1, \ldots, a_m$, respectively, from each of m shipping origins, and received in amounts $b_1, \ldots, b_n$, respectively, by each of n shipping destinations. The problem is to determine the amounts x_{ij} to be shipped from origin i to destination j.

The constraints of this problem are:

$$\begin{aligned} \sum_{j=1}^{n} x_{ij} &= a_i;\ i = 1, \ldots, m, \\ \sum_{i=1}^{m} x_{ij} &= b_j;\ j = 1, \ldots, n, \\ x_{ij} &\geq 0;\ i = 1, \ldots, m;\ j = 1, \ldots, n. \end{aligned} \tag{11.32}$$

The first set of conditions states that the total amount shipped from origin i must be equal to the sum of the amounts going from it to all destinations $j = 1, \ldots, n$.

The second set of conditions state that the total amount received at destination j must be equal to the sum of all amounts shipped to that destination from all origins $i = 1, \ldots, m$.

Apart from other conditions, the compatibility of the above system leads to

$$\sum_{i=1}^{m}\sum_{j=1}^{n} x_{ij} = \sum_{i=1}^{m} a_i = \sum_{j=1}^{n} b_j, \tag{11.33}$$

which states that the total amount shipped from all origins must be equal to the total amount received at all destinations.

Example 11.1 (Transportation problem). Consider the transportation problem in Fig. 11.14, where there are $m = 3$ shipping origins and $n = 3$ destinations, and

$$a_1 = 2, a_2 = 3, a_3 = 4; b_1 = 5, b_2 = 2, b_3 = 2.$$

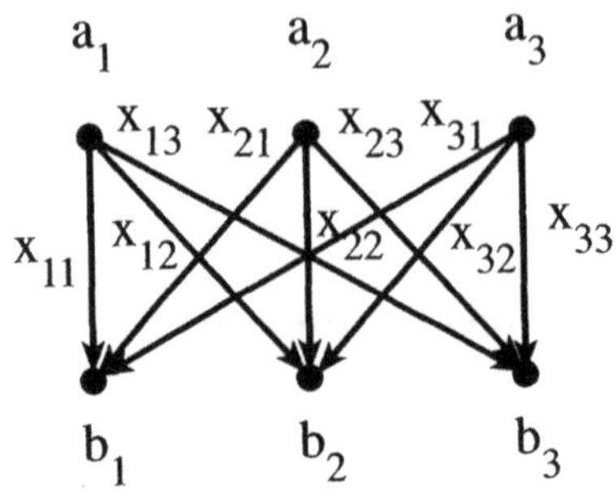

Fig. 11.14 Illustration of the transportation problem.

In this case, system (11.32) becomes

$$\mathbf{Ax} = \begin{pmatrix} 1 & 1 & 1 & 0 & 0 & 0 & 0 & 0 & 0 \\ 0 & 0 & 0 & 1 & 1 & 1 & 0 & 0 & 0 \\ 0 & 0 & 0 & 0 & 0 & 0 & 1 & 1 & 1 \\ 1 & 0 & 0 & 1 & 0 & 0 & 1 & 0 & 0 \\ 0 & 1 & 0 & 0 & 1 & 0 & 0 & 1 & 0 \\ 0 & 0 & 1 & 0 & 0 & 1 & 0 & 0 & 1 \end{pmatrix} \begin{pmatrix} x_{11} \\ x_{12} \\ x_{13} \\ x_{21} \\ x_{22} \\ x_{23} \\ x_{31} \\ x_{32} \\ x_{33} \end{pmatrix} = \begin{pmatrix} a_1 \\ a_2 \\ a_3 \\ b_1 \\ b_2 \\ b_3 \end{pmatrix} = \mathbf{b}, \qquad (11.34)$$

$$x_{ij} \geq 0; i, j = 1, 2, 3.$$

Since we have a system of linear equations with non-negative variables of the type (8.7), according to (8.8), system (11.34) has solution iff

$$\mathbf{V}^T\mathbf{b} = \mathbf{0} \wedge \mathbf{W}^T\mathbf{b} \leq \mathbf{0}, \qquad (11.35)$$

where $\Omega(\mathbf{A}_\pi) \equiv \mathbf{V}_\rho + \mathbf{W}_\pi$ is the dual of the cone $\mathbf{A}_\pi$ generated by the columns of $\mathbf{A}$.

Thus, we use the Γ-algorithm and obtain Table 11.6. Then (11.35) leads to

$$\begin{array}{lcl} \mathbf{V}^T\mathbf{b} = 0 & \Leftrightarrow & -a_1 - a_2 - a_3 + b_1 + b_2 + b_3 = 0, \\ \mathbf{W}^T\mathbf{b} \leq 0 & \Leftrightarrow & a_1, a_2, a_3, b_1, b_2, b_3 \geq 0, \end{array} \qquad (11.36)$$

where we have used the equality constraint $\mathbf{v}_1 \bullet \mathbf{b} = 0$ to modify the constraint $\mathbf{w}_5 \bullet \mathbf{b} \leq 0$ to its equivalent $b_3 \geq 0$.

The equality constraint only says that the total amount shipped from the three origins must coincide with the total amount received at the three destinations. The inequality constraints only state the non-negativity of $a_1, a_2, a_3, b_1, b_2, b_3$.

To obtain the solution of system (11.34) we use the results of Section 8.3. Initially, we introduce a new auxiliary variable $y_0 = 1$ and write system (11.34)

Table 11.6 Calculating the dual of Example 11.1.

Dual						
$\mathbf{w}_1$	$\mathbf{w}_2$	$\mathbf{w}_3$	$\mathbf{w}_4$	$\mathbf{w}_5$	$\mathbf{w}_6$	$\mathbf{v}_1$
0	–1	0	0	–1	0	–1
–1	0	0	0	–1	0	–1
0	0	0	0	–1	–1	–1
0	0	0	–1	1	0	1
0	0	–1	0	1	0	1
0	0	0	0	0	0	1

in its homogeneous form as

$$\begin{pmatrix} 1 & 1 & 1 & 0 & 0 & 0 & 0 & 0 & 0 & -a_1 \\ 0 & 0 & 0 & 1 & 1 & 1 & 0 & 0 & 0 & -a_2 \\ 0 & 0 & 0 & 0 & 0 & 0 & 1 & 1 & 1 & -a_3 \\ 1 & 0 & 0 & 1 & 0 & 0 & 1 & 0 & 0 & -b_1 \\ 0 & 1 & 0 & 0 & 1 & 0 & 0 & 1 & 0 & -b_2 \\ 0 & 0 & 1 & 0 & 0 & 1 & 0 & 0 & 1 & -b_3 \end{pmatrix} \begin{pmatrix} x_{11} \\ x_{12} \\ x_{13} \\ x_{21} \\ x_{22} \\ x_{23} \\ x_{31} \\ x_{32} \\ x_{33} \\ y_0 \end{pmatrix} = \begin{pmatrix} 0 \\ 0 \\ 0 \\ 0 \\ 0 \\ 0 \end{pmatrix}$$

$$\begin{pmatrix} -1 & 0 & 0 & 0 & 0 & 0 & 0 & 0 & 0 & 0 \\ 0 & -1 & 0 & 0 & 0 & 0 & 0 & 0 & 0 & 0 \\ 0 & 0 & -1 & 0 & 0 & 0 & 0 & 0 & 0 & 0 \\ 0 & 0 & 0 & -1 & 0 & 0 & 0 & 0 & 0 & 0 \\ 0 & 0 & 0 & 0 & -1 & 0 & 0 & 0 & 0 & 0 \\ 0 & 0 & 0 & 0 & 0 & -1 & 0 & 0 & 0 & 0 \\ 0 & 0 & 0 & 0 & 0 & 0 & -1 & 0 & 0 & 0 \\ 0 & 0 & 0 & 0 & 0 & 0 & 0 & -1 & 0 & 0 \\ 0 & 0 & 0 & 0 & 0 & 0 & 0 & 0 & -1 & 0 \\ 0 & 0 & 0 & 0 & 0 & 0 & 0 & 0 & 0 & -1 \end{pmatrix} \begin{pmatrix} x_{11} \\ x_{12} \\ x_{13} \\ x_{21} \\ x_{22} \\ x_{23} \\ x_{31} \\ x_{32} \\ x_{33} \\ y_0 \end{pmatrix} \le \begin{pmatrix} 0 \\ 0 \\ 0 \\ 0 \\ 0 \\ 0 \end{pmatrix}, \tag{11.37}$$

where

$$y_0 = 1.$$

Thus, using Algorithm 6.1, we obtain the dual of the cone generated by the rows of the matrices in (11.37), obtaining (in its last phase) Table 11.7, which after forcing $y_0 = 1$ gives Table 11.8. From it, we get

Table 11.7 Last tableau in the Γ-algorithm.

Dual										
$\mathbf{w}_1$	$\mathbf{w}_2$	$\mathbf{w}_3$	$\mathbf{w}_4$	$\mathbf{w}_5$	$\mathbf{w}_6$	$\mathbf{w}_7$	$\mathbf{w}_8$	$\mathbf{w}_9$	$\mathbf{w}_{10}$	$\mathbf{w}_{11}$
0	0	0	1	3/2	3/2	0	3/2	3/4	3/2	9/8
0	0	1	0	0	0	3/2	0	3/4	0	0
1	3/2	0	0	0	0	0	0	0	0	9/8
3/2	3/4	3/2	3/2	0	3/4	3/4	3/4	0	0	0
0	3/2	0	0	3/2	3/2	0	0	3/4	3/4	9/4
0	0	0	0	3/4	0	3/2	3/2	3/2	3/2	9/8
1	3	1	0	9/4	3/2	3	3/2	3	9/4	9/2
1	0	0	1	0	0	0	3/2	0	3/4	0
0	0	1	1	3/4	3/2	0	0	0	0	0
1/2	3/4	1/2	1/2	3/4	3/4	3/4	3/4	3/4	3/4	9/8

Table 11.8 Last tableau in the Γ-algorithm after forcing $y_0 = 1$.

Normalized dual										
$\mathbf{w}_1$	$\mathbf{w}_2$	$\mathbf{w}_3$	$\mathbf{w}_4$	$\mathbf{w}_5$	$\mathbf{w}_6$	$\mathbf{w}_7$	$\mathbf{w}_8$	$\mathbf{w}_9$	$\mathbf{w}_{10}$	$\mathbf{w}_{11}$
0	0	0	2	2	2	0	2	1	2	1
0	0	2	0	0	0	2	0	1	0	0
2	2	0	0	0	0	0	0	0	0	1
3	1	3	3	0	1	1	1	0	0	0
0	2	0	0	2	2	0	0	1	1	2
0	0	0	0	1	0	2	2	2	2	1
2	4	2	0	3	2	4	2	4	3	4
2	0	0	2	0	0	0	2	0	1	0
0	0	2	2	1	2	0	0	0	0	0
1	1	1	1	1	1	1	1	1	1	1

$$
\begin{aligned}
\mathbf{V} &\equiv \{\},\\
\mathbf{W} &\equiv \{\},\\
\mathbf{Q} &\equiv \{(0,0,2,3,0,0,2,2,0)^T,(0,0,2,1,2,0,4,0,0)^T,(0,2,0,3,0,0,2,0,2)^T,\\
&\quad (2,0,0,3,0,0,0,2,2)^T,(2,0,0,0,2,1,3,0,1)^T,(2,0,0,1,2,0,2,0,2)^T,\\
&\quad (0,2,0,1,0,2,4,0,0)^T,(2,0,0,1,0,2,2,2,0)^T,(1,1,0,0,1,2,4,0,0)^T,\\
&\quad (2,0,0,0,1,2,3,1,0)^T,(1,0,1,0,2,1,4,0,0)^T\}.
\end{aligned}
\tag{11.38}
$$

Thus, the general solution of (11.34) is a linear convex combination of the vectors in **Q**.

Note that each vector in **Q** is a solution of the initial system (11.34). ■

11.3.2 The Optimization Problem

In the transportation problem we are normally interested in minimizing cost, that is,

$$\sum_{i=1}^{m}\sum_{j=1}^{n} c_{ij}x_{ij}, \tag{11.39}$$

where c_{ij} is the cost associated with sending a unit from origin i to destination j, which is a linear programming problem. Assuming some particular values for c_{ij}, this becomes:

Maximize

$$-x_{11} - 2x_{12} - 3x_{13} - 2x_{21} - x_{22} - 2x_{23} - 3x_{31} - 2x_{32} - x_{33} \tag{11.40}$$

subject to

$$\begin{pmatrix} 1 & 1 & 1 & 0 & 0 & 0 & 0 & 0 & 0 \\ 0 & 0 & 0 & 1 & 1 & 1 & 0 & 0 & 0 \\ 0 & 0 & 0 & 0 & 0 & 0 & 1 & 1 & 1 \\ 1 & 0 & 0 & 1 & 0 & 0 & 1 & 0 & 0 \\ 0 & 1 & 0 & 0 & 1 & 0 & 0 & 1 & 0 \\ 0 & 0 & 1 & 0 & 0 & 1 & 0 & 0 & 1 \\ -2 & -2 & -2 & -2 & -2 & -2 & -2 & -2 & -2 \end{pmatrix} \begin{pmatrix} x_{11} \\ x_{12} \\ x_{13} \\ x_{21} \\ x_{22} \\ x_{23} \\ x_{31} \\ x_{32} \\ x_{33} \end{pmatrix} \le \begin{pmatrix} 2 \\ 3 \\ 4 \\ 5 \\ 2 \\ 2 \\ -18 \end{pmatrix},$$

$$x_{ij} \ge 0; i, j = 1, 2, 3,$$

where the six equalities of (11.34) have been written as seven inequalities (see Proposition 10.1).

Using the EPM, we obtain a maximum value of -14 for the objective function, which implies a minimum cost of 14 units, at point $(2, 0, 0, 1, 2, 0, 2, 0, 2)$.

11.4 PRODUCTION–SCHEDULING PROBLEMS

A manufacturer produces a product, whose demand fluctuates in time, in a form similar to that in Fig. 11.15. The manufacturer must always meet the monthly requirements established by demand. In general, any scheduling problem has many different schedules that will satisfy the requirements. There are two possible alternatives:

1. **Variable production**: The manufacturer could produce each month the exact number of units required by demand. However, since a fluctuating production schedule is costly to maintain, because of overtime cost in high–production months, and the costs associated with the release of personnel and machinery in low–production months, this type of production schedule is not efficient.

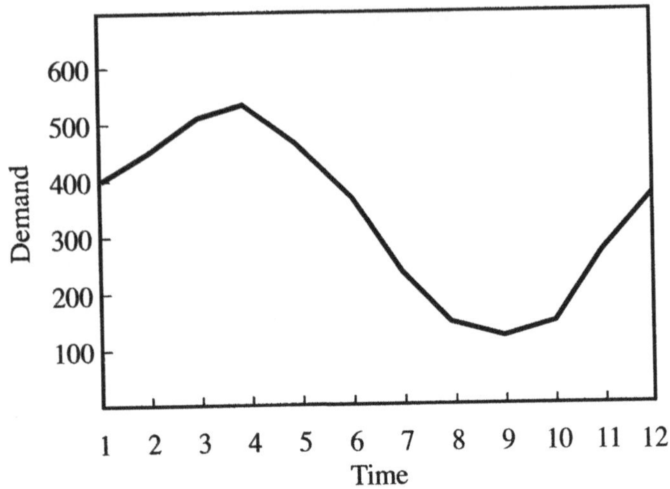

Fig. 11.15 Demand function over time.

2. **Constant production**: The manufacturer faced with fluctuating requirements could overproduce in periods of low requirements and store the surplus to be used in periods of high demand. Thus, the production can be made constant, compensating underproduction with past overproduction. However, because of the cost of keeping a manufactured item in storage, such a solution may be undesirable if it yields comparatively large monthly surpluses.

Problems of this nature illustrate the difficulties that arise when conflicting objectives are inherent in a given system. Our objective is to design a production schedule that maximizes benefit after consideration of the costs of production fluctuations and inventories.

The mathematical statement of this problem is as follows: Let s_0 be the amount of an item available in storage at the beginning of the considered period. Let

x_t be the number of units produced in month t,

y_t be the number of units required in month t,

s_t be the number of units in storage in month t.

Since the storage in month t contains the storage of month $t-1$ plus the production x_t minus the demand y_t, we have

$$\begin{aligned} s_t &= s_{t-1} + x_t - y_t; t = 1, 2, \ldots \\ s_t, x_t, y_t &\geq 0. \end{aligned} \tag{11.41}$$

Table 11.9 Demand as a function of time in Example 11.2.

Time	Demand
1	2
2	3
3	6
4	1

Example 11.2 (Production-scheduling problem). In this example we illustrate the production-scheduling problem. Consider the demand function in Table 11.9 and assume that the initial storage is $s_0 = 2$. Then system (11.41) becomes

$$\mathbf{Ax} = \begin{pmatrix} -1 & 0 & 0 & 0 & 1 & 0 & 0 & 0 \\ 1 & -1 & 0 & 0 & 0 & 1 & 0 & 0 \\ 0 & 1 & -1 & 0 & 0 & 0 & 1 & 0 \\ 0 & 0 & 1 & -1 & 0 & 0 & 0 & 1 \end{pmatrix} \begin{pmatrix} s_1 \\ s_2 \\ s_3 \\ s_4 \\ x_1 \\ x_2 \\ x_3 \\ x_4 \end{pmatrix} = \begin{pmatrix} 0 \\ 3 \\ 6 \\ 1 \end{pmatrix} = \mathbf{b}, \tag{11.42}$$

$$s_t, x_t \geq 0.$$

According to (8.8), system (11.42) has solution iff

$$\begin{array}{rcl} \mathbf{V}^T\mathbf{b} & = & \mathbf{0} \\ \mathbf{W}^T\mathbf{b} & \leq & \mathbf{0}, \end{array} \tag{11.43}$$

where $\Omega(\mathbf{A}_\pi) \equiv \mathbf{V}_\rho + \mathbf{W}_\pi$.

Thus, we need to obtain the linear space and cone components of the dual of the cone generated by the columns of the matrix $\mathbf{A}$ in the left–hand side of (11.42). Using the Γ–algorithm we obtain

$$\begin{array}{rcl} \mathbf{V} & \equiv & \emptyset; \\ \mathbf{W} & \equiv & \emptyset. \end{array} \tag{11.44}$$

Equation (11.42) together with the compatibility condition (11.43) implies that the above system has a solution. In fact, the production-scheduling problem always has a solution.

Table 11.10 Last tableau in the Γ-algorithm.

Dual											
$\mathbf{w}_1$	$\mathbf{w}_2$	$\mathbf{w}_3$	$\mathbf{w}_4$	$\mathbf{w}_5$	$\mathbf{w}_6$	$\mathbf{w}_7$	$\mathbf{w}_8$	$\mathbf{w}_9$	$\mathbf{w}_{10}$	$\mathbf{w}_{11}$	$\mathbf{w}_{12}$
0	0	1	0	3	0	1	0	10	0	10/3	0
0	0	0	0	2	2	0	0	7	7	10/3	10/3
0	1	0	0	0	0	1/3	1/3	1	1	10/3	10/3
1	1	0	0	0	0	0	0	0	0	10/3	10/3
0	0	1	0	3	0	1	0	10	0	10/3	0
0	0	0	1	0	3	0	1	0	10	0	10/3
0	1	2	2	0	0	7/3	7/3	0	0	0	0
1	0	1/3	1/3	1/3	1/3	0	0	0	0	0	0
0	0	1/3	1/3	1/3	1/3	1/3	1/3	1	1	0	0

To obtain the solution of system (11.42), we first transform this system, introducing the auxiliary variable $y_0 = 1$, to get

$$\mathbf{Ax} = \begin{pmatrix} -1 & 0 & 0 & 0 & 1 & 0 & 0 & 0 & 0 \\ 1 & -1 & 0 & 0 & 0 & 1 & 0 & 0 & -3 \\ 0 & 1 & -1 & 0 & 0 & 0 & 1 & 0 & -6 \\ 0 & 0 & 1 & -1 & 0 & 0 & 0 & 1 & -1 \end{pmatrix} \begin{pmatrix} s_1 \\ s_2 \\ s_3 \\ s_4 \\ x_1 \\ x_2 \\ x_3 \\ x_4 \\ y_0 \end{pmatrix} = \begin{pmatrix} 0 \\ 0 \\ 0 \\ 0 \end{pmatrix},$$

$$\mathbf{Bx} = \begin{pmatrix} -1 & 0 & 0 & 0 & 0 & 0 & 0 & 0 & 0 \\ 0 & -1 & 0 & 0 & 0 & 0 & 0 & 0 & 0 \\ 0 & 0 & -1 & 0 & 0 & 0 & 0 & 0 & 0 \\ 0 & 0 & 0 & -1 & 0 & 0 & 0 & 0 & 0 \\ 0 & 0 & 0 & 0 & -1 & 0 & 0 & 0 & 0 \\ 0 & 0 & 0 & 0 & 0 & -1 & 0 & 0 & 0 \\ 0 & 0 & 0 & 0 & 0 & 0 & -1 & 0 & 0 \\ 0 & 0 & 0 & 0 & 0 & 0 & 0 & -1 & 0 \\ 0 & 0 & 0 & 0 & 0 & 0 & 0 & 0 & -1 \end{pmatrix} \begin{pmatrix} s_1 \\ s_2 \\ s_3 \\ s_4 \\ x_1 \\ x_2 \\ x_3 \\ x_4 \\ y_0 \end{pmatrix} \le \begin{pmatrix} 0 \\ 0 \\ 0 \\ 0 \\ 0 \\ 0 \\ 0 \\ 0 \\ 0 \end{pmatrix}, \tag{11.45}$$

where

$$y_0 = 1.$$

Next, we use Algorithm 6.1 with the rows of matrices **A** and **B** in (11.45) and obtain, in its last phase, Table 11.10, which, after forcing $y_0 = 1$, gives Table 11.11. From it, we get

Table 11.11 Last tableau in the Γ-algorithm after forcing $y_0 = 1$.

Normalized dual											
$\mathbf{w}_1$	$\mathbf{w}_2$	$\mathbf{w}_3$	$\mathbf{w}_4$	$\mathbf{w}_5$	$\mathbf{w}_6$	$\mathbf{w}_7$	$\mathbf{w}_8$	$\mathbf{w}_9$	$\mathbf{w}_{10}$	$\mathbf{w}_{11}$	$\mathbf{w}_{12}$
0	0	3	0	9	0	3	0	10	0	10/3	0
0	0	0	0	6	6	0	0	7	7	10/3	10/3
0	1	0	0	0	0	1	1	1	1	10/3	10/3
1	1	0	0	0	0	0	0	0	0	10/3	10/3
0	0	3	0	9	0	3	0	10	0	10/3	0
0	0	0	3	0	9	0	3	0	10	0	10/3
0	1	6	6	0	0	7	7	0	0	0	0
1	0	1	1	1	1	0	0	0	0	0	0
0	0	1	1	1	1	1	1	1	1	0	0

$$
\begin{aligned}
\mathbf{V} &\equiv \{\}, \\
\mathbf{W} &\equiv \{(0,0,0,1,0,0,0,1)^T, (0,0,1,1,0,0,1,0)^T, \\
&\quad (10/3,10/3,10/3,10/3,10/3,0,0,0)^T, \\
&\quad (0,10/3,10/3,10/3,0,10/3,0,0)^T\}, \\
\mathbf{Q} &\equiv \{(3,0,0,0,3,0,6,1)^T, (0,0,0,0,0,3,6,1)^T, \\
&\quad (9,6,0,0,9,0,0,1)^T, (0,6,0,0,0,9,0,1)^T, \\
&\quad (3,0,1,0,3,0,7,0)^T, (0,0,1,0,0,3,7,0)^T, \\
&\quad (10,7,1,0,10,0,0,0)^T, (0,7,1,0,0,10,0,0)^T\}.
\end{aligned}
\tag{11.46}
$$

Thus, the general solution is a non-negative linear combination of the vectors in **W** plus a linear convex combination of the vectors in **Q**.

Note that:

1. Each vector in **Q** is a solution of (11.42). These solutions represent the minimum production required to satisfy the demand. They lead to a zero storage in at least one time step.

2. Each vector in **W** is a solution of the associated homogeneous system. These solutions represent the production excesses over demand and only increase the storage, making the production policy more robust against possible demand increments.

Table 11.12 Dual of the compatibility problem.

Dual						
$\mathbf{w}_1$	$\mathbf{w}_2$	$\mathbf{w}_3$	$\mathbf{w}_4$	$\mathbf{w}_5$	$\mathbf{w}_6$	$\mathbf{w}_7$
0	0	-1	0	0	-1	0
0	0	0	0	-1	-1	0
0	0	0	0	0	0	0
0	0	0	0	0	0	0
0	-1	-1	0	0	0	0
0	0	0	-1	-1	-1	0
0	0	0	0	0	0	-1
-1	0	0	0	0	0	0
$\mathbf{w}_8$	$\mathbf{w}_9$	$\mathbf{w}_{10}$	$\mathbf{w}_{11}$	$\mathbf{w}_{12}$	$\mathbf{w}_{13}$	$\mathbf{w}_{14}$
0	0	-1	0	0	0	-1
0	-1	-1	0	0	-1	-1
-1	-1	-1	0	-1	-1	-1
0	0	0	-1	-1	-1	-1
0	0	0	0	0	0	0
0	0	0	0	0	0	0
-1	-1	-1	0	0	0	0
0	0	0	-1	-1	-1	-1

Consider now that we have a limit on storage of 4 units. Then, introducing four auxiliary slack variables z_1, z_2, z_3 and z_4, system (11.42) becomes

$$\begin{pmatrix} -1 & 0 & 0 & 0 & 1 & 0 & 0 & 0 & 0 & 0 & 0 & 0 \\ 1 & -1 & 0 & 0 & 0 & 1 & 0 & 0 & 0 & 0 & 0 & 0 \\ 0 & 1 & -1 & 0 & 0 & 0 & 1 & 0 & 0 & 0 & 0 & 0 \\ 0 & 0 & 1 & -1 & 0 & 0 & 0 & 1 & 0 & 0 & 0 & 0 \\ 1 & 0 & 0 & 0 & 0 & 0 & 0 & 0 & 1 & 0 & 0 & 0 \\ 0 & 1 & 0 & 0 & 0 & 0 & 0 & 0 & 0 & 1 & 0 & 0 \\ 0 & 0 & 1 & 0 & 0 & 0 & 0 & 0 & 0 & 0 & 1 & 0 \\ 0 & 0 & 0 & 1 & 0 & 0 & 0 & 0 & 0 & 0 & 0 & 1 \end{pmatrix} \begin{pmatrix} s_1 \\ s_2 \\ s_3 \\ s_4 \\ x_1 \\ x_2 \\ x_3 \\ x_4 \\ z_1 \\ z_2 \\ z_3 \\ z_4 \end{pmatrix} = \begin{pmatrix} 0 \\ 3 \\ 6 \\ 1 \\ 4 \\ 4 \\ 4 \\ 4 \end{pmatrix},$$

$$s_t, x_t, z_t \geq 0. \tag{11.47}$$

Thus, to obtain the new compatibility conditions, we obtain the dual of the cone generated by the columns of the matrix on the left-hand side of (11.47), using the Γ-algorithm, and obtain Table 11.12.

To obtain the solution of system (11.47), we first transform it to a homogeneous system, introducing the auxiliary variable $y_0 = 1$, to get

$$\begin{pmatrix} -1 & 0 & 0 & 0 & 1 & 0 & 0 & 0 & 0 & 0 & 0 & 0 & 0 \\ 1 & -1 & 0 & 0 & 0 & 1 & 0 & 0 & 0 & 0 & 0 & 0 & -3 \\ 0 & 1 & -1 & 0 & 0 & 0 & 1 & 0 & 0 & 0 & 0 & 0 & -6 \\ 0 & 0 & 1 & -1 & 0 & 0 & 0 & 1 & 0 & 0 & 0 & 0 & -1 \\ 1 & 0 & 0 & 0 & 0 & 0 & 0 & 0 & 1 & 0 & 0 & 0 & -4 \\ 0 & 1 & 0 & 0 & 0 & 0 & 0 & 0 & 0 & 1 & 0 & 0 & -4 \\ 0 & 0 & 1 & 0 & 0 & 0 & 0 & 0 & 0 & 0 & 1 & 0 & -4 \\ 0 & 0 & 0 & 1 & 0 & 0 & 0 & 0 & 0 & 0 & 0 & 1 & -4 \end{pmatrix} \begin{pmatrix} s_1 \\ s_2 \\ s_3 \\ s_4 \\ x_1 \\ x_2 \\ x_3 \\ x_4 \\ z_1 \\ z_2 \\ z_3 \\ z_4 \\ y_0 \end{pmatrix} = \begin{pmatrix} 0 \\ 0 \\ 0 \\ 0 \\ 0 \\ 0 \\ 0 \\ 0 \end{pmatrix},$$

$$s_t, x_t, z_t, y_0 \geq 0$$

$$y_0 = 1. \tag{11.48}$$

Next, we use Algorithm 6.1 with the rows of the matrix in (11.48) and the corresponding rows associated with the non-negativity conditions for the variables s_t, x_t, z_t, and y_0 and, after forcing $y_0 = 1$, we get

$$\begin{array}{rcl} \mathbf{V} & \equiv & \emptyset, \\ \mathbf{W} & \equiv & \emptyset, \\ \mathbf{Q} & \equiv & \{(3,0,0,0,3,0,6,1,1,4,4,4)^T, (0,0,0,0,0,3,6,1,4,4,4,4)^T, \\ & & (3,0,1,0,3,0,7,0,1,4,3,4)^T, (0,0,1,0,0,3,7,0,4,4,3,4)^T, \\ & & (4,1,0,0,4,0,5,1,0,3,4,4)^T, (4,1,1,0,4,0,6,0,0,3,3,4)^T, \\ & & (0,4,0,0,0,7,2,1,4,0,4,4)^T, (0,4,1,0,0,7,3,0,4,0,3,4)^T, \\ & & (4,4,0,0,4,3,2,1,0,0,4,4)^T, (4,4,1,0,4,3,3,0,0,0,3,4)^T, \\ & & (3,0,4,3,3,0,10,0,1,4,0,1)^T, (0,0,4,3,0,3,10,0,4,4,0,1)^T, \\ & & (4,1,4,3,4,0,9,0,0,3,0,1)^T, (0,4,4,3,0,7,6,0,4,0,0,1)^T, \\ & & (4,4,4,3,4,3,6,0,0,0,0,1)^T, (3,0,0,4,3,0,6,5,1,4,4,0)^T, \\ & & (0,0,0,4,0,3,6,5,4,4,4,0)^T, (4,1,0,4,4,0,5,5,0,3,4,0)^T, \\ & & (0,4,0,4,0,7,2,5,4,0,4,0)^T, (4,4,0,4,4,3,2,5,0,0,4,0)^T, \\ & & (3,0,4,4,3,0,10,1,1,4,0,0)^T, (0,0,4,4,0,3,10,1,4,4,0,0)^T, \\ & & (4,1,4,4,4,0,9,1,0,3,0,0)^T, (0,4,4,4,0,7,6,1,4,0,0,0)^T, \\ & & (4,4,4,4,4,3,6,1,0,0,0,0)^T\}. \end{array} \tag{11.49}$$

Thus, the general solution is a linear convex combination of the vectors in **Q** with their last fourth components removed.

Table 11.13 Example of an input–output table.

Sales	Railroad	Steel	Coal	Other	Final demand	Total
Railroad	x_{11}	x_{12}	x_{13}	x_{14}	y_1	x_1
Steel	x_{21}	x_{22}	x_{23}	x_{24}	y_2	x_2
Coal	x_{31}	x_{32}	x_{33}	x_{34}	y_3	x_3
Other	x_{41}	x_{42}	x_{43}	x_{44}	y_4	x_4

Note that now $\mathbf{W} \equiv \emptyset$; that is, production is limited by the limited storage. Note also that all solutions in (11.46) with their first four components smaller than 4 (the storage limit) are included in (11.49). All other solutions in (11.49) have at least one of their first four components equal to 4 (the storage limit). ■

11.5 THE INPUT–OUTPUT TABLES

By way of introduction to the input–output tables, consider an economy that consists of only three basic industries – railroad, steel, and coal – and a fourth category of all other industries. The aim of the problem consists of analyzing the interrelationships of these industries in terms of sales to each other and to other elements of the economy, during a given period of time. This analysis can be done by means of an input–output table.

Each element in Table 11.13 represents a total sales activity of each industry during the considered time period. For example, the first row of the table describes the sales of the railroad industry to each of the other industries.

Let

$x_i > 0$ be the total output of industry i during the base period;

$x_{ij} \geq 0$ be the total sales of industry i to industry j during the base period;

$y_i \geq 0$ be the amount of final demand for industry i.

The interrelationships of the industries in our economy can be written as

$$x_i - \sum_{j=1,\ldots,n} x_{ij} = y_i; i = 1, \ldots, n. \tag{11.50}$$

In the 1930s Leontief assumed that the linear model of the economy for a base period could be used to predict the structure of the economy for a future period, and called $a_{ij} = x_{ij}/x_i$, the *input–output coefficient*, which is the amount of industry i necessary to produce one unit of commodity j. Thus

Table 11.14 Example of an input–output table.

Sales	Railroad	Steel	Coal	Other	Final demand	Total
Railroad	2	2	6	4	6	20
Steel	5	1	2	4	8	20
Coal	3	4	2	5	6	20
Other	6	6	3	6	9	30

Table 11.15 Example of an input–output coefficients table.

Sales	Railroad	Steel	Coal	Other	Final demand
Railroad	0.10	0.10	0.30	0.20	0.30
Steel	0.25	0.05	0.10	0.20	0.40
Coal	0.15	0.20	0.10	0.25	0.30
Other	0.20	0.20	0.10	0.20	0.30

system (11.50) becomes

$$\begin{array}{rcl} (1-a_{ii})x_i - \sum\limits_{j\neq i} a_{ij}x_j & = & y_i; i=1,\ldots,n \\ x_i & \geq & 0; i=1,\ldots,n. \end{array} \tag{11.51}$$

Example 11.3 (Input-output table). Consider Table 11.14, an input–output table, and its corresponding coefficients in Table 11.15. In this case, system (11.51) becomes

$$\begin{array}{c} \begin{pmatrix} 0.90 & -0.10 & -0.30 & -0.20 \\ -0.25 & 0.95 & -0.10 & -0.20 \\ -0.15 & -0.20 & 0.90 & -0.25 \\ -0.20 & -0.20 & -0.10 & 0.80 \end{pmatrix} \begin{pmatrix} x_1 \\ x_2 \\ x_3 \\ x_4 \end{pmatrix} = \begin{pmatrix} 6 \\ 8 \\ 6 \\ 9 \end{pmatrix}, \\ x_i \geq 0; i=1,2,3,4. \end{array} \tag{11.52}$$

The compatibility conditions of system (11.52) are obtained by finding the dual of the cone generated by its associated columns, which is given in Table 11.16. Thus, it is compatible iff

$$\begin{pmatrix} -1.481 & -0.426 & -0.615 & -0.669 \\ -0.568 & -1.323 & -0.403 & -0.599 \\ -0.534 & -0.504 & -1.423 & -0.704 \\ -0.579 & -0.500 & -0.432 & -1.655 \end{pmatrix} \begin{pmatrix} d_1 \\ d_2 \\ d_3 \\ d_4 \end{pmatrix} \leq \begin{pmatrix} 0 \\ 0 \\ 0 \\ 0 \end{pmatrix}, \tag{11.53}$$

where d_1, d_2, d_3, d_4 are the final demands of the four industries.

Note that for non-negative final demands the system always has a solution, but there are other solutions.

Table 11.16 Dual of the cone generated by the columns of matrix in (11.52).

Dual			
$\mathbf{w}_1$	$\mathbf{w}_2$	$\mathbf{w}_3$	$\mathbf{w}_4$
-1.481	-0.568	-0.534	-0.579
-0.426	-1.323	-0.504	-0.500
-0.615	-0.403	-1.423	-0.432
-0.669	-0.599	-0.704	-1.655

If the input–output tables are correctly defined, this problem always has a solution for positive final demands.

To obtain its general solution we solve the system of inequalities using the Γ-algorithm, using the rows of the matrix of the homogeneous system associated with (11.53), that is, including an auxiliary variable $x_5 = 1$, and obtain

$$\begin{aligned} \mathbf{V} &\equiv \{\}, \\ \mathbf{W} &\equiv \{\}, \\ \mathbf{Q} &\equiv \left\{ \left(\frac{3125}{142}, \frac{3095}{142}, \frac{1570}{71}, \frac{3545}{142} \right)^T \right\}, \end{aligned} \tag{11.54}$$

which gives a unique solution. ■

11.6 THE DIET PROBLEM

11.6.1 The Feasible Solution Set Problem

The nutrient contents of a number of different foods are given. The minimum daily requirements for each nutrient are also given. The problem consists of determining all diets satisfying the requirements. Let

m be the number of nutrients,

n be the number of different foods,

a_{ij} be the amount of nutrient i in one unit of food j,

b_i be the minimum amount of nutrient i required,

x_j be the amount of food j to be purchased.

Table 11.17 Nutrient contents of five different foods: (DN) digestible nutrients, (DP) digestible protein, (Ca) calcium, and (Ph) phosphorus, as given by Goldstein.

Nutrient	Required amount	Corn	Oats	Milo maize	Bran	Linseed meal
DN	74.2	78.6	70.1	80.1	67.2	77.0
DP	14.7	6.5	9.4	8.8	13.7	30.4
Ca	0.14	0.02	0.09	0.03	0.14	0.41
Ph	0.55	0.27	0.34	0.30	1.29	0.86

Since the total amount of a given nutrient i is the sum of the amounts of the same nutrient in all foods, we have

$$\begin{aligned} \sum_{j=1}^{n} a_{ij}x_j &\geq b_i; i = 1, \ldots, m \\ x_j &\geq 0; i = 1, \ldots, n. \end{aligned} \tag{11.55}$$

Example 11.4 (Diet problem). Consider the five nutrients and the minimum requirements of digestible nutrients (DN), digestible protein (DP), calcium (Ca), and phosphorus (Ph) shown in Table 11.17.

Equation (11.55) becomes

$$\begin{aligned} \begin{pmatrix} 78.6 & 70.1 & 80.1 & 67.2 & 77.0 \\ 6.50 & 9.40 & 8.80 & 13.7 & 30.4 \\ 0.02 & 0.09 & 0.03 & 0.14 & 0.41 \\ 0.27 & 0.34 & 0.30 & 1.29 & 0.86 \end{pmatrix} \begin{pmatrix} x_1 \\ x_2 \\ x_3 \\ x_4 \\ x_5 \end{pmatrix} &\geq \begin{pmatrix} 74.2 \\ 14.7 \\ 0.14 \\ 0.55 \end{pmatrix}, \\ x_1, x_2, x_3, x_4, x_5 &\geq 0. \end{aligned} \tag{11.56}$$

Using auxiliary slack variables $z_1, \ldots, z_4$, system (11.56) becomes

$$\begin{pmatrix} 78.6 & 70.1 & 80.1 & 67.2 & 77.0 & -1 & 0 & 0 & 0 \\ 6.50 & 9.40 & 8.80 & 13.7 & 30.4 & 0 & -1 & 0 & 0 \\ 0.02 & 0.09 & 0.03 & 0.14 & 0.41 & 0 & 0 & -1 & 0 \\ 0.27 & 0.34 & 0.30 & 1.29 & 0.86 & 0 & 0 & 0 & -1 \end{pmatrix} \begin{pmatrix} x_1 \\ x_2 \\ x_3 \\ x_4 \\ x_5 \\ z_1 \\ z_2 \\ z_3 \\ z_4 \end{pmatrix} = \begin{pmatrix} 74.2 \\ 14.7 \\ 0.14 \\ 0.55 \end{pmatrix},$$

$$x_1, x_2, x_3, x_4, x_5, z_1, z_2, z_3, z_4 \geq 0. \tag{11.57}$$

This system is compatible because the dual associated with the columns of the matrix in (11.57) is empty.

The general solution is obtained using the Γ-algorithm with the rows of the matrix associated with the homogeneous system of (11.56) and the rows associated with the non-negativity conditions of $x_1, \ldots, x_5$: see Table 11.18. Note that this general solution consists of a non-negative linear combination of the vectors in **W** plus a convex linear combination of the vectors in **Q**. The meanings of both parts are clear:

- **Q** contains the minimum amounts of nutrients required to satisfy the minimum requirements.
- **W** contains any amount of nutrient leading to an excess of requirements.

■

11.6.2 The Optimization Problem

In the diet problem we are interested in minimizing the cost of the nutrients, that is,

$$\text{Minimize} \sum_{j=1}^{n} c_j x_j, \tag{11.58}$$

where c_j is the unit cost of food j.

Example 11.5 (Diet problem). Consider the following cost coefficients for the diet problem in Example 11.4:

$$c_1 = 1; \; c_2 = 0.5; \; c_3 = 2; \; c_4 = 1.2; \; c_5 = 3.$$

Thus, we have the following LPP:

$$\text{Maximize} \quad -x_1 - 0.5x_2 - 2x_3 - 1.2x_4 - 3x_5 \tag{11.59}$$

subject to (11.56).

Using the EPM we obtain the maximum value of the objective function of -0.792769, which implies a minimum cost of 0.792769 units, at the point $(0, 1.53026, 0, 0.0230316, 0)$. ■

11.7 NETWORK FLOW PROBLEMS

Assume a transportation network (pipeline system, railroad, highway, communication, etc.) through which we wish to send a homogeneous commodity (oil, grain, cars, messages, etc.) from particular points of the network, called *source nodes*, to particular destination points, called *sink nodes*. In addition to source or sink nodes, the network can contain intermediate nodes, where no input or output of commodity occurs. Let us denote the flow going from node i to node j by x_{ij} (positive in the direction $i \to j$, and negative otherwise).

Table 11.18 Nutrient content of ten different foods. (DN) digestible nutrients, (DP) digestible protein, (Ca) calcium and (Ph) phosphorus, as given by Goldstein.

	Corn	Oats	Milo	Bran maize	Linseed meal
			W		
$\mathbf{w}_1$	1	0	0	0	0
$\mathbf{w}_2$	0	1	0	0	0
$\mathbf{w}_3$	0	0	1	0	0
$\mathbf{w}_4$	0	0	0	1	0
$\mathbf{w}_5$	0	0	0	0	1
			Q		
$\mathbf{q}_1$	0	0	0	1.10417	0
$\mathbf{q}_2$	0	0	0	0	0.963636
$\mathbf{q}_3$	0.0448423	0	0	1.05172	0
$\mathbf{q}_4$	0	0.0873163	0	1.01308	0
$\mathbf{q}_5$	0	0	0.0567182	1.03656	0
$\mathbf{q}_6$	7.	0	0	0	0
$\mathbf{q}_7$	0	0	4.66667	0	0
$\mathbf{q}_8$	0.220126	0	0	0.968553	0
$\mathbf{q}_9$	0	0	0.170524	0.963459	0
$\mathbf{q}_{10}$	0	1.61765	0	0	0
$\mathbf{q}_{11}$	0.458531	0	0	0	0.495577
$\mathbf{q}_{12}$	0	0.629273	0	0	0.390752
$\mathbf{q}_{13}$	0	0	0.468746	0	0.476019
$\mathbf{q}_{14}$	0.544348	0	0	0.096701	0.323583
$\mathbf{q}_{15}$	0	1.53026	0	0.023032	0
$\mathbf{q}_{16}$	0	0.736664	0	0.088176	0.216031
$\mathbf{q}_{17}$	0	0	0.574136	0.116168	0.265004
$\mathbf{q}_{18}$	0.10857	1.53143	0	0	0
$\mathbf{q}_{19}$	1.12406	0	0	0	0.286631
$\mathbf{q}_{20}$	0	1.51786	0.11309	0	0
$\mathbf{q}_{21}$	0	0	1.08128	0	0.262346
$\mathbf{q}_{22}$	0.022243	1.51631	0	0.022055	0
$\mathbf{q}_{23}$	0.814007	0	0	0.070968	0.277523
$\mathbf{q}_{24}$	0	1.51325	0.017539	0.023436	0
$\mathbf{q}_{25}$	0	0	0.598964	0.114763	0.258449

By imposing the conservation of flow condition at all nodes, the following constraints are obtained:

$$\begin{aligned}\sum_j (x_{ij} - x_{ji}) &= f_i; \quad i = 1, \ldots, n; \\ -f_{ij} \le x_{ij} &\le f_{ij}; \quad \forall i < j,\end{aligned} \tag{11.60}$$

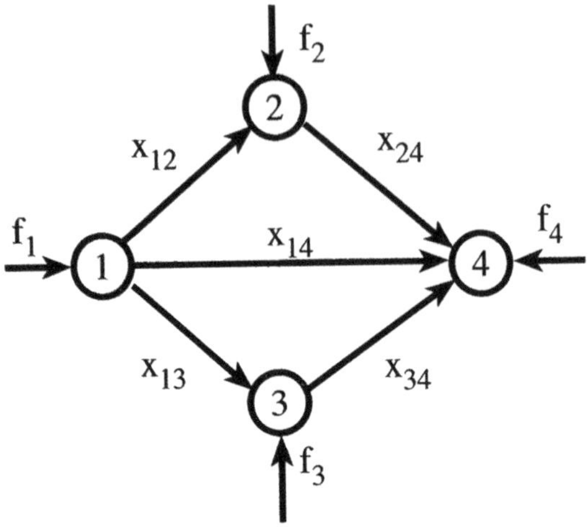

Fig. 11.16 Illustration of a network flow problem.

where n is the number of nodes in the network, f_i is the flow entering node i (if the flow leaves the node f_i must be negative), and f_{ij} is the maximum flow capacity of link $i \to j$.

Problems of flow networks are very common in engineering. In fact the water supply systems, the network communications systems, and others lead to network flow problems as those described above. In addition to solving the optimization associated problems one can be interested in analyzing the solution set, and how this set changes when some elements in the system fail. The following example discusses one of such problems.

Example 11.6 (Network flow problem). Consider the network flow problem in Fig. 11.16.

Equation (11.60) becomes

$$\begin{pmatrix} 1 & 1 & 1 & 0 & 0 \\ -1 & 0 & 0 & 1 & 0 \\ 0 & -1 & 0 & 0 & 1 \\ 0 & 0 & -1 & -1 & -1 \end{pmatrix} \begin{pmatrix} x_{12} \\ x_{13} \\ x_{14} \\ x_{24} \\ x_{34} \end{pmatrix} = \begin{pmatrix} f_1 \\ f_2 \\ f_3 \\ f_4 \end{pmatrix}, \tag{11.61}$$

$$\begin{array}{rcl} x_{ij} & \leq & f_{ij}; \forall i,j, \\ -x_{ij} & \leq & f_{ij}; \forall i,j. \end{array}$$

Since system (11.61) is not in the form of (8.6), we define the new variables

$$y_{ij} = x_{ij} + f_{ij}; \forall i,j,$$

and then system (11.61) transforms to

$$
\begin{pmatrix}
1 & 1 & 1 & 0 & 0 & 0 & 0 & 0 & 0 & 0\\
-1 & 0 & 0 & 1 & 0 & 0 & 0 & 0 & 0 & 0\\
0 & -1 & 0 & 0 & 1 & 0 & 0 & 0 & 0 & 0\\
0 & 0 & -1 & -1 & -1 & 0 & 0 & 0 & 0 & 0\\
1 & 0 & 0 & 0 & 0 & 1 & 0 & 0 & 0 & 0\\
0 & 1 & 0 & 0 & 0 & 0 & 1 & 0 & 0 & 0\\
0 & 0 & 1 & 0 & 0 & 0 & 0 & 1 & 0 & 0\\
0 & 0 & 0 & 1 & 0 & 0 & 0 & 0 & 1 & 0\\
0 & 0 & 0 & 0 & 1 & 0 & 0 & 0 & 0 & 1
\end{pmatrix}
\begin{pmatrix} y_{12}\\ y_{13}\\ y_{14}\\ y_{24}\\ y_{34}\\ z_{12}\\ z_{13}\\ z_{14}\\ z_{24}\\ z_{34}\end{pmatrix}
=
\begin{pmatrix} f_1^*\\ f_2^*\\ f_3^*\\ f_4^*\\ 2f_{12}\\ 2f_{13}\\ 2f_{14}\\ 2f_{24}\\ 2f_{34}\end{pmatrix},
\qquad
\begin{aligned}
y_{ij} &\geq 0; \forall i,j,\\
z_{ij} &\geq 0; \forall i,j,
\end{aligned}
\tag{11.62}
$$

where $z_{ij} \geq 0; \forall i,j$ are auxiliary slack variables used to convert the system of inequalities (11.61) into the system of equalities (11.62), and

$$
\begin{pmatrix} f_1^*\\ f_2^*\\ f_3^*\\ f_4^*\end{pmatrix}
=
\begin{pmatrix} f_1 + f_{12} + f_{13} + f_{14}\\ f_2 - f_{12} + f_{24}\\ f_3 - f_{13} + f_{34}\\ f_4 - f_{14} - f_{24} - f_{34}\end{pmatrix}.
\tag{11.63}
$$

An analysis of the compatibility of the system requires obtaining the dual of the cone generated by the columns of the matrix in (11.62), which is given in Table 11.19. It can be seen that there is one single **V** vector and seventeen **W** vectors, which lead to one compatibility equation (the first in Table 11.20) and 17 compatibility inequalities (the remaining rows in Table 11.20)

The compatibility equation

$$f_1 + f_2 + f_3 + f_4 = 0, \tag{11.64}$$

only says that the conservation of flow must be satisfied globally, that is, input flow to the system and output flow from the system must be equal (lack of losses is assumed).

The inequality conditions can be divided into two groups:

- *Non-negative inequality constraints*, which only express that the capacities must be non-negative.

- *Maximum flow inequality constraints*, which require that the input or output flow cannot exceed the maximum capacity associated with given sections (indicated in Fig. 11.17). For example, in Section S_1, the maximum flow is given by $f_{12} + f_{13} + f_{14}$ (the maximum flow of the links associated with node 1), which is an upper bound for either f_1 or $-f_1$ (flow can go in either direction). Similar arguments explain the physical meaning of the remaining compatibility conditions, which

Table 11.19 Final tableau of the Γ-algorithm for the compatibility problem.

Final Tableau

w_1	w_2	w_3	w_4	w_5	w_6	w_7	w_8	w_9
0	0	0	0	0	0	0	1	1
1	0	0	–1	0	0	0	1	0
0	0	0	0	0	1	0	1	1
0	0	0	0	0	0	0	0	0
0	0	–1	–1	0	0	0	0	–1
0	0	0	0	0	0	0	0	0
0	0	0	0	–1	0	0	–1	–1
–1	–1	0	0	0	0	0	–1	0
0	0	0	0	0	–1	–1	–1	–1

w_{10}	w_{11}	w_{12}	w_{13}	w_{14}	w_{15}	w_{16}	w_{17}	v_1
0	–1	–1	0	–1	–1	1	1	1
0	0	–1	0	0	–1	1	0	1
0	0	0	–1	–1	–1	0	0	1
0	0	0	0	0	0	0	0	1
0	0	0	0	0	0	0	–1	0
–1	0	0	–1	0	0	–1	–1	0
0	0	0	0	0	0	–1	–1	0
0	0	0	0	0	0	–1	0	0
0	0	0	0	0	0	0	0	0

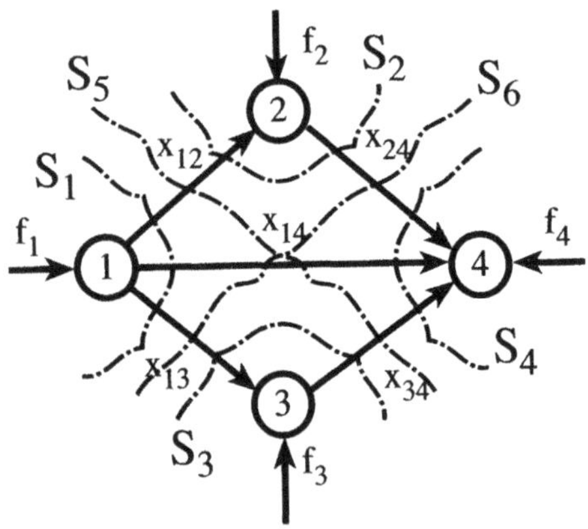

Fig. 11.17 Sections associated with the inequality constraints of Example 11.6.

are flow balances in different parts of the network assuming maximum allowed flow to occur in their links.

Table 11.20 Compatibility conditions.

Vector	Constraint	Section
Global flow balance equality constraints		
$(1,1,1,1,0,0,0,0,0)^T$	$f_1 + f_2 + f_3 + f_4 = 0$	
Non-negative inequality constraints		
$(0,0,0,0,-1,0,0,0,0)^T$	$f_{12} \geq 0$	
$(0,0,0,0,0,-1,0,0,0)^T$	$f_{13} \geq 0$	
$(0,0,0,0,0,0,-1,0,0)^T$	$f_{14} \geq 0$	
$(0,0,0,0,0,0,0,-1,0)^T$	$f_{24} \geq 0$	
$(0,0,0,0,0,0,0,0,-1)^T$	$f_{34} \geq 0$	
Maximum flow inequality constraints		
$(-1,0,0,0,0,0,0,0,0)^T$	$-f_1 \leq f_{12} + f_{13} + f_{14}$	S_1
$(1,0,0,0,-1,-1,-1,0,0)^T$	$f_1 \leq f_{12} + f_{13} + f_{14}$	S_1
$(0,1,0,0,0,0,0,-1,0)^T$	$f_2 \leq f_{12} + f_{24}$	S_2
$(0,-1,0,0,-1,0,0,0,0)^T$	$-f_2 \leq f_{12} + f_{24}$	S_2
$(0,0,1,0,0,0,0,0,-1)^T$	$f_3 \leq f_{13} + f_{34}$	S_3
$(0,0,-1,0,0,-1,0,0,0)^T$	$-f_3 \leq f_{13} + f_{34}$	S_3
$(1,1,1,0,0,0,-1,-1,-1)^T$	$f_1 + f_2 + f_3 \leq f_{14} + f_{24} + f_{34}$	S_4
$(-1,-1,-1,0,0,0,0,0,0)^T$	$-f_1 - f_2 - f_3 \leq f_{14} + f_{24} + f_{34}$	S_4
$(1,0,1,0,-1,0,-1,0,-1)^T$	$f_1 + f_3 \leq f_{12} + f_{14} + f_{34}$	S_5
$(-1,0,-1,0,0,0,0,0,0)^T$	$-f_1 - f_3 \leq f_{12} + f_{14} + f_{34}$	S_5
$(-1,-1,0,0,0,0,0,0,0)^T$	$-f_1 - f_2 \leq f_{13} + f_{14} + f_{24}$	S_6
$(1,1,0,0,0,-1,-1,-1,0)^T$	$f_1 + f_2 \leq f_{13} + f_{14} + f_{24}$	S_6

Table 11.21 Final tableau of the Γ-algorithm for the solution problem.

Dual			
$\mathbf{w}_1$	$\mathbf{w}_2$	$\mathbf{w}_3$	$\mathbf{w}_4$
4	8	4	8
8	8	7	3
7	3	8	8
0	4	0	4
7	7	6	2
4	0	4	0
0	0	1	5
1	5	0	0
8	4	8	4
1	1	2	6
1	1	1	1

Now system (11.62) is solved for the particular case $(f_1, f_2, f_3, f_4)^T = (7, -4, -1, -2)^T$ and $f_{ij} = 4; \forall i, j$. Using an auxiliary unknown $y_0 = 1$ the system can be written as

$$\begin{pmatrix} 1 & 1 & 1 & 0 & 0 & 0 & 0 & 0 & 0 & 0 & -19 \\ -1 & 0 & 0 & 1 & 0 & 0 & 0 & 0 & 0 & 0 & 4 \\ 0 & -1 & 0 & 0 & 1 & 0 & 0 & 0 & 0 & 0 & 1 \\ 0 & 0 & -1 & -1 & -1 & 0 & 0 & 0 & 0 & 0 & 14 \\ 1 & 0 & 0 & 0 & 0 & 1 & 0 & 0 & 0 & 0 & -8 \\ 0 & 1 & 0 & 0 & 0 & 0 & 1 & 0 & 0 & 0 & -8 \\ 0 & 0 & 1 & 0 & 0 & 0 & 0 & 1 & 0 & 0 & -8 \\ 0 & 0 & 0 & 1 & 0 & 0 & 0 & 0 & 1 & 0 & -8 \\ 0 & 0 & 0 & 0 & 1 & 0 & 0 & 0 & 0 & 1 & -8 \end{pmatrix} \begin{pmatrix} y_{12} \\ y_{13} \\ y_{14} \\ y_{24} \\ y_{34} \\ z_{12} \\ z_{13} \\ z_{14} \\ z_{24} \\ z_{34} \\ y_0 \end{pmatrix} = \begin{pmatrix} 0 \\ 0 \\ 0 \\ 0 \\ 0 \\ 0 \\ 0 \\ 0 \\ 0 \end{pmatrix}$$

$$\begin{pmatrix} -1 & 0 & 0 & 0 & 0 & 0 & 0 & 0 & 0 & 0 & 0 \\ 0 & -1 & 0 & 0 & 0 & 0 & 0 & 0 & 0 & 0 & 0 \\ 0 & 0 & -1 & 0 & 0 & 0 & 0 & 0 & 0 & 0 & 0 \\ 0 & 0 & 0 & -1 & 0 & 0 & 0 & 0 & 0 & 0 & 0 \\ 0 & 0 & 0 & 0 & -1 & 0 & 0 & 0 & 0 & 0 & 0 \\ 0 & 0 & 0 & 0 & 0 & -1 & 0 & 0 & 0 & 0 & 0 \\ 0 & 0 & 0 & 0 & 0 & 0 & -1 & 0 & 0 & 0 & 0 \\ 0 & 0 & 0 & 0 & 0 & 0 & 0 & -1 & 0 & 0 & 0 \\ 0 & 0 & 0 & 0 & 0 & 0 & 0 & 0 & -1 & 0 & 0 \\ 0 & 0 & 0 & 0 & 0 & 0 & 0 & 0 & 0 & -1 & 0 \\ 0 & 0 & 0 & 0 & 0 & 0 & 0 & 0 & 0 & 0 & -1 \end{pmatrix} \begin{pmatrix} y_{12} \\ y_{13} \\ y_{14} \\ y_{24} \\ y_{34} \\ z_{12} \\ z_{13} \\ z_{14} \\ z_{24} \\ z_{34} \\ y_0 \end{pmatrix} \leq \begin{pmatrix} 0 \\ 0 \\ 0 \\ 0 \\ 0 \\ 0 \\ 0 \\ 0 \\ 0 \\ 0 \\ 0 \end{pmatrix}, \tag{11.65}$$

and using the Γ-algorithm we obtain the dual of the cone generated by the rows of the matrices in (11.65), which is given in Table 11.21:

$$
\begin{aligned}
\mathbf{V} &\equiv \{\}, \\
\mathbf{W} &\equiv \{\}, \\
\mathbf{Q} &\equiv \{(4,8,7,0,7,4,0,1,8,1)^T, (8,8,3,4,7,0,0,5,4,1)^T, \\
&\quad (4,7,8,0,6,4,1,0,8,2)^T, (8,3,8,4,2,0,5,0,4,6)^T\}.
\end{aligned} \tag{11.66}
$$

Returning to the initial x variables we obtain the final solution:

$$
\begin{aligned}
\mathbf{x} &= \lambda_1(0,4,3,-4,3)^T + \lambda_2(4,4,-1,0,3)^T \\
&\quad +\lambda_3(0,3,4,-4,2)^T + \lambda_4(4,-1,4,0,-2)^T\}, \\
1 &= \lambda_1 + \lambda_2 + \lambda_3 + \lambda_4.
\end{aligned} \tag{11.67}
$$

Note that the four vectors in (11.67) satisfy the initial system (11.61). ■

Exercises

11.1 Consider the network flow problem in Fig. 11.18. Show that for this case, Equation (11.60) becomes

$$
\begin{pmatrix}
1 & 1 & 0 & 0 & 0 & 0 & 0 & 0 \\
-1 & 0 & 1 & 1 & 0 & 0 & 0 & 0 \\
0 & -1 & 0 & 0 & 1 & 1 & 0 & 0 \\
0 & 0 & -1 & 0 & -1 & 0 & 1 & 0 \\
0 & 0 & 0 & 0 & 0 & -1 & 0 & 1 \\
0 & 0 & 0 & -1 & 0 & 0 & -1 & -1
\end{pmatrix}
\begin{pmatrix} x_{12} \\ x_{13} \\ x_{24} \\ x_{26} \\ x_{34} \\ x_{35} \\ x_{46} \\ x_{56} \end{pmatrix}
=
\begin{pmatrix} f_1 \\ f_2 \\ f_3 \\ f_4 \\ f_5 \\ f_6 \end{pmatrix},
$$

$$
\begin{aligned}
x_{ij} &\le f_{ij}; \forall i,j, \\
-x_{ij} &\le f_{ij}; \forall i,j.
\end{aligned} \tag{11.68}
$$

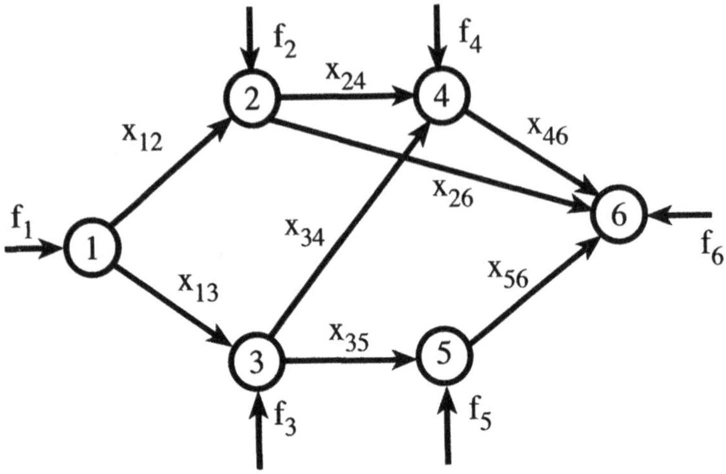

Fig. 11.18 A network flow problem.

Defining the new variables $y_{ij} = x_{ij} + f_{ij}; \forall i, j$, show that system (11.68) transforms to

$$\begin{pmatrix} 1 & 1 & 0 & 0 & 0 & 0 & 0 & 0 \\ -1 & 0 & 1 & 1 & 0 & 0 & 0 & 0 \\ 0 & -1 & 0 & 0 & 1 & 1 & 0 & 0 \\ 0 & 0 & -1 & 0 & -1 & 0 & 1 & 0 \\ 0 & 0 & 0 & 0 & 0 & -1 & 0 & 1 \\ 0 & 0 & 0 & -1 & 0 & 0 & -1 & -1 \end{pmatrix} \begin{pmatrix} y_{12} \\ y_{13} \\ y_{24} \\ y_{26} \\ y_{34} \\ y_{35} \\ y_{46} \\ y_{56} \end{pmatrix} = \begin{pmatrix} f_1^* \\ f_2^* \\ f_3^* \\ f_4^* \\ f_5^* \\ f_6^* \end{pmatrix},$$

$$\begin{array}{rcl} y_{ij} + z_{ij} & = & 2f_{ij}; \forall i, j, \\ y_{ij} & \geq & 0; \forall i, j, \\ z_{ij} & \geq & 0; \forall i, j, \end{array} \tag{11.69}$$

where $z_{ij} \geq 0; \forall i, j$ are auxiliary slack variables used to convert the system of inequalities (11.68) into the system of equalities (11.69), and

$$\begin{pmatrix} f_1^* \\ f_2^* \\ f_3^* \\ f_4^* \\ f_5^* \\ f_6^* \end{pmatrix} = \begin{pmatrix} f_1 - f_{12} - f_{13} \\ f_2 + f_{12} - f_{24} - f_{26} \\ f_3 + f_{13} - f_{34} - f_{35} \\ f_4 + f_{24} + f_{34} - f_{46} \\ f_5 + f_{35} - f_{56} \\ f_6 + f_{26} + f_{46} + f_{56} \end{pmatrix}. \tag{11.70}$$

Obtain the compatibility conditions and the general solution of the problem.

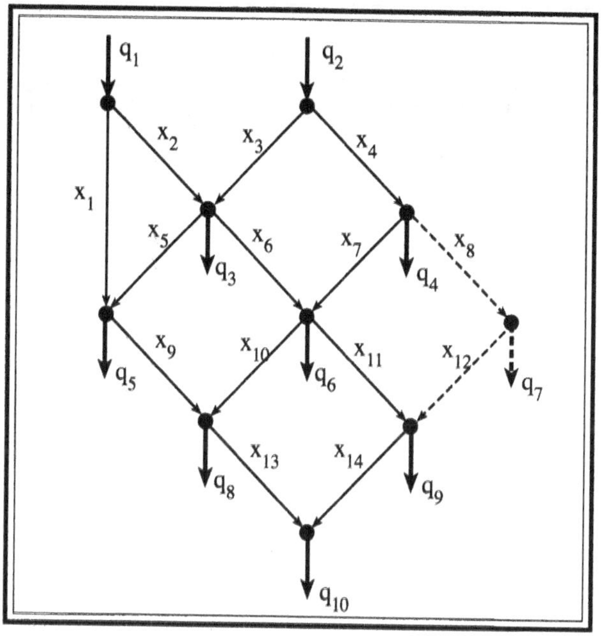

Fig. 11.19 Water supply system showing the flows in all conduits.

11.2 Consider the water supply network system in Fig. 11.19, where $\{q_1, q_2\}$ are the supply flows, $\{q_3, \ldots, q_{10}\}$ are the consumption flows at the nodes of the network, and $\{x_1, \ldots, x_{14}\}$ are the corresponding flows for the required supply. By convention, we consider $q_1, q_2 < 0$ and $q_i \geq 0$ for $i = 3, \ldots, 14$.

(a) Write the balance of flow in each of the nodes to obtain the system of equations for the water supply system.

(b) Obtain the compatibility conditions and give them a physical interpretation.

(c) Assume that the supplies satisfy $q_1 = q_2$ and the consumptions are $q_3 = q_4 = q_5 = q_6 = q_7 = q_8 = q_9 = q_{10} = 1\ \text{m}^3/\text{sec}$. What is the commom value of q_1 and q_2?

(d) Obtain the general solution of the system of equations in this case.

(e) Consider the following two situations:

i. Consumer q_7 has a normal supply, but its consumption is null.

ii. We cut off the water supply to consumer q_7.

Obtain the new solutions based on the old ones.

11.3 Given the structure in Fig. 11.4, write the stiffness matrices for different cases of support.

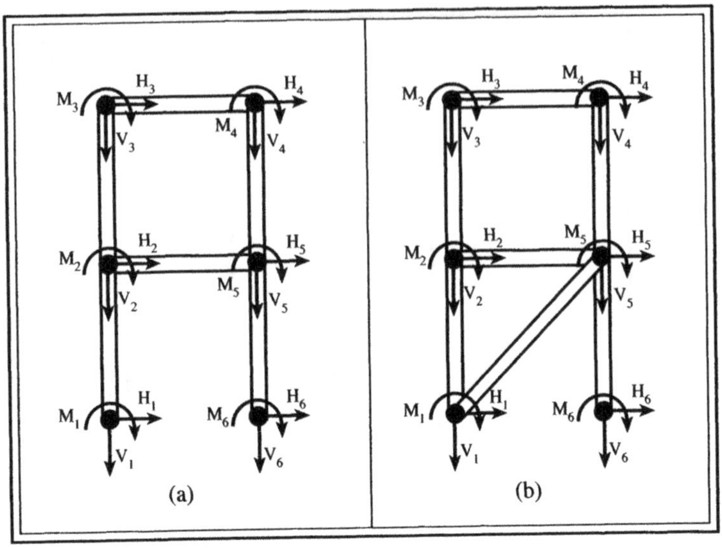

Fig. 11.20 Two two-level structures.

11.4 Given the structure in Fig. 11.20 , write the stiffness matrices for different cases of support.

11.5 Consider again the transportation problem in Section 11.3.2 and assume now that the system has another alternative shipping origin, such that the cost function to be minimized becomes

$$x_{11} + 3x_{12} + 2x_{13} + 1x_{21} + 2x_{22} + 2x_{23} + 1x_{31} + 2x_{32} + 3x_{33}. \quad (11.71)$$

Determine which of the two solutions – the original or this one – is more economical.

11.6 Consider the network flow problem in Fig. 11.21.

(a) Write the system of equations associated with this problem.

(b) Obtain the compatibility conditions and give them a physical interpretation.

(c) Obtain the set of feasible solutions and explain the result.

(d) Obtain the optimal solution for unit costs and $f_i = 10; \forall i$.

(e) Assuming that the shipping associated with x_{47} fails; redo parts (a)–(d) without starting from scratch.

11.7 Consider the demand graph in Fig. 11.22 and assume an initial stock of two units.

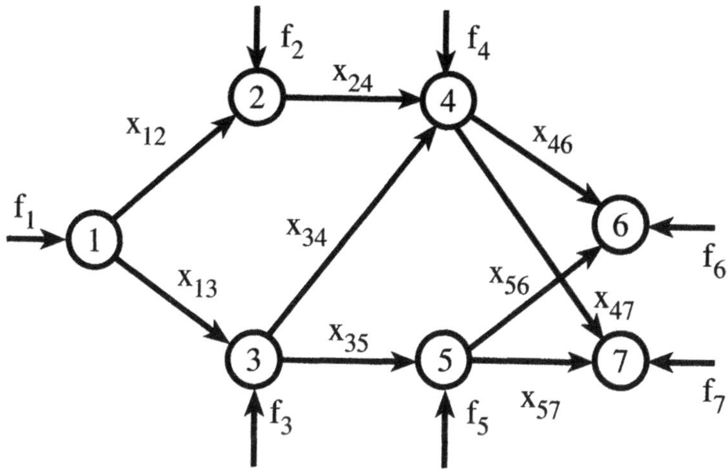

Fig. 11.21 Network flow problem.

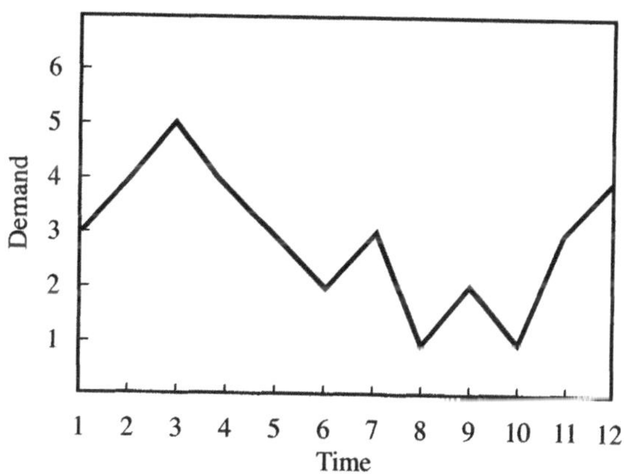

Fig. 11.22 Production-scheduling problem.

(a) Write the system of equations associated with this problem.

(b) Obtain the compatibility conditions and give them a physical interpretation.

(c) Obtain the set of feasible solutions and explain the result.

(d) Fix the required costs.

(e) Obtain the optimal solution for these costs.

Part V

Appendices

Appendix A: A Java Application

In this appendix we present a Java application for solving some problems of linear algebra using orthogonalization techniques. This application can be obtained in the following URL:

`http://ccaix3.unican.es/~AIGroup/Orthogonal.html`,

where an *Applet* version has also been included for its direct use through Internet.

The application can be used to solve the following problems:

1. Obtaining the orthogonal decomposition of a linear space with respect to a linear space in another linear space.

2. Obtaining the inverse of a square matrix.

3. Obtaining the determinant of a square matrix.

4. Obtaining the rank of a matrix.

5. Analyzing the compatibility of a linear system of equations.

6. Solving a linear system of equations.

7. Checking the equivalence of two systems of equations.

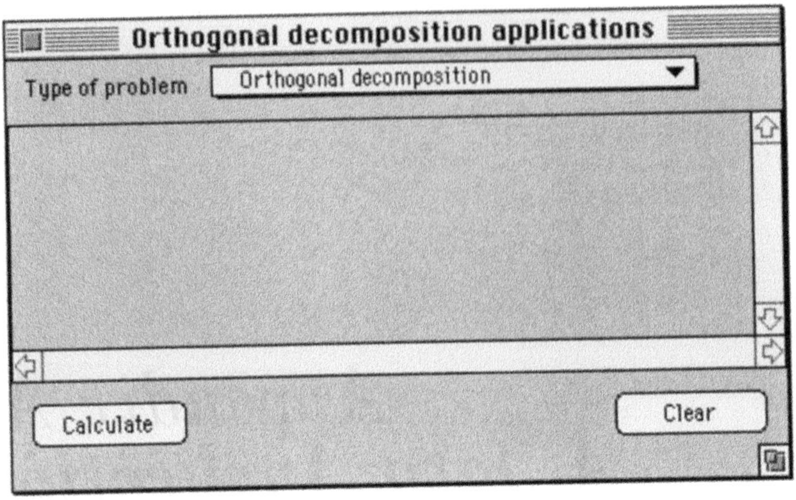

Fig. A.1 The program's main window.

8. Solving a system of linear equations in some selected variables.

9. Obtaining the reciprocal image of a linear space by a given linear transformation.

10. Obtaining the intersection of two linear spaces.

A.1 HOW TO USE THE PROGRAM

Fig. A.1 shows the application's main window, which has the following components:

- A choice list that allows selecting the type of problem to be solved (see Fig. A.2).

- A scrollable area to show the results.

- A Calculate button to initiate the process for solving the problem.

- A Clear button to reset the data and solve a new problem.

The appearance of the menu bar depends on the option selected from the problems choice list. In the following subsections we describe how to solve each of the preceding problems.

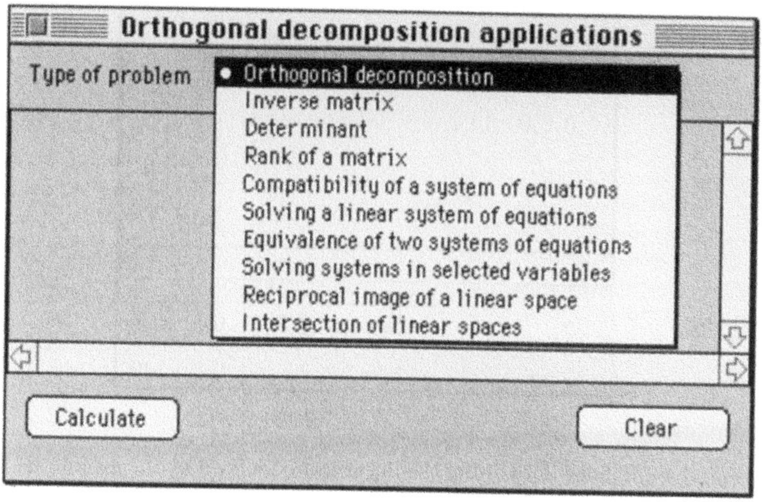

Fig. A.2 Pop-up menu showing the types of problems that can be solved.

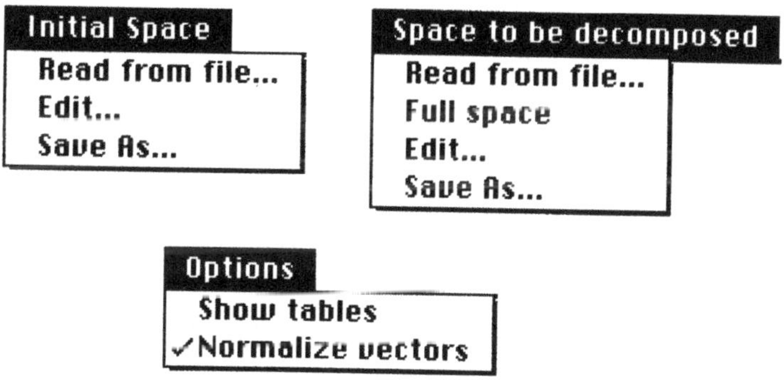

Fig. A.3 Submenus appearing in the main menu when the orthogonal decomposition option is selected.

A.1.1 Performing Orthogonal Decompositions

The application is initially prepared to perform the orthogonal decomposition of a linear space **V** with respect to another linear space **U**. The menu bar contains the submenus shown in Fig. A.3.

- The **Initial Space** menu is used for reading the linear space **U**. You can read the linear space from a text file or use the window as in Fig. A.4 to introduce or modify the data.

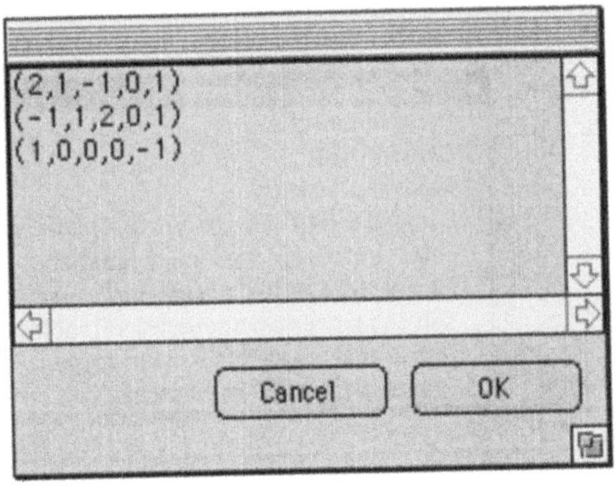

Fig. A.4 Window for editing the linear space, defined by its generators.

- The **Read from file...** option of this menu permits selecting a text file and reading from it. This text file must contain the generators of the linear space with the following structure:

```
(2,1,-1,0,1)
(-1,1,2,0,1)
(1,0,0,0,-1)
```

  that is, a generator per line with each defined by its corresponding list of components separated by commas and between parentheses.
- The **Edit...** option allows typing the generators directly in the window of Fig. A.4 or modifying the generators previously read from a file space.
- With the **Save As...** option the generators of the space can be saved in a text file, so that it can be read and used later.

• The **Space to be decomposed** menu is used for reading the linear space **V**. In addition to the three options of the previous menu, it has the additional **Full space** option, which is selected when the space to be decomposed is the whole $\mathbb{R}^n$ space.

• The **Options** menu has the following options:

  - **Show tables**: When this option is chosen the application shows the iterations of the algorithm, one by one, in a tabular form. By default, this option is disabled.
  - **Normalize vectors**: When this option is enabled, the orthogonal decomposition algorithm normalizes vectors; otherwise, it doesn't.

Once the linear spaces **U** and **V** have been read, the **Calculate** button of the main window gets activated. After pressing this button the orthogonal

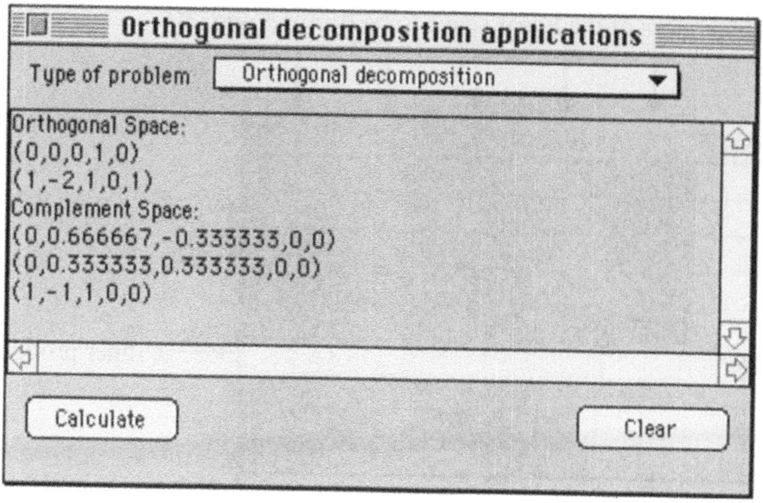

Fig. A.5 Orthogonal decomposition showing the generators of the orthogonal set and its complement.

decomposition is performed, and the output is shown in the main window. For example, Fig. A.5 shows the orthogonal decomposition of the $\mathbb{R}^5$ space with respect to the space generated by the vectors $(2,1,-1,0,1)^T$, $(-1,1,2,0,1)^T$, and $(1,0,0,0,-1)^T$.

If the **Show tables** option is activated, before obtaining the result in Fig A.5, a new window with a tabular form is shown, where all the iterations are presented. Initially, this window contains the generators of the linear space to be decomposed, and the first generator of the **U** space (see Fig. A.6). The last row contains the inner products of this first generator by the generators of the **V** space. The pivot column is distinguished with a different color. If the **Next Table** button is pressed, the algorithm performs a new iteration using the pivot column.

Figures A.7–A.9 show all iterations of the algorithm for the proposed example and the corresponding pivot columns. In each step, the generators of the orthogonal set to the set generated by the already introduced vectors –those shown in the first column– and its complement are shown in different colors.

The last table shows the generators of the orthogonal space and its complement in different colors. Note that the **Next Table** button in this last table is disabled because the end of the orthogonalization process has already been reached.

If the decomposition is performed with the **Normalize vectors** option disabled, the last table will be different (see Fig. A.10). In this case, the last element of the first column is the common denominator of all cells.

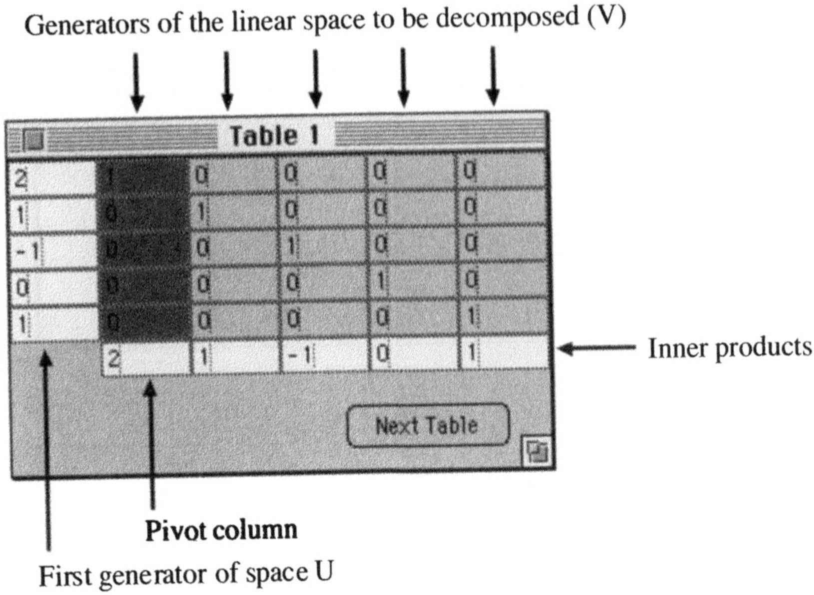

Fig. A.6 First table showing the generators of the space to be decomposed and the first generator of the **U** space.

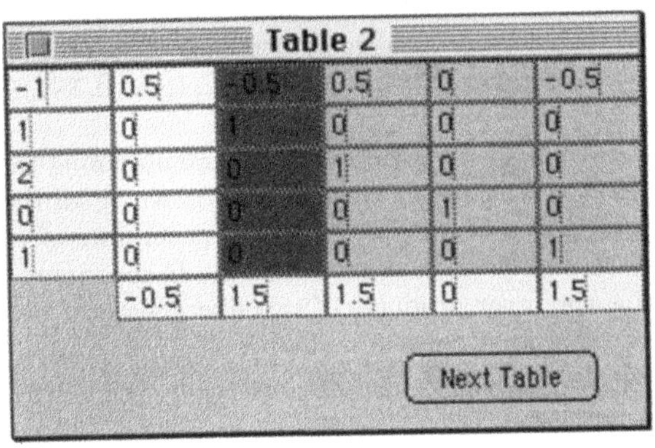

Fig. A.7 Table corresponding to the second iteration showing the pivot column, and the generators of the orthogonal set (i.e., orthogonal to the set generated by the already introduced vectors) and its complement.

A.1.2 Calculating the Determinant, Inverse, and Rank of a Matrix

If you want to calculate the determinant, the inverse matrix, or the rank of a matrix, select the corresponding option in the pop-up menu of the main window (see Fig. A.2). Once the option is selected, the menu bar will display

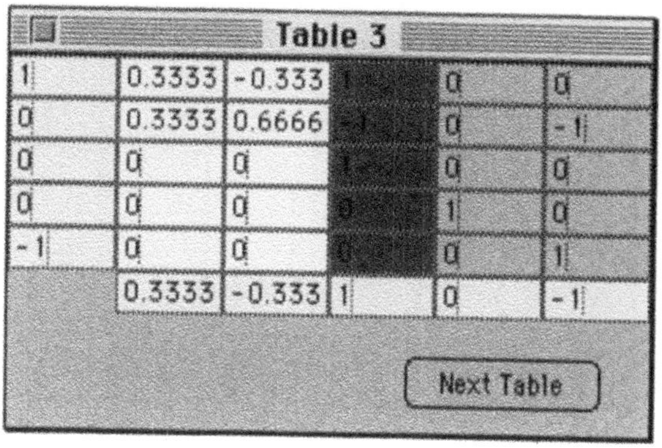

Fig. A.8 Table corresponding to the third iteration showing the pivot column, and the generators of the orthogonal set (i.e., orthogonal to the set generated by the already introduced vectors) and its complement.

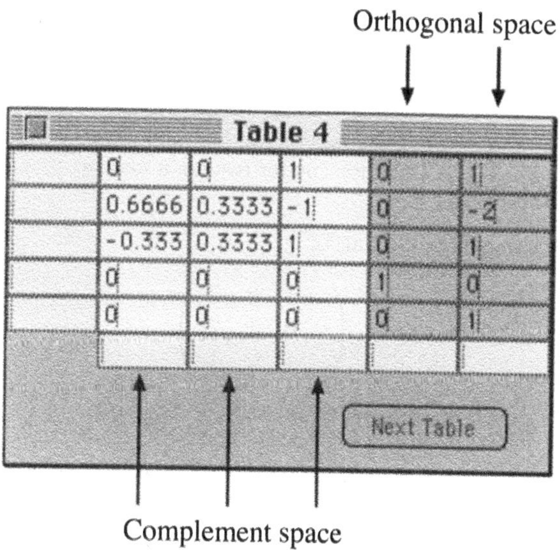

Fig. A.9 Table corresponding to the last iteration showing the generators of the orthogonal set (i.e., orthogonal to the set generated by the already introduced vectors) and its complement.

the submenu shown in Fig. A.11. With this submenu you can read the row vectors of a matrix from a file or type them in the editing window.

Once the matrix has been read, the **Calculate** button becomes activated. After pressing this button the output is shown in the main window. For

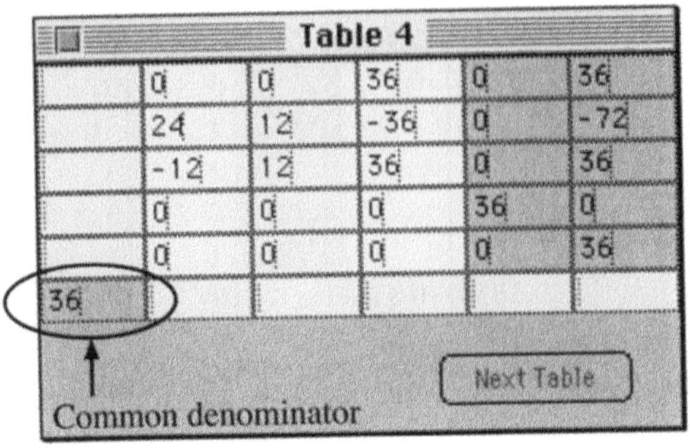

Fig. A.10 Last table with the normalize option disabled.

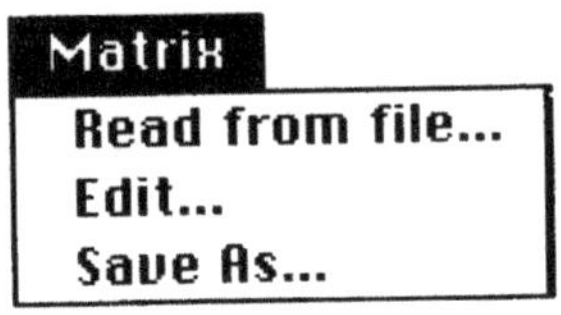

Fig. A.11 Submenu for reading a matrix.

example, if we first enter the matrix

$$\mathbf{A} = \begin{pmatrix} -1 & 2 & 3 \\ 0 & 1 & 2 \\ 1 & 2 & 1 \end{pmatrix},$$

then Fig. A.12 shows the inverse of $\mathbf{A}$.

A.1.3 Linear Systems of Equations

The program can be used for:

1. Analyzing the compatibility of a linear system of equations.
2. Solving a linear system of equations.
3. Checking the equivalence of two systems of equations.
4. Solving a system of linear equations in some selected variables.

Each problem in the preceding list has a corresponding option in the pop-up menu of the main window (see Fig. A.2). In these cases, the menu bar will contain the submenu in Fig. A.13 for reading the system of equations. In the

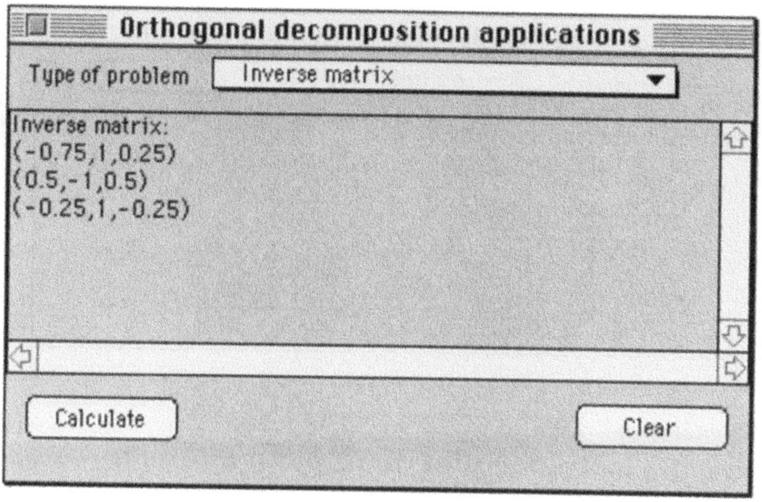

Fig. A.12 The inverse matrix.

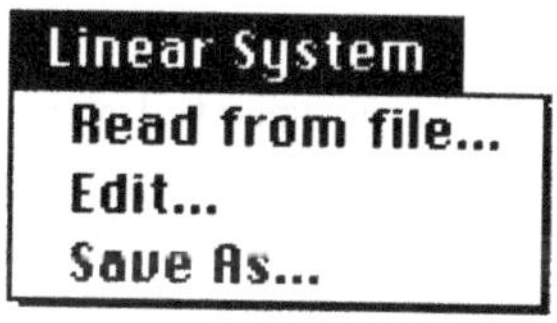

Fig. A.13 Submenu for reading a linear system of equations.

case of equivalence of two systems, the menu bar will contain two identical submenus for reading the two systems.

For example, the linear system

$$\begin{array}{rrrrrrcl} -x_1 & +2x_2 & +3x_3 & +4x_4 & +x_5 & +2x_6 & = & 3 \\ 2x_1 & +3x_2 & +4x_3 & +5x_4 & +6x_5 & +x_6 & = & 1 \\ 3x_1 & +2x_2 & +x_3 & +3x_4 & & +5x_6 & = & 2 \end{array} \tag{A.1}$$

is first read from a text file with the following structure:

```
(-1,2,3,4,1,2)
(2,3,4,5,6,1)
(3,2,1,3,0,5)
END
(3,1,2)
```

The first three lines contain the coefficients of the unknowns in the equations, and the last row contains the list of independent terms. The word `END` is a separator that indicates the end of the equation coefficients. When we click on the Calculate button, the solution will be displayed, as in Fig. A.14.

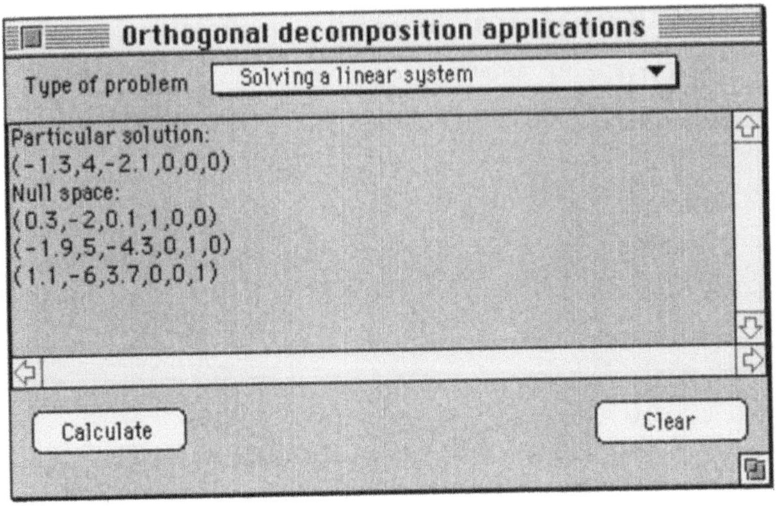

Fig. A.14 The solution of a linear system.

When solving a system in some selected variables, these variables have to be given in the data file below the independent terms, using again the separator END. For example, if we are interested in solving the system (A.1) in the variables x_1, x_3, x_4, x_5, and x_6, we use the following text file:

```
(-1,2,3,4,1,2)
(2,3,4,5,6,1)
(3,2,1,3,0,5)
END
(3,1,2)
END
1,3,4,5,6
```

Fig. A.15 shows the solution for this example.

A.1.4 Intersection of Two Linear Spaces

The program can also calculate the intersection of two linear spaces. In this case, the menu bar contains two identical submenus for reading the space generators. You can read them from a file or write them in the editing window as described above. For example, Fig. A.16 shows the intersection of the following linear spaces:

$$\begin{aligned} U &\equiv \mathcal{L}\left\{(2,1,-3)^T,(1,0,2)^T\right\}, \\ V &\equiv \mathcal{L}\left\{(4,5,-1)^T,(-1,1,4)^T\right\}. \end{aligned}$$

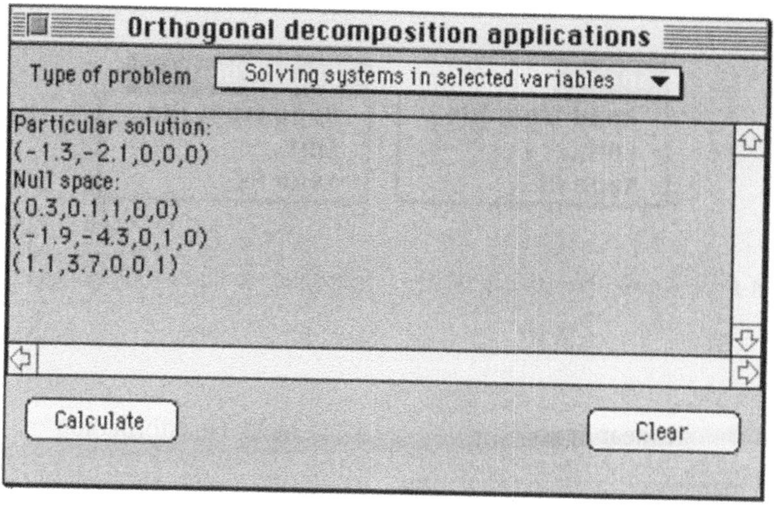

Fig. A.15 The solution of a linear systems in some selected variables.

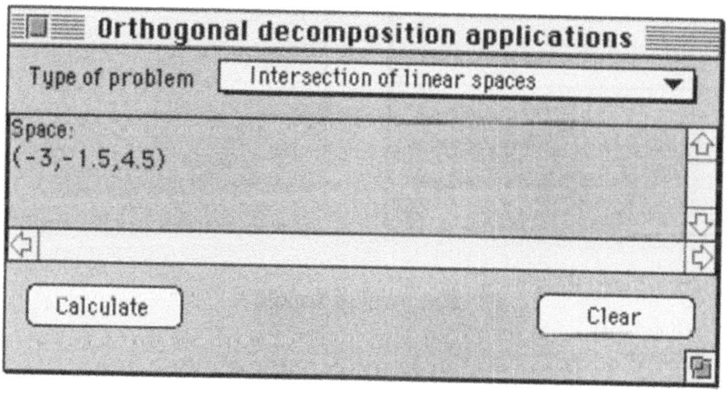

Fig. A.16 Intersection of two linear spaces.

A.1.5 Reciprocal Image of a Linear Space by a Linear Transformation

The orthogonalization techniques are also useful for obtaining the reciprocal image of a linear space by a linear transformation. To this end, select the corresponding option in the pop-up menu and read the linear space generators and the matrix of the linear transformation with the submenus in Fig. A.17.

Fig. A.18 shows an example of reciprocal image of a linear space generated by two vectors with respect to a linear transformation in $\mathbb{R}^4$.

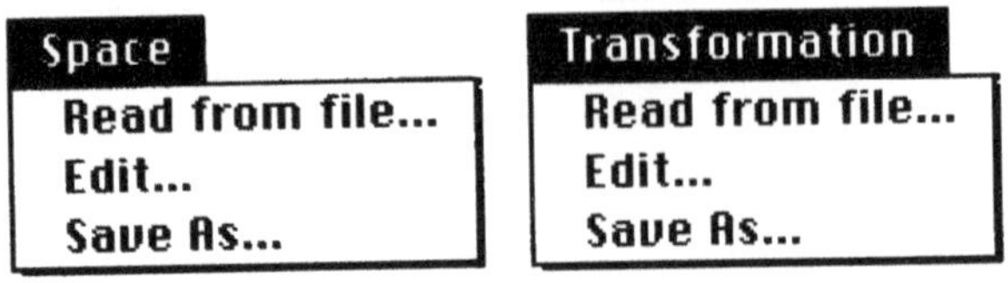

Fig. A.17 Submenus for reading the linear space and the linear transformation.

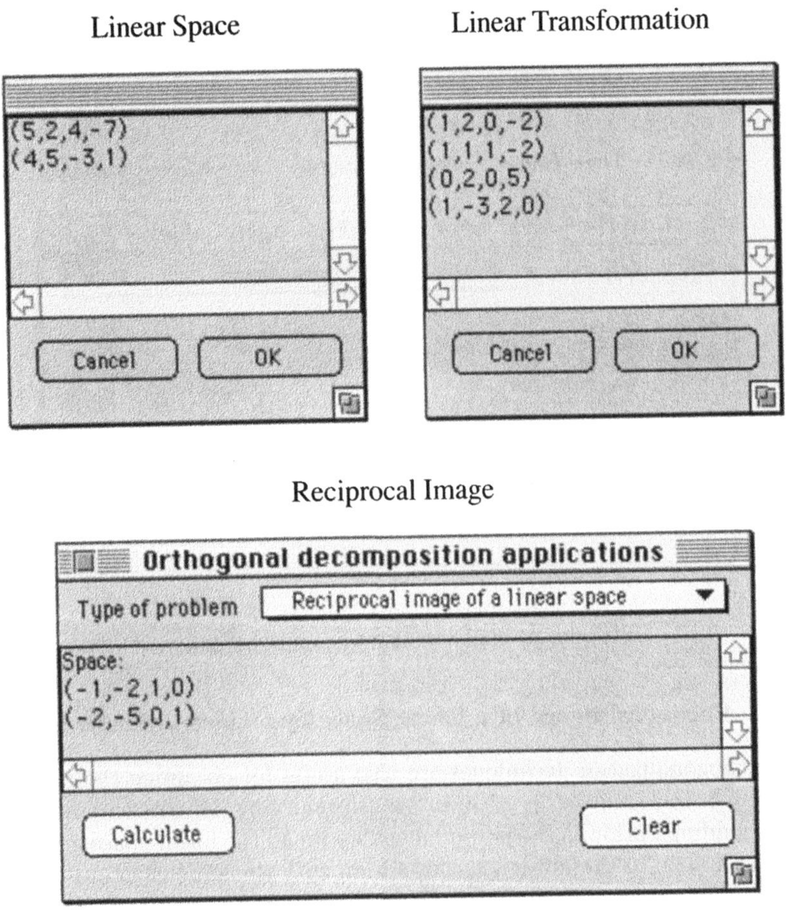

Fig. A.18 Reciprocal image of a linear space by a linear transformation.

Appendix B: List of Notation

In this appendix we list the main notation used in this book. We have attempted to keep the notation as consistent as possible throughout the book. The notation common to all chapters is listed first, followed by the notation mainly specific to each chapter.

Notation Common to All Chapters

$\mathbf{A}$	Matrix of real numbers
$-\mathbf{A}$	Negative of $\mathbf{A}$
$\mathbf{A}^T$	Transpose of $\mathbf{A}$
$\mathbf{A}^{-1}$	Inverse of a nonsingular matrix $\mathbf{A}$
$A \supseteq B$	A is a superset of or coincides with B
$A \subseteq B$	A is a subset of or coincides with B
$A \subset B$	A is a proper subset of B
$\bar{A}$	Complement of A
$\mathbf{a}_i$	ith column of matrix $\mathbf{A}$
$\mathbf{a}_i^T$	ith row of matrix $\mathbf{A}$
a_{ij}	Element (i, j) of matrix $\mathbf{A}$
$card(A)$	Number of elements of set A
$dim(\mathcal{L}(\cdot))$	Dimension of the linear space $\mathcal{L}(\cdot)$
$d(\mathbf{x}, \mathbf{y})$	Euclidean distance between vectors $\mathbf{x}$ and $\mathbf{y}$

E^n	n-dimensional Euclidean space
$\mathbf{I}_n$	Identity matrix of order n
$\mathbb{R}$	Set of real numbers
$\mathbb{R}^+$	Set of non-negative real numbers
$\mathbb{R}^{++}$	Set of positive real numbers
$\mathbb{R}^n$	Set of vectors of n real numbers
$rank(\mathbf{A})$	Rank of the matrix $\mathbf{A}$
$\|a\|$	Absolute value of a
$\|\mathbf{A}\|$	Determinant of matrix $\mathbf{A}$
$\\|\mathbf{x}\\|$	Euclidean norm of the vector $\mathbf{x}$
$\mathbf{x} \bullet \mathbf{y}$	Dot product of vectors $\mathbf{x}$ and $\mathbf{y}$
$X \setminus Y$ or $X - Y$	X minus Y (set difference)
$\mathcal{L}(\cdot), \mathcal{U}, \mathcal{V}$	Linear spaces
WWW	World Wide Web
$\mathbf{0}$	Zero vector, $\mathbf{0} = (0, \ldots, 0)$
$\cap$	Set intersection
$\cup$	Set union
$\emptyset$	Empty set
$\exists$	Exists at least one
$\forall$	For all
$\iff$ or iff	If and only if
$\in$	Belongs to
$\prod_{i=1}^n x_i$	$x_1 \times \cdots \times x_n$
$\sum_{i=1}^n x_i$	$x_1 + \cdots + x_n$
$\wedge$	And
$\vee$	Or
$\oplus$	Direct sum
δ_{ij}	Kronecker deltas
$\equiv$	Equality sign for sets

Chapter 1

$A \times B$	Cartesian product of A and B
$\mathbf{A}$	Ordered set of k vectors $\{\mathbf{a}_1, \ldots, \mathbf{a}_k\}$
$\mathbf{A}_\rho$	$\{\rho_1 \mathbf{a}_1 + \cdots + \rho_k \mathbf{a}_k \mid \rho_i \in \mathbb{R}, \forall i\}$
$\mathbf{A}_\xi$	$\{\xi_1 \mathbf{a}_1 + \cdots + \xi_k \mathbf{a}_k \mid \xi_i \in \mathbb{R} \setminus \{0\}, \forall i\}$
$(\mathbf{A} : \mathbf{b})$	Matrix with the columns of $\mathbf{A}$ and $\mathbf{b}$
$\mathbf{e}_j$	jth unit vector
$\{\mathbf{e}_1, \ldots, \mathbf{e}_n\}$	Canonical basis in $\mathbb{R}^n$
H	Hyperplane, $\{\mathbf{x} \in E^n \mid \mathbf{a} \bullet \mathbf{x} = b\}$
H^-	$\{\mathbf{x} \in E^n \mid \mathbf{a} \bullet \mathbf{x} \leq b\}$
H^+	$\{\mathbf{x} \in E^n \mid \mathbf{a} \bullet \mathbf{x} > b\}$
$\mathcal{K}$	A field
$\mathcal{P}(\mathbb{R})$	Set of polynomials with coefficients in $\mathbb{R}$

$\mathcal{P}_n(\mathbb{R})$ .... $\{a_0 + a_1x^1 + \cdots + a_nx^n \mid a_i \in \mathbb{R},\ i = 1, \ldots, n\}$
$p_a(x), p_b(x)$ ...... Elements of $\mathcal{P}_n(\mathbb{R})$
$\mathbf{u}^\perp$ .... Linear space orthogonal to vector $\mathbf{u}$
$\mathcal{U}^\perp$ .... Linear space orthogonal to space $\mathcal{U}$
$(\mathcal{V}, +, *)$ .... Vector or linear space
$\mathcal{W}_r$ .... $\{p(x) \in \mathcal{P}(\mathbb{R}) \mid p(r) = 0\}$
$\mathbf{x}_h$ .... General solution of a homogeneous system
$\mathbf{x}_0$ .... Particular solution of a system
α, β, ρ .... Real numbers
ξ .... Non–null real number ($\xi \in \mathbb{R} \setminus \{0\}$)
θ .... Angle between two vectors
$+$ .... Internal operation
$*$ .... External operation

Chapter 2

$\mathcal{L}\{\mathbf{v}_1^j, \ldots, \mathbf{v}_n^j\}$ .... Linear space spanned by $\{\mathbf{v}_1^j, \ldots, \mathbf{v}_n^j\}$
$\mathbf{M}_j^{-1}$ .... Matrix $\left(\mathbf{e}_1, \ldots, \mathbf{e}_{j-1}, t_j^{*T}, \mathbf{e}_{j+1}, \ldots, \mathbf{e}_n\right)^T$
t_j^* .... $\left(-t_1^j/t_j^j, \ldots, -t_{j-1}^j/t_j^j, 1/t_j^j, \ldots, -t_n^j/t_j^j\right)$
t_k^j .... Real number; normally $\mathbf{u}_j \bullet \mathbf{v}_k^{j-1}$
$\mathbf{u}_\mathbf{V}^\perp$ .... Linear space orthogonal to the vector $\mathbf{u}$ on $\mathbf{V}$
$\mathbf{V}^j$ .... Linear space spanned by $\{\mathbf{v}_1^j, \ldots, \mathbf{v}_n^j\}$
$\mathbf{V}^\perp$ .... Space orthogonal to $\mathbf{V}$
$\mathbf{v}^j$ .... Vector $\mathbf{v}$ at the jth step
$\mathbf{v}_k^j$ .... kth component of $\mathbf{v}$ at the jth step

Chapter 3

$\mathbf{A}^*$ .... Modified matrix $\mathbf{A}$
$\mathbf{A}_{ij}$ .... Matrix partition ij of matrix $\mathbf{A}$
a_{ij}^* .... Element (i, j) of $\mathbf{A}^{-1}$
$B(i, j)$ .... Element (i, j) of matrix $\mathbf{B}$
$\mathbf{C}$ .... Matrix obtained from $\mathbf{A}$ by modifying one row
$\mathbf{P}_{ij}$ .... Permutation matrix
$*$ .... Matrix product

Chapter 4

$\Pi_1(t), \Pi_2(t)$ ...... Planes in $\mathbb{R}^3$
ρ_i .... Real number
t .... Time parameter

$\mathbf{x}_a, \mathbf{x}_b$ .... Associated vectors with $\mathbf{A}$ and $\mathbf{B}$ respectively, such that $\mathbf{A}\mathbf{x}_a = \mathbf{B}\mathbf{x}_b$

Chapter 5

$C^2[a, b]$........... Set of continuous functions in the interval $[a, b]$ with first and second continuous derivatives
F_i................ Weight component in the movement direction
I................. Current intensity
K................ Coulomb
m................ Number of equations
mA .... Milliampere
n .... Number of unknowns
R................ Electric resistance
R_i .... Friction force
T_i................ String tension
t_j^i .... Dot products $a_i \bullet v_j$
t_j^{*i} .... Updated dot products
$\mathbf{U}$................ Coefficient matrix of the linear system
u_{ij} .... Real number
V .... Voltage
$\hat{\mathbf{w}}$................ Vector $\mathbf{w}$ after removing its last component
$\mathbf{X}^m$ .... Set of solution vectors at step m
$\mathbf{x}^*$ .... Solution vector
$\mathbf{x}_a$ .... $(x_1, \ldots, x_r, x_{n+1})$
$\mathbf{x}_b$................ $(x_{r+1}, \ldots, x_n)$
x_+ .... Positive part
$\phi(j)$.............. Associates to each column $\mathbf{w}^j$ of pivot equation
$\phi^*(j)$............. Updated $\phi(j)$ function
$\psi(j)$ .... List of nonbasic equations that are not orthogonal to $\mathbf{w}^j$
$\psi^*(j)$ .... Updated $\psi(j)$ function
$\partial S/\partial a_i$........... Partial derivative

Chapter 6

$\mathbf{A}_\lambda$............... Set of vectors that are convex linear combinations of vectors of $\mathbf{A}$
$\mathbf{A}_\pi$............... Cone generated by the columns of matrix $\mathbf{A}$
$\mathbf{A}_\sigma$............... Set of vectors that are all positive linear combinations of vectors of $\mathbf{A}$
$\mathbf{A}^\mathbf{x}$............... Binding set of vector $\mathbf{x}$ in the cone $\mathbf{A}_\pi$
$\mathbf{A}_\pi^{bd}$ .... Boundary of cone $\mathbf{A}_\pi$

$\mathbf{A}_\pi^{ex}$ .... Exterior of cone $\mathbf{A}_\pi$
$\mathbf{A}_\pi^{in}$ .... Interior of cone $\mathbf{A}_\pi$
$(\mathbf{A}:\mathbf{B})_\pi$ .... Cone generated by $\mathbf{A}$ and $\mathbf{B}$
$\mathbf{A}^i$ .... Matrix with the first i columns of $\mathbf{A}$
$\mathbf{A}_\pi^h, \mathbf{V}_\rho^h, \mathbf{W}_\pi^h$ ..... $\mathbf{A}_\pi, \mathbf{V}_\rho, \mathbf{W}_\pi$ at the hth step of the Γ–algorithm
$A(\mathbf{W}_c^{-+})$......... $\{A_{\mathbf{w}_j} \cap A_{\mathbf{w}_k} \mid j \in I(\mathbf{W}^-), k \in I(\mathbf{W}^+)\}$
$A(\mathbf{w})$ .... $\{\mathbf{a}_j \mid \mathbf{w} \bullet \mathbf{a}_j = 0\}$
$A_{\mathbf{w}_j}$ .... $\{j \mid \mathbf{a}_j \in A(\mathbf{w}_j^h)\}$
$A(\mathbf{U})$ .... $\{A_{\mathbf{u}_j} \mid j = 1, \ldots n\}$
$\mathbf{B}$................ $\{\mathbf{a}_i \mid -\mathbf{a}_i \in \mathbf{A}_\pi\}$
$\mathbf{B}^{\sim b_i}$ .... Matrix $\mathbf{B}$ after removing its column $\mathbf{b}_i$
$\mathbf{C}$................ $\{\mathbf{a}_i \mid -\mathbf{a}_i \notin \mathbf{A}_\pi\}$
$I(\mathbf{W}^-)$.......... $\{i \leq s \mid t_i < 0\}$
$I(\mathbf{W}^+)$.......... $\{i \leq s \mid t_i > 0\}$
$I(\mathbf{W}^0)$.......... $\{i \leq s \mid t_i = 0\}$
n_b................ Dimension of linear space $\mathbf{B}_\rho$
n_c................ Dimension of linear space $\mathbf{C}_\pi$
t_k .... $\mathbf{w}_k^{h-1} \bullet \mathbf{a}_h$
$\mathbf{V}_\rho^*$.............. $\mathbf{V}_\rho$ component of the bidual
$\mathbf{W}(\mathbf{x})$ .... Orthogonal set to $\mathbf{x}$
$\mathbf{W}_\pi^*$.............. $\mathbf{W}_\pi$ component of the bidual
$\mathbf{W}^-$ .... $\{\mathbf{w}_i^{h-1} \mid \mathbf{w}_i^{h-1} \bullet \mathbf{a}_h < 0\}$
$\mathbf{W}^+$.............. $\{\mathbf{w}_i^{h-1} \mid \mathbf{w}_i^{h-1} \bullet \mathbf{a}_h > 0\}$
$\mathbf{W}^0$ .... $\{\mathbf{w}_i^{h-1} \mid \mathbf{w}_i^{h-1} \bullet \mathbf{a}_h = 0\}$
$\mathbf{W}^{-+}$ .... $\{\mathbf{w} = t_j\mathbf{w}_i^{h-1} - t_i\mathbf{w}_j^{h-1} \mid \mathbf{w}_i \in \mathbf{W}^-, \mathbf{w}_j \in \mathbf{W}^+\}$
$\mathbf{W}_c^{-+}$ .... $\{t_k\mathbf{u}_j - t_j\mathbf{u}_k \mid j \in I(\mathbf{W}^-), k \in I(\mathbf{W}^+)\}$
λ................. Real number in the interval $[0,1]$
π .... Nonnegative real number
σ .... Positive real number
$\Omega(\mathbf{A}_\pi)$ .... Dual or polar cone of $\mathbf{A}_\pi$
$\Omega(\mathbf{A}_\pi^h)$ .... Dual at h-th step of the Γ-algorithm
$\Omega(\Omega(\mathbf{A}_\pi))$....... Bidual of $\mathbf{A}_\pi$

Chapter 7

$\mathbf{P}_\lambda$............... Polytope
$\mathbf{P} = \{p_1, \ldots p_n\}$ .. Set of generators of $\mathbf{P}_\lambda$
$\overline{O_3O_1}$ .... Vector defined by the points O_3 and O_1
$\begin{pmatrix}\mathbf{P}\\ \mathbf{1}\end{pmatrix}_\pi$ .... Cone associated with the polytope $\mathbf{P}_\lambda$

Chapter 8

$\mathbf{B}$ .... Matrix coefficients of the unrestricted variables
$\mathbf{C}$ .... Matrix coefficients of the non-negative variables
$\mathbf{D}$ .... Matrix coefficients of the positive variables
K_V .... $\{k \mid v_{nk} \neq 0\}$
K_W .... $\{k \mid w_{nk} \neq 0\}$
v_{nk} .... (n, k) component of matrix $\mathbf{V}$
w_{nk} .... (n, k) component of matrix $\mathbf{W}$
$\mathbf{x}_\rho$ .... Vector with unrestricted variables
$\mathbf{x}_\pi$ .... Vector with non-negative variables
$\mathbf{x}_\sigma$ .... Vector with positive variables

Chapter 9

$a_{i1}, \ldots, a_{in}$ .... Coefficients of initial variables
b_i .... Independent terms
$f(\mathbf{x})$ .... Linear objective function
$g_i(\mathbf{x}), h_i(\mathbf{x})$ .... Linear functions
LPP .... Linear programming problem
m .... Dimension of the slackness variables set
n .... Dimension of initial variables set
Q .... Point load
X .... Moment at the supported end
x_i .... Variables
$\mathbf{x}_\mathbf{B}$ .... Basic variables

Chapter 10

$\mathbf{A}^{(t)}$ .... Partition of $\mathbf{U}^{(t)}$– matrix associated with sets $\mathbf{x}_{I_t(\mathcal{P})} \times \mathbf{x}_{J_t(\mathcal{S})}$
$\mathbf{A}^{(0)}$ .... Partition of $\mathbf{U}^{(0)}$– matrix associated with sets $\mathbf{x}_{J_t(\mathcal{S})} \times \mathbf{x}_{I_t(\mathcal{P})}$
$a_{ij}^{(0)}, a_{ij}^{(t)}$ .... Elements (i, j) of $\mathbf{A}^{(0)}$ and $\mathbf{A}^{(t)}$, respectively
$\mathbf{B}^{(t)}$ .... Partition of $\mathbf{U}^{(t)}$– matrix associated with sets $\mathbf{x}_{I_t(\mathcal{P})} \times \mathbf{x}_{J_t(\mathcal{P})}$
$\mathbf{B}^{(0)}$ .... Partition of $\mathbf{U}^{(0)}$– matrix associated with sets $\mathbf{x}_{J_t(\mathcal{S})} \times \mathcal{S}\mathbf{x}_{J_t(\mathcal{P})}$
$b_{ij}^{(0)}, b_{ij}^{(t)}$ .... Elements (i, j) of $\mathbf{B}^{(0)}$ and $\mathbf{B}^{(t)}$, respectively
$\mathbf{C}^{(t)}$ .... Partition of $\mathbf{U}^{(t)}$– matrix associated with sets $\mathbf{x}_{I_t(\mathcal{S})} \times \mathbf{x}_{J_t(\mathcal{S})}$
$\mathbf{C}^{(0)}$ .... Partition of $\mathbf{U}^{(0)}$– matrix associated

with sets $\mathbf{x}_{I_t(\mathcal{S})} \times \mathcal{S}\mathbf{x}_{I_t(\mathcal{P})}$
$c_{ij}^{(0)}, c_{ij}^{(t)}$ .... (i,j) elements of $\mathbf{C}^{(0)}$ and $\mathbf{C}^{(t)}$, respectively
$\mathbf{D}^{(t)}$ .... Partition of $\mathbf{U}^{(t)}-$ matrix associated with sets $\mathbf{x}_{I_t(\mathcal{S})} \times \mathbf{x}_{J_t(\mathcal{P})}$
$\mathbf{D}^{(0)}$ .... Partition of $\mathbf{U}^{(0)}-$ matrix associated with sets $\mathbf{x}_{I_t(\mathcal{S})} \times \mathbf{x}_{J_t(\mathcal{P})}$
$d_{ij}^{(0)}, d_{ij}^{(t)}$ .... (i,j) elements of $\mathbf{D}^{(0)}$ and $\mathbf{D}^{(t)}$, respectively
EPM.............. Exterior point method
I_t .... $\{j \in \mathcal{R} \mid j \notin J_t\}$. Basic variables set at state t
$I_t(\mathcal{P})$ .... Indices of I_t associated with basic variables
$I_t(\mathcal{S})$ .... Indices of I_t associated with secondary variables
J_t .... $\{j \mid \mathbf{u}_j^{(t)} = \mathbf{e}_j\}$; nonbasic variables set at state t
$J_t(\mathcal{P})$ .... Indices of J_t associated with basic variables
$J_t(\mathcal{S})$ .... Indices of J_t associated with secondary variables
$\mathcal{P}$ .... Set of indices of principal variables
$\mathcal{R}$ .... $\{\mathcal{P}, \mathcal{S}\}$
$\mathbf{U}^{(t)}$ .... Matrix $\mathbf{U}$ used in the EPM-algorithm, at step t
$\mathbf{U}_{\mathcal{S}}$ .... Partition of $\mathbf{U}-$matrix associated with $\mathcal{S}$
$\mathbf{U}_{\mathcal{P}}$ .... Partition of $\mathbf{U}-$matrix associated with $\mathcal{P}$
$u_{\alpha 0}^{(t)}$ .... $\min\{u_{i0}^{(t)} \mid u_{i0}^{(t)} < 0\}$
$\mathcal{S}$ .... Set of indices of secondary variables
$X^{(t)}$ .... Feasible region at step t
$\mathbf{x}_{\mathcal{P}}$ .... Initial variables
$\mathbf{x}_{\mathcal{S}}$ .... Slackness variables
x_f .... Objective function $f(\mathbf{x})$
$\mathbf{x}^*$ .... Solution of the LPP
$\Delta_\alpha^{(t)}$ .... $\max\{\Delta_i^{(t)} \mid u_{i0}^{(t)} < 0\}$
$\Delta_i^{(t)}$ .... $\dfrac{u_{f\beta}^{(t)}}{u_{i\beta}^{(t)}} - \max\limits_{u_{ij}^{(t)}>0} \dfrac{u_{fj}^{(t)}}{u_{ij}^{(t)}}$

Chapter 11

A .... Cross section
E .... Young's modulus
f_i .... Flow entering node i
f_{ij} .... Maximum flow capacity of link ij
H_i .... Horizontal force applied to node i
h_i .... Horizontal displacement of node i
I .... Moment of inertia of the cross section of the piece
K .... Stiffness matrix
K_i .... Element of the stiffness matrix
L .... Length
M_i .... Moment applied to node i

S_i .... Section associated with node i in the network
V_i .... Vertical force applied to node i
v_i .... Vertical displacement of node i
θ_i .... Rotation of node i

References

Aitken, A. C. *Determinants and Matrices*, Interscience Publishers, New York, 1954.

Banett, R. *Templates for the Solution of Linear Systems*, SIAM, Philadelphia, 1994.

Bloom, D. M. *Linear Algebra and Geometry*, Cambridge University Press, New York, 1979.

Burden, R. I. and Faires, J. D. *Numerical Analysis*, PWS, E.E.U.A, Boston, 1985.

Calver, J. and Voxman, W. *Linear Programming*, Academic Press, Orlando, FL, 1989.

Castillo, E., Cobo, A., Fernández-Canteli, A., Jubete, F., and Pruneda, R. E. Updating Inverses in Matrix Analysis of Structures, *International Journal for Numerical Methods in Engineering (in press)*.

Castillo, E., Cobo, A., Gutiérrez, J. M., and Pruneda, R. E. Working with Differential, Functional and Difference Equations Using Functional Networks. *Applied Mathematical Modeling (in press)*.

Ciarlet, P. G. *An Introduction to Numerical Linear Algebra and Optimization*, Cambridge University Press, New York, 1989.

Contesse, L. *Méthode de double description pour la résolution complète de systèmes d'Inégalités linéaires avec elimination des inégalités redondantes*, Ph D. Thesis, Universite de Lille I, 1977.

Dantzig, G. B. *Linear Programming and Extensions*, Princeton University Press, Princeton, NJ, 1963.

Díaz–Hernando, J. A. *Álgebra, Geometría, Cálculo*, Vol. 1 and 2, Tebar Flores, Madrid, 1985.

Dwyer, P. S. A Matrix Presentation of Least Squares and Correlation Theory with Matrix Justification of Improved Methods of Solutions, *Annals of Mathematical Statistics*, **15**, 82–29, 1944.

Dyer, M. D. and Proll, A. Eliminating Extraneous Edges in Greenberg's Algorithm, *Mathematical Programming*, Vol. 19, Academic Press, Orlando, FL, 1980.

Farkas, J. Uber der einfachen Ungleichungen, *Journal fuer die Reine und Angewandte Mathematik*, **124**, 1–27, 1902.

Forrest, J. J. H. and Tomlin, J. A. Implementing Interior Point Linear Programming Methods for the Optimization Subroutine Library, *IBM Systems Journal* **31**, No. 1, 26–38, 1992.

Gass, S. I. *Linear Programming. Methods and Applications*, McGraw-Hill, New York, 1969.

Gill, P. E., Golub, G. H., Murray, W., and Saunders, M. A. Methods for Modifying Matrix Factorizations, *Mathematics and Computation*, **28**, 505–535, 1974.

Gill, P. E., Murray, W., Saunders, M. A., Tomlin, J. A., and Wright, M. H. On Projected Newton Barrier Methods for Linear Programming and an Equivalence to Karmarkar's Projective Method, *Mathematical Programming*, **36**, 183–209, 1986.

Gill, P. E., Murray, W., and Wright, M. H. *Numerical Linear Algebra and Optimization*, Addison-Wesley, Reading, MA, 1991.

Golub, G. H. and Van Loan C. F. (1989), *Matrix Computations*, 2nd ed., John Hopkins University Press, Baltimore, 1989.

Greenberg, H. *Integer Programming*, Mathematics in Science and Engineering, Vol. 19, Academic Press, Orlando, FL, 1971.

Greenberg, H. An Algorithm for Determining Redundant Inequalities and all Solutions to Convex Polyhedra, *Numerische Mathematik*, **24**, 19–26, 1975.

Grossman, S. I. *Elementary Linear Algebra with Applications*, McGraw-Hill, 1994.

Horn, R. A. and Charles, R. J. *Matrix Analysis*, Cambridge University Press, New York, 1991.

IBM Optimization Subroutine Library (OSL), Release 2, Publication number SC23-0519-03, 1992.

Hadi, A. S. *Matrix Algebra as a Tool*, Duxbury Press, Boston, 1996.

Jubete, F. *Programación no lineal convexa. El método polar*, CIS, Santander, Spain, 1987.

Jubete, F. *El cono poliédrico convexo. Su incidencia en el álgebra lineal y la programación no lineal*, CIS, Santander, Spain, 1991.

Kuenzi, H. P., Tzchach, H. G., and Zehnder, C. A. *Numerical Methods of Mathematical Optimization*, Academic Press, Orlando, FL, 1971.

Kuiper, N. H. *Numerical Linear Algebra and Geometry*, North Holland Publishing, Amsterdam, 1965.

Mansfield, L. E. *Linear Algebra with Geometric Applications*, Marcel Dekker, New York, 1976.

Mehrotra, S. On the Implementation of a (Primal Dual) Interior Point Method.Technical Report 90–03, Department of Industrial Engineering and Management Sciences, Northwestern University, Evanston, Illinois, 1990.

Minkowski, H. *Gesammelte Abhandlugen*, Teubner, Berlin, 1911.

Motzkin, T.S., Raiffa, H., Thompson, T. S., and Thrall, R. M. Double Description Method. *Contributions to the Theory of Games*, Vol. 19, 51–73, Princeton University Press, Princeton, NJ, 1966.

Muir, T. *A Treatise on the Theory of Determinants*, Dover Publications, Mineola, NY, 1960.

Nashed, M. *Generalized Inverses and Applications*, Academic Press, Orlando, FL, 1980.

Press, W. H., Teukolsky, S. A., Vetterling, W. T., and Flannery, B. P. *Numerical Recipes in C. The Art of Scientific Computing*, Cambridge University Press, New York, 1985.

Raiffa, H., Thompson, G. L., and Thrall, R. M. An Algorithm for the Determination of all Solutions of a Two-Person Zero-Sum Game with a Finite Number of Strategies: Double Descriptive Method, *Directorate of Management Analysis*, **122**, 100–114, 1953.

Samelson, H. *An Introduction to Linear Algebra,* Wiley, New York, 1974.

Shere, W. and Love, G. *Applied Mathematics for Engineering and Science,* Prentice-Hall, Englewood Cliffs, NJ, 1969.

Strong, G. *Linear Algebra and its Applications,* Academic Press, Orlando, FL, 1980.

Trefethen, L. N. and Bau, L. N. *Numerical Linear Algebra,* SIAM, 1997.

Wreub, W. H. *Linear Algebra,* Springer-Verlag, New York, 1963.

Index

A

Adding equations, 102
Adding one unknown, 119
Algorithm
  orthogonalization, 32
  solving a system of linear equations, 92
All positive linear combination, 149
Angle definition, 12
Applications, 345
Artificial variable, 278
Associative property, 6

B

Basic variable, 279
Basis definition, 11, 171
Binding set, 176
  definition, 176
Block-matrix inverse, 57
Boundary of a cone definition, 181
Boundary set of a cone definition, 181

C

Calculating the determinant of a matrix, 44
Coefficient matrix, 20
Columns, 16
Combinations
  all positive linear, 149
  convex linear, 149
  linear, 9
  types of linear, 148
Commutative property, 6
Compatibility
  of a system of linear equations algorithm, 88
  of a system of linear equations, 86
  restricted variables, 217
  strictly restricted variables, 218
  unrestricted variables homogeneous, 216
  unrestricted variables, 216
Compatible equation, 91
Complexity
  of a matrix inverse, 50
  orthogonalization method, 95
Cone associated with a polytope, 193
Cone generators, 151
Cone
  boundary, 181
  extended, 174
  exterior, 181
  general form, 155
  in a cone, 155
  interior, 181
  notable, 175
  polar, 156
  reduced, 175
  set of vectors in a, 155
  vector in a, 154
  vectors in a, 157
Convex polyhedron, 191
Convex set definition, 148
Convex

linear combination, 149, 191
Cube example, 228

D

Definition
angle, 12
basis, 11, 171
binding set, 176
boundary of a cone, 181
boundary set of a cone, 181
convex set, 148
determinant, 19
edge, 212
Euclidean n-space, 12
Euclidean norm, 12
extended cone, 174
exterior of a cone, 181
feasible solution, 256
general form of a cone, 155
hyperplane, 14
identity matrix, 17
interior of a cone, 181
interior set of a cone, 181
Linear space, 6
linear dependency, 10
linear programming problem, 256
matrix, 16
minimum set of generators, 171
nonpositive dual, 156
notable cone, 175
optimal solution, 256
orthogonal subspaces, 14
orthogonal, 12
orthonormal, 12
pivoting transformation, 24
polar cone, 156
polyhedron, 207
polytope, 191
rank, 19
reciprocal image, 77
reduced cone, 175
set of generators, 11
subspace, 8
vertex, 207
Determinant of a matrix
algorithm, 46
calculation, 45–46
Determinant, 18
definition, 19
of row-modified matrices, 51, 53
Determining the rank of a matrix, 43
Diet problem, 378
Direct sum, 15–16
Dual cone, 156
Dynamically updating the solution of a system of equations, 114

E

Edge definition, 212
Equivalence
of two systems of equations, 96
Euclidean n-space definition, 12
Euclidean dot product, 12
Euclidean norm definition, 12
Euclidean space, 11
Example from matrix analysis of structures, 346
Extended cone definition, 174
Exterior of a cone definition, 181

F

Farkas theorem, 157
Feasible region, 276
Feasible
region, 256
solution, 256
Field, 6
Free vectors, 7

G

Gamma process, 159
Gamma-algorithm, 171
complete, 182
Mathematica programs, 184
Gamma-process, 157
General form of a cone definition, 155

H

Halfspaces, 14
Homogeneous, 20
Hyperplane
definition, 14
Hyperplanes, 14

I

Identity matrix definition, 17
Incompatible system of equations, 91
Infeasible region, 276
Input-Output tables, 376
Interior of a cone definition, 181
Interior set of a cone definition, 181
Intersection of two linear subspaces, 73
Inverse element, 6
Inverse matrix, 18
Inverse of a matrix, 42
algorithm, 46
calculation, 46
complexity, 50
Inverse, 18
of a block-matrix, 57

of a matrix after removing rows and columns, 62
of a partitioned matrix, 60
of a row-modified matrix, 51
of a symbolic matrix, 55
of matrices after adding rows and columns, 60

J

JAVA programs, 395

K

Kronecker delta definition, 12

L

Linear combination, 9
non-null, 9
nonnegative, 149
types of, 148
Linear dependency definition, 10
Linear independence and cones, 170
Linear programming problem definition, 256
Linear space as a cone, 151
Linear space definition, 6
Linear spaces, 6
Linear subspace
orthogonal to a linear subspace, 15
orthogonal to a vector, 15
Linear systems of equations, 402
Linear transformation, 77
Linearity, 12
LPP, 256

M

Main variable, 278
Mathematica program
equivalence of two systems, 136
compatibility of a system of equations, 135
determinant of a matrix, 67–68
Gamma-algorithm, 184
inverse of a matrix, 67
inverse of a row modified matrix, 68
matrices, 67, 81
orthogonalization, 36
rank of a matrix, 67
solve a system of linear equations in some selected variables, 136
solve a system of linear equations, 135
systems, 135
Mathematica programs
orthogonal set, 37
Matrices, 16
Mathematica programs, 67, 81
Matrix definition, 16
Minimum set of generators
definition, 171
obtaining, 171
Modifying systems of equations, 101
Moving polytopes
three-dimensional, 204
two-dimensional, 197
Multiplicative unit, 6

N

Network flow problem, 380
Non-null linear combination, 9
Nonbasic variable, 278
Nonnegative linear combination, 149
Nonpositive dual definition, 156
Nonsingular matrix, 18
Notable cone definition, 175

O

Optimal solution definition, 256
Order of a matrix, 16
Orthogonal decomposition using Java programs, 395
Orthogonal decomposition with respect to a vector, 16
Orthogonal decomposition, 24
with respect to a linear subspace, 30
with respect to a vector, 26
Orthogonal definition, 12
Orthogonal matrix, 21
Orthogonal sets and decompositions, 14
Orthogonal sets and inequalities, 215
Orthogonal sets and miscellanea, 73
Orthogonal sets and systems of equations, 85
Orthogonal subspaces definition, 14
Orthogonalization method
Complexity, 95
Orthogonalization module, 31
Orthogonalization
algorithm, 32
Mathematica programs, 36
Orthonormal definition, 12
Other applications, 79

P

Parallelogram law, 7
Partitioned matrices, 18
Pivoting transformation
definition, 24
normalization step, 24
pivoting process, 24
Polar cone definition, 156
Polar cone, 156
Polyhedra, 191
Polyhedral Convex Cones, 147

Polyhedral convex cone, 151
Polyhedral convex cones, 151
Polyhedron definition, 207
Polytope, 191
  defined by inequalities, 192
  definition, 191
Polytopes, 191
Position vectors, 7
Positive definiteness, 11
Process Gamma, 157, 159
Product of two matrices, 17
Production-scheduling problem, 369

R

Rank, 18
  algorithm, 46
  definition, 19
  of a matrix, 46
Reciprocal image definition, 77
Reciprocal image of a linear subspace by a linear transformation, 77
Reduced cone
  definition, 175
Redundant equations
  have previously been removed, 108
  remain in the system, 108
Regular matrix, 18
Removing equations, 108
Removing unknowns, 104
Restricted variables
  compatibility, 217
  uniqueness, 217
Rows, 16

S

Scalars, 6
Secondary variable, 278
Selecting a linear combination, 150
Set of generators definition, 11
Set of vectors in a cone, 155
Simplex method, 275
Singular matrix, 18
Slack variable, 278
Solving a system in some selected variables, 99
Solving a system of linear equations, 90
  algorithm, 92
Some extensions matrices, 56
Square matrix, 16
Strictly restricted variables
  compatibility, 218
  uniqueness homogeneous, 219
  uniqueness, 218
Subspace definition, 8
Sum of matrices, 16
Sum, 6
Symmetry, 12
System of equations
  compatibility, 86
  incompatible, 91
Systems with restricted variables, 217
Systems with strictly restricted variables, 218
Systems with unrestricted variables, 216
Systems
  Mathematica programs, 135
  of linear equations, 20

T

Theorem of Farkas, 157
Transportation problem, 365
Transpose matrix, 17
Types of linear combinations, 148

U

Uniqueness
  restricted variables, 217
  strictly restricted variables homogeneous, 219
  strictly restricted variables, 218
  unrestricted variables, 216
Unit coordinate vector, 11
Unrestricted variables
  compatibility homogeneous, 216
  compatibility, 216
  uniqueness, 216

V

Variable
  artificial, 278
  basic, 279
  main, 278
  nonbasic, 278
  secondary, 278
  slack, 278
Vector of independent terms, 20
Vector
  in a cone, 154, 157
  in a space, 169
Vectors, 6
Vertex definition, 207

PURE AND APPLIED MATHEMATICS

A Wiley-Interscience Series of Texts, Monographs, and Tracts

ADÁMEK, HERRLICH, and STRECKER—Abstract and Concrete Catetories
ADAMOWICZ and ZBIERSKI—Logic of Mathematics
AKIVIS and GOLDBERG—Conformal Differential Geometry and Its Generalizations
ALLEN and ISAACSON—Numerical Analysis for Applied Science
*ARTIN—Geometric Algebra
AZIZOV and IOKHVIDOV—Linear Operators in Spaces with an Indefinite Metric
BERMAN, NEUMANN, and STERN—Nonnegative Matrices in Dynamic Systems
BOYARINTSEV—Methods of Solving Singular Systems of Ordinary Differential Equations
BURK—Lebesgue Measure and Integration: An Introduction
*CARTER—Finite Groups of Lie Type
CASTILLO, COBO, JUBETE and PRUNEDA—Orthogonal Sets and Polar Methods in Linear Algebra: Applications to Matrix Calculations, Systems of Equations, Inequalities, and Linear Programming
CHATELIN—Eigenvalues of Matrices
CLARK—Mathematical Bioeconomics: The Optimal Management of Renewable Resources, Second Edition
COX—Primes of the Form $x^2 + ny^2$: Fermat, Class Field Theory, and Complex Multiplication
*CURTIS and REINER—Representation Theory of Finite Groups and Associative Algebras
*CURTIS and REINER—Methods of Representation Theory: With Applications to Finite Groups and Orders, Volume I
CURTIS and REINER—Methods of Representation Theory: With Applications to Finite Groups and Orders, Volume II
*DUNFORD and SCHWARTZ—Linear Operators
Part 1—General Theory
Part 2—Spectral Theory, Self Adjoint Operators in Hilbert Space
Part 3—Spectral Operators
FOLLAND—Real Analysis: Modern Techniques and Their Applications
FRÖLICHER and KRIEGL—Linear Spaces and Differentiation Theory
GARDINER—Teichmüller Theory and Quadratic Differentials
GREENE and KRATZ—Function Theory of One Complex Variable
*GRIFFITHS and HARRIS—Principles of Algebraic Geometry
GROVE—Groups and Characters
GUSTAFSSON, KREISS and OLIGER—Time Dependent Problems and Difference Methods
HANNA and ROWLAND—Fourier Series, Transforms, and Boundary Value Problems, Second Edition
*HENRICI—Applied and Computational Complex Analysis
Volume 1, Power Series—Integration—Conformal Mapping—Location of Zeros
Volume 2, Special Functions—Integral Transforms—Asymptotics—Continued Fractions
Volume 3, Discrete Fourier Analysis, Cauchy Integrals, Construction of Conformal Maps, Univalent Functions

*HILTON and WU—A Course in Modern Algebra
*HOCHSTADT—Integral Equations
JOST—Two-Dimensional Geometric Variational Procedures
*KOBAYASHI and NOMIZU—Foundations of Differential Geometry, Volume I
*KOBAYASHI and NOMIZU—Foundations of Differential Geometry, Volume II
LAX—Linear Algegra
LOGAN—An Introduction to Nonlinear Partial Differential Equations
McCONNELL and ROBSON—Noncommutative Noetherian Rings
NAYFEH—Perturbation Methods
NAYFEH and MOOK—Nonlinear Oscillations
PANDEY—The Hilbert Transform of Schwartz Distributions and Applications
PETKOV—Geometry of Reflecting Rays and Inverse Spectral Problems
*PRENTER—Splines and Variational Methods
RAO—Measure Theory and Integration
RASSIAS and SIMSA—Finite Sums Decompositions in Mathematical Analysis
RENELT—Elliptic Systems and Quasiconformal Mappings
RIVLIN—Chebyshev Polynomials: From Approximation Theory to Algebra and Number Theory, Second Edition
ROCKAFELLAR—Network Flows and Monotropic Optimization
ROITMAN—Introduction to Modern Set Theory
*RUDIN—Fourier Analysis on Groups
SENDOV—The Averaged Moduli of Smoothness: Applications in Numerical Methods and Approximations
SENDOV and POPOV—The Averaged Moduli of Smoothness
*SIEGEL—Topics in Complex Function Theory
Volume 1—Elliptic Functions and Uniformization Theory
Volume 2—Automorphic Functions and Abelian Integrals
Volume 3—Abelian Functions and Modular Functions of Several Variables
SMITH and ROMANOWSKA—Post-Modern Algebra
STAKGOLD—Green's Functions and Boundary Value Problems, Second Editon
*STOKER—Differential Geometry
*STOKER—Nonlinear Vibrations in Mechanical and Electrical Systems
*STOKER—Water Waves: The Mathematical Theory with Applications
WESSELING—An Introduction to Multigrid Methods
WHITMAN—Linear and Nonlinear Waves
†ZAUDERER—Partial Differential Equations of Applied Mathematics, Second Edition

*Now available in a lower priced paperback edition in the Wiley Classics Library.
†Now available in paperback.

Printed and bound by CPI Group (UK) Ltd, Croydon, CR0 4YY
16/09/2026
14963441-0005